Geosciences: Novel Concepts, Technologies and Applications

Geosciences: Novel Concepts, Technologies and Applications

Editor: Jacques Howard

R CALLISTO REFERENCE

www.callistoreference.com

Callisto Reference,
118-35 Queens Blvd., Suite 400,
Forest Hills, NY 11375, USA

Visit us on the World Wide Web at:
www.callistoreference.com

ISBN: 978-1-63239-867-3 (Hardback)

The publisher's policy is to use permanent paper from mills that operate a sustainable forestry policy. Furthermore, the publisher ensures that the text paper and cover boards used have met acceptable environmental accreditation standards.

Trademark Notice: Registered trademark of products or corporate names are used only for explanation and identification without intent to infringe.

Printed in the United States of America.

Cataloging-in-publication Data

Geosciences : novel concepts, technologies and applications / edited by Jacques Howard.
   p. cm.
Includes bibliographical references and index.
ISBN 978-1-63239-867-3
1. Earth sciences. 2. Geology. 3. Environmental sciences. 4. Physical sciences. I. Howard, Jacques.
QE33 .G46 2017
550--dc23

Table of Contents

Permissions

List of Contributors

Index

Preface

Geoscience is an interdisciplinary field of study. It employs tools from varied fields such as mathematics, physics etc. to form a better understanding of the diverse aspects such as biosphere, hydrosphere, atmosphere and other related aspects. This book contains some path-breaking studies in the field of geoscience. It traces the progress of this field and highlights some of its key concepts and applications. It strives to provide a fair idea about this discipline and to help develop a better understanding of the latest advances within this field. This book is appropriate for student seeking detailed information in this area as well as for experts. The readers would gain knowledge that would broader their perspective about geoscience.

The researches compiled throughout the book are authentic and of high quality, combining several disciplines and from very diverse regions from around the world. Drawing on the contributions of many researchers from diverse countries, the book's objective is to provide the readers with the latest achievements in the area of research. This book will surely be a source of knowledge to all interested and researching the field.

In the end, I would like to express my deep sense of gratitude to all the authors for meeting the set deadlines in completing and submitting their research chapters. I would also like to thank the publisher for the support offered to us throughout the course of the book. Finally, I extend my sincere thanks to my family for being a constant source of inspiration and encouragement.

Editor

Integration of onshore and offshore seismic arrays to study the seismicity of the Calabrian Region: a two steps automatic procedure for the identification of the best stations geometry

A. D'Alessandro[1,2], **I. Guerra**[3], **G. D'Anna**[1], **A. Gervasi**[1], **P. Harabaglia**[4], **D. Luzio**[2], and **G. Stellato**[3]

[1]Istituto Nazionale di Geofisica e Vulcanologia, Centro Nazionale Terremoti, Rome, Italy
[2]Università di Palermo, Dipartimento delle Scienze della Terra e del Mare, Palermo, Italy
[3]Università della Calabria, Dipartimento di Fisica, Arcavacata di Rende (Cosenza), Italy
[4]Università della Basilicata, Scuola di Ingegneria, Potenza, Italy

Correspondence to: A. D'Alessandro (antonino.dalessandro@ingv.it)

Abstract. We plan to deploy in the Taranto Gulf some Ocean Bottom broadband Seismometer with Hydrophones. Our aim is to investigate the offshore seismicity of the Sibari Gulf. The seismographic network optimization consists in the identification of the optimal sites for the installation of the offshore stations, which is a crucial factor for the success of the monitoring campaign. In this paper, we propose a two steps automatic procedure for the identification of the best stations geometry. In the first step, based on the application of a set of a priori criteria, the suitable sites to host the ocean bottom seismic stations are identified. In the second step, the network improvement is evaluated for all the possible stations geometries by means of numerical simulation. The application of this procedure allows us to identify the best stations geometry to be achieved in the monitoring campaign.

1 Introduction

The Pollino Massif (Southern Italy) is a stocky mountain chain, triangle-shaped and E–W oriented which marks the transition from the Southern Apennines to the Calabrian Arc. On the western side it is characterized by a moderate seismicity (9 $M_L > 4$ events in the last 50 years, Fig. 1a), rather well documented in the last 400 years (Peresan and Panza, 2002; Castello et al., 2006; Luzi et al., 2008; Rovida et al., 2011; Iside catalog). The Moment Tensor Solutions (MTS) available in this area (Fig. 1b, European-Mediterranean CMT catalog; Ekström et al., 2012) mainly yields normal faults with coherent Southern Apenninic trend. This remains true also for several tens of Fault Plane Solutions of the dense seismic sequence, which interested the western Pollino area in the years 2010–2012 (Totaro et al., 2013). South of the Massif, in most of the Sibari plane, seismic activity is very scarce, while it is again rather intense in its southeastern corner, both onshore and offshore. There are however only a few MTS in this south-eastern part of the area; two of them show the right strike slip kinematics of the associated events, with of the possible fault planes coherent with the Southern Apenninic trend, while the third one derives from a thrust event in the perpendicular direction. It is also noteworthy that at least a couple of MTS around the Sila Massif still yield a Southern Apenninc trend. The morphology also presents some Southern Apenninic trend: the Pollino Massif crests and valley do show it, as well as the shore direction from the Sibari Plain up to Cirò Marina.

The above observations point to the perspective that the stress field of a vast portion of Northern Calabria still resembles that of the Southern Apennines (Guerra et al., 2005). In this frame, it becomes important to investigate the offshore seismicity of the Sibari Gulf and the deformation pattern within the Sibari Plane. The latter might function as a hinge to transfer the deformation of the extensional fault system in the Pollino area to a different offshore fault system. Since return times of largest events might be very long, we need to investigate the true seismic potential of the offshore faults and to verify whether they are truly strike slip or if they could

(a)   (b)

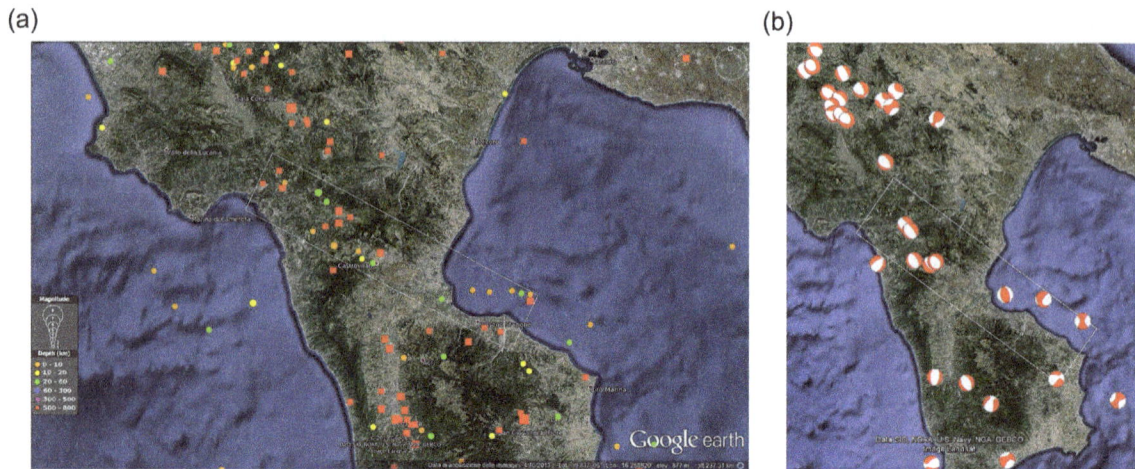

Figure 1. (a) Instrumental (full circle) and historical seismicity (red square) of the study area (earthquake with $M > 4$, data from UCI2001 (1960–1980), CSI (1981–2002), ITACA (2003–2005/04/15), ISIDE 2005/04/16-today and CPTI11 (1000–1959) catalogs; **(b)** Moment Tensor solutions from Global Centroid-Moment-Tensor and European-Mediterranean Regional Centroid Moment Tensor catalogs; the white rectangle in both the figures indicates the area of main interest.

involve relevant thrust or normal components, that would add to the risk of potentially associated tsunamis.

The seismicity of the Calabrian area is monitored by the Italian National Seismic Network (INSN) managed by Istituto Nazionale di Geofisica e Vulcanologia and by the Calabrian University Seismic Network (CUSN) managed by the University of Calabria (D'Alessandro et al., 2013a). Both network comprise only on-land seismic stations (Fig. 2a). The lack of offshore stations does not allow accurate determination of the hypocentral parameters also for moderate–strong earthquakes that occur in the Calabrian offshore (D'Alessandro et al., 2013a). Figure 2b shows the spatial distribution of the location uncertainty in the study area determined integrating INSN and CUSN.

The location uncertainty has been determined for $M_L =$ 1.5 and hypocentral depth of 10 km using the SNES method (D'Alessandro et al., 2011a, 2013a). Figure 2b report the Radius of the Equivalent Sphere (RES) parameter, which is the radius of the sphere whose volume equals that of the 95 % confidence ellipsoid of the hypocentral parameters (D'Alessandro et al., 2011a). D'Alessandro et al. (2013a), observed that only few stations will detect small magnitude earthquakes in the offshore area of the Sibari Gulf, with resulting azimuthal gap exceeding 180° and location errors of several kilometers (Fig. 2b). The lack of offshore seismic stations also does not allow the accurate determination of hypocentral coordinates and focal parameters also for even the largest earthquakes commonly observed in the investigated area.

With the aim of investigating the near shore seismicity in the Sibari offshore and its eventual relationship with the Pollino activity, we plan to deploy some OBS/H in the Taranto Gulf. The monitoring campaign is planned for the

2015, and it will last about a year. The equipment will consist in five OBS/H, each equipped with a broadband seismometer and a hydrophone.

The stations will be designed, assembled and deployed by the Gibilmanna OBSLab, a laboratory created in 2005 by INGV to address the offshore extension of the Italian seismic network (Mangano et al., 2011) and for the development of seismic station based on Micro Electro-Mechanical Systems (MEMS) technology (D'Alessandro and D'Anna, 2013). Several seismic monitoring experiments, conducted in the Mediterranean Sea have already resulted in a better understanding of the seismo-tectonic and seismo-volcanic activity of some submarine seismogenic districts (D'Alessandro et al., 2009, 2012a, 2013b, D'Alessandro, 2014; Adelfio et al., 2012).

Due to the high costs and limited available resources, the network optimization, consisting in the identification of the optimal sites for the installation of the offshore stations, is a crucial factor for the success of the monitoring campaign. In the following, we analyze, by means of numerical simulations, the effect of the OBS/H stations in terms of network coverage and of hypocentral localization improvement. The results of the simulations are critically analyzed in order to identify the best OBS/H geometry, which will be realized in the monitoring campaign.

2 Evaluation of the best OBS/H array geometry

In planning a monitoring campaign, it is necessary to find an optimal set of observation sites. They will ensure the widest and most homogeneous coverage of the area of interest, with a significant improvement of hypocenter estimation. They normally be selected among a larger set of possible candidate

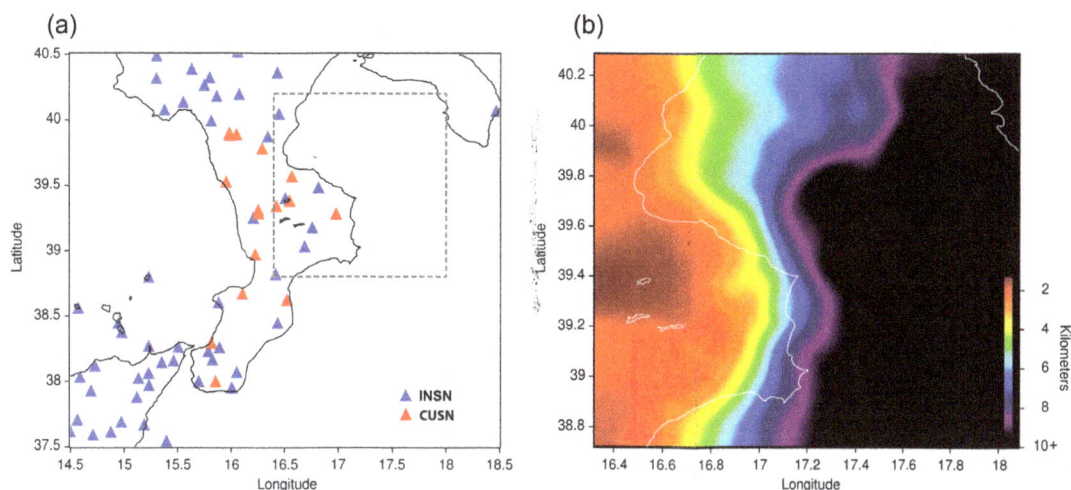

Figure 2. (a) distribution of the seismic stations on the Calabrian territory; **(b)** RES maps at $M_L = 1.5$, hypocentral depth of $10\,km$ and confidence level of $95\,\%$, for INSN+CUSN (D'Alessandro et al., 2013a).

locations and must ensure integration with the onshore stations.

The candidate sites to host an OBS/H station must satisfy the following requirements:

– $6\,km$ maximum water depth;

– $5°$ maximum average slope in an circular area of radius of $5\,km$;

– $10\,km$ minimum distance from the coastline;

– $10\,km$ minimum distance among the stations.

The first constrain originates from the maximum operating depth of the OBS/H of $6\,km$ (Mangano et al., 2011). The second one from the need that each OBS/H is deployed in a flat nearly horizontal area wide enough to prevent station or sensor overturnings. Even areas of moderate slope can compromise the sensor leveling and therefore the quality of the acquired signals. The third criterion is necessary in order to reduce the seismic noise power on the OBS/H signals. It is well known that wave breaking on the coastline is a strong source of noise. In addition, human activities near the coasts are generally intense and could be an additional source of noise. Finally, the fourth point is related to the ration between the extension of the study area and the number of measurement points and to a reasonable minimum stations spacing.

Figure 3 shows the bathymetric maps of the Taranto Gulf. The locations of the onshore seismic stations and of the candidate sites for OBS/H installation are indicated. Because the maximum depth in the Taranto Gulf is about $3\,km$, the candidate sites were identified only on the basis of the three remaining criteria. Starting from the whole set of sites, which fulfill the condition 2, an automatic search algorithm identified seven sites that would be suitable to host an OBS/H station (Fig. 3).

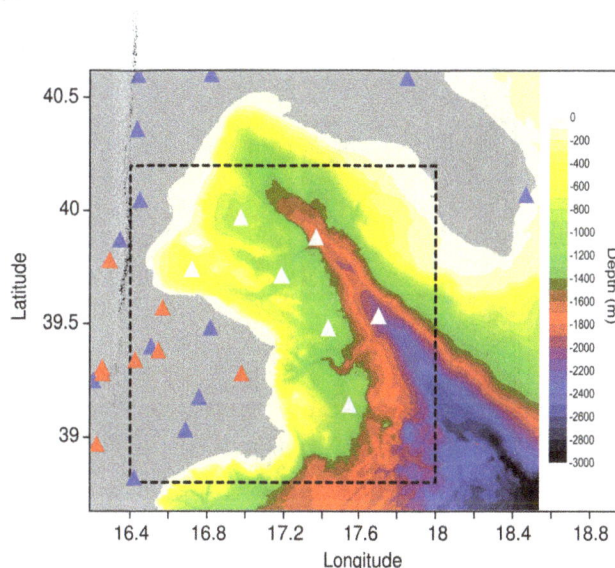

Figure 3. Bathymetric map of the Taranto Gulf; the dashed square indicates the area of main interest; blue triangles are INSN stations, red triangles are CUSN stations, white triangles are candidate sites for OBS/H installation.

Since the possible sites are seven but there are only five available OBS/H, the number of possible stations geometries is 21. Therefore, it becomes crucial to identify an optimal OBS/H geometry, capable to ensure the best coverage of the area of interest, in view of both the limited number of suitable sites and of OBS/H.

To this purpose, we have simulated all the possible different OBS/H geometries and we analyzed them by means of the SNES method (D'Alessandro et al., 2011a). This method has been extensively used for seismic networks performance evaluation (D'Alessandro et al., 2011b, 2012b, c, 2013c; D'Alessandro and Rupert, 2012) and optimization

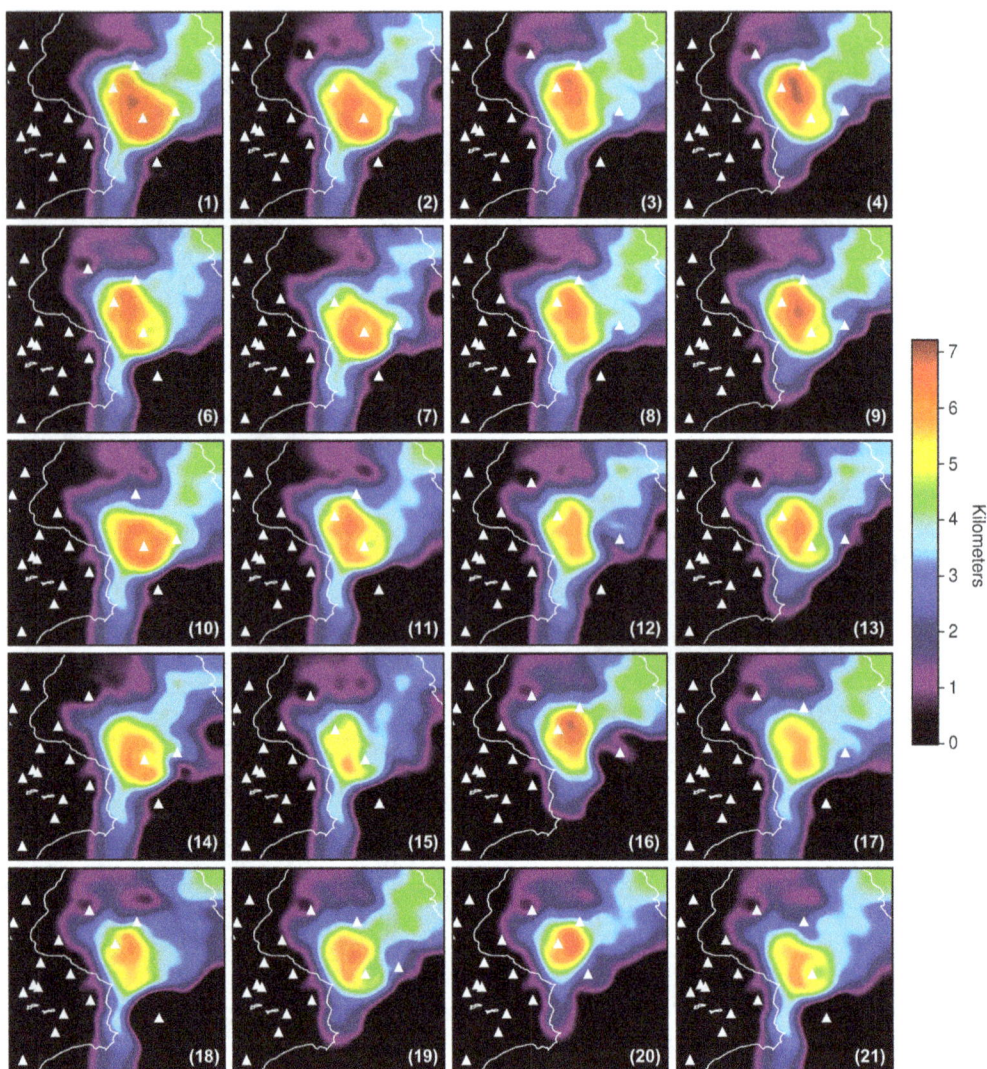

Figure 4. Improvement maps for the area of interest as average RES over all the investigated hypocenter depths, for different stations geometry; white triangles indicates the stations position; geometry 5 is showed in Fig. 6.

(D'Alessandro et al., 2013d). The SNES method allows to determine, as a function of magnitude, hypocentral depth and confidence level, the spatial distribution of the following parameters: magnitude detection threshold, number of stations active in the location procedure, azimuthal gap and confidence levels of hypocentral parameters. Details on the method and on the computation algorithms can be found in D'Alessandro et al. (2011a).

On the basis of the results of D'Alessandro et al. (2013a) and of the features of the local seismicity, the SNES maps were determined for $M_L = 1.5$ and hypocentral depths of 5, 10, 15, 20, 25 and 30 km, with 95 % confidence intervals. In our evaluation, we considered only the RES parameter because it takes into account both epicentral and hypocentral depth errors, and is therefore the best parameter to quantify the performance of a seismic network. To quantify the improvement of the network due to the addition of five OBS/H in the candidate sites, we have determined, for each possible stations geometry, the difference between the RES maps with and without the offshore stations.

Figure 4 shows, for the area of interest and for all the possible stations geometries, the improvement maps as average RES determined over all the investigated hypocenter depths. The area of greatest improvement is located in the southwestern part of Taranto Gulf, where the RES reduction reaches values of about 7 km.

The shape and the extension of the improvement area is clearly dependent on the OBS/H geometry. It is very difficult to identify the best OBS/H geometry by a simple visual inspection. This means that we need objective quantitative criteria.

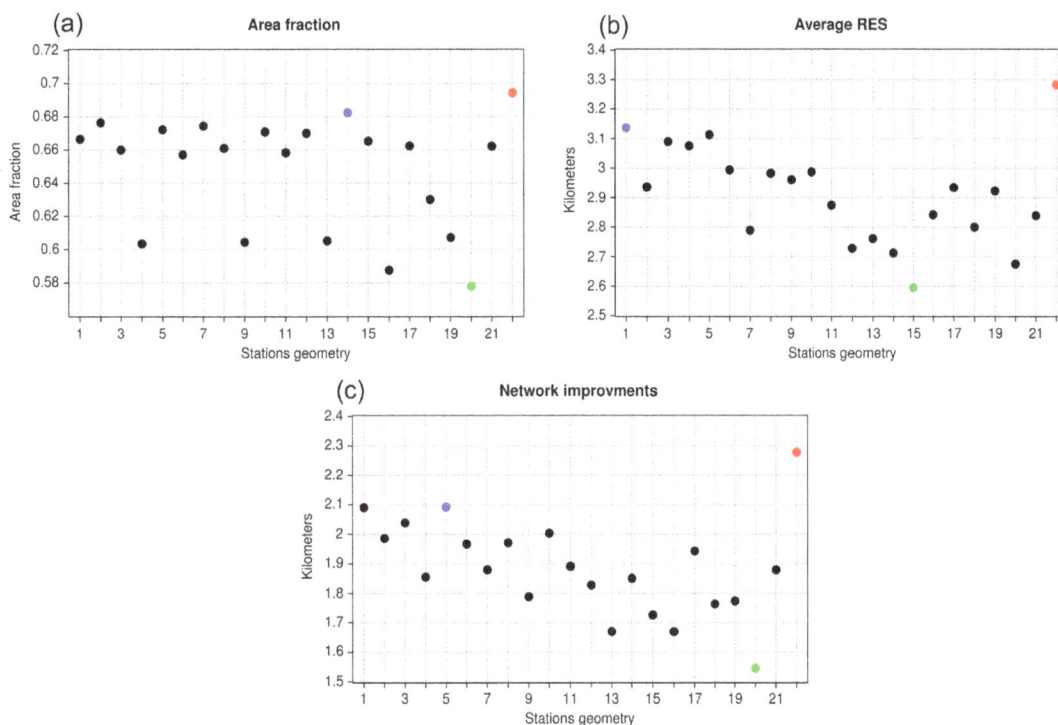

Figure 5. (a) A_{RES}, (b) $\overline{RES}$ and (c) $I\left(A_{RES}, \overline{RES}\right)$ as function of OBS/H geometry of Fig. 4; station geometry 22 refers to that assuming seven OBS/H (Fig. 6).

Therefore, to identify the optimal OBS/H geometry we determined:

- fraction of area in which there is a significant reduction of the hypocentral error (A_{RES});

- average error reduction in this area ($\overline{RES}$);

- improvement index ($I\left(A_{RES}, \overline{RES}\right)$).

A_{RES} is determined as the ratio between the area where RES reduction is more than 0.5 km (A_{IMP}) and the area of interest (A_{TOT}); $\overline{RES}$ is determined as average RES in A_{IMP}; $I\left(A_{RES}, \overline{RES}\right)$ is determined as $A_{RES} \cdot \overline{RES}$.

Figure 5 shows the distribution of these parameters as functions of the network geometry. In Fig. 6 we compare network geometry number 5, our preferred one, with a hypothetical configuration with all sites occupied by OBS/H (stations geometry 22).

On the basis of Fig. 5, we can see that the fraction of area covered using only five OBS/H ranges between 0.682 (stations geometry 14) and 0.578 (stations geometry 20) and would be 0.694 using all the seven candidate sites. The average error reduction $\overline{RES}$ ranges between 3.136 km (stations geometry 1) and 2.595 km (stations geometry 15), and would be 3.282 km using all the seven candidate sites. The improvement index ranges between 2.092 km (stations geometry 5) and 1.545 km (stations geometry 20), and would be 2.278 km using all the seven candidate sites.

3   Discussion and conclusion

The seismicity and the seismogenic volumes of the Sibari Gulf are not well characterized, despite their importance in the understanding of the seismotectonic processes in the Southern Apenninic – Calabrian Arc border and surrounding areas. The main reason is the poor distribution of the seismic network sensors due to the elongated shape of the Calabrian arc and in the lack of offshore stations.

With the aim of investigating the near shore seismicity in the Sibari offshore and its eventual relationship with the Pollino activity, we plan to deploy several OBS/H in the Taranto Gulf. The monitoring campaign, planned for the 2015, will allow for the acquisition of a large amount of data. Their integration with those acquired by the onshore permanent networks will help in characterizing the seismicity and the seismogenetic volume of this area.

A careful assess stations geometry is of primary importance in any monitoring campaign. This is especially true when submarine stations are involved. This assessment must take into account both the presence of pre-existing networks and the logistical problems due to the deployment of the OBS/H. An optimal stations geometry must be achieved, that is the one with the largest are coverage and the greatest reduction in the hypocentral error.

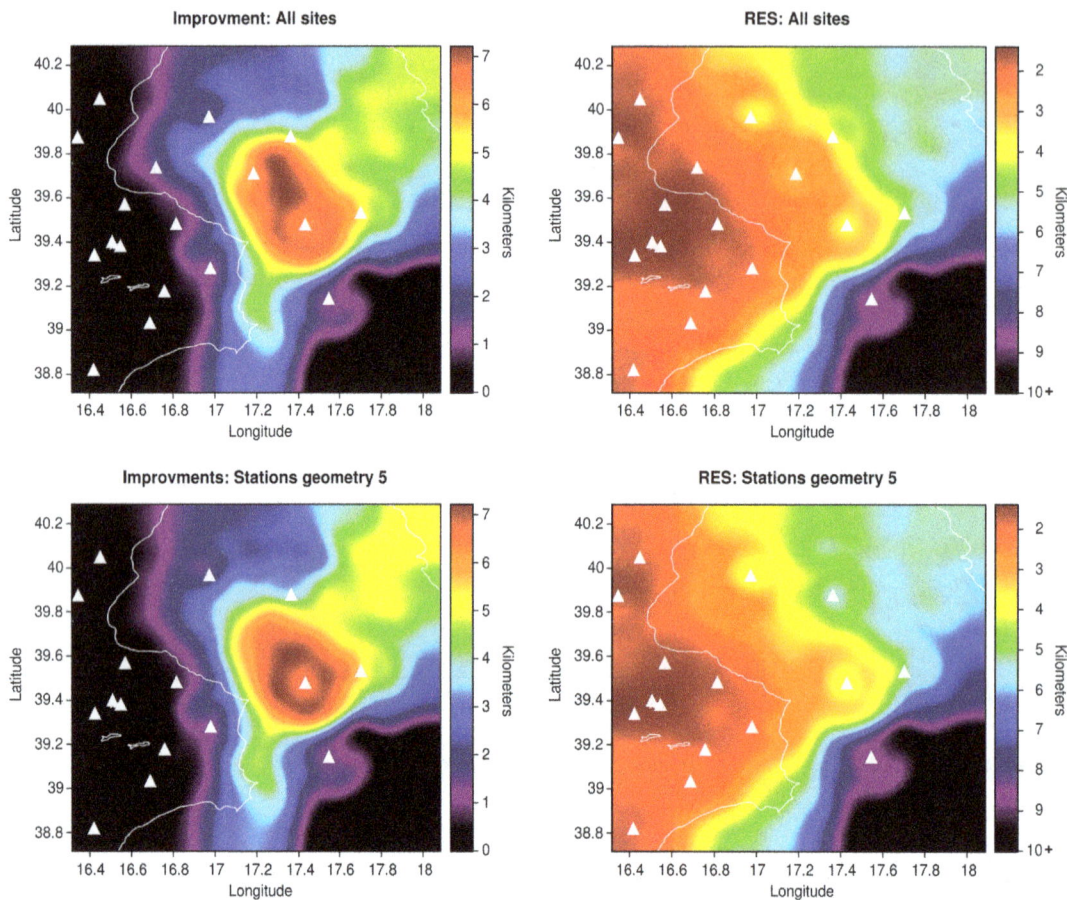

Figure 6. Improvement and RES maps using the stations geometry 5 and assuming seven OBS/H.

In this paper, we propose a two steps automatic procedure for the identification of the best stations geometry. In the first step, we identify suitable sites to host OBS/H stations, based on some a priori criteria. In the second step we evaluate the network improvement for all the possible stations geometries. We specifically identify seven candidate site suitable to host the five OBS/H planned for the monitoring campaign; this led to a total of 21 possible stations geometries.

The results of our simulation shows that for $M_L = 1.5$, on the basis of A_{RES}, the OBS/H geometry 14 should be surely the best, although many geometries could ensure a very similar coverage. However, seven geometries have to be excluded (4, 9, 13, 16, 18, 19, 20) because they do not provide a sufficient extension of the area covered by the network. On the basis of $\overline{RES}$ the best station geometry should be the array 1, even though similar values are obtained for the geometries 3, 4 and 5. However these parameters, separately, do not permit to univocally determine the best OBS/H array. The analysis of the improvement index $I\left(A_{RES}, \overline{RES}\right)$ instead, allow us to identify in geometry 5 (Fig. 6) the best one and, as second choice, geometry 1. The array number 20 is that with the minimum $I\left(A_{RES}, \overline{RES}\right)$ and than should be avoided.

Figure 6 compares the improvement and RES maps relative to geometry 5 and to the ideal seven OBS/H geometry. We can see that the use of geometry 5 greatly improves the location performance in most of the area of our interest and by using two further OBS/H's we will not obtain significantly better results.

It is clear that the results of the optimization procedure are highly dependent on: criteria for the choice of suitable sites, shape and size of the area of interest, magnitudes and hypocenters depths of interest and optimization criteria. Clearly, these should be evaluated from time to time according to the specific needs and goals we want to achieve.

Acknowledgements. We are grateful to the anonymous reviewers and to the editor Damiano Pesaresi for their constructive comments and suggestions.

References

Adelfio, G., Chiodi, M., D'Alessandro, A., Luzio, D., D'Anna, G., and Mangano, G.: Simultaneous seismic wave clustering and registration, Comput. Geosci., 44, 60–69, doi:10.1016/j.cageo.2012.02.017, 2012.

Castello B., Selvaggi G., Chiarabba C., and Amato A.: CSI – Catalogo della sismicità italiana 1981–2002, versione 1.1. INGV-CNT, Roma, available at: http://www.ingv.it/CSI (last access: 1 August 2014), 2006.

D'Alessandro, A.: The Marsili Seamount, the biggest European volcano, could be still active!, Current Science, 106, p. 1339, 2014.

D'Alessandro, A. and D'Anna, G.: Suitability of low cost 3 axes MEMS accelerometer in strong motion seismology: tests on the LIS331DLH (iPhone) accelerometer, Bull. Seismol. Soc. Am., 103, 2906–2913, doi:10.1785/0120120287, 2013.

D'Alessandro, A. and Stickney, M.: Montana Seismic Network Performance: an evaluation through the SNES method, Bull. Seismol. Soc. Am., 102, 73–87, doi:10.1785/0120100234, 2012.

D'Alessandro, A., D'Anna, G., Luzio, D., and Mangano, G.: The INGV's new OBS/H: analysis of the signals recorded at the Marsili submarine volcano, J. Volcanol. Geoth. Res., 183, 17–29, doi:10.1016/j.jvolgeores.2009.02.008, 2009.

D'Alessandro, A., Luzio, D., D'Anna, G., and Mangano, G.: Seismic Network Evaluation through Simulation: An Application to the Italian National Seismic Network, Bull. Seismol. Soc. Am., 101, 1213–1232, doi:10.1785/0120100066, 2011a.

D'Alessandro, A., Papanastassiou, D., and Baskoutas, I.: Hellenic Unified Seismological Network: an evaluation of its performance through SNES method, Geophys. J. Int., 185, 1417–1430, doi:10.1111/j.1365-246X.2011.05018.x, 2011b.

D'Alessandro, A., Mangano, G., and D'Anna, G.: Evidence of persistent seismo-volcanic activity at Marsili seamount, Ann. Geophys., Scientific News, 55, 213–214, doi:10.4401/ag-5515, 2012a.

D'Alessandro, A. and Ruppert, N.: Evaluation of Location Performance and Magnitude of Completeness of Alaska Regional Seismic Network by SNES Method, Bull. Seismol. Soc. Am., 102, 2098–2115, doi:10.1785/0120110199, 2012b.

D'Alessandro, A., Danet, A., and Grecu, B.: Location Performance and Detection Magnitude Threshold of the Romanian National Seismic Network, Pure Appl. Geophys., 169, 2149–2164, doi:10.1007/s00024-012-0475-7, 2012c.

D'Alessandro, A., Gervasi, A., and Guerra, I.: Evolution and strengthening of the Calabrian Regional Seismic Network, Adv. Geosci., 36, 11–16, doi:10.5194/adgeo-36-11-2013, 2013a.

D'Alessandro, A., Mangano, G., D'Anna, G., and Luzio, D.: Waveforms clustering and single-station location of microearthquake multiplets recorded in the northern Sicilian offshore region, Geophys. J. Int., 194, 1789–1809, doi:10.1093/gji/ggt192, 2013b.

D'Alessandro, A., Badal, J., D'Anna, G., Papanastassiou, D., Baskoutas, I., and Özel, M. M.: Location Performance and Detection Threshold of the Spanish National Seismic Network, Pure Appl. Geophys., 170, 1859–1880, doi:10.1007/s00024-012-0625-y, 2013c.

D'Alessandro, A., Scarfì, L., Scaltrito, A., Di Prima, S., and Rapisarda, S.: Planning the improvement of a seismic network for monitoring active volcanic areas: the experience on Mt. Etna, Adv. Geosci., 36, 39–47, doi:10.5194/adgeo-36-39-2013, 2013d

Ekström, G., Nettles, M., and Dziewonski, A. M.: The global CMT project 2004-2010: Centroid-moment tensors for 13,017 earthquakes, Phys. Earth Planet. Inter., 200–201, 1–9, doi:10.1016/j.pepi.2012.04.002, 2012.

European-Mediterranean CMT catalog: European-Mediterranean Regional Centroid-Moment Tensors Catalog, available at: http://www.bo.ingv.it/RCMT, last access: 1 August 2014.

Guerra, I., Harabaglia, A., Gervasi, A., and Rosa, A. B.: The 1998–1999 Pollino (Southern Apennines, Italy) seismic crisis: Tomography of a sequence, Ann. Geophys., 48, 995–1007, 2005, http://www.ann-geophys.net/48/995/2005/.

ISIDE catalog, Italian Seismic Instrumental and parametric DatabasE, available at: http://iside.rm.ingv.it/iside, last access: 1 August 2014.

Luzi, L., Hailemikael, S., Bindi, D., Pacor, F., Mele, F., and Sabetta, F.: ITACA (ITalian ACcelerometric Archive): A Web Portal for the Dissemination of Italian Strong motion Data, Seismol. Res. Lett., 79, 716–722, doi:10.1785/gssrl.79.5.716, 2008.

Mangano, G., D'Alessandro, A., and D'Anna, G.: Long-term underwater monitoring of seismic areas: design of an Ocean Bottom Seismometer with Hydrophone and its performance evaluation, OCEANS 2011 IEEE Conference, 6–9 June, Santander, Spain, in: OCEANS 2011 IEEE Conference Proceeding, 9, doi:10.1109/Oceans-Spain.2011.6003609, 2011.

Peresan, A. and Panza, G. F.: UCI2001: The Updated Catalogue of Italy, The Abdus Salam International Centre for Theoretical Physics, ICTP, Miramare, Trieste, Italy, Internal report IC/IR/2002/3, 2002.

Rovida, A., Camassi, R., Gasperini, P., and Stucchi, M. (Eds.): CPTI11, the 2011 version of the Parametric Catalogue of Italian Earthquakes, Milano, Bologna, available at: http://emidius.mi.ingv.it/CPTI, doi:10.6092/INGV.IT-CPTI11, 2011.

Totaro, C., Presti, D., Billi, A., Gervasi, A., Orecchio, B., Guerra, I., and Neri, G.: The ongoing seismic crisis of Pollino Mts in Southern Italy, Seismol. Res. Lett., 84, 955–962, 2013.

Improving seismic networks performances: from site selection to data integration (EGU2014 SM1.2/GI3.7 session)

D. Pesaresi[1]**, J. Clinton**[2]**, and H. Pedersen**[3]

[1]OGS (Istituto Nazionale di Oceanografia e di Geofisica Sperimentale), Trieste, Italy
[2]ETHZ, Zurich, Switzerland
[3]RESIF, Grenoble, France

Correspondence to: D. Pesaresi (dpesaresi@inogs.it)

Abstract. The number and quality of seismic stations and networks in Europe continually improves, nevertheless there is always scope to optimize their performance. In this EGU2014 SM1.2/GI3.7 session we welcomed contributions from all aspects of seismic network installation, operation and management. This includes site selection; equipment testing and installation; planning and implementing communication paths; policies for redundancy in data acquisition, processing and archiving; and integration of different datasets including GPS and OBS.

1 Introduction

The history of seismic networks sessions at European Geosciences Union (EGU) General Assemblies started in 2010 with the SM1.3 "Seismic Centers Data Acquisition" session (Pesaresi, 2011), where the Convener Damiano Pesaresi supported by the Orfeus Data Center (ODC) Director Co-Convener Reinoud Sleeman chaired a session of 7 oral and 16 posters. A similar session was later the same year held at the XXXII European Seismological Commission (ESC) General Assembly: "SD1, 3 Seismic centers data acquisition", conveners D. Pesaresi and R. Sleeman, with 15 oral presentations.

The history continued in 2011 with the EGU2011 SM1.3/G3.8/GD3.7/GI-19/TS8.7 "Improving seismic networks performances: from site selection to data integration" session (EGU2011 SM1.3/G3.8/GD3.7/GI-19/TS8.7 Improving seismic networks performances: from site selection to data integration, 2011) where the Convener Damiano Pesaresi supported by the Co-Conveners John Clinton and Robert Busby chaired a session of 9 oral and 20 posters, in 2012 with the EGU2012 SM1.3/GI1.7 "Improving seismic

networks performances: from site selection to data integration" session (Pesaresi and Vernon, 2013) where the Convener Damiano Pesaresi supported by the Co-Convener Frank Vernon chaired a session of 6 oral and 22 posters, and in 2013 with the SM1.4/GI1.6 "Improving seismic networks performances: from site selection to data integration" session (Pesaresi and Busby, 2013) where the Convener Damiano Pesaresi supported by the Co-Convener Robert Busby chaired a session of 6 oral and 13 posters.

2 The EGU2014 SM1.2/GI3.7 session

In the EGU2014 SM1.2/GI3.7 "Improving seismic networks performances: from site selection to data integration" session (EGU2014 SM1.2/GI3.7 Improving seismic networks performances: from site selection to data integration, 2014) the Convener Damiano Pesaresi supported by the Co-Conveners John Clinton and Helle Pedersen chaired a session (Fig. 1) of 12 oral (Table 1) and 27 posters (Table 2).

The 39 presentations come from 16 countries (USA, Norway, Germany, France, Canada, Italy, Poland, Finland, Taiwan, Austria, Romania, Malta, Spain, Algeria, Switzerland, UK) from 4 different continents (North America, Europe, Asia, and Africa), which well fits the goals of the European Geosciences Union.

Solicited presentations in this session were:

i. "Improvements in Data Quality, Integration and Reliability: New Developments at the IRIS DMC" by Tim Ahern, Rick Benson, Rob Casey, Chad Trabant, and Bruce Weertman (Ahern et al., 2014);

ii. "Seismic Sensor orientation by complex linear least squares" by Francesco Grigoli, Simone Cesca, Lars Krieger, Manuel Olcay, Carlos Tassara,

Figure 1. EGU2014 SM1.2/GI3.7 session (from EGU2014 homepage).

Monika Sobiesiak, and Torsten Dahm (Grigoli et al., 2014);

iii. "Detecting and locating teleseismic events with using USArray as a big antenna" by Lise Retailleau, Nikolaï Shapiro, Jocelyn Guilbert, Michel Campillo, and Philippe Roux (Retailleau et al., 2014);

iv. "A high-resolution ambient seismic noise model for Europe", by Toni Kraft (Kraft, 2014);

v. "Improving Station Performance by Building Isolation Walls in the Tunnel", by Yan Jia, Nikolaus Horn, and Roman Leohardt (Jia et al., 2014);

vi. "Romanian Data Center: A modern way for seismic monitoring", by Cristian Neagoe, Liviu Marius Manea, and Constantin Ionescu (Neagoe et al., 2014);

vii. "Comparison Study Between Vault Seismometers and Posthole Seismometers", by Neil Spriggs, Geoffrey Bainbridge, and Wesley Greig (Spriggs et al., 2014);

viii. "RESIF national datacentre: new features and upcoming evolutions", by Pierre Volcke, Catherine Pequegnat, Benjamin Brichet-Billet, Albanne Lecointre, David Wolyniec, and Philippe Guéguen (Volcke et al., 2014);

ix. "Testing various modes of installation for permanent broadband stations in open field environment", by Jérôme Vergne, Olivier Charade, Benoît Arnold, and Thierry Louis-Xavier (Vergne et al., 2014);

x. "Data Quality Control of the French Permanent Broadband Network in the RESIF Framework", by Marc Grunberg, Sophie Lambotte, Fabien Engels, Remi Dretzen, and Alain Hernandez (Grunberg et al., 2014);

xi. "Enhancement of Network Performance through Integration of Borehole Stations", by Edith Korger, Katrin Plenkers, John Clinton, Toni Kraft, Tobias Diehl, Stephan Husen, and Michael Schnellmann (Korger et al., 2014);

xii. "Testing the Lower Thresholds of Broadband Seismometers", by Nathan Pearce, Cansun Guralp, Murray Mcgowan, and Horst Rademacher (Pearce et al., 2014).

The papers published in these proceedings of the EGU2014 SM1.2/GI3.7 session are:

I. "Integration of onshore and offshore seismic stations to study the seismicity of the Calabrian Region: a two steps automatic procedure for the identification of the best stations geometry" by A. D'Alessandro, I. Guerra,

Table 1. Oral programme EGU2014 SM1.2/GI3.7 session.

EGU abstract ref.	Title	Authors
EGU2014-1576	Improvements in Data Quality, Integration and Reliability: New Developments at the IRIS DMC	Tim Ahern, Rick Benson, Rob Casey, Chad Trabant, and Bruce Weertman
EGU2014-2030	Combination of High Rate, Real-time GNSS and Accelerometer Observations – Preliminary Results Using a Shake Table and Historic Earthquake Events	Michael Jackson, Paul Passmore, Leonid Zimakov, and Jared Raczka
EGU2014-2136	Noise and detection levels for the Norwegian National Seismic Network	Andrea Demuth and Lars Ottemoller
EGU2014-2282	Seismic Sensor orientation by complex linear least squares	Francesco Grigoli, Simone Cesca, Lars Krieger, Manuel Olcay, Carlos Tassara, Monika Sobiesiak, and Torsten Dahm
EGU2014-5850	Detecting and locating teleseismic events with using USArray as a big antenna	Lise Retailleau, Nikolaï Shapiro, Jocelyn Guilbert, Michel Campillo, and Philippe Roux
EGU2014-6361	Microseismic Network Performance Estimation: Comparing Predictions to an Earthquake Catalogue	Wesley Greig and Nick Ackerley
EGU2014-6735	Improvements of the Regional Seismic network of Northwestern Italy in the framework of ALCoTra program activities	Fabrizio Bosco
EGU2014-8497	"13 BB star" – broadband seismic array at the edge of East European Craton in Poland	Marcin Polkowski, Marek Grad, Monika Wilde-Piórko, Jerzy Suchcicki, and Tadeusz Arant
EGU2014-11707	Northern Finland Seismological Network: a tool to analyse long-period seismological signals	Elena Kozlovskaya and Riitta Hurskainen
EGU2014-12407	A high-resolution ambient seismic noise model for Europe	Toni Kraft
EGU2014-13054	Testing the "PRESTo" early warning algorithm with OGS, ARSO and ZAMG seismic data: first results	Luca Elia, Andrej Gosar, Wolfgang Lenhardt, Marco Mucciarelli, Damiano Pesaresi, Matteo Picozzi, Mladen Živčić, and Aldo Zollo
EGU2014-13735	VADASE: a new approach for real-time fast displacement detection – First application to Taiwan High-Rate GNSS Network	Huang-Kai Hung, Ruey-Juin Rau, Gabriele Colosimo, Elisa Benedetti, Mara Branzanti, Mattia Crespi, and Augusto Mazzoni

Table 2. Poster programme EGU2014 SM1.2/GI3.7 session.

EGU abstract ref.	Title	Authors
EGU2014-3312	Improving Station Performance by Building Isolation Walls in the Tunnel	Yan Jia, Nikolaus Horn, and Roman Leohardt
EGU2014-3320	Romanian Data Center: a modern way for seismic monitoring	Cristian Neagoe, Liviu Marius Manea, and Constantin Ionescu
EGU2014-3622	Implementation of a new picking procedure in the Antelope software	Lara Tiberi, Giovanni Costa, and Daniele Spallarossa
EGU2014-4793	An improved real-time seismic network in the Central Mediterranean	Matthew Agius, Pauline Galea, and Sebastiano D'Amico
EGU2014-4953	The performance of the stations of the Romanian seismic network in monitoring the local seismic activity	Luminita Angela Ardeleanu and Cristian Neagoe
EGU2014-5066	Small instrument to volcanic seismic signals	Normandino Carreras, Spartacus Gomariz, and Antoni Manuel
EGU2014-5571	A multiple-criteria network optimization	Anna Tramelli, Giuseppe De Natale, Claudia Troise, and Massimo Orazi
EGU2013-7906	Borehole prototype for seismic high-resolution exploration	Rüdiger Giese, Katrin Jaksch, Felix Krauß, Kay Krüger, Marco Groh, and Andreas Jurczyk
EGU2014-5971	"SeismoSAT" project state of the art: connecting seismic data centres via satellite	Damiano Pesaresi, Wolfgang Lenhardt, Markus Rauch, Mladen Zivcic, Rudolf Steiner, and Michele Bertoni
EGU2014-6441	Comparison Study Between Vault Seismometers and Posthole Seismometers	Neil Spriggs, Geoffrey Bainbridge, and Wesley Greig
EGU2014-6547	Data quality control of ADSN Broadband stations	Azouaou Alili, Abd el karim Yelles-chaouche, Toufik Allili, and Walid Messemen
EGU2014-9623	Using Antelope and Seiscomp in the framework of the Romanian Seismic Network	George Marius Craiu, Andreea Craiu, Alexandru Marmureanu, and Cristian Neagoe
EGU2014-9750	The 2013 Earthquake Series in the Southern Vienna Basin: Location	Maria-Theresia Apoloner, Irene Bianchi, Götz Bokelmann, Ewald Brückl, Helmut Hausmann, Stefan Mertl, and Rita Meurers

Table 2. Continued.

EGU abstract ref.	Title	Authors
EGU2014-9931	Retrieve Ocean Bottom and Downhole Seismic sensors orientation using integrated low cost gyroscope and direct rotation measurements	Antonino D'Alessandro and Giuseppe D'Anna
EGU2014-10064	Urban MEMS based seismic network for post-earthquakes rapid disaster assessment	Antonino D'Alessandro, Dario Luzio, and Giuseppe D'Anna
EGU2014-10230	Investigating active faults in SE Iberia: borehole and surface seismic monitoring	Maria Jose Jurado, Jose Crespo, Teresa Teixido, and Carlos Viñolo
EGU2014-12270	RESIF national datacentre: new features and upcoming evolutions	Pierre Volcke, Catherine Pequegnat, Benjamin Brichet-Billet, Albanne Lecointre, David Wolyniec, and Philippe Guéguen
EGU2014-12337	Testing various modes of installation for permanent broadband stations in open field environment	Jérôme Vergne, Olivier Charade, Benoît Arnold, and Thierry Louis-Xavier
EGU2014-13911	The Central and Eastern European Earthquake Research Network – CE3RN	Pier Luigi Bragato, Giovanni Costa, Antonella Gallo, Andrej Gosar, Nikolaus Horn, Wolfgang Lenhardt, Marco Mucciarelli, Damiano Pesaresi, Rudolf Steiner, Peter Suhadolc, Lara Tiberi, Mladen Živčić, and Giuliana Zoppé
EGU2014-14138	Data Quality Control of the French Permanent Broadband Network in the RESIF Framework	Marc Grunberg, Sophie Lambotte, Fabien Engels, Remi Dretzen, and Alain Hernandez
EGU2014-14311	The Italian Strong Motion Network (RAN)	Giovanni Costa, Alfredo Ammirati, Rita de Nardis, Luisa Filippi, Antonella Gallo, Giusy Lavecchia, Sebastiano Sirignano, Elisa Zambonelli, and Mario Nicoletti
EGU2014-14529	Enhancement of Network Performance through Integration of Borehole Stations	Edith Korger, Katrin Plenkers, John Clinton, Toni Kraft, Tobias Diehl, Stephan Husen, and Michael Schnellmann
EGU2014-14796	How to create a very-low cost, very-low-power, credit-card-sized and real-time ready datalogger	Maxime Bès de Berc, Marc Grunberg, and Fabien Engels

Table 2. Continued.

EGU abstract ref.	Title	Authors
EGU2014-14918	Seismic catalog condensation with applications to multifractal analysis of South Californian seismicity	Yavor Kamer, Guy Ouillon, Didier Sornette, and Jochen Wössner
EGU2014-14987	Significant breakthroughs in monitoring networks of the volcanological and seismological French observatories	Arnaud Lemarchand, André Anglade, Jean-Marie Saurel and the arnaudl@ipgp.fr Team
EGU2014-15285	Testing the Lower Thresholds of Broadband Seismometers	Nathan Pearce, Cansun Guralp, Murray Mcgowan, and Horst Rademacher
EGU2014-15311	A new Shallow Water Cabled OBS System off California	Horst Rademacher, Chris Pearcey, Giorgio Mangano, Cansun Guralp, and Nathan Pearce
EGU2014-15722	Integration of onshore and offshore seismological data to study the seismicity of the Calabrian Region	Antonino D'Alessandro, Ignazio Guerra, Giuseppe D'Anna, Anna Gervasi, Paolo Harabaglia, Dario Luzio, and Gilda Stellato

G. D'Anna, A. Gervasi, P. Harabaglia, D. Luzio, and G. Stellato;

II. "The 2013 Earthquake Series in the Southern Vienna Basin: Location" by M.-T. Apoloner, I. Bianchi, G. Bokelmann, E. Brückl, H. Hausmann, S. Mertl, and R. Meurers;

III. "Urban MEMS based seismic network for post-earthquakes rapid disaster assessment" by A. D'Alessandro, D. Luzio, and G. D'Anna, which shows the usage of the new MEMS low dimensions-low cost devices in monitoring urban areas;

IV. "Retrieve Ocean Bottom and Downhole Seismic sensors orientation using integrated MEMS gyroscope and direct rotation measurements" by A. D'Alessandro and G. D'Anna, which again shows the usage of the new MEMS low dimensions-low cost devices, here remote sensors orientation;

V. "Detecting and locating seismic events using US-Array as a large antenna", by L. R. Retailleau, N. M. S. Shapiro, J. G. Guilbert, M. C. Campillo, and P. R. Roux;

VI. "Significant breakthroughs in monitoring networks of the volcanological and seismological French observatories", by A. A. Anglade, A. L. Lemarchand, and J. M. S. Saurel;

VII. "How to create a very-low cost, very-low-power, credit-card-sized and real-time ready datalogger", by M. Bès de Berc, M. Grunberg, and F. Engels;

VIII. "SeismoSAT project state of the art: connecting seismic data centres via satellite" by D. Pesaresi, W. Lenhardt, M. Rauch, M. Živčić, R. Steiner, and M. Bertoni;

IX. "Improvements in Data Quality, Integration and Reliability: New Developments at the IRIS DMC", by T. Ahern, R. Benson, R. Casey, C. Trabant, and B. Weertman;

X. "Trans-National Earthquake Early Warning (EEW) in North-Eastern Italy, Slovenia and Austria: First Experience with PRESTo at the CE3RN Network", by M. Picozzi, L. Elia, D. Pesaresi, A. Zollo, M. Mucciarelli, A. Gosar, W. Lenhardt, and M. Živčić.

3 Conclusions

The quality and quantity of presentations made at the EGU2014 SM1.2/GI3.7 session well satisfied the expectations of the Convener and Co-Conveners, and well fitted the goals of the European Geosciences Union.

The increasing number of presentations at such yearly seismic networks sessions encourage the conveners that the path they followed in organizing such sessions is a valid one, and that there is need in the seismological community worldwide to present and discuss different solutions to common problems in running seismic networks.

Acknowledgements. Authors wish to thank the EGU2014 SM1.2/GI3.7 session presentations authors, especially those who made the effort to publish their presentations in these proceedings on Advances in Geosciences: A. D'Alessandro, M.-T. Apoloner, L. Retailleau, A. Lemarchand, M. Bès de Berc, D. Pesaresi and M. Picozzi. Authors also especially thank the EGU Seismology Division President Charlotte Krawczyk for her continuous strong support to the seismic networks sessions at EGU.

References

Ahern, T., Benson, R., Casey, R., Trabant, C., and Weertman, B.: Improvements in Data Quality, Integration and Reliability: New Developments at the IRIS DMC, EGU General Assembly, EGU2014-1576, 27 April–2 May 2014, Vienna, Austria, 2014.

EGU2011 SM1.3/G3.8/GD3.7/GI-19/TS8.7 Improving seismic networks performances: from site selection to data integration, http://meetingorganizer.copernicus.org/EGU2011/session/7340 (last access: 8 August 2014), 2011.

EGU2014 SM1.2/GI3.7 Improving seismic networks performances: from site selection to data integration, http://meetingorganizer.copernicus.org/EGU2014/session/14856, last access: 8 August 2014.

Grigoli, F., Cesca, S., Krieger, L., Olcay, M., Tassara, C., Sobiesiak, M., and Dahm, T.: Seismic Sensor orientation by complex linear least squares, EGU General Assembly, EGU2014-2282, 27 April–2 May 2014, Vienna, Austria, 2014.

Grunberg, M., Lambotte, S., Engels, F., Dretzen, R., and Hernandez, A.: Data Quality Control of the French Permanent Broadband Network in the RESIF Framework, EGU General Assembly, EGU2014-14138, 27 April–2 May 2014, Vienna, Austria, 2014.

Jia, Y., Horn, N., and Leohardt, R.: Improving Station Performance by Building Isolation Walls in the Tunnel, EGU General Assembly, EGU2014-3312, 27 April–2 May 2014, Vienna, Austria, 2014.

Korger, E., Plenkers, K., Clinton, J., Kraft, T., Diehl, T., Husen, S., and Schnellmann, M.: Enhancement of Network Performance through Integration of Borehole Stations, EGU General Assembly, EGU2014-14529, 27 April–2 May 2014, Vienna, Austria, 2014.

Kraft, T.: A high-resolution ambient seismic noise model for Europe, EGU General Assembly, EGU2014-2282, 27 April–2 May 2014, Vienna, Austria, 2014.

Neagoe, C., Manea, L. M., and Ionescu, C.: Romanian Data Center: A modern way for seismic monitoring, EGU General Assembly, EGU2014-3320, 27 April–2 May 2014, Vienna, Austria, 2014.

Pearce, N., Guralp, C., Mcgowan, M., and Rademacher, H.: Testing the Lower Thresholds of Broadband Seismometers, EGU General Assembly, EGU2014-15285, 27 April–2 May 2014, Vienna, Austria, 2014.

Pesaresi, D.: The EGU2010 SM1.3 Seismic Centers Data Acquisition session: an introduction to Antelope, EarthWorm and SeisComP, and their use around the World, Ann. Geophys., 54, 1–7, doi:10.4401/ag-4972, 2011.

Pesaresi, D. and Busby, R.: EGU2013 SM1.4/GI1.6 session: "Improving seismic networks performances: from site selection to data integration", Adv. Geosci., 36, 1–5, doi:10.5194/adgeo-36-1-2013, 2013.

Pesaresi, D. and Vernon, F.: EGU2012 SM1.3/GI1.7 session: "Improving seismic networks performances: from site selection to data integration", Adv. Geosci., 34, 1–4, doi:10.5194/adgeo-34-1-2013, 2013.

Retailleau, L., Shapiro, N., Guilbert, J., Campillo, M., and Roux, P.: Detecting and locating teleseismic events with using USArray as a big antenna, EGU General Assembly, EGU2014-5850, 27 April–2 May 2014, Vienna, Austria, 2014.

Spriggs, N., Bainbridge, G., and Greig, W.: Comparison Study Between Vault Seismometers and Posthole Seismometers, EGU General Assembly, EGU2014-6441, 27 April–2 May 2014, Vienna, Austria, 2014.

Vergne, J., Charade, O., Arnold, B., and Louis-Xavier, T.: Testing various modes of installation for permanent broadband stations in open field environment, EGU General Assembly, EGU2014-12337, 27 April–2 May 2014, Vienna, Austria, 2014.

Volcke, P., Pequegnat, C., Brichet-Billet, B., Lecointre, A., Wolyniec, D., and Guéguen, P.: RESIF national datacentre : new features and upcoming evolutions, EGU General Assembly, EGU2014-12270, 27 April–2 May 2014, Vienna, Austria, 2014.

3

Relative role of bed roughness change and bed erosion on peak discharge increase in hyperconcentrated floods

W. Li[1], Z. B. Wang[1], D. S. van Maren[1], H. J. de Vriend[1], and B. S. Wu[2]

[1]Faculty of Civil Engineering and Geosciences, Delft University of Technology, Delft, the Netherlands
[2]State Key Laboratory of Hydroscience and Engineering, Tsinghua University, Beijing, China

Correspondence to: W. Li (w.li@tudelft.nl)

Abstract. River floods are usually featured by a downstream flattening discharge peak whereas a downstream increasing discharge peak is observed at a rate exceeding the tributary discharge during highly silt-laden floods (hyperconcentrated floods) in China's Yellow River. It entails a great challenge in the downstream flood defence and the underlying mechanisms need to be unravelled. Previous study on this issue only focuses on one possible mechanism, while the present work aims to reveal the relative importance of bed roughness change and bed erosion in the hyperconcentrated flood. Using a newly developed fully coupled morphodynamic model, we have conducted a numerical study for the 2004 hyperconcentrated flood in the Xiaolangdi-Jiahetan reach of the Lower Yellow River. In order to focus on the physical mechanism and to reduce uncertainty from low-resolution topography data, the numerical modeling was carried out in a schematized 1-D channel of constant width. The basic understanding that bed roughness decreases with concentration at moderate concentrations (e.g. several 10 s to 100 s g L^{-1}) was incorporated by a simple power-law relation between Manning roughness coefficient and sediment concentration. The feedback between the bed deformation and the turbid flow, however, was fully accounted for, in the constituting equations as well as in the numerical solutions. The model successfully reproduced the downstream flood peak increase for the 2004 flood when considering the hyperconcentration-induced bed roughness reduction. As the hyperconcentration lags shortly behind the flood peak, later parts of the flood wave may experience less friction and overtake the wave front, leading to the discharge increase. In comparison, bed erosion is much less important to the discharge increase, at least for hyperconcentrated flood of moderate sediment concentration.

1 Introduction

The Yellow River, the second longest river in China, is famous for the high sediment load in its middle and lower reaches. However, dam construction, water-soil conservation and water diversions for irrigation and other purposes (in addition to climate change) have altered the flow regime and sediment load considerably (Wu et al., 2008a). At the Huayuankou hydrological station where the discharge and sediment load represent those entering the lower reach, the average annual runoff and suspended sediment load were 48.6×10^9 m^3 and 1.56×10^9 tonnes respectively, prior to the construction of the Sanmenxia dam in 1960 (Wu et al., 2008a). Yet in the 1990s, they reduced dramatically to 25.7×10^9 m^3 and 0.7×10^9 tonnes respectively. After the construction of the Xiaolangdi Reservoir (in October 1999), the reduction continued, leading to 20.8×10^9 m^3 and 0.13×10^9 tonnes for the average annual runoff and suspended sediment load respectively in the 2000s.

As a result, the hyperconcentrated flood, which is defined as a water-sediment mixture with sediment concentrations higher than 100–200 kg m^{-3} typically in the Yellow River (Wan and Wang, 1994; He et al., 2012), exhibits obviously different behaviors in different periods, among which the phenomenon of downstream peak discharge increase is the focus of this paper. Before 2000, a downstream increasing peak discharge was occasionally observed during a hyperconcentrated flood that inundated floodplains. The increase was relatively small and the maximal increasing rate was 30 %. After 2004 when the water-sediment regulation by the Xiaolangdi Reservoir became operational, an increasing peak discharge of the hyperconcentrated flood was frequently observed between the Xiaolangdi and the

Huayuankou hydrological stations (in the Lower Yellow River), which are approximately 125.8 km apart. In this period, the average increasing rate is as high as 50 %, with floods of moderate concentrations (i.e., 100 to 400 kg m^{-3}) mostly conveyed inside the main channel. This considerable increase in the peak discharge greatly increases the flood risk in the lower reach, along with the severe sedimentation due to runoff reduction. Therefore, it is of great importance to unravel the mechanisms underlying the peak discharge increase and to find out solutions for mitigating potential damages.

The peak discharge increase may be related to rapid morphological changes, to a modified bed roughness, or to bed erosion related increasing flow volume. In a hyperconcentrated flood with floodplain inundation, the channel morphological change from wide and shallow to narrow and deep may increase the flow velocity, thus probably causing a downstream increasing peak discharge when the flood peak is caught up by the successive flood waves (Wang et al., 2009). It is also suggested that the intensive sediment transport due to strong bed erosion may contribute to a peak discharge increase by increasing the flow volume of the water-sediment mixture (Cao et al., 2006; Qi et al., 2010). In addition, the bed roughness change in high concentration conditions could also be a main factor for the discharge increase. Field observations and laboratory experiments show that the bed roughness decreases with sediment concentration at moderate concentrations (Jiang et al., 2006; Zhu and Hao, 2008). The decreasing roughness may accelerate flow propagation leading to a downstream peak discharge increase. Jiang et al. (2006) numerically reproduced the peak discharge increase of the 2004 hyperconcentrated flood at Huayuankou by considering the effect of sediment concentration on bed roughness. Based on theoretical analysis, Li (2008) also suggested that the discharge increase should be attributed to a roughness reduction.

So far, there has been no consensus, however, on which mechanism contributes most to the peak discharge increase. Focusing on the floods mostly conveyed inside the main channel, this paper aims to reveal the relative importance of bed roughness change and bed erosion to the peak discharge increase. The 2004 flood in the Xiaolangdi-Jiahetan reach is revisited by a numerical study using a newly developed morphodynamic model (Li et al., 2013). The effect of bed erosion on increasing flow volume is distinguished by two model versions (a fully coupled version and a partially coupled version). A simple power-law relation is used to represent a decreasing roughness with concentration.

2 Mathematical model

The numerical study is conducted using a coupled morphodynamic model based on the finite volume method and the 2nd order extension of upwind-biased First Order Centered (UFORCE, see Stecca et al., 2010) scheme. This model is

second order accurate in space and time, and validated by a series of dam-break tests (Li et al., 2013). In this paper, only the basic formulations for the 1-D modeling are introduced; see details in Li et al. (2013).

2.1 Governing equations

In a 1-D coupled morphodynamic model of non-capacity sediment transport, the governing equations consist of the mass and momentum conservation equations for sediment-laden flow, the mass conservation equation for sediment in motion, and a bed update equation (Cao et al., 2004). For a fully coupled version (the effects of bed deformation and sediment density on the flow are fully considered), the vector form of the governing equations can be written, following the methods of Cao et al. (2004) and Li and Duffy (2011):

$$\frac{\partial \boldsymbol{U}}{\partial t} + \frac{\partial \boldsymbol{F}}{\partial x} = \boldsymbol{R} \tag{1}$$

$$\boldsymbol{U} = \begin{bmatrix} h \\ hu \\ hc \\ \varphi \end{bmatrix} \tag{2}$$

$$\boldsymbol{F} = \begin{bmatrix} hu \\ hu^2 + 0.5gh^2 \\ huc \\ huc \end{bmatrix} \tag{3}$$

$$\boldsymbol{R} = \begin{bmatrix} \frac{E-D}{1-p} \\ gh(S_0 - S_f) - \frac{(\rho_s - \rho_w)gh^2}{2\rho}\frac{\partial c}{\partial x} - \frac{(\rho_0 - \rho)(E-D)}{\rho(1-p)}u \\ E - D \\ 0 \end{bmatrix} \tag{4}$$

where $\boldsymbol{U}$ = vector of conservative variables; $\boldsymbol{F}$ = vector of flux variables; $\boldsymbol{R}$ = vector of source terms for the fully coupled model; t = time; x = horizontal coordinate; h = water depth; u = depth-averaged flow velocity in x direction; c = depth averaged volumetric sediment concentration; z = bed elevation; E, D = sediment entrainment and deposition fluxes respectively; $S_0 = -\partial z/\partial x$ = bed slope in x direction; S_f = friction slope; $\rho_s = 2650$ kg m^{-3} = sediment density; $\rho_w = 1000$ kg m^{-3} = water density; $\rho = \rho_w(1-c) + \rho_s c$ = density of sediment-laden flow; $\rho_0 = \rho_w p + \rho_s(1-p)$ = density of saturated bed; p = bed porosity; $g = 9.8$ m s^{-2} = acceleration of gravity; $\varphi = (1-p)z + hc$ = newly-constructed conservative variable. In order to reveal the contribution of bed erosion to the discharge increase, a partially coupled model, which neglects the effect of bed deformation on increasing flow volume (i.e., mass conservation), is also used as a comparison. It differs from the fully coupled version in the source term,

$$\boldsymbol{R}' = \begin{bmatrix} 0 \\ gh(S_0 - S_f) - \frac{(\rho_s - \rho_w)gh^2}{2\rho}\frac{\partial c}{\partial x} - \frac{(\rho_s - \rho_w)(E-D)u}{\rho} \\ E - D \\ 0 \end{bmatrix} \tag{5}$$

where R' = vector of source terms for the partially coupled model.

2.2 Empirical relations

The friction slope is estimated using Manning roughness n

$$S_f = \frac{n^2 u^2}{h^{4/3}} \tag{6}$$

Following Cao et al. (2004), the sediment entrainment and deposition are estimated by the adaptation coefficient α, as

$$E - D = \alpha \omega_s (c_* - c) \tag{7}$$

where ω_s = effective sediment settling velocity (m s^{-1}); c_* = sediment transport capacity ($-$); the coefficient α is calculated by an empirical formula that is widely used in the Yellow River (Wang and Xia, 2001),

$$\alpha = \begin{cases} 0.001/\omega_s^{0.3}, & c > c_* \\ 0.001/\omega_s^{0.7}, & c \leq c_* \end{cases} \tag{8}$$

The effective sediment settling velocity is given by Richardson-Zaki formula,

$$\omega_s = \omega_0 (1 - \frac{c}{1-p})^5 \tag{9}$$

where ω_0 = sediment settling velocity in clear and still water, which is computed by Zhang and Xie (1993)'s formula. The sediment transport capacity is estimated by Wu et al. (2008b),

$$c_* = \frac{K}{\rho_s} \left(\frac{\rho}{\rho_s - \rho} \frac{u^3}{gh\omega_s} \right)^m \tag{10}$$

where $K = 0.4515$; $m = 0.7414$.

Based on the previous experimental findings covering sediment concentration lower than 300 kg m^{-3} (Zhu and Hao, 2008), we use a power law relation to estimate the Manning roughness:

$$n = n_r (1 + c_r - c)^\beta \tag{11}$$

where n_r = reference roughness; c_r = reference concentration; β = power exponent (>0). The reference roughness and concentration are set to the initial values (n_0, c_0) for each numerical case.

3 Numerical simulations

3.1 Numerical cases and model settings

In the 2004 hyperconcentrated flood, the measured peak discharge was 2690 m^3 s^{-1} at Xiaolangdi. About 16 h later, the peak discharge was 3990 m^3 s^{-1} at Huayuankou. At both stations, the sediment peak (around 350 kg m^{-3}) arrived later

than the flood peak. The tributaries only contribute to a very small discharge (200 m^3 s^{-1}), which are therefore neglected in the present numerical study.

This flood has been numerically investigated in a 1-D schematic channel of constant width. The channel is 226.6 km long and 1000 m wide with an initial bed slope of 2.55×10^{-4}, representing the Xiaolangdi-Jiahetan reach of the Lower Yellow River. The Huayuankou hydrological station is almost half way of this reach ($x = 125.8$ km). Uniform sediment is considered: sediment median diameter $d_{50} = 0.02$ mm, bed porosity $p = 0.45$. At the upstream boundary, the discharge and concentration measured at Xiaolangdi are prescribed; at the downstream boundary, the stage-discharge relationship at Jiahetan is used. The initial steady-uniform flow condition is assumed: $h_0 = 0.96$ m, $u_0 = 0.6688$ m s^{-1}, $n_0 = 0.02328$. The initial sediment concentration is assumed at the capacity state as $c_0 = 0.00438$.

Three cases are simulated to investigate the relative role of bed roughness change and bed deformation on peak discharge increase. Using a fully coupled model, Case 1 considers the contribution of bed deformation on the flow volume (Eq. 4) and a varied bed roughness with sediment concentration (Eq. 11). Case 2 also uses a fully coupled model to include the bed deformation contribution while deploying a constant bed roughness. In Case 3, a varied bed roughness is considered while the effect of bed deformation on the flow volume is neglected by using a partially coupled model (Eq. 5).

3.2 Numerical results and discussions

3.2.1 Discharge hydrographs and concentrations

Figure 1 shows the discharge hydrographs at distinct locations for the three cases. The results of Cases 1 and 3 are obtained by setting the parameter $\beta = 3$ in the roughness equation (Eq. 11). When considering a varied roughness with concentration, Cases 1 and 3 compute a downstream increasing peak discharge (see the first flood peak) in the upper reach ($x = 0$ km to $x = 125.8$ km). During the first flood peak, the sediment peak (Fig. 2), which is slightly behind the peak discharge, results in a considerable roughness reduction. Therefore, the later parts of the flood wave may experience less friction and overtake the wave front causing the peak discharge to increase. In the lower reach (> 125.8 km), the peak discharge is still larger than that at Xiaolangdi ($Q = 2690$ m^3 s^{-1} at Xiaolangdi) but decreases downstream probably due to a more diffusive roughness effect with distance increase. Qualitatively, the numerical results of Cases 1 and 3 are in line with the observations that peak discharge increase only occurs in the upper reach for the 2004 flood (Jiang et al., 2006).

Moreover, the contribution of bed deformation to the flow volume is trivial for the 2004 flood as the difference between Cases 1 and 3 is small. Only incorporating the bed

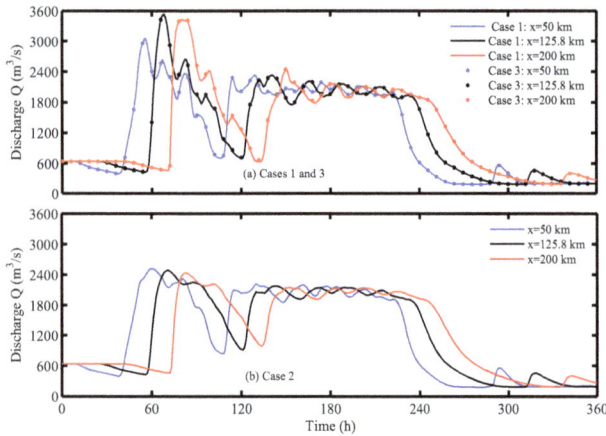

Fig. 1. Discharge hydrographs at distinct locations for **(a)** Cases 1 and 3, **(b)** Case 2.

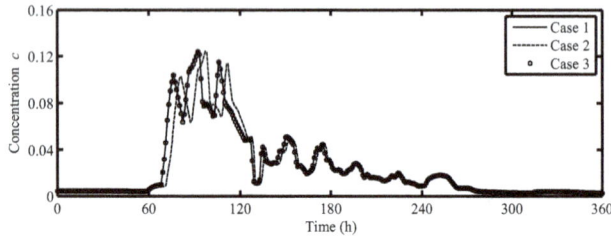

Fig. 2. Sediment concentrations at the Huayuankou station ($x = 125.8\,\text{km}$).

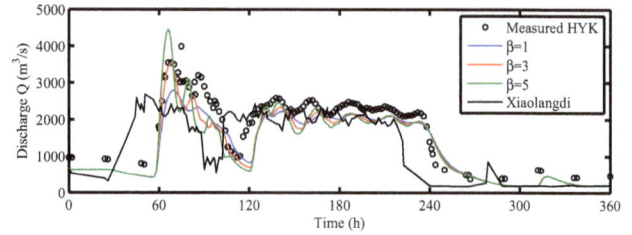

Fig. 3. Computed and measured discharge at Huayuankou, with reference to the discharge at Xiaolandi.

deformation effect, Case 2 fails to compute a downstream increasing peak discharge when using a constant bed roughness. Therefore, the effect of bed roughness reduction with concentration is more important to the peak discharge increase than the bed deformation contribution for the 2004 flood.

In addition, there is no peak discharge increase during the second flood peak (after $t = 120\,\text{h}$) in all the three cases. This is mainly because the sediment concentration is small and changes very slowly (compared to the first flood peak) in this period (Fig. 2), and the effect of roughness reduction cannot be sufficiently strong to induce a discharge increase. Therefore, the difference of the second flood peak between Case 2 and Cases 1–3 is not as significant as the difference of the first flood peak.

3.2.2 Influence of downstream stage-discharge relationship

It should be noted that an unjustified downstream boundary condition could cause errors to the computational region. It is unclear whether the outlet stage-discharge relationship drawn from the Jiahetan measurement will affect the findings of downstream peak discharge increase in our 1-D simplified channel. Therefore, an auxiliary computation has been con-

ducted to remove this impact by re-simulating Case 1 in a sufficiently long channel of 800 km (about the full length of the Lower Yellow River) with a constant water level at the downstream boundary. The results (not shown) illustrate that at the two locations ($x = 125.8\,\text{km}$ and $x = 226.6\,\text{km}$), the differences of discharge hydrographs are negligible between the short reach (226.6 km) and the long reach (800 km), though the computed water level has some small discrepancy at $x = 226.6\,\text{km}$ where the downstream boundary of the short reach is located. This implies the stage-discharge relationship based on the Jiahetan measurement is justified for the 1-D simplified short reach and the computed peak discharge increase is not forced by a downstream boundary condition.

3.2.3 Comparisons of the computation and measurement

At $x = 125.8\,\text{km}$ where the Huayuankou hydrological station is located, the computed results with $\beta = 3$ (in Case 1) can well reproduce the measured discharge hydrographs (Fig. 3). For the first flood peak, the computed discharge peak is about 32 % larger than the discharge at Xiaolangdi, which is comparable to the measured data of 48 %.

As the present work focuses on the effects of roughness change and bed erosion on the peak discharge increase, sensitivity analysis has been done for the parameters that directly influence these processes. Different values of the parameter β (Eq. 11), and the different sediment transport capacity formulae that are suitable for the conditions of the hyperconcentrated flow in the Yellow River, are analyzed. The results show a downstream peak discharge increase can be reproduced when β is larger than 1 and the magnitude of discharge peak increases with β (Fig. 3). This is valid when different formulae of the sediment transport capacity are used (not shown). In this paper, we present and analyze the computed results by $\beta = 3$ because best agreement between the computed and measured discharge peak (also the roughness reduction) is obtained with this value.

4 Conclusions

The relative importance of bed roughness reduction and bed erosion-related flow volume increase to the downstream peak discharge increase during hyperconcentrated floods is numerically investigated by a newly developed coupled morphodynamic model. The role of bed erosion is highlighted by comparing two model versions: a fully coupled model and a partially coupled model. The effect of bed roughness reduction is considered by a power law relation between the sediment concentration and Manning roughness. The 2004 hyperconcentrated flood with moderate concentrations is modeled in a 1-D schematic Xiaolangdi-Jiahetan reach of the Lower Yellow River.

When considering a decreasing bed roughness with concentration, a downstream peak discharge increase can be computed for the 2004 flood, no matter whether the effect of bed erosion is included or not. While only incorporating the bed erosion effect, the computed peak discharge decreases downstream. Therefore, the effect of bed roughness reduction with concentration is more important than the bed erosion contribution to the peak discharge increase of the 2004 flood. The rapid and considerable roughness reduction immediately behind the flood peak may accelerate the later parts of the flood waves thus leading to the downstream peak discharge increase.

Quantitative uncertainties are embedded in the empirical relations, parameters and simplified topography. Best agreement between the computed and measured results is obtained by setting the parameter $\beta = 3$ in the roughness equation. Improvements can be made by using a more realistic topography and advancing the basic understanding of the sediment concentration effects on bed roughness.

Acknowledgements. This research is supported by the China Scholarship Council (2008621194) and the Sino-Dutch collaboration project (08-PSA-E-01).

References

Cao, Z. X., Pender, G., Wallis, S., and Carling, P.: Computational dam-break hydraulics over erodible sediment bed, J. Hydraul. Eng.-ASCE, 130, 689–703, 2004.

Cao, Z. X., Pender, G., and Carling, P.: Shallow water hydrodynamic models for hyperconcentrated sediment-laden floods over erodible bed, Adv. Water Res., 29, 546–557, 2006.

He, L., Duan, J. G., Wang, G. Q., and Fu, X. D.: Numerical simulation of unsteady hyperconcentrated sediment-laden flow in the Yellow River, J. Hydraul. Eng.-ASCE, 138, 958–969, 2012.

Jiang, E. H., Zhao, L. J., and Wei, Z. L.: Mechanism of flood peak increase along the Lower Yellow River and its verification, J. Hydraul. Eng., 37, 1454–1459, 2006 (in Chinese).

Li, G. Y.: Analysis on mechanism of peak discharge increasing during flood routing in lower reaches of Yellow River, J. Hydraul. Eng., 39, 511–517, 2008 (in Chinese).

Li, S. C. and Duffy, C. J.: Fully coupled approach to modeling shallow water flow, sediment transport, and bed evolution in rivers, Water Resour. Res., 47, W03508, doi:10.1029/2010WR009751, 2011.

Li, W., de Vriend, H. J., Wang, Z., and van Maren, D. S.: Morphological modeling using a fully coupled, total variation diminishing upwind-biased centered scheme, Water Resour. Res., 49, 3547–3565, 2013.

Qi, P., Sun, Z. Y., and Qi, H. H.: Flood discharge and sediment transport potentials of the Lower Yellow River and development of an efficient flood discharge channel, Yellow River Hydraulics Publisher, Zhengzhou, China, 2010 (in Chinese).

Stecca, G., Siviglia, A., and Toro, E .F.: Upwind-biased FORCE schemes with applications to free-surface shallow flows, J. Comput. Phys., 229, 6362–6380, 2010.

Wan, Z. and Wang, Z. Y.: Hyperconcentrated flow. IAHR monograph series, Balkema, Rotterdam, the Netherlands, 1994.

Wang, G. Q. and Xia, J. Q.: Channel widening during the degradation of alluvial rivers, Int. J. Sediment Res., 16, 139–149, 2001.

Wang, Z. Y., Qi, P., and Melching, C. S.: Fluvial hydraulics of hyperconcentrated Floods in Chinese rivers, Earth Surf. Process. Land., 34, 981–993, 2009.

Wu, B. S., Wang, G. Q., Xia, J. Q., Fu, X. D., and Zhang, Y. F.: Response of bankfull discharge to discharge and sediment load in the Lower Yellow River, Geomorphology, 100, 366–376, 2008a.

Wu, B. S., van Maren, D. S., and Li, L. Y.: Predictability of sediment transport in the Yellow River using selected transport formulas. Int. J. Sediment Res., 23, 283–298, 2008b.

Zhang, R. J. and Xie, J. H.: Sedimentation research in China-systematic selections, China Water Power Press, Beijing, 1993.

Zhu, C. J. and Hao, Z. C.: A study on the resistance reduction of flows with hyper-concentration in open channel, in: International Workshop on Education Technology and Training & 2008 International Workshop on Geoscience and Remote Sensing, Shanghai, China, 141–144, 2008.

4

How kilometric sandy shoreline undulations correlate with wave and morphology characteristics: preliminary analysis on the Atlantic coast of Africa

D. Idier[1] and A. Falqués[2]

[1]R3C, DRP, BRGM, Orléans, France
[2]Applied Physics Department, UPC, Barcelona, Spain

Correspondence to: D. Idier (d.idier@brgm.fr)

Abstract. Sandy coasts are characterized by a number of rhythmic patterns like, amongst others, shoreline undulations or sandwaves at a kilometric scale. One hypothesis for their formation is that high angle waves (large incidence angle with respect to shore normal) could induce an instability of the shoreline (Ashton et al., 2001). More recently, a scaling for their wavelength has also been proposed (van den Berg et al., 2014). The existing studies rely mainly on modelling but quantitative field tests are lacking. We aim at investigating how both the formation hypothesis of these shoreline undulations and the theoretical scaling do fit with nature at a global scale. The first step, which is the goal of this paper, is to set up the methodology by analyzing the Atlantic African coast as test site. First, based on global databases, shoreline wavelength L_S, wave characteristics (obliquity θ_W and wavelength λ_W) and mean shoreface slope β are determined. Then the wave obliquity is confronted with the presence of shoreline undulations. Finally the values of the ratio $\beta L_S / \lambda_W$ are estimated and discussed in comparison with the estimate of van den Berg et al. (2014). It is found that the correlation between shoreline sandwave occurrence and wave obliquity is very good, allowing the identification of 5 new potential unstable shoreline stretches, whereas the results on the scaling are not conclusive and deserve further investigations.

1 Introduction

Sandy coasts are characterized by a number of rhythmic patterns like, for instance, cusps (metric scale), megacusps (hundreds of meters) and shoreline sandwaves (kilometric scale). The processes involved in the formation of cusps and megacusps have already been studied by field observation and modelling showing that cusps are related to swash zone processes and megacusps are related to surf zone processes. However, shoreline sandwaves are less known and according to some modelling studies (see, e.g., Falqués and Calvete, 2005 and Falqués et al., 2011) they would be mainly controlled by shoaling area processes.

On one hand it is theoretically clear that a rectilinear shoreline can be unstable in case of high wave incidence angle (High Angle Wave Instability: HAWI) and that from such instability a number of large scale shoreline features may appear (Ashton et al., 2001), including these kilometric scale shoreline sandwaves. This is now widely supported by a number of modelling studies, e.g., Ashton et al. (2006a), van den Berg et al. (2012), Kaergaard and Fredsoe (2013a), showing that the instability develops for deepwater wave incidence θ_W larger than about $42°$, with respect to shoreline normal. In case of using the CERC formula for the sediment transport, other formulae giving a range between $35°$ and $50°$.

On the other hand, looking at a global scale, it seems that kilometric sandy shoreline undulations do occur on many coasts: Fig. 1 shows 29 identified shoreline sandwave sites. 15 have been already identified and investigated in previous studies whereas the 14 others have been identified by eye by the authors based on punctual Google Earth visit, i.e.,

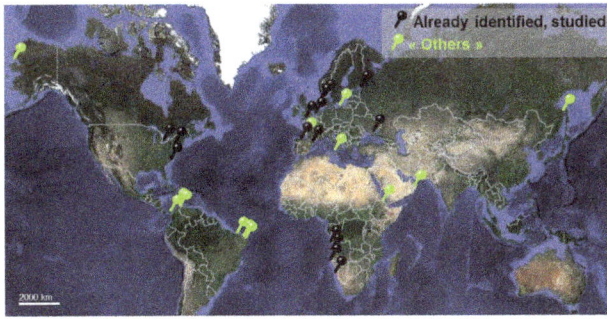

Fig. 1. Map of identified kilometric shoreline sandwave sites. Background image: Google Earth.

satellite images. No exhaustive identification has been done yet, so that the existence of many other kilometric shoreline undulations can be suspected.

A first research question (Q1) is therefore: how does the hypothesis of HAWI origin of shoreline undulations fit with nature? Presently, it is still not clear if the observed shoreline sandwaves result from HAWI. At some sites it seems that HAWI could be responsible for the origin and persistence of such undulations. For example, Ashton and Murray (2006b) studied the local wave climate north shore of Lake Erie (Canada) suggesting that unstable sandwaves have shaped the spit there (see also Davidson-Arnott and van Heyningen, 2003). Falqués (2006) investigated whether the subtle but systematic shoreline sandwaves along the Dutch coast (Ruessink and Jeuken, 2002) could be related to HAWI. That coast appeared to be at the threshold for instability and can be stable/unstable depending of some of the hypothesis of the study. Kaergaard and Fredsoe (2013b) investigated possible HAWI occurrence on the West coast of Denmark. The mean wave approach is quite oblique there, from the NW, but the angle with the shore normal is nearly at the threshold of instability. Kaergaard and Fredsoe (2013b) have done the corresponding stability analysis and concluded that only if the large storm waves were excluded from the wave climate the coastline would be unstable and shoreline sandwaves with the observed wavelength would emerge. Kaergaard et al. (2012) investigated also the cross-shore extent of coastline undulations on a site located on the West coast of Denmark, based on specific bathymetric surveys providing temporal and spatial data. This field approach allows a better understanding, but cannot be used for a global analysis, because of the lack of such bathymetric surveys. Thus, no extensive data analysis in a large scale environment has been done.

In addition, recent model experiments propose that the wavelength L_S of sandwaves initially emerging from high angle wave instability scales with the wavelength of surface waves λ_W (in deep water) divided by the mean shoreface slope β (van den Berg, 2012; van den Berg et al., 2014): $L_S \approx c\lambda_W/\beta$, with $c = O(1)$. A second research question

(Q2) is then: how does this model based scaling fit with nature?

The present research aims at investigating kilometric sandy coastline undulations in a wide geographical perspective (global scale) within an effort to answer to the two questions Q1 and Q2. The present paper sets up the methodology for such analysis, based on the two following investigations: (1) correlation between wave obliquity and shoreline sandwave occurrence, (2) correlation between shoreline sandwave wavelength and wave and bathymetric parameters. As a first step, we focus on the Atlantic coast of Africa, a "natural environment", exposed to energetic wave conditions.

2   Method and data

From the literature, HAWI's occurrence relates with wave obliquity θ_W, whereas HAWI wavelength L_S should scale with wavelength λ_W and mean shoreface slope β. Thus, the 4 quantities should be estimated: L_S, θ_W, λ_W and β. Such estimate requires shoreline (for L_S), wave (for θ_W, λ_W) and bathymetric data (for β). Within an effort to set up a global method, such data analysis is based on the joint use of global shoreline, wave and bathymetric databases.

The selected shoreline database is the WVS® (NGA) database. The shoreline corresponds to high water contour. It is obtained from satellite images LANDSAT 2000. The accuracy is comprised between 250 and 50 m, whereas the spatial resolution is about 100 m. This shoreline data is processed in order to provide shoreline orientation (needed to estimate wave obliquity) and shoreline wavelengths. The shoreline wavelengths are obtained by analysing 40 km long sections of shoreline. Figure 2, which, for sake of clarity, shows only a portion of the entire processed Atlantic African coast, illustrates the type of processing which is done. First the raw data are spatially filtered to remove very long undulations of several hundreds of kilometres (these long oscillations can be seen on Fig. 2, middle panel). Then, for every 40 km section, a Fast Fourier Transformation is done focusing on the wavelengths ranging between 1 and 20 km. This FFT provides several amplitude peaks. In the analysis, among the 4 largest ones, we keep only the one which has the largest amplitude (in blue) and the one which has the smallest wavelength (in red), respectively called L_{SD} and L_{SS}.

The wave data comes from the IOWAGA project (Rascle and Ardhuin, 2013). Within that project global and local wave hindcasts have been done, using the WW3 model and the CEP wind data. In the present study we use the global wave hindcast, and, to set up the method, we focus on the 2012 year. From this database, we obtain the three following yearly averaged wave characteristics: significant wave height H_S, peak period T_p and direction α. They are used not only to estimate the parameters θ_W, λ_W, but also to characterise

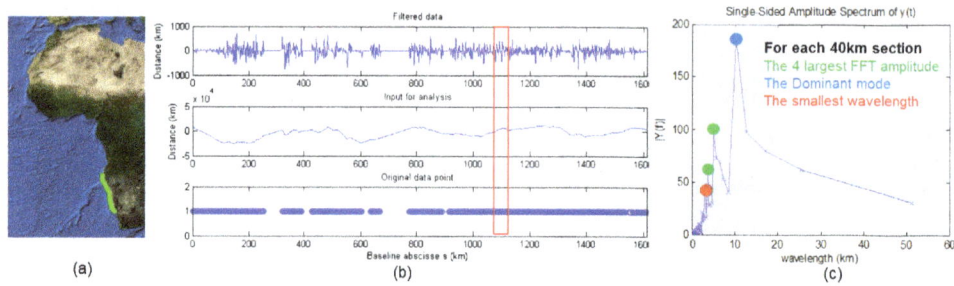

Fig. 2. Shoreline data analysis for the south portion of the Atlantic coast of Africa **(a)**: **(b)** input, filtered and original point data, **(c)** Fast Fourier Transformation shoreline amplitude on the 40 km section indicated by the red area. The points on Fig. 2c indicate the local maxima of FFT amplitude.

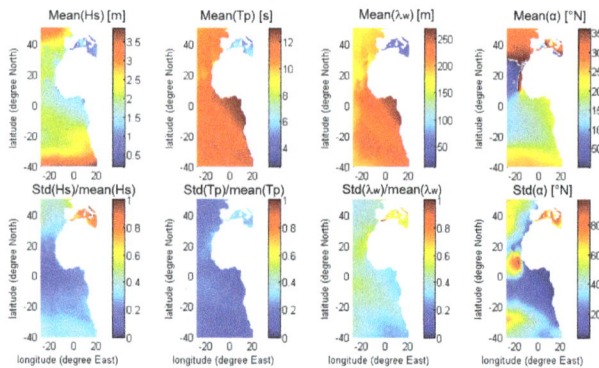

Fig. 3. 2012 Wave Climate. Yearly averaged values of: significant wave height H_s (m), peak direction T_p (s), wavelength λ_w (m), wave direction α (°, clockwise from North), together with the normalized standard deviation. Source: IOWAGA wave hindcast database (Rascle and Ardhuin, 2013).

the wave climate of the study area (see Fig. 3). The spatial resolution is 0.5°.

For the bathymetry, the GEBCO (NOAA) database is used (Amante and Eakins, 2009), with a spatial resolution of $1'$ (~ 1.9 km in the study area). This database allows estimating the slope β. In the present analysis, this slope is computed based on the distance between the shoreline and the 100 m bathymetric contour. It should be said that the ocean wave wavelength is quite large in this area (about 200 m, Fig. 3), such that the computed slope corresponds to the slope between the bathymetric contour of the ocean wave base ($\lambda_W/2$) and the shoreline. The bathymetric data is also of use for intermediate step, like estimating the location of the limit of wave action on the sea-bed (wave base) (wave parameters must be estimated there, and not too far or too close from the coast).

3   Wave obliquity and Kilometric coastline undulation

Figure 3 shows the main characteristics of the wave climate over the 2012 year. First, we can notice that the yearly aver-

aged significant wave height ranges from about 1.5 to 3 m, with larger values at the North and South of the Atlantic Africa coast. The peak period are comprised between 10 and 13s, with the largest values in the central part of the coast. The wave direction comes mainly from the North – North West ($\alpha \sim 350°$) in the northern part of the studied area, and from the South – South West ($\alpha \sim 200°$) in the southern part. At this stage, even without any wave obliquity computation, regarding the shoreline orientation, we can expect areas of strong wave obliquity. These values are yearly averaged values, such that it is worthwhile to analyse the standard deviation of these parameters. Figure 3 shows that these parameters are quite constant (small standard deviation), except for the wave direction at the 10° latitude. This can be explained by the fact that, in this area, the wave climate is characterised by two dominant wave directions: one from the North and one from the South (see wave rose on Fig. 4). This is consistent with the location of the limit between the north wave dominated areas, and the south wave dominated areas.

The next step is to estimate quantitatively the wave obliquity θ_W along the entire coast and to determine whether or not wave obliquity is correlated with the existence of shoreline undulations. Figure 4 shows the areas of high-angle wave incidence (blue and red, such that $|\theta_W| > 45°$) and low-angle wave incidence (grey, such that $|\theta_W| \leq 45°$). There are two main areas subject to high wave obliquity (one at the North (blue) and one at the South (red)), and one area subject to low angle waves. As a preliminary investigation, a crude analysis of the shoreline is done, based on Google Earth in order to identify large shoreline undulations (satellite images) but also the sediment composition of the coast (photos) and the presence of anthropic modifications (hard defences, harbour, ...). The type of the coast has also been validated using local studies (e.g. Fayet, 2010). Figure 4 shows the roughly estimated sandy areas, as well as satellite images in the three main areas. In the Northern area, subject to high obliquity from the North, large kilometric shoreline undulations are observed in three main locations (green marker, Fig. 4), but not systematically along the coast. All the observed undulations are asymmetric towards the South, indicating a

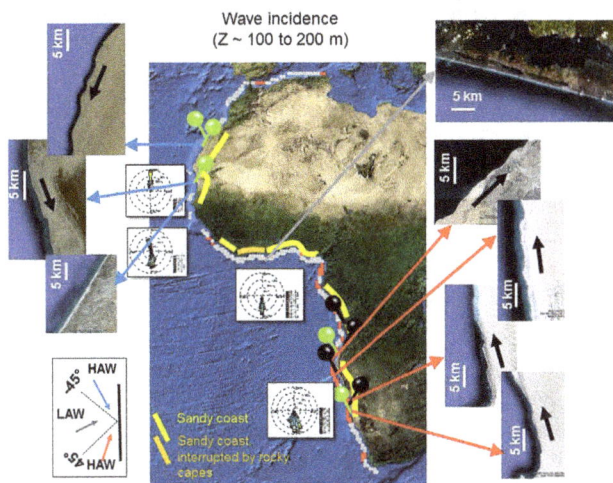

Fig. 4. Wave incidence (2012) and shoreline undulations. Incidence larger than $45°$ and smaller than $-45°$ are indicated resp. in red and blue on the shoreline, whereas low angle waves are indicated in grey. HAW means High Angle Wave. Green symbols indicate shoreline undulations not identified as such before this study. Black ones indicate already identified shoreline sandwaves. Wave roses illustrate the wave direction. Satellite images (Google Earth) illustrate the undulated or un-undulated character of the shoreline.

southward migration. This is consistent with the mean wave direction oriented toward the South. These are new potential sites for HAWI that had not been previously identified. In the area subject to low wave obliquity (grey arrow), the shoreline is highly straight, without any undulation. In the southern part (red), subject to highly oblique waves from the South, many shoreline undulations are observed. In these areas, van den Berg (2012a) already observed shoreline undulations probably related to HAWI's (black markers), whereas the present study has identified a few more potential sites for HAWI (green markers). Again, the shoreline undulations are asymmetric, now toward the North, which is consistent with the mean wave direction (oriented towards the North too).

4   Shoreline wavelength, wave and morphology

From the databases analysis, the following parameters have been computed: L_{SD}, L_{SS}, θ_W, λ_W and β. Figure 5 shows the longshore variation of the shoreline wavelengths of the dominant mode (L_{SD}) and the smallest wavelength (L_{SS}), as explained in Sect. 2, for 2 shoreline sections exposed to high angle wave incidence: one at the South (Fig. 5, left panel), one at the North (Fig. 5, right panel). These sections are respectively 1600 and 3100 km long. In the southern area the dominant mode has a wavelength varying by a factor 2, with a mean wavelength of about 10 km, whereas the smallest wavelength varies between 4 and 6 km. At the North, the dominant mode has a larger wavelength, in average equal to

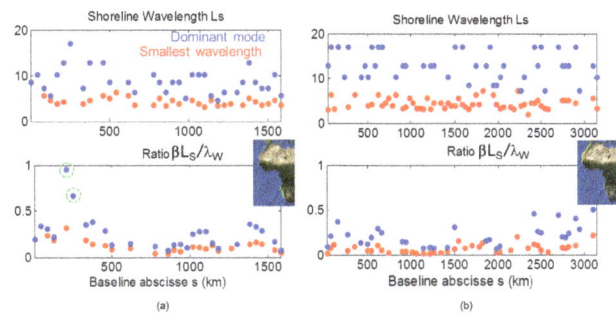

Fig. 5. Shoreline wavelength and computed ratio for (a) the South (red area) and (b) North part (yellow area) of the Atlantic coast of Africa.

12 km, with spatial variations of a factor 2 too, whereas the smallest wavelength has a mean value of about 4 km.

As explained in the introduction, van den Berg et al. (2014) proposed a scaling of the shoreline wavelength with the wave wavelength divided by the slope. Thus, the ratio $\beta L_S / \lambda_W$, is computed for every 40 km long sub-sections. Figure 5 (bottom panel) shows the spatial variations of this ratio. For the southern part, which is characterised by a wave period of about 12 s, a significant wave height of 2.5 m, and a slope β smaller than 0.02, if we exclude two outliers (green circles), the ratio is comprised in the following ranges: [0.1–0.4] for the dominant wavelength L_{SD}, [0.05–0.3] for the smallest wavelength L_{SS}. Thus, whatever the wavelength type, the ratio is in the range [0.05–0.4]. In the Northern part, which is characterised by wave period of about 10 s, a significant wave height of 2 m, and a slope β smaller than 0.005, the ratio is in the range [0.02–0.5] ([0.1–0.5] for the dominant wavelength and [0.02–0.2] for the smallest wavelength). Table 1 summarizes the ratio variations and the wave and slope conditions. As it can be seen, the parameter variations are not very large (large wave height, large wave period, small slope), such that no significant correlation can be found. Coming back to the model experiment of van den Berg et al. (2014), we can notice that the experiment has been done for a Dean profile, a shoreface slope β ranging from 0.004 to 0.013 (with a mean value of 0.09) and a wave height of 1 m. However, using these results for the wave period range corresponding to the ones along the two studied shoreline portions, we can deduce from the model a ratio value ranging between 0.1 and 0.4. Table 1 summarizes the comparison between the ratio obtained from data, and the ones obtained from the van den Berg (2012) study. As a first draft comparison, the data provides a ratio ranging between 0.02 and 0.5, whereas the model provides values varying between about 0.1 and 0.4. Thus, the results are of the same order of magnitude, roughly consistent but not conclusive.

Several limitations make the data – model scaling comparison not straightforward. First, the model of van den Berg et al. (2014) provides the wavelength of the initial shoreline

Table 1. Ratio $\beta L_S / \lambda_W$ obtained from data analysis (this paper) and the modelling results (van den Berg et al., 2014), with the corresponding wave (for the data analysis: averaged over year 2012) and morphological characteristics.

	Data analysis	Modelling results
Ratio $\beta L_S / \lambda_W$	[0.02–0.5]	[0.1–0.4]
Wave height H_S	[2–3] m	1 m
Wave period T_p	[10–12.5] s	[10–12.5] s
Slope β	<0.02	[0.004–0.013]
Bathymetry	Complex	Dean profile
Nature of L_S	Largest amplitude (L_{SD}) and smallest wavelength (L_{SS})	Fastest growing mode in the linear regime (initial wavelength)

undulations. However, in nature, non-linear processes occur, such that the shoreline is characterised by many wavelengths at the same locations. Also, larger wavelengths than the initial one can form with time. At the end, it is difficult to determine which wavelength corresponds to the initial one. Second, the bathymetry of the study area is complex and not longshore uniform whereas the one used in the model analysis was based on a longshore uniform Dean profile. Thus, in reality there is no longshore uniform "basic state" and high angle waves are modified by updrift bathymetry with a cross-shore profile that can be different from the local one. Third, the Atlantic coast of Africa is exposed to larger wave height and period than the ones used in the model analysis. Forth, as suggested by Kaergaard and Fredsoe (2013b), the smallest waves within the wave spectrum arrive with a larger obliquity at the coast and can contribute to the shoreline instability even more than the dominant waves. Finally, the sensitivity of the ratio results should be analysed taking into account the data quality.

5 Conclusions

The present paper sets up a global scale methodology to answer to the two questions Q1 and Q2: "how does the hypothesis of HAWI origin of shoreline undulations fit with nature?", "how does the model based wavelength scaling fit with nature?". The proposed methodology is based on three databases: WVS for the shoreline, IOWAGA for the waves and GEBCO for the bathymetry. These databases allow estimating the parameters required for the analysis: shoreline wavelength L_S, wave obliquity θ_W and wavelength λ_W, and the mean shoreface slope β. The application on the Atlantic coast of Africa provides some preliminary results. First, a good qualitative correlation has been found between wave obliquity and shoreline sandwave occurrence (Q1). Areas of high wave obliquity are characterized by the presence of shoreline undulations while areas of low obliquity are characterized by straight shoreline. Even new shoreline sandwave sites not previously described were discovered from the knowledge of the wave climate (i.e., wave obliquity on a sandy coast seems to be a good predictor of shoreline undula-

tions). These first results are promising for the identification of potential HAWI sites at the global scale. Within future explorations, using wavelet analysis would be worthwhile to investigate shoreline undulation (Tebbens et al., 2002; Lazarus et al., 2011) and automatically detect potential HAWI sites. Second the analysis of the ratio between the shoreline wavelength L_S, the mean shoreface slope β and wave wavelength λ_W, in comparison with the scaling proposed by van den Berg et al. (2014), does not provide conclusive results (Q2). Indeed, the data analysis provides a ratio ($\beta L_S / \lambda_W$) ranging from 0.02 to 0.5, which is only "roughly" consistent with the tested scaling. This illustrates that the comparison between data analysis results and model based formula is not straightforward. There are several reasons (complex bathymetry, several shoreline wavelengths, wave climate, data quality) for such inconclusive results. Further investigations should be done, based on global scale analysis, together with a wider model exploration.

Acknowledgements. Funding from BRGM (CCRC and ShorelineSW projects), the ANR-Carnot institution and the Spanish government (CTM2009-11892/IMNOBE project and TM2006-08875/MAR project) are acknowledged. The authors are grateful to the providers of databases: NOAA (GEBCO and WVS) and F. Ardhuin (IOWAGA). The authors like also to thank D. Calvete, N. Van den Berg and F. Ribas for fruitful discussions, as well as the reviewers of this paper (K. Ells and anonymous) for their comments.

References

Amante, C. and Eakins, B. W.: ETOPO1 1 arc-minute global relief model: Procedures, data sources and analysis, NOAA Technical Memorandum NESDIS NGDC-24, 19 pp., 2009.

Ashton, A. D. and Murray, A. B.: High-angle wave instability and emergent shoreline shapes: 1. Modeling of sand waves, flying spits, and capes, J. Geophys. Res., 111, F04011, doi:10.1029/2005JF000422, 2006a.

Ashton, A. D. and Murray, A. B.: High-angle wave instability and emergent shoreline shapes: 2. Wave climate analysis and comparisons to nature, J. Geophys. Res., 111, F04012, doi:10.1029/2005JF000423, 2006b.

Ashton, A. D., Murray, A. B., and Arnault, O.: Formation of coastline features by large-scale instabilities induced by high-angle waves, Nature, 414, 296–300, 2001.

Davidson-Arnott, R. G. D. and van Heyningen, A.: Migration and sedimentology of longshore sand waves, Long Point, Lake Erie, Canada, Sedimentology, 50, 1123–1137, 2003.

Falqués, A.: Wave driven alongshore sediment transport and stability of the Dutch coastline, Coast. Eng., 53, 243–254, 2006.

Falqués, A. and Calvete, D.: Large scale dynamics of sandy coastlines: Diffusivity and instability, J. Geophys. Res., 110, C03007, doi:10.1029/2004JC002587, 2005.

Falqués, A., Calvete, D. and Ribas, F.: Shoreline Instability due to Very Oblique Wave Incidence: Some Remarks on the Physics, J. Coastal Res., 27, 291–295, 2011.

Fayet, I. B. N.: Dynamique du trait de côte sur les littoraux sableux de la Mauritanie à la Guinée-Bissau (Afrique de l'Ouest): Approches régionale et locale par photointerprétation, traitement d'images et analyse de cartes anciennes, PhD Thesis, UBO, 2010.

Kaergaard, K. and Fredsoe, J.: Numerical modeling of shoreline undulations part 1: Constant wave climate, Coast. Eng., 75, 64–76, 2013a.

Kaergaard, K. and Fredsoe, J.: Numerical modeling of shoreline undulations part 2: Varying wave climate and comparison with observations, Coast. Eng., 75, 77–90, 2013b.

Kaergaard, K., Fredsoe, J., and Knudsen, S. B.: Coastline undulations on the West Coast of Denmark: Offshore extent, relation to breaker bars and transported sediment volume, Coast. Eng., 60, 109–122, 2012.

Rascle, N. and Ardhuin, F.: A global wave parameter database for geophysical applications. Part 2: Model validation with improved source term parameterization, Ocean Model., 70, 174–188, 2013.

Ruessink, B. G. and Jeuken, M. C. J. L.: Dunefoot dynamics along the dutch coast, Earth Surf. Proc. Land., 27, 1043–1056, doi:10.1002/esp.391, 2002.

van den Berg, N.: Modelling the dynamics of large scale shoreline sand waves. PhD Thesis, Applied Physics Department, Universitat Politecnica de Catalunya, 2012.

van den Berg, N., Falqués, A., and Ribas, F.: Modelling large scale shoreline sand waves under oblique wave incidence, J. Geophys. Res., 117, F03019, doi:10.1029/2011JF002177, 2012.

van den Berg, N., Falqués, A., Ribas, F., and Caballeria, M.: On the mechanism of wavelength selection of self-organized shoreline sand waves, J. Geophys. Res., doi:10.1002/2013JF002751, accepted, 2014.

5

Urban MEMS based seismic network for post-earthquakes rapid disaster assessment

A. D'Alessandro[1,2], D. Luzio[2], and G. D'Anna[1]

[1]Istituto Nazionale di Geofisica e Vulcanologia, Centro Nazionale Terremoti, Rome, Italy
[2]Università degli Studi di Palermo, Dipartimento di Scienze della Terra e del Mare, Palermo, Italy

Correspondence to: A. D'Alessandro (antonino.dalessandro@ingv.it)

Abstract. In this paper, we introduce a project for the realization of the first European real-time urban seismic network based on Micro Electro-Mechanical Systems (MEMS) technology. MEMS accelerometers are a highly enabling technology, and nowadays, the sensitivity and the dynamic range of these sensors are such as to allow the recording of earthquakes of moderate magnitude even at a distance of several tens of kilometers. Moreover, thanks to their low cost and smaller size, MEMS accelerometers can be easily installed in urban areas in order to achieve an urban seismic network constituted by high density of observation points. The network is being implemented in the Acireale Municipality (Sicily, Italy), an area among those with the highest hazard, vulnerability and exposure to the earthquake of the Italian territory. The main objective of the implemented urban network will be to achieve an effective system for post-earthquake rapid disaster assessment. The earthquake recorded, also that with moderate magnitude will be used for the effective seismic microzonation of the area covered by the network. The implemented system will be also used to realize a site-specific earthquakes early warning system.

1 Introduction

In recent decades, the population growth and the consequent expansion of urban centers, often close to major industrial areas, have led to a significant increase in exposure of the urban areas to the risk induced by earthquakes. When a strong earthquake occurs, the loss of human lives depends primarily on the intensity of the shaking, on the vulnerability of the buildings and on the effectiveness of the rescue operations in the immediate post-quake. The impact of a strong earthquake on an urban center can be considerably reduced by an emergency management center, through timely and targeted actions in the immediate post-earthquake.

A real-time Urban Seismic Network (USN), consisting of a high density of stations installed on the urban center to monitoring, could provide immediate alert and post-earthquake information summarized in maps of ground motion parameters, which might allow to greatly improving the effectiveness of rescue operations. The centers for post-earthquake emergency management could use this information to decide the action priorities in order to minimize the loss of human lives, with optimally managing the available resources. Unfortunately, the high costs associated with the construction and installation of traditional seismic stations has made impossible (until now) the realization of a seismic network at urban scale.

The recent technological developments in the field of MEMS (Micro Electro-Mechanical Systems) sensors, can now allow the creation of a USN at low cost. MEMS device are a highly enabling technology with a huge commercial potential. In the 90s, MEMS sensors revolutionized the automotive airbag system and are today widely used in laptops, games controllers and mobile phones. Thanks to the great commercial success, the research and development of MEMS technology actively continues all over the world.

Due to their versatility, MEMS sensors are increasingly being used in a wide field of science, including the physical, engineering and medical one. In the last decade, a number of research institutes in the fields of geophysics and seismology gained interest in this promising technology. Nowadays, the sensitivity and the dynamic range of these sensors are such as to allow the recording of earthquakes of moderate magnitude even at a distance of several tens of kilometers (D'Alessandro and D'Anna, 2013; Evans et al., 2014). Moreover, because of their low cost and small size, MEMS accelerometers can be easily installed in urban areas in order to achieve an USN constituted by very densely spaced stations. In California the development of seismic networks consisting of MEMS sensors has already started, including the Quake-Catcher Network (Cochran et al., 2009), operated by Stanford University, and the Community Seismic Network (Clayton et al., 2011) operated by the California Institute of Technology. However, the aforementioned networks are still at an experimental stage as they are built on the basis of the activities of volunteers. It is clear however, that the international community of seismology is focusing on that technology that could revolutionize in a short time the way to monitor earthquakes (Chung et al., 2011; Cochran et al., 2012; Kohler et al., 2013; Lawrence et al., 2014; D'Alessandro, 2014b).

At the beginning of 2014, under an agreement between the University of Palermo, the National Institute of Geophysics and Volcanology and the Department of Sicilian Civil Protection (Italy) has been funded a project called MEMS (Monitoring of Earthquakes through MEMS Sensors). The MEMS project involves the realization of an USN constituted of stations based on MEMS technology. As a pilot site for the realization of the MEMS project was chosen the municipality of Acireale (Sicily, Italy), whose urban areas are among those with the highest seismic hazard and vulnerability to the earthquake effects in Italy (Azzaro et al., 2010, 2013).

The network that is being achieved will be able to measure both the translational and the rotational component of the wave field generated by an earthquake. The MEMS stations will be located inside strategic and sensitive buildings (i.e. characterized by high vulnerability and exposure) such as schools, hospitals, public buildings and places of worship. The recorded waveforms will be processes in real-time by an automated system to determine, following a moderate to strong earthquake, several shaking parameters that will be used to create shake maps at the urban scale. These maps can be used by the competent authorities, like the Civil Protection, for the optimization of the immediate post-quake interventions.

In the following, after the characterization of Acireale in terms of historical seismicity and related seismic hazard, we describe as we are realizing the USN and the post-earthquakes rapid disaster assessment system. Both performances, potentialities and possible future developments will be discussed.

2 Acireale setting

The area selected for the realization of the USN is the Acireale municipality (Italy). Acireale is located in the middle of the Ionian coast of the Sicily, at the southeastern slopes of the Etna volcano (Fig. 1). The municipality covers an area of about $40 \, \text{km}^2$ with a population in excess of 50 000

(population density of 1300 inhabitants per km^2). The city settled in the current area as early as the fourteenth century. The town is sited on a lava plateau called Timpa, with an average altitude of about 160 m.

The urban plan is typical of late-medieval cities of Sicily. Acireale is regarded as an important economic, cultural and artistic heritage. The center of Acireale has many historic buildings and places of worship of great artistic value, which are considered to be sites of national heritage. The National Institute for Conservation and Restoration has identified more than a hundred of historic buildings of great cultural and artistic value having high seismic vulnerability. Acireale, which has grown considerably in the second half of the nineteenth century as the "City of the Studies", now gathers a large number of primary, secondary and higher secondary, public and private schools, used by most students of neighboring municipalities. Figure 2 shows the location of more than 200 sites with high exposure and vulnerability to the earthquake identified among the historic buildings and schools of the Acireale municipality.

The town of Acireale is exposed to the effects of damage of both regional earthquakes and local volcano-tectonic events (Fig. 1). Among the first one there are several strong earthquakes such as those of 1169 and 1693, located in the southeastern Iblean, and that of 1908, with epicenter in the Strait of Messina. The earthquake of 4 February 1169, hit the entire eastern Sicily and the southern part of Calabria causing major destruction. This earthquake shook all the villages that arose in the current area of Acireale. The earthquake of 1169 caused the almost total destruction of the villages and the dispersion of the population in the hinterland with the birth of the various Aci villages. In 1693, on two occasions, on 9 and 11 January, two earthquakes hit a vast area of eastern Sicily hard. The effects were catastrophic in about 40 towns in the Noto Valley. The damage caused by the two earthquakes were enormous, since it can be considered the greatest catastrophe in the history of the Italian seismicity. Acireale was devastated, more than sixty per cent of its built heritage was destroyed. There were 739 victims on a population of nearly 13 000 inhabitants. The earthquake of 28 December 1908, had its epicenter in the Messina Straits and caused very serious damages and collapses in all towns that stood on the northern and eastern slopes of Mount Etna, also taking several deaths in Acireale.

To such catastrophic events, also some destructive local earthquakes must be added, as that of 1818 ($M = 6.2$) and the numerous volcano-tectonic events generated by the Etna volcano. Even if such events have generally low magnitude ($M < 5$), because of its shallow depth ($< 2 \, \text{km}$), in the past have caused very serious damage and even destruction with maximum epicentral intensity of X of the European Macroseismic Scale (EMS-98, Grünthal, 1998).

Figure 1a shows the distribution of earthquakes that occurred in the Etna region from 1669 to 2008. In less than 200 years over 190 earthquakes occurred that

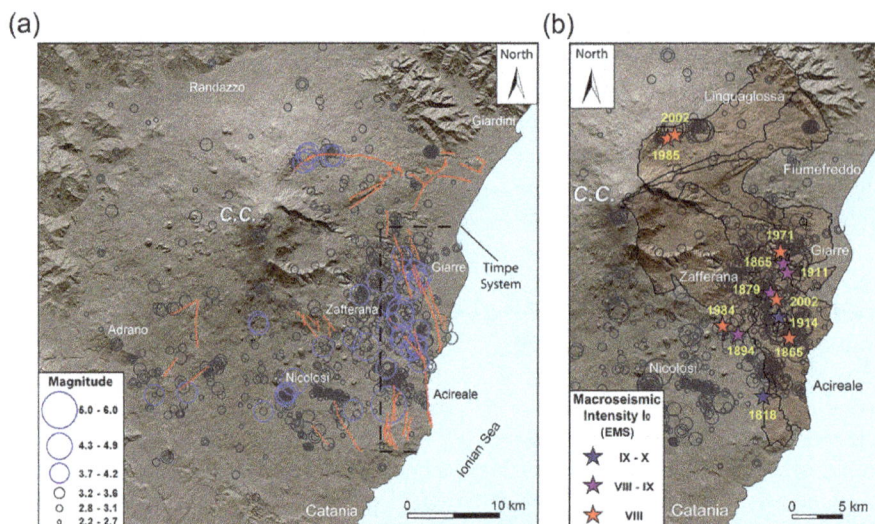

Figure 1. (a) Distribution of Etna earthquakes occurred between 1669 and 2008 (Azzaro et al., 2010). In blue the epicenters of earthquakes of magnitude $M \geq 3.7$, corresponding to an epicentral intensity $I_0 \geq$ VII EMS, in gray smaller events ($I_0 \leq$ VII EMS). The red lines indicate major faults outcropping (dashes on the low side); **(b)** earthquakes with epicentral intensity $I_0 \geq$ VIII EMS-98, corresponding to a magnitude $M \geq 4.1$, occurred from 1669 to 2008 in the Acireale and surrounding areas (Azzaro et al., 2010).

caused damage, almost one every year. These earthquakes have caused severe damage ($I_0 =$ VII–VIII EMS-98) every 15 years and destruction ($I_0 =$ IX–X EMS-98) with loss of human lives every 30 years (Azzaro et al., 2010). Figure 1b instead shows the earthquakes with epicentral intensity $I_0 \geq$ VIII EMS-98, corresponding to magnitudes $M \geq 4.1$, occurred from 1669 to 2008 in the Acireale and surrounding areas (Azzaro et al., 2010). Most of these events were generated by the Timpe fault system on which stands the Acireale town. This system is characterized by a set of tectonic structures with complex failure mechanisms, strongly heterogeneous from the point of view of seismo-tectonic behavior.

In recent times, several neighborhoods of the Acireale town were damaged by the mainshock (M_L 4.4) of the seismic sequence that started on 29 October 2002. Following this earthquake, more than 400 buildings of Acireale were declared uninhabitable.

Recent seismic microzonation studies (Azzaro et al., 2010) have highlighted the extreme variability of the Acireale substrate foundation (Fig. 2). This shows a great lithological variability, showing rapid lateral and vertical variations from over-consolidated clays, to massive lava, slag or anthropogenic carryovers. The strong geological variations have made the characterization of Acireale in seismic perspective very difficult and the seismic microzonation unsatisfactory. These abrupt changes of the geology (in particular the presence of low S wave velocity body in contact with a high S wave velocity body) may generate strong site effects that can significantly amplify the intensity of shaking during an earthquake. Current and recent floods in contact with old rocks are present in the Acireale urban area (Fig. 2). A

detailed microzonation of the urban center of Acireale, based on direct observation of the shaking caused by an earthquake, would therefore be highly desirable, especially looking at the results of recent studies of seismic hazard of the eastern slope of the Etna (Azzaro et al., 2013).

3 Instruments, methods and performance

MEMS devices have been recognized as one of the most promising technologies of the XXI century, able to revolutionize both the industrial world and that of the consumer products. MEMS devices have dimensions on the order of microns and are made directly on a silicon substrate. The construction technique is fairly simple and therefore economical, because is the same of any integrated circuit (photolithographic process). The electromechanical microsystems are nothing more than a set of devices of various kinds (mechanical, electrical and electronic) highly miniaturized integrated on the same silicon substrate, combining electrical properties of the semiconductor integrated with the opto-mechanical one. Therefore, these "intelligent" systems combine electronic and mechanical functions in a very small space.

The operation of MEMS can be described considering the integrated circuit as the "brain" of the system that makes possible the monitoring of the surrounding environment through the other devices (sensors) found on the same chip. A MEMS device is therefore a system capable of receiving information from the environment by translating the physical quantities into electrical impulses. The sensors can measure phenomena of various kinds: mechanical (sound, acceleration

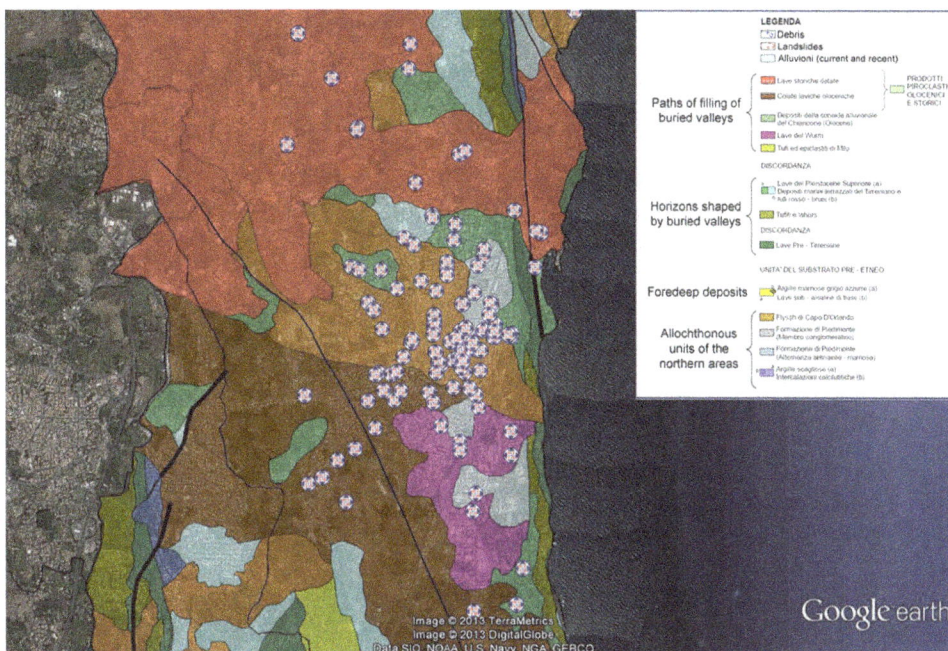

Figure 2. Geological map of Acireale and surrounding (Azzaro et al., 2010) with located more than 200 candidate sites identified among historical buildings and schools with high seismic vulnerability and exposure.

and pressure), thermal (temperature and heat flux), biological (cell potential), chemical (pH), optical (intensity of light radiation, spectroscopy), magnetic (intensity of flow).

Nowadays, the sensitivity and the dynamic range of several low-cost MEMS accelerometers are such as to allow the recording of earthquakes of moderate magnitude even at a distance of several tens of kilometers (D'Alessandro and D'Anna, 2013; Evans et al., 2014). The MEMS device used for the construction of the stations that will constitute the USN, is the model 1044_0 (3/3/3 PhidgetSpatial Precision High Resolution) produced by the Canadian company Phidget Inc. The sensor was chosen from among hundreds of MEMS devices on the market based on its performance, cost and in relation to the objectives that the MEMS project aims to achieve.

The MEMS device model 1044_0 (Fig. 3) integrates a three-axis capacitive accelerometer and is also equipped with a gyroscope and a magnetometer, both tri-axial. The circuit of transduction is internal to the device and is of the digital type, for which the outputs are already in a digital format and proportional to the measured quantity. The integrated accelerometer is able to measure both constant accelerations (usable as a tilt sensor) or variable in time (used to measure the oscillation induced by an earthquake). It features three sensing elements oriented along three mutually orthogonal axes able to measure the acceleration vector of the wavefield generated by an earthquake. All accelerometers manufactured by Phidgets are individually calibrated by absolute measurements of gravity along the three axes, and this guarantees the high accuracy of the sensors. The recent paper

Figure 3. MEMS model 1044_0 (3/3/3 PhidgetSpatial Precision High Resolution) produced by the Canadian company Phidget Inc.; the reported measures are in inches.

published by Evans et al. (2014) and further laboratory tests, confirms the high performance of this MEMS accelerometer and its suitability for the earthquakes monitoring.

The integrated accelerometer operates in high sensitivity mode for acceleration less of ± 2 g (g $= 9.80665$ m s^{-2}); in this range, the sensor provides a linear output signal. This value guarantees that the sensor is able to record even very strong accelerations, such as those induced by a catastrophic earthquake, remaining in linear regime. The resolution or sensitivity of the sensor is 76.3μg, which implies a dynamic range greater than 44 dB. The bandwidth, which is the frequency band in which the system has linear response, is 0–497 Hz. As well known the frequencies generated by an earthquake are at most of the order of tens of Hz. Those of interest for earthquake engineering, i.e. those that can create more damage to buildings triggering resonance phenomena, are included in a more narrow range (approximately 0.1–20 Hz) and therefore largely contained in the frequency band of this MEMS accelerometer. The MEMS accelerometer self-noise has standard deviation of 280 μg. It is clear that any seismic acceleration less than this value would be hardly detectable in the signals acquired by the sensor. However, this value of standard deviation does not constitute a limitative factor for the aim of the MEMS project because is of the same order of that expected for the seismic noise in urban areas.

In order to evaluate the performance of the USN in terms of detection magnitude, was determined the Power Spectrum Density (PSD) of the self-noise generated by the tri-axial accelerometer integrated into the MEMS device model 1044_0. The self-noise noise was determined by acquiring 72 h of signal, using six MEMS 1044_0 devices, at a site characterized by very low seismic noise. The acceleration PSD's calculated for all three components of the six sensors are very similar; therefore their average value has been calculated (Fig. 4a). In the frequency range 0.05–40 Hz, the PSD lies between approximately −58 and −70 dB $(m\,s^{-2})^2\,Hz^{-1}$ and shows a monotonically decreasing trend as a function of frequency. For a comparison, in Fig. 4a is also shown the upper Peterson (1993) seismic noise reference PSD. This PSD indicates the maximum expected power for the seismic noise of sites where the bedrock outcrop, located far from urban center or industrial areas, and refers only to the vertical component. The self-noise produced by the MEMS sensor is quite a bit higher than the Peterson (1993) reference curve, but low enough to allow recording of earthquakes of moderate magnitude, and therefore suitable for the purposes of the project.

Moreover, as well known, human and industrial activities generate an intense noise in a wide range of frequencies, which can exceed of 20–30 dB the upper Peterson (1993) curve. In addition, the ambient noise recorded on the horizontal components, is generally much larger than that recorded on the vertical component, in particular in the presence of site effects. For the aims of the MEMS project, the stations will be installed in the urban center, sampling all outcropping lithologies. For these reasons, would be useless to use accelerometers with better performance but much more expensive. Figure 4b shows a local earthquake of M_L 4 recorded by a MEMS 1044_0 devices and by a professional force balance accelerometer at the epicentral distance of about 35 km; the PSD of this earthquake is reported in Fig. 4a. It is possible to observe that, in a relative wide range of frequencies, the spectrum of the earthquake exceeds in power that of the self-noise of the MEMS accelerometer.

In order to evaluate the performance of such MEMS sensor in detecting the PGA on the horizontal components generated by an earthquake, we used the empirical attenuation law obtained by Ambraseys et al. (1996) and Simpson (1996). This attenuation law has been obtained from the analysis of 422 European earthquakes with hypocentral depth less than 30 km, magnitude between 4 and 7.9 and epicentral distance

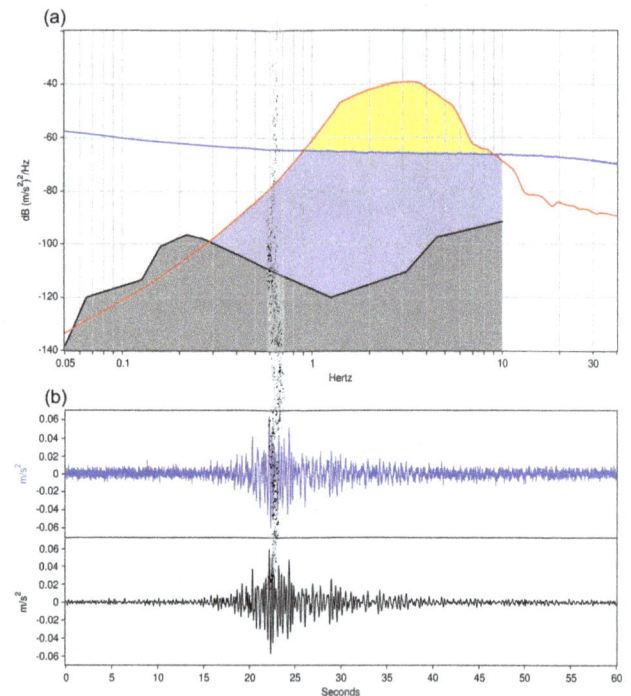

Figure 4. (a) Average PSD of the self-noise produced by the MEMS 1044_0 device accelerometer (blue line) with over plotted the mean horizontal earthquake spectra (red line) of a shallow local seismic event with M_L 4 recorded at the epicentral distance of about 35 km by the same device. For comparison it is also over plotted the upper Peterson (1993) seismic noise reference PSD (black line); **(b)** raw NS component waveform of the local event described in **(a)** as recorded by a MEMS 1044_0 device accelerometer (blue line) and a professional force balance accelerometer (black line).

between 0 and 260 km. Using this empirical law we determined the PGA values (horizontal components) generated by earthquakes of magnitude (M_S) between 3 and 7.8 and epicentral distance between 0 and 300 km. These PGA values were compared with the values of standard deviation of the self-noise of the tri-axial accelerometer integrated in the MEMS device model 1044_0. Figure 5 shows the values of magnitude and hypocentral distance for which the MEMS accelerometer would be able to record the PGA of the horizontal components, with a signal to noise ratio greater than 10. The figure shows how the PGA generated by events of magnitude 3 may be recorded up to hypocentral distances of about 70 km, in the case of bedrock outcrops, and of about 130 km, in the case of soft soil. The PGA generated by earthquakes of greater magnitude can be clearly recorded also at hypocentral distances significantly higher.

In addition, the Single Board Computer (SBC) used for the realization of the station (described follow) has various digital inputs that may allow the addition of other MEMS sensors. Adding additional MEMS accelerometers to a single station could improve the signal to noise ratio by means of the synchronously stacking of the signals. Assuming that

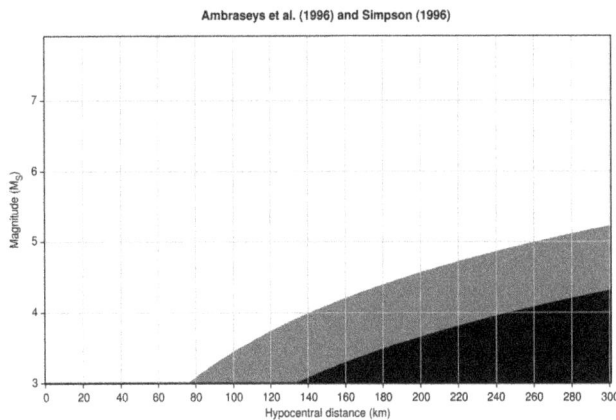

Ambraseys et al. (1996) and Simpson (1996)

Figure 5. The graph shows, as a function of the magnitude (M_S) and of the hypocentral distance, the detection performance of the accelerometer integrated into MEMS device model 1044_0. The chart was obtained comparing the values of standard deviation of the noise self-produced by the MEMS sensor with the PGA values, expected on the horizontal components, determined by the empirical law of Ambraseys et al. (1996) and Simpson (1996). The white and gray areas are those where the expected PGA exceeds the standard deviation of the MEMS accelerometer self-noise by a factor of 10, for sites with soft sediments and bedrock outcrops, respectively.

the coherent signal (in our case the earthquake waveform) is the same in the N recordings, while the self-noise is not correlated, the synchronously stacking of the signals would allow an increase in the signal to noise ratio equal to $\sqrt{N}$.

However, a tri-axial accelerometer alone does not allow the full characterization of the ground motion generated by an earthquake. The motion of the medium sampled by the wavefield generated by an earthquake can be described splitting the wave field into two parts: translational and rotational. Because of the difficulties associated with the realization of highly sensitive instruments capable of measuring with accuracy and precision the rotational motion generated by an earthquake, the modern seismology is based on the measurement of the translational component only. However, numerous scientific papers report observations of rotational motions generated by earthquakes (Lee et al., 2009a, b). In the last decade a new field of seismology has developed, called Rotational Seismology, which studies all the aspects related to rotation motions that can be induced by earthquakes. The two recent monographs Teisseyre et al. (2006, 2008) and an entire special issue devoted to this topic by the Seismological Society of America (Lee et al., 2009a) confirm the importance of this rapidly developing branch of seismology. Therefore, the complete characterization of the ground shaking caused by an earthquake should include both translational and rotational measures.

As previously stated, the MEMS device model 1044_0 also includes a tri-axial gyroscope, which is an instrument capable of measuring the angular velocities of all three axes (the same of the accelerometer). In precision mode, the

gyroscope seems to have good performance (resolution of 0.02 and 0.013° s⁻¹ on horizontals and vertical axis, respectively, self-noise standard deviation of 0.095° s⁻¹, and maximum measurable speed > 300° s⁻¹). Unfortunately, since the still small number of experimental observations made in the field of Rotational Seismology, it is not possible to estimate the maximum possible rotations expected due to an earthquake and therefore determine the effectiveness of this sensor for the monitoring of earthquakes.

The MEMS device has a low power consumption, estimated at a maximum current consumption of 55 mA. Power can be supplied via a simple USB connection with a potential difference of between 4.4 and 5.3 V. The sampling step can be set between a minimum of 4 ms to a maximum of 1 s, in local acquisition mode, and in a minimum of 12 ms for a maximum of 1 s, in remote acquisition mode. These sampling steps, according to the Nyquist theorem, allow reconstruction without aliasing of frequencies up to 125 and 41 Hz in local or remote mode, respectively. These values ensure the possibility to properly observe all frequencies of interest in the field of earthquake seismology. The analog to digital conversion is done automatically by means of a 16-bit digital converter. Furthermore, the temperature range of operation is extremely large (−40 to 85 °C) ensuring the operation of the device even in the most extreme temperature conditions. All specifications above described, together with the small size of the device (3 × 3.5 × 0.4 cm) and its low cost, make this MEMS device suitable for the implementation of the USN.

Another sensor included in the MEMS device is the tri-axial magnetometer. This magnetometer can be used as an electronic compass facilitating, during the installation of a station, the correct orientation of the horizontal components with respect to the North. Such information, together with the values of the acceleration of gravity, allow the correct installation of the MEMS station without any external reference. In addition, once station is installed it is possible to remotely control any changes in these parameters caused by accidental movement of the device.

For the realization of the MEMS stations we also used the SBC 1073_0 even produced by the Phidgets Inc (Fig. 6). The SBC handles the pre-processing of the acquired signals and their routing and forwarding, via TCP/IP, to a server properly laid out for the collection and analysis of data recorded by all stations that will constitute the USN. The SBC is provided with a network adapter with an Ethernet port and WiFi antenna for the connection to the network. In addition to the MEMS device described above, the SBC manages the data acquired from a GPS antenna, necessary for the time synchronization of the acquired signals. The SBC is provided with several input and output ports that can easily allow the integration of other sensors in a technological upgrade.

An appropriately sized buffer battery, capable of powering the station in case of electrical blackout, accompany the described system. All the elements described so far, are integral part of the seismic station and included in an appropriate

Figure 6. SBC model 1073_0 produced by the Canadian company Phidget Inc.; measurements are in inches.

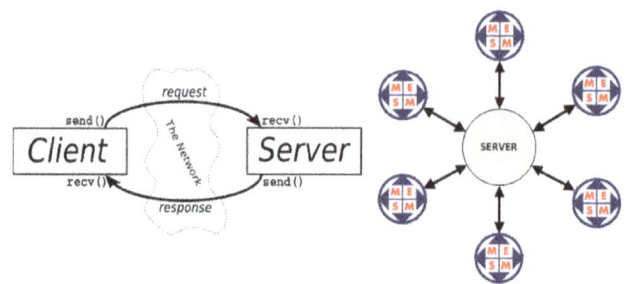

Figure 7. Schematic example of the client-server system that will be implemented under the MEMS project.

case that protects the electro-mechanical devices from strong shocks or the presence of water.

The total cost of realization of a MEMS station is estimated to be less of EUR 600. This cost is much lower therefore that of a standard strong motion station and allows the realization of a high density of stations USN at very low costs.

Suitable sites will be sought between the strategic and sensitive buildings (i.e. those with high exposure and vulnerability to earthquake), such as schools, hospitals, public buildings and places of worship, already highlighted for the Acireale urban center in Fig. 2. Additional sites will also be identified and evaluated during the project. Station will be located among them, within the same buildings or in areas close to these locations. During the identification of the suitable sites for installation of the MEMS stations, we will assess the structural nature of the building and the ability to access the foundation plans. For each building, we will carefully evaluate possible soil-structure interaction and the stations will be installed at the foundation of structures. The final geometry of the network will provide the most comprehensive and homogeneous coverage of the urban area, taking into account the local geological conditions and the values of exposure and vulnerability of the identified sites.

The data acquired by the MEMS network will be sent in real time to a Server that will handle their collection and processing. The system will be built according to the traditional Client-Server scheme (Fig. 7); the individual MEMS stations will be the Clients while a properly sized computer, will act as a Server. The signals will be transmitted using the TCP-IP Internet Protocol. The Server will be installed at the INGV seismic room, where specialized staff, present 24 h on 24 and 7 days a week, can monitor the operation of the system.

The server will process the signals transmitted by different stations in order to extract the different parameters that can be used to assess the potential damage due to of an earthquake. The potential damage will be estimated by an automatic system, appropriately implemented, that will determine some parameters, such as the maximum of the PGA and the Arias (1970) intensity of an earthquake, measured by each station of the network. These parameters will be used to

construct, in almost real-time, shake maps for the area covered by the network. Suitable algorithms will be developed for avoiding false alarms that may be generated during anthropogenic transients generated in proximity of one or more MEMS station. The implemented algorithms will allow the creation of a shake map only if a significant percentage of the MEMS stations simultaneously detect an event. In case an earthquake is established, the server will send the information and the shake maps produced to the competent authorities for the actions in the immediate post-earthquake, such as Civil Protection.

4 Conclusion and general remarks

The realization of the USN would have a high socio-economic impact, providing a useful tool to reduce the seismic risk, capable of increasing the safety of the population of the urban area covered by the network. The impact of a strong seismic event on an urban community can be reduced with a well-planned and timely action by the centers for the emergency management. The MEMS project will lead to the realization, in the Acireale Municipality, of the first real-time USN for the post-earthquakes rapid disaster assessment. It would be an important tool for civil protection authorities by providing timely alerts that allow a rapid and effective mobilization of the post-earthquake interventions. The implemented system will be able to timely provide the distribution of the intensity of the ground shaking due to an earthquake. The shaking maps would allow the involve centers, the optimal management of priorities and the allocation of resources so as to achieve a significant reduction in the number of victims following an earthquake. The rescue operations and verification of damage to buildings could then be carried out according to a logic of priority on the basis of the highest urban shaking measured by the seismic network. It would minimize secondary effects induced by an earthquake allowing the protection of critical infrastructure and may allow the economic and social structures to return to the "normal" as quickly as possible.

The MEMS project would lead to the realization of the first network able to record both translation and rotational component generated by a strong earthquake, leading to enormous advances in the field of Rotational Seismology. The data collected could be used for the accurate reconstruction of the seismogenic processes and seismogenetic structures that generated the earthquakes recorded. This would lead to enormous advances in the understanding of the earthquake generation phenomena.

The data acquired, also after a low magnitude earthquake, will provide important information for seismic microzonation of the Acireale Municipality. From Fig. 2 we can see that the candidate sites are located on lithologies with very different mechanical behavior that may generate strong site effects. The use of direct measurements of ground motion generated by an earthquake could help to significantly improve the Acireale seismic microzonation and critically evaluate the results of the previous investigations obtained by empirical or indirect methods. This could possibly lead to the development and validation of new techniques and new approaches to the assessment of the site effects.

The project would lead to the development of innovative algorithms for real-time analysis of accelerometric signals and for the construction of shake maps at urban scale. Great progress could be made in the field of the Earthquake Early Warning System (EEWS). The extreme characteristics of the ground motion generated by a strong earthquake could be estimated in the first few seconds of the initial P wave allow realizing a site-specific EEWS. This EEWS could be usable to enhance in real-time the safety margin of specific critical engineered systems mitigating the seismic risk by reducing the exposure of the facility by automated safety actions.

The prototypes of seismic stations, designed and built with the latest MEMS technology, will lead the acquisition of an important technological know-how. This know-how could be used to improve the performance of existing seismic networks (D'Alessandro et al., 2011a, b, 2012b, 2013a, c; D'Alessandro and Stickney, 2012; D'Alessandro and Ruppert, 2012) or for the realization of highly miniaturized sensors usable in extreme environmental conditions, like in volcanic areas (D'Alessandro et al., 2013d) or in ocean bottoms (D'Alessandro et al., 2009, 2012a, 2013b; Mangano et al., 2011; D'Alessandro, 2014a). Such knowledge could be also used for the further development of MEMS stations suitable to monitor also other environmental parameters of great scientific interest. The implemented system could revolutionize the way to monitor earthquakes and quickly be extended to all areas with high seismic risk in Italy and other countries.

Acknowledgements. We are grateful to the anonymous reviewers, to Elizabeth Cochran and the Editor Damiano Pesaresi for their constructive comments and suggestions. We want to thank Giuseppe Chiarenza, Calogero Foti and the Dipartimento Regionale della Protezione Civile (Sicily), for the technical and logistical support. Special thanks to Chester Fitchett and the Phidgets Inc. (Canada) without which this project would not have been achieved.

References

Ambraseys, N. N., Simpson, K. A., and Bommer, J. J.: Prediction of horizontal response spectra in Europe, Earthq. Eng. Struct. D., 25, 371–400, 1996.

Arias, A.: A measure of earthquake intensity, in: Seismic design of nuclear power plants, The MIT Press, 438–468, 1970.

Azzaro, R., Carocci, C. F., Maugeri, M., and Torrisi, A.: Microzonazione sismica del versante orientale dell'Etna, Studi di primo livello, Regione Siciliana, Dipartimento della Protezione Civile, Le Nove Muse Editrice, 184 pp., 2010.

Azzaro, R., D'Amico, S., Peruzza, L., and Tuvè, T.: Probabilistic seismic hazard at Mt. Etna (Italy): The contribution of local fault activity in mid-term assessment, J. Volcanol. Geoth. Res., 251, 158–169, doi:10.1016/j.jvolgeores.2012.06.005, 2013.

Chung, A. I., Neighbors, C., Belmonte, A., Miller, M., Sepulveda, H. H., Christensen, C., Jakka, R. S., Cochran, E. S., and Lawrence, J. F.: The Quake-Catcher Network Rapid Aftershock Mobilization Program Following the 2010 M 8.8 Maule, Chile Earthquake, Seismol. Res. Lett., 82, 526–532, doi:10.1785/gssrl.82.4.526, 2011.

Clayton, R. W., Heaton, T., Chandy, M., Krause, A., Kohler, M., Bunn, J., Guy, R., Olson, M., Faulkner, M., Cheng, M., Strand, L., Chandy, R., Obenshain, D., Liu, A., and Aivazis, M.: Community Seismic Network, Ann. Geophys.-Italy, 54, 6, doi:10.4401/ag-5269, 2011.

Cochran, E. S., Lawrence, J. F., Christensen, C., and Jakka, R. S.: The Quake-Catcher Network: Citizen Science Expanding Seismic Horizons, Seismol. Res. Lett., 80, 26–30, doi:10.1785/gssrl.80.1.26, 2009.

Cochran, E. S., Lawrence, J. F., Kaiser, A., Fry, B., Chung, A., and Christensen, C.: Comparison between low-cost and traditional MEMS accelerometers: a case study from the M 7.1 Darfield, New Zealand, aftershock deployment, Ann. Geophys.-Italy, 54, 728–737, doi:10.4401/ag-5268, 2012.

D'Alessandro, A.: The Marsili Seamount, the biggest European volcano, could be still active!, Curr. Sci., 106, p. 1339, 2014a.

D'Alessandro, A.: Monitoring of earthquakes using MEMS sensors, Curr. Sci., 107, 733–734, 2014b.

D'Alessandro, A. and D'Anna, G.: Suitability of low cost 3 axes MEMS accelerometer in strong motion seismology: tests on the LIS331DLH (iPhone) accelerometer, B. Seismol. Soc. Am., 103, 2906–2913, doi:10.1785/0120120287, 2013.

D'Alessandro, A. and Ruppert, N.: Evaluation of Location Performance and Magnitude of Completeness of Alaska Regional Seismic Network by SNES Method, B. Seismol. Soc. Am., 102, 2098–2115, doi:10.1785/0120110199, 2012.

D'Alessandro, A. and Stickney, M.: Montana Seismic Network Performance: an evaluation through the SNES method, B. Seismol. Soc. Am., 102, 73–87, doi:10.1785/0120100234, 2012.

D'Alessandro, A., D'Anna, G., Luzio, D., and Mangano, G.: The INGV's new OBS/H: analysis of the signals recorded at the Marsili submarine volcano, J. Volcanol. Geoth. Res., 183, 17–29, doi:10.1016/j.jvolgeores.2009.02.008, 2009.

D'Alessandro, A., Luzio, D., D'Anna, G., and Mangano, G.: Seismic Network Evaluation through Simulation: An Application to the Italian National Seismic Network, B. Seismol. Soc. Am., 101, 1213–1232, doi:10.1785/0120100066, 2011a.

D'Alessandro, A., Papanastassiou, D., and Baskoutas, I.: Hellenic Unified Seismological Network: an evaluation of its performance through SNES method, Geophys. J. Int., 185, 1417–1430, doi:10.1111/j.1365-246X.2011.05018.x, 2011b.

D'Alessandro, A., Mangano, G., and D'Anna, G.: Evidence of persistent seismo-volcanic activity at Marsili seamount, Ann. Geophys.-Italy, Scientific News, 55, 213–214, doi:10.4401/ag-5515, 2012a.

D'Alessandro, A., Danet, A., and Grecu, B.: Location Performance and Detection Magnitude Threshold of the Romanian National Seismic Network, Pure Appl. Geophys., 169, 2149–2164, doi:10.1007/s00024-012-0475-7, 2012b.

D'Alessandro, A., Gervasi, A., and Guerra, I.: Evolution and strengthening of the Calabrian Regional Seismic Network, Adv. Geosci., 36, 11–16, doi:10.5194/adgeo-36-11-2013, 2013a.

D'Alessandro, A., Mangano, G., D'Anna, G., and Luzio, D.: Waveforms clustering and single-station location of microearthquake multiplets recorded in the northern Sicilian offshore region, Geophys. J. Int., 194, 1789–1809, doi:10.1093/gji/ggt192, 2013b.

D'Alessandro, A., Badal, J., D'Anna, G., Papanastassiou, D., Baskoutas, I., and Özel, M. M.: Location Performance and Detection Threshold of the Spanish National Seismic Network, Pure Appl. Geophys., 170, 1859–1880, doi:10.1007/s00024-012-0625-y, 2013c.

D'Alessandro, A., Scarfì, L., Scaltrito, A., Di Prima, S., and Rapisarda, S.: Planning the improvement of a seismic network for monitoring active volcanic areas: the experience on Mt. Etna, Adv. Geosci., 36, 39–47, doi:10.5194/adgeo-36-39-2013, 2013d.

Evans, J. R., Allen, R. M., Chung, A. I., Cochran, E. S., Guy, R., Hellweg, M., and Lawrence, J. F.: Performance of Several Low-Cost Accelerometers, Seismol. Res. Lett., 85, 147–158, doi:10.1785/0220130091, 2014.

Grünthal, G.: European Macroseismc Scale 1998 (EMS-98), European Seismological Commission, subcommission on Engineering Seismology, working Groupo Macroseismic Scales, Conseil de l'Europe, Cahiers du Centre Euroéen de Géodynamique et de Séismologie, 15, Luxembourg, 99 pp., 1998.

Kohler, M. D., Heaton, T. H., and Cheng, M.-H.: The community seismic network and quake-catcher network: enabling structural health monitoring through instrumentation by community participants, Proc. SPIE 8692, Sensors and Smart Structures Technologies for Civil, Mechanical, and Aerospace Systems 2013, 86923X (19 April 2013), doi:10.1117/12.2010306, 2013.

Lawrence, J. F., Cochran, E. S., Chung, A., Kaiser, A., Christensen, C. M., Allen, R., Baker, J. W., Fry, B., Heaton, T., Kilb, D., Kohler, M. D., and Taufer, M.: Rapid Earthquake Characterization Using MEMS Accelerometers and Volunteer Hosts Following the M 7.2 Darfield, New Zealand, Earthquake, B. Seismol. Soc. A., 104, 184–192, 2014.

Lee, W. H. K., Celebi, M., Todorovska, M. I., and Igel, H.: Introduction to the special issue on rotational seismology and engineering applications, B. Seismol. Soc. Am., 99, 945–957, 2009a.

Lee, W. H. K., Igel, H., and Trifunac, M. D.: Recent Advances in Rotational Seismology, Seismol. Res. Lett., 80, 479–490, doi:10.1785/gssrl.80.3.479, 2009b.

Mangano, G., D'Alessandro, A., and D'Anna, G.: Long-term underwater monitoring of seismic areas: design of an Ocean Bottom Seismometer with Hydrophone and its performance evaluation, OCEANS 2011 IEEE Conference, 6–9 June, Santander, Spain, in OCEANS 2011 IEEE Conference Proceeding, 9 pp., doi:10.1109/Oceans-Spain.2011.6003609, 2011.

Peterson, J.: Observation and modelling of background seismic noise, U.S. Geol. Surv. Open-File Rept., Albuquerque, New Mexico, 93–322, 1993.

Simpson, K. A.: Attenuation of strong ground-motion incorporating near-surface foundation conditions, Ph.D. thesis, University of London, 1996.

Teisseyre, R., Takeo, M., and Majewski, E. (Eds.): Earthquake Source Asymmetry, Structural Media and Rotation Effects, Berlin: Springer, 2006.

Teisseyre, R., Nagahama, H., and Majewski, E. (Eds.): Physics of Asymmetric Continua: Extreme and Fracture Processes: Earthquake Rotation and Soliton Waves, Berlin & Heidelberg: Springer-Verlag, 2008.

6

Coral-rubble ridges as dynamic coastal features – short-term reworking and weathering processes

Michaela Spiske[1,2]

[1]Universität Trier, Geozentrum, Behringstr. 21, 54296 Trier, Germany
[2]Westfälische Wilhelms-Universität, Institut für Geologie und Paläontologie, Corrensstr. 24, 48149 Münster, Germany

Correspondence to: Michaela Spiske (spiske@uni-muenster.de)

Abstract. A coral-rubble ridge built by storm waves at Anegada (British Virgin Islands) underwent remarkable changes in shape and weathering in a 23-month period. The ridge is located along the island's north shore, in the lee of a fringing reef and a reef flat. This coarse-clast ridge showed two major changes between March 2013, when first examined, and February 2015, when revisited. First, a trench dug in 2013, and intentionally left open for further examination, was found almost completely infilled in 2015, and the ridge morphology was modified by slumping of clasts down the slope and by reworking attributable to minor storm waves. In size, composition and overall condition, most of the clasts that filled the trench resemble reworked clasts from the ridge itself; only a small portion had been newly brought ashore. Second, a dark gray patina formed on the whitish exteriors of the carbonate clasts that had been excavated in 2013. These biologically weathered, darkened clasts had become indistinguishable from clasts that had been at the ridge surface for a much longer time.

The findings have two broader implications. First, coastal coarse-clast ridges respond not solely to major storms, but also to tropical storms or minor hurricanes. The modification and reworking of the ridge on Anegada most probably resulted from hurricane Gonzalo which was at category 1–2 as it passed about 60 km north of the island in October 2014. Second, staining of calcareous clasts by cyanobacteria in the supralittoral zone occurs within a few months. In this setting, the degree of darkening quickly saturates as a measure of exposure age.

1 Introduction

This note documents short-term modifications of modern coarse-clast storm deposits that line part of the north shore of Anegada, British Virgin Islands. Anegada is a low-lying Caribbean island that faces the Puerto Rico Trench (Fig. 1). Nowhere more than 9 m a.s.l. (above sea level), the island is composed mainly of Pleistocene limestone and is fringed by sandy beach ridges and, locally, by a coral-rubble ridge that forms the island's coarsest deposits of modern storms. The rubble ridge hugs the island's central north shore, which faces the open Atlantic Ocean and is sheltered by a fringing reef which reduces the height of swell significantly. While the sandy shores on the western side of the island are subject to human modification, e.g. construction of beach houses, the central and eastern part of Anegada's north shore is very remote, can only be reached by foot, and is untouched by any anthropogenic activities. The ridge was found to extend discontinuously for 1.5–1.8 km along the shore (for detailed maps see Spiske and Halley, 2014). Its sedimentology was first surveyed in March 2013 along two transects (Spiske and Halley, 2014). A trench along one of these transects (transect II) was intentionally left open to be able to better detect any, even small-scaled, changes of ridge morphology or any other type of modification by surface processes, weathering or later inundation events (Fig. 2a and b).

Nearly two years later, in February 2015, additional field work revealed that the trench had been partly filled with coral rubble (Fig. 2c and d). This was unexpected because no strong hurricane and related surge had affected Anegada in the interim.

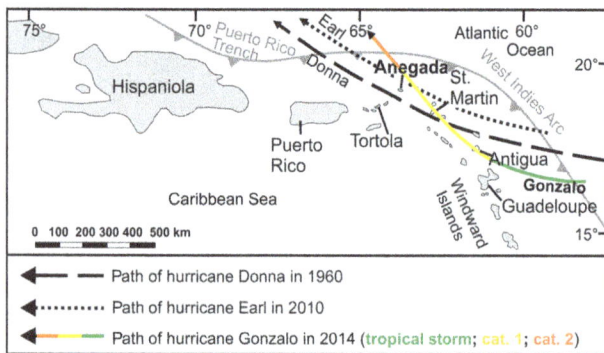

Figure 1. Location of Anegada and tracks of hurricane Donna (1960), Earl (2010) and Gonzalo (2014) that affected the island with different consequences in terms of storm surge inundation and sediment emplacement.

Here, the material that newly infills the trench is documented and compared with the composition of the ridge in 2013. It is asked if the clast size, type, and source of the new material are similar to the ridge composition in 2013. In addition, the degree and rapidness of surface weathering is addressed. These findings, give insights into the magnitude of inundation processes that can entrain and transport the respective material, and modify an existing ridge or build a new ridge.

2   Hurricanes on Anegada

The risk of hurricane-related storm surges inundating the low-lying coast of Anegada is discussed by Spiske and Halley (2014) and Atwater et al. (2012, 2014). Most notable surges and large swell that affected Anegada in the last decades were related to hurricane Donna in 1960 and Earl in 2010 (Fig. 1; Table 1).

Hurricane Donna passed about 15 km south of Anegada on 5 September 1960 with wind speeds of 115–120 knots, attaining category 3–4 (Dunn, 1961; National Oceanic and Atmospheric Administration, 2012; Fig. 1; Table 1). Eyewitness accounts suggest that Donna's trailing-left quadrant produced a 2.5 m storm surge on Anegada's south shore (Atwater et al., 2012). The storm's effects on the north shore are unknown.

Hurricane Earl passed about 30 km north of Anegada on 30 August 2010 as a category 4 hurricane with wind speeds of up to 115 knots (Cangialosi, 2011; Fig. 1; Table 1). Wrack lines were surveyed in 2011 at maximum elevations of 1.5 m a.s.l. from surge near the low-lying south shore and 2.0 m a.s.l. from surge and wash on the north shore (Atwater et al., 2014). The storm also suspended microbial matter in salt ponds of the island's interior. On the south shore, sand and lime mud as much as 10 cm thick were deposited on a sandy spillover fan that extends a few tens of meters inland

Table 1. Parameters of hurricanes Donna, Earl and Gonzalo at their closest position to Anegada.

Name	Date	Distance from Anegada	Category	Wind speed (knots)
Donna	5 Sep 1960	15 km south	3–4	115–120
Earl	30 Aug 2010	30 km north	4	115
Gonzalo	14 Oct 2014	60 km north	1–2	90

(Atwater et al., 2014). Along the north shore sand was eroded and transferred into the sea.

The only noteworthy event that happened between the surveys in March 2013 and February 2015 was hurricane Gonzalo (Fig. 1; Table 1). It passed Anegada with wind speeds of 90 knots, as a category 1–2 hurricane, ~60 km to its north on 14 October 2014 (Brown, 2015). Before reaching Anegada, Gonzalo made landfall on Antigua and St. Martin where waves and storm surge caused major damage to harbors and coastal structures (Brown, 2015). No severe impact was reported from Anegada.

3   Methods

The ridge was re-surveyed using the methods described in Spiske and Halley (2014). The trench was hand-dug from the mean tide level to the landward limit of the ridge-forming rubble, which apparently terminates at dense vegetation (Fig. 2a). For each of the 708 clasts in the former trench, long, intermediate, and short axes were measured, the clast type was determined, and angularity, roundness, karstification, encrustation, and borings were documented. After clearing the trench from any newly deposited clasts, the ridge thickness was measured with a tape measure in order to detect thickness changes between 2013 and 2015 (Fig. 3). The thickness of the infill was calculated by subtracting the height difference of the surface of the ridge filling from the ridge surface at the sides of the trench.

4   Characteristics of the ridge in 2013

When surveyed in March 2013 (Spiske and Halley, 2014), the ridge at transect II started at a distance of 8.6 m from the mean tide level and continued to 23.5 m inland, for a width of 14.9 m. The maximum thickness at the ridge crest was 0.8 m. The seaward side was steep. The ridge was composed of well-rounded clasts (96 %) with an average clast size of $16 \times 11 \times 4$ cm, the biggest clast was $75 \times 40 \times 5$ cm. Main components (Fig. 4a) of the clast-supported ridge were corals (61 %), reef rock (28 %), conch shells (5 %), serpulite rock (4 %), beach rock (1 %), and Pleistocene limestone (1 %). The coral species present (Fig. 4b) were *Acropora* (49 %; *A. palmata* and *A. cervicornis*), *Diploria* (26 %; *Diploria* sp. and *D. labyrinthiformis*), *Montastrea* (17 %;

Figure 2. Comparison of transect II (**a, b**) in 2013 and (**c, d**) in 2015. (**a, b**) The trench is free of any clasts in 2013. Clasts that were moved when the trench was dug are covering the ridge to its left and right side, depicting a fresh bright surface (area is circled in white and red) in contrast to the clasts that were exposed to surface weathering processes before 2013. (**c, d**) Infill of the trench as encountered in 2015 (white and green lines confine the sides of the former trench). (**c**) Dark grey surface of the ridge in 2015; the bright material moved in 2013 is no longer distinguishable, i.e. already strongly weathered within the elapsed two years. (**d**) Detailed view of the material that newly infills the trench, leaving only a linear depression that coincides with the former trench. The blue arrows indicate slumping of the upper parts of the former walls into the trench. For orientation, the location of a large *Acropora* coral that has not been moved, is marked by a yellow circle in all photos.

Montastrea sp. and *M. cavernosa*), *Porites* (6 %) and *Mille-porida* (2 %). About 80 % of the clasts were encrusted with *Homotrema rubrum* (Lamarck) or bored. Sand was only present in the lowermost ∼ 5 cm of the ridge. Some flattened clasts were imbricated in a direction that suggested emplacement as bed load during unidirectional landward flow.

Spiske and Halley (2014) referred the emplacement of the ridge to hurricanes stronger than hurricane Earl (category 4) in 2010 because no new ridge was formed, and the preexisting ridge was not significantly altered during the event. Earl was only capable of slightly reworking the lower sea-

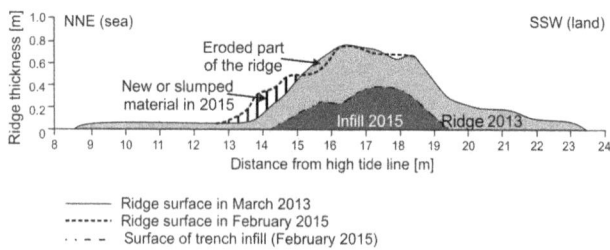

Figure 3. Comparison of the ridge thickness and morphology (at transect II) in March 2013 and February 2015. In the interim the trench dug at transect II had been partly refilled with $\sim 0.74\,\mathrm{m}^3$ of material.

ward parts of the ridge and of transporting few small pieces of coral rubble onshore.

5 Characteristics of the ridge in 2015

The position of the seaward onset and landward termination of the ridge had not changed since 2013. Its maximum thickness was still 0.8 m at a distance of ~ 16 m from the mean tide limit. However, the morphology of the seaward slope and top of the ridge had changed (Fig. 3). Where the slope was relatively uniformly steep in 2013, the slope in 2015 exhibited two obvious steps (at ~ 13.9 and 14.8 m from the shore). Close to its top (at ~ 15.5 m from the shore), the thickness of the ridge in 2015 was ~ 10 cm less than in 2013. In contrast, the thickness in the lower half of the seaward side (~ 13.5–14.8 m from the shore) locally increased by 20–25 cm.

The material found infilling the trench started at ~ 9.2 m from the shore, about 0.6 m farther inland than the ridge itself (Fig. 3). The maximum thickness of the infilled material was 0.4 m at ~ 17.5 m from the shore. The morphology of the infill paralleled the surface of the current ridge, with a break in slope roughly in the middle of the seaward flank. In total 708 clasts were measured and the total volume of the infilled clasts amounts to $\sim 0.74\,\mathrm{m}^3$. The average clast size was $15 \times 10 \times 5$ cm, and the biggest clast in the infill was a piece of *Acropora palmata* ($86 \times 60 \times 15$ cm). About 63 % of the clasts were encrusted with *Homotrema rubrum* (Lamarck) and 96 % were well-rounded. The main components (Fig. 4a) were corals (78 %), reef rock (7 %), conch shells (6 %), Pleistocene limestone (4 %), serpulite rock (2 %), beach rock (1 %), gastropods (1 %) and other material (1 %; e.g. wood, flip flops). The coral species represented (Fig. 4b) were *Acropora* sp. (74 %), *Diploria* sp. (12 %), *Porites* (7 %), *Montastrea* sp. (2 %), *Siderastrea* (2 %), *Milleporida* (2 %) and *Gorgonia* (1 %). No sand was present in the interstices of the coarse-clast framework of the ridge.

Neither sedimentary structures, nor systematic vertical or lateral trends in particle size were observed in the trench infill. However, low-density clasts such as serpulite rock,

wood, flip flops, *Gorgonia* and *Milleporida* were found higher and farther inland, on average, than were the relatively dense clasts of Pleistocene limestone, beach rock and *Montastrea*.

6 Discussion

Field observations and analytical data clearly show a modification of the ridge and the infilling of the trench during the 23 months that elapsed since the initial survey by Spiske and Halley (2014). However, these changes were unexpected considering the short time period and the fact that the island was not affected by any severe wave conditions in the meantime. Thus, modifications seem to underlie much weaker energetic processes, and bioweathering of fresh clast surfaces seems to already occur within a few months. There are no signs of any human modification of the ridge.

6.1 Bioweathering

The material that was manually moved during the excavation of the trench was apparent because of its light color in 2013 (Fig. 2a and b). The light color was a consequence of burial that had inhibited weathering by surface processes such as sea spray or lithobiontic weathering (Spiske and Halley, 2014). In 2015 these manually moved clasts were no longer distinguishable (Fig. 2c) because weathering had already darkened their surfaces.

Colonization by epi- and endolithic cyanobacteria (blue-green algae) is an inevitable process that affects coastal sites (Golubic et al., 1980). The rubble ridge at Anegada is positioned in the supralittoral zone, often being wetted by sea spray, as well as by wave splash during storms. This ecological niche is a preferred living environment of cyanobacteria (Radkte et al., 1997; Spencer and Viles, 2002; Gómez-Pujol et al., 2006). The rate of colonization is a function of the temperature, and of the exposure to insolation, waves, spray or splash water (e.g. Folk et al., 1973; Radkte et al., 1997; Schneider and Le Campion-Alsumard, 1999). The activity of the cyanobacteria causes a dark grey to black coating (biopatina) on the clast surfaces and a few micrometers to millimeters deep (e.g. Folk et al., 1973). Porous, bored or karstified rocks, such as the coral and limestone clasts on Anegada, promote the colonization by endolithics (Hoppert et al., 2004). However, endolithic colonization rates are much slower compared to epilithics (e.g. Viles, 1987; Hoppert et al., 2004; McNamara et al., 2006). Under ideal conditions, colonization rates can be extremely rapid with initial occupation of fresh surfaces within 8–9 days of exposure and a complete overgrowth in three weeks to less than four months (Spencer (1988); and references therein). Spencer (1988) notes that colonization leaves the initially fresh surface visually indistinguishable from the surrounding surfaces. This indeed applies for the surfaces of the clasts on Anegada that

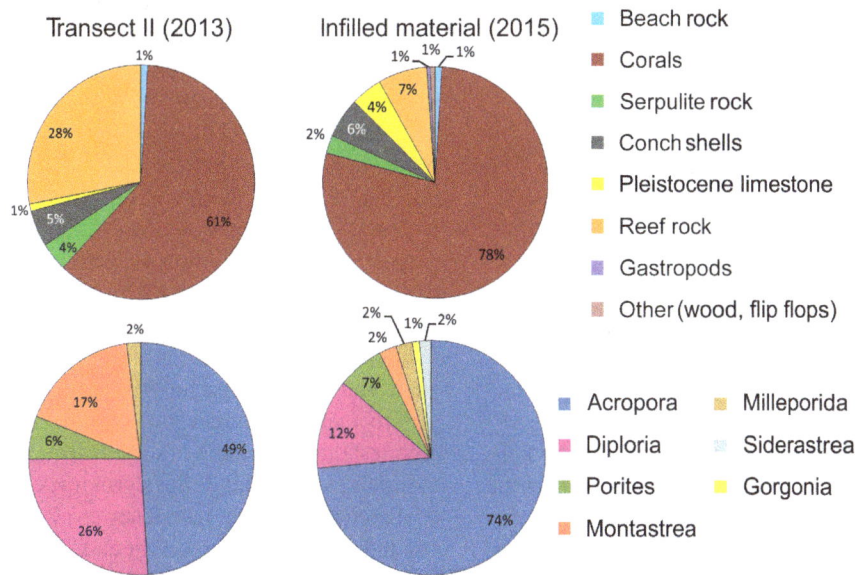

Figure 4. (a) Percentage of each component and **(b)** percentage of each coral species of the trench in 2013 and the new trench infill in 2015.

were manually moved and put onto the ridge in 2013 that were no longer distinguishable 23 months later (Fig. 2a–c). Warscheid and Braams (2000) reported that freshly quarried stone could be covered by biofilms within months. On the Aldabra Atoll (Indian Ocean), a tropical setting similar to Anegada, Whitton and Potts (1979) documented that supralittoral beach rock which was newly exposed by coastal erosion already had a light steel-blue color caused by blue-green algae within two weeks after exposure. This report is underlined by Folk et al. (1973) who state that algae coatings grow fastest and cover is more dense and abundant in tropical humid climates with high temperatures and in settings with constant wetting by sea spray or splash. Thus, it is not surprising after all, that algae would have darkened the light-colored clasts removed from the trench the next two years. In that case, the surface color of calcareous clasts in tropical climate settings, like Anegada, quickly saturates as a guide to the duration of clasts exposure, i.e. relative dating of ridge emplacement.

In close-up view, the infilled material and the ridge surface in 2015 show a mixture of clasts with light and dark gray (bioweathered) surfaces (Fig. 2d). This distribution results from (i) mixing of weathered and fresh rubble and (ii) overturning of the ridge-derived clasts during transport. The ridge-derived clasts exhibit at least one darkened side, previously being exposed at the ridge surface, and brighter sides that were not directly exposed to spray, splash and insolation. Assuming that hurricane Gonzalo emplaced the clasts in mid October 2014, four months passed until the survey in February 2015. At the time of the survey the clast surfaces of the moved material were still light-colored and did not yet show signs of darkening by biofilms. Consequently, at the

given conditions on Anegada, the minimum time for algal colonization is > 4 months. Contrariwise, the maximum colonization time is less than 23 months because clasts that were artificially moved when the trench was dug in March 2013 were no longer distinguishable from longer exposed clasts in February 2015.

6.2 Ridge composition

Most obviously the trench dug in 2013 was nearly completely infilled (Fig. 2c) and the minor depression as the only visible remain of the previous trench (Fig. 2d) was hard to detect in 2015. Detailed analyses of the trench infill document an average clast size of $15 \times 10 \times 5$ cm, which is very similar to the average clast size in 2013 ($16 \times 11 \times 4$ cm). Likewise, the degree of reworking and encrustation (up to 96 %) of the clasts is very high both in 2013 and 2015. The main components that constitute the ridge in 2013 and its infill in 2015 are corals, conch shells, Pleistocene limestone, serpulite rock, and minor portions of reef rock and beach rock (Fig. 4a). The same applies for the main coral species (*Acropora*, *Diploria*, *Porites*, *Montastrea*; Fig. 4b). The percentages of each component and coral species shifted, but the overall composition did not change significantly, even though some minor constituents like *Gorgonia*, *Siderastrea* and gastropods were only counted during one of the surveys. The deviation in percentage and the occurrence of some species in just one of the surveys may be the result of spatial variations in ridge composition. Spiske and Halley (2014) presented these variations when comparing transects I and II.

The composition of the ridge and the infilling of the trench do not significantly differ. This implies that (i) either the source of the material remained the same and still provided sufficient material for the infilling, or (ii) the ridge itself provided (portions of) the material as it was partly reworked by swell, most probably related to the passage of hurricane Gonzalo in October 2014.

6.3 Ridge modification and emplacement

Infilling of the trench was only possible during a storm-induced surge that allowed waves to overtop the ~ 1–1.5 m high coastal platform along Anegada's central north coast (for a detailed view of the platform see Figs. 2a and 3a in Spiske and Halley, 2014). Even during rough sea conditions, with offshore wave heights > 2.5 m, as observed during the survey in 2015, the fringing reef and the shallow lagoon effectively decrease wave heights, not allowing for an inundation of the coastal platform. Consequently, only a storm surge, i.e. an elevated wedge of water pushed towards the coast, on top of which the storm waves can propagate, enables waves to cross the shoreline and the small limestone platform.

The clast-supported framework of the ridge with interlocking components is quite stable. However, storm wave action will at least rework the loose clasts on the surface of the seaward flank of the ridge. At the transect site, material from the uppermost parts of the ridge was eroded and either transported downslope or slumped into the excavated trench. Slumping of clasts from the upper trench walls into the trench may vaguely be observed in Fig. 2d. Nevertheless, it remains unknown which portions of the ~ 0.74 m^3 of infilled material were newly entrained in the shallow marine environment and which derive from reworking of the ridge. If a significant number of clasts was entrained and transported onshore, clasts with all surfaces being unweathered should be found all along the ridge. However, as nearly all clasts have one dark side, it is assumed that reworking of ridge material is the dominant process.

Spiske and Halley (2014) assumed that the rubble ridge was emplaced during a hurricane-related inundation event which was most probably more energetic in terms of surge height, wave speed and wave amplitude, than hurricanes Earl and Donna. However, less energetic storm conditions account for the infilling of the trench that occurred between March 2013 and February 2015. The largest storm at Anegada in that interval was hurricane Gonzalo, which passed about 60 km north of Anegada in October 2014 (category 1–2). Nevertheless, since most of the infill seems to represent eroded ridge material and only a small portion of fresh material was added, it can be referred that such a weak event may not create a ridge, due to the lack of energy to transport clasts onshore, but at least modify a preexisting ridge. Consequently, coarse-clast ridges are vulnerable to even small storms and thus ridge systems are highly dynamic. This conclusion is underlined by a recent study of Xu et al. (2015) who sampled the coarse-clast ridge on Anegada to create a 20th century coral calibration. They counted the growth

bands and used uranium-series dating to determine the duration of coral growth represented by two coral clast samples. The end of growth, i.e. timing of the entrainment of the coral and its subsequent death, were dated as AD 1953.3 $\pm$ 0.72 and AD 1989.5 $\pm$ 0.58, respectively. Of course, these are only two samples, however they show that clasts were most probably added to the preexisting ridge at least for decades. No notable hurricanes are documented to have passed Anegada at the respective times, again pointing to the fact that even minor storm events are capable of adding material to the ridge.

Future storm impact on Anegada should be monitored in greater detail to get information on parameters that can govern the characteristics of hurricane-related inundations, amongst others, the role of the hurricane intensity, distance to the island, alignment of the storm track relative to the island, or direction of swell (Gardener et al., 2005). These factors can influence the surge height and the clast transport capacity of the waves, and under particular conditions even tropical storms or category 1–2 hurricanes may cause severe coastal damage, whereas in turn the effectiveness of category > 2 storm may be attenuated by factors such as swell direction or alignment of the storm track relative to the island.

7 Conclusions

The coral-rubble ridge along Anegada's central north coast that was initially surveyed in 2013 underwent morphological changes in the subsequent 23 months. A trench that has been dug through the ridge and left open in 2013 was found nearly completely infilled with clasts in 2015. Light-colored clasts placed on the ridge surface in 2013 were found bioweathered and darkened in 2015. These comparisons suggest:

– modification and emplacement of coarse-clast coastal ridges occurs already at lower wave energy levels;

– minor hurricane-induced surges mainly rework preexisting ridge structures, adding only few freshly entrained marine components;

– under the prevailing (supralittoral) conditions, colonization by cyanobacteria that cause staining of the clast surfaces takes at least 4 months, but less than 23 months;

– in a tropical climate context, blackening of calcareous clasts by biopatina saturates in less than two years as a measure of exposure time of calcareous deposits.

Acknowledgements. This study is part of the US Geological Survey's "Tsunami Hazards Potential in the Caribbean" project. I thank Brian Atwater, Anna Lisa Cescon, Robert Halley and Jean Roger for support in the field and during manuscript preparation. J. Roger helped to measure > 700 clasts.

References

Atwater, B. F., ten Brink, U. S., Buckley, M., Halley, R. B., Jaffe, B. E., López-Venegas, A. M., Reinhardt, E. G., Tuttle, M. P., Watt, S., and Wei, Y.: Geomorphic and stratigraphic evidence for an unusual tsunami or storm a few centuries ago at Anegada, British Virgin Islands, Nat. Hazards, 63, 51–84, 2012.

Atwater, B. F., Fuentes, Z., Halley, R. B., Ten Brink, U. S., and Tuttle, M. P.: Effects of 2010 Hurricane Earl amidst geologic evidence for greater overwash at Anegada, British Virgin Islands, Adv. Geosci., 38, 21–30, doi:10.5194/adgeo-38-21-2014, 2014.

Brown, D. P.: Tropical cyclone report Hurricane Gonzalo (AL082014) 12–19 October 2014: National Hurricane Center, p. 30, http://www.nhc.noaa.gov/data/tcr/AL082014_Gonzalo.pdf, last access: August 2015.

Cangialosi, J. P.: Tropical cyclone report Hurricane Earl (AL072010) 25 August–4 September 2010: National Hurricane Center, p. 29, http://www.nhc.noaa.gov/data/tcr/AL072010_Earl.pdf (last access: August 2015), 2011.

Dunn, G. E.: The hurricane season of 1960, Mon. Weather Rev., 89, 99–108, 1961.

Folk, R. L., Roberts, H. H., and Moore, C. H.: Black phytokarst from Hell, Cayman Islands, British West Indies, Geol. Soc. Am. Bull., 84, 2351–2360, 1973.

Gardner, T. A., Côté, I. M., Gill, J. A., Grant, A., and Watkinson, A. R.: Hurricanes and Caribbean Coral Reefs: Impacts, Recovery Patterns, and Role in Long-Term Decline, Ecology, 1, 174–184, 2005.

Golubic, S., Friedmann, E. I., and Schneider, J.: The lithobiontic ecological niche, with special reference to microorganisms, J. Sediment. Petrol., 51, 475–478, 1980.

Gómez-Pujol, L., Fornós, J. J., and Swantesson, J. O. H.: Rock surface millimetre-scale roughness and weathering of supratidal Mallorcan carbonate coasts (Balearic Islands), Earth Surf. Proc. Land., 31, 1792–1801, 2006.

Hoppert, M., Flies, C., Pohl, W., Günzl, B., and Schneider, J.: Colonization strategies of lithobiontic microorganisms on carbonate rocks, Environ. Geol., 21, 183–191, 2004.

McNamara, C. J., Perry, T. D., Bearce, K. A., Hernandez-Duque, G., and Mitchell, R.: Epilithic and endolithic bacterial communities in limestone from a Maya archaeological site, Microb. Ecol., 51, 51–64, 2006.

National Oceanic and Atmospheric Administration: Historical hurricane tracks, http://www.csc.noaa.gov/hurricanes/# (last access: August 2015), 2012.

Radtke, G., Le Campion-Alsumard, T., and Golubic, S.: Microbial assemblages involved in tropical coastal bio-erosion: an Atlantic–Pacific comparison, Proceedings of the 8th International Coral Reef Symposium, 24–29 June 1996, Panama City, 1825–1830, 1997.

Schneider, J. and Le Campion-Alsumard, T.: Construction and destruction of carbonates by marine and freshwater Cyanobacteria, Eur. J. Phycol., 34, 417–42, 1999.

Spencer, T.: Limestone coastal morphology: the biological contribution, Prog. Phys. Geogr., 12, 66–101, 1988.

Spencer, T. and Viles, H.: Bioconstruction, bioerosion and disturbance on tropical coasts: coral reefs and rocky limestone shores, Geomorphology, 48, 23–50, 2002.

Spiske, M. and Halley, R. B.: A coral-rubble ridge as evidence for hurricane overwash, Anegada (British Virgin Islands), Adv. Geosci., 38, 9–20, doi:10.5194/adgeo-38-9-2014, 2014.

Viles, H. A.: Blue-green algae and terrestrial limestone weathering on Aldabra Atoll: a SEM and light microscope study, Earth Surf. Proc. Land., 12, 319–330, 1987.

Warscheid, T. and Braams, J.: Biodeterioration of stone: a review, Int. Biodeterio. Biodegrad., 46, 343–368, 2000.

Whitton, B. A. and Potts, M.: Blue-green algae (cyanobacteria) of the oceanic coast of Aldabra, Atoll Res. Bull., 238, 1–9, 1979.

Xu, Y., Pearson, S., and Kilbourne, K.: 2015, Assessing coral Sr/Ca–SST calibration techniques using the species Diploria strigosa, Palaeogeogr. Palaeocl., 440, 353–362, 2015.

7

Detecting and locating seismic events with using USArray as a large antenna

L. Retailleau[1,2]**, N. M. Shapiro**[1]**, J. Guilbert**[2]**, M. Campillo**[3]**, and P. Roux**[3]

[1]Institut de Physique du Globe de Paris, Sorbonne Paris Cité, CNRS (UMS 7154), Paris, France
[2]CEA/DAM/DIF, F-91297 Arpajon, France
[3]Institut des Sciences de la Terre, CNRS, Université Joseph Fournier, Grenoble, France

Correspondence to: L. Retailleau (retailleau.lise@gmail.com)

Abstract. We design an earthquake detection and location algorithm that explores coherence and characteristic behavior of teleseismic waves recorded by a large-scale seismic network. The procedure consists of three steps. First, for every tested source location we construct a time-distance gather by computing great-circle distances to all stations of the network and aligning the signals respectively. Second, we use the constructed gather to compute a Tau-P transform. For waves emitted by teleseismic sources, the amplitude of this transform has a very characteristic behavior with maxima corresponding to different seismic phases. Relative location of these maxima on the time-slowness plane strongly depends on the distance to the earthquake. To explore this dependence, in a third step, we convolve the Tau-P amplitude with a time-slowness filter whose maxima are computed based on prediction of a global travel-time calculator. As a result of this three-step procedure, we obtain a function that characterizes a likelihood of occurrence of a seismic event at a given position in space and time. We test the developed algorithm by applying it to vertical-component records of USArray to locate a set of earthquakes distributed around the Globe with magnitudes between 6.1 and 7.2.

mic events with using a network of seismic receivers as an antenna. The records by large regional-scale seismic networks installed during recent decades such as USArray in the United States, HI-Net Array in Japan or VEBSN in Europe have been recently used to study sources of large earthquakes such as the 2004 Sumatra-Andaman event (Ishii et al., 2005), the 2011 Tohoku earthquake (e.g., Satriano et al., 2014) and the 2012 Wharton Basin earthquake (Satriano et al., 2012) based on beam forming algorithms. More recently, Retailleau et al. (2014) used a network-based analysis of the USArray data to study the antipodal focusing of seismic waves emitted by an earthquake in the Indian Ocean. In these studies, the authors focused on characterization of details of the source processes of large earthquakes with a priory known locations. In the present paper, we explore another approach when records from large networks are used to detect and to locate epicenters of seismic events without a priory knowledge about their occurrence. We first test the method with a set of synthetic seismograms computed in the spherically symmetric Earth model PREM (Dziewonski and Anderson, 1981) and then apply it to a few real earthquakes.

1 Introduction

Traditional earthquake detection and location algorithms use the arrival times of short period body wave phases such as *P* or *S*. The waves are identified at every individual station when the wavefront generated by the source passes through it. The measured arrival times are then grouped and inverted to locate the event. An alternative approach is to study seis-

2 Data

We use vertical-component records from the Transportable Array (TA) component of the USArray, which contained roughly 400 receivers during 2010–2011. We selected 200 closely located stations forming a compact group that can be used as an antenna (Fig. 1). A quality check was performed to suppress data with high amplitude glitches. We selected five earthquakes, described in Table 1, to illustrate the method.

Table 1. Description of the 4 earthquakes on which the method was tested. These characteristics were obtained from the CMT catalog (Dziewonski and Anderson, 1981; Ekström et al., 2012).

Event	Latitude	Longitude	Depth (km)	Source date	Centroid Time	Magnitude	Distance to the network
1. Vanuatu	− 13.81	166.65	42.9	2010/05/27	17:14:55.3 GMT	7.2	99.8°
2. Japan	25.86	128.61	18	2010/02/26	20:31:29.7 GMT	7.0	99.5°
3. Argentina	−27.02	−63.21	586	2011/01/01	09:57:03.8 GMT	7.0	74.4°
4. Turkey	38.82	40.04	15.1	2010/03/08	02:32:37.4 GMT	6.1	95.2°

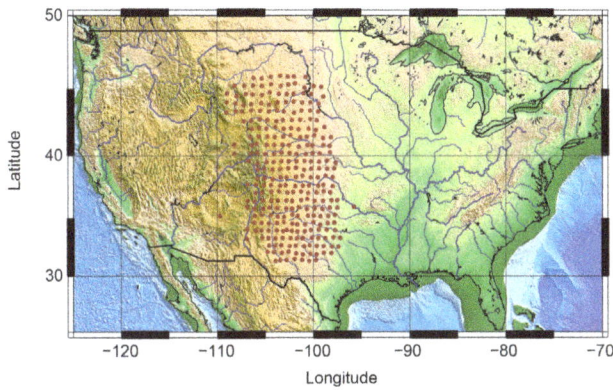

Figure 1. Map of the United States where the red circles show locations of the USArray Transportable seismic stations used in this study.

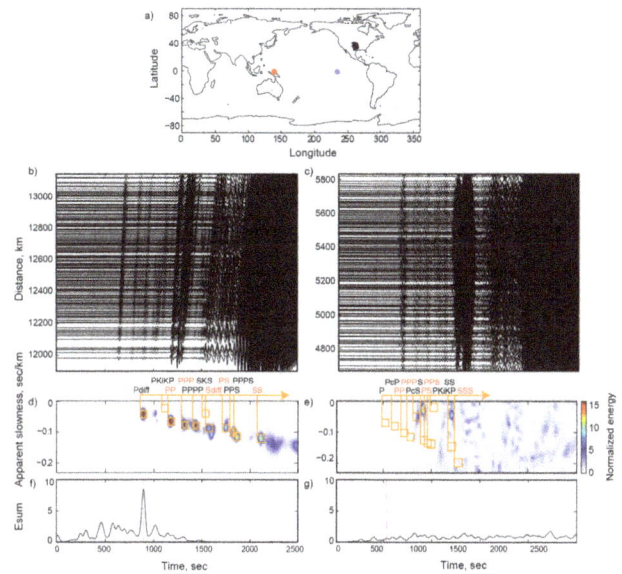

Figure 2. (a) Map showing the network used and two points showing the correct location (in red) and a wrong one (in blue). Details of the method for two potential locations, the correct (**b, d, f**) and a wrong one (**c, e, g**). (**b**) and (**c**) show the alignment of the seismograms with respect to their respective potential source. (**d**) and (**e**) Tau-P transforms and theoretical masks (in orange) generated for each case. (**f**) and (**g**) The energy functions E_{sum} generated. The theoretical arrival of the first phase is shown with a dashed line.

For every earthquake we used records of 4 h starting at the source time of the event.

We also created a synthetic dataset to test the method. For this purpose, we used the focal mechanism from the Global CMT catalog for event 1 as our source. Its location was set to longitude 140 and latitude 0 as shown in Fig. 2a. The synthetic seismograms were computed with the method AXISEM, a 2-D spectral-element solver for 3-D elastodynamics in global, spherically symmetric background models (Nissen-Meyer et al., 2007), using the isotropic version of the PREM model (Dziewonski and Anderson, 1981) and with including the viscoelastic attenuation. Both the real signals and the synthetic seismograms were bandpassed in the period band between 20 and 50 s.

3 Method

The basic idea of our method is to use simultaneously distinctive characteristics of different seismic body-wave phases recorded by an array of receivers. Every seismic phase is characterized by its slowness and back-azimuth that can be measured with an array-based analysis. For body waves emitted by the same earthquake, the back-azimuths measured at a given time are similar, because the seismic structure of the Earth is well approximated by a spherically symmet-

ric model. The slowness of every seismic phase is varying with the epicentral distance. When using N seismic phases, the combination of their slownesses is becoming a N values vector whose dependence on the epicentral distance is much more selective comparing to a single phase.

Main steps of our method are illustrated in Fig. 2 with using the synthetic dataset. First we select a tested source position (Fig. 2a) and construct a time-distance gather by computing great-circle distances to all stations of the network and aligning the signals respectively (Fig. 2b and c). Then, we compute a Tau-P transform to present the gather energy as function of travel time and apparent slowness (Fig. 2d and e).

The Tau-P transform is smoothed in time with a 30 s sliding window to suppress high frequency variations. Then, we

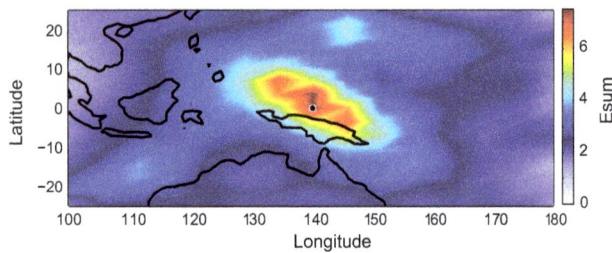

Figure 3. Application of the described network-based location method to the synthetic dataset. Colors show the maximum of E_{sum} at every location. The black point shows the correct location of the source.

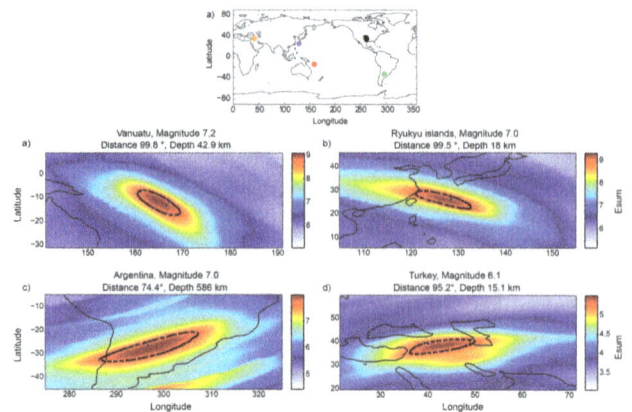

Figure 4. (a) Results of the location method for four earthquakes described in Table 1, which locations are represented in (a). The dashed black line shows the 90 % energy around the maximum of E_{sum}. (b) Vanuatu 27 May 2010 magnitude 7.2 earthquake, (c) Ryukyu islands (Japan) 26 February 2010 magnitude 7.0 earthquake, (d) Santiago del Estero (Argentina) 1 January 2011 magnitude 7.0 earthquake, and (e) Turkey 8 March 2010 magnitude 6.1 earthquake. Table 1 describes the events.

decided to emphasize the energy of the first arriving body waves that is lower than the energy of the latter arrivals (especially the surface waves). The early body-wave phases have narrow maxima on the time-slowness plane and are, therefore, more useful for determining the distance with our method. To increase their relative contribution, we apply a normalization, consisting in dividing the Tau-P transform by its average over all slownesses smoothed in time with a 600 s sliding window.

Finally, we design a time-slowness filter whose maxima are computed based on predictions of the global travel-time calculator of Crotwell et al. (1999).

This filter consists of a set of rectangles centered at arrival times and slowness of selected phases as illustrated by yellow lines in Fig. 2d and e. The filter is continuously shifted along the time coordinate and the time-slowness energy within the boxes is summed. When the tested source position coincides with the epicenter and the time coincides with the arrival time of all phases, the summed energy E_{sum} shows a clear maximum (Fig. 2f). If the tested source position is far from the epicenter, the values of E_{sum} are relatively weak (Fig. 2g). We perform a grid search and test different source positions located on a geographical grid. The results of this grid search for the synthetic test are shown in Fig. 3 and show a clear maximum located very close to the true source position.

4 Discussion

After testing the method with the synthetic seismograms, we applied it to records of real earthquakes shown in Table 1. The results shown in Fig. 4 demonstrate the example of three events of magnitude 7 and one of magnitude 6 which were well located. These events occurred at different locations on Earth with different depths and magnitudes providing a first view of the abilities of the method.

Further assessment of the method performance requires its testing with more events and with signals filtered in different frequency bands. Indeed these different frequency bands should highlight different events in the signals and a higher frequency would locate the events more precisely.

Another important issue that remains out of scope of this short paper is the possible depth sensitivity of the proposed methods. In the presented tests, the source depth used to compute travel times and to design the time-slowness-filters was fixed to their depth (obtained from the Global CMT catalog, Table 1). We expect that for shallow events (above 50 km) the difference in depth should not have much consequences in the shape of the time slowness-filters and the resulting E_{sum} values. However, the depth might become important for deep earthquakes (such as event 2 with its depth of 586 km) when the pP and sP phases appear. Including these phases into the analysis may help to discriminate the depth.

Methods to study the slowness and back azimuth of a source in the time domain have been developed at Norwegian Seismic Array (NORSAR) long ago (King et al., 1976). Their method generates a function of energy of the waves as opposed to time, azimuth, and slowness. Nonetheless our method also takes into account the distance of the source to the network, and thus the location of the source of signal. Indeed our mask study obtains the decays of arrivals between the different phases, which give a good idea of the location of the event. We are able to get more information from the Tau-P transform.

Moreover, we do not only focus on a precise location as Ishii et al. (2005) did to study the Sumatra rupture. Adding several phases increases the precision and enables us to observe events spread worldwide, in all the range of distances.

Basically our method does not need any a priori information such as the time and location of the event we want to detect. It is possible to detect two events that occurred at the same time in different places. The phases also do not need to be selected and identified, nor their time arrivals measured a priori.

Since this approach does not need any information about the events it should lead to the detection of events we do not expect. For this purpose, we intend to apply the method to a few years of continuous USArray records to evaluate its capacity for automatic detection and location of different types of seismic events.

Acknowledgements. We thank Alexandre Fournier and Tarje Nissen-Meyer for their help about the computing of synthetics with Axisem. Data from the TA network are made freely available as part of the EarthScope USArray facility, operated by Incorporated Research Institutions for Seismology (IRIS). This work was supported by the Commissariat à l'Energie Atomique (CEA, France), by the European Research Council under the contract FP7 ERC Advanced grant no. 227507 (WHISPER), and by the project DataScale funded by the French BigData program. The work of NMS was supported by the Russian Science Foundation (grant no. 14-47-00002).

References

Crotwell, H., Owens, T., and Ritsema, J.: The TauP Toolkit: Flexible seismic travel-time and ray-path utilities, Seismol. Res., 70, 154–160, 1999.

Dziewonski, A. and Anderson, D.: Preliminary reference Earth model, Phys. Earth Planet. In., 25, 297–356, 1981.

Ekström, G., Nettles, M., and Dziewoński, A.: The global CMT project 2004–2010: Centroid-moment tensors for 13,017 earthquakes, Phys. Earth Planet. In., 200–201, 1–9, doi:10.1016/j.pepi.2012.04.002, 2012.

Ishii, M., Shearer, P. M., Houston, H., and Vidale, J. E.: Extent, duration and speed of the 2004 Sumatra-Andaman earthquake imaged by the Hi-Net array., Nature, 435, 933–6, doi:10.1038/nature03675, 2005.

King, D. W., Husebye, E. S., and Haddon, R. A. W.: Processing of seismic precursor data, Phys. Earth Planet. In., 12, 128–134, 1976.

Nissen-Meyer, T., Fournier, A., and Dahlen, F. A.: A two-dimensional spectral-element method for computing spherical-earth seismograms – I. Moment-tensor source, Geophys. J. Int., 168, 1067–1092, doi:10.1111/j.1365-246X.2006.03121.x, 2007.

Retailleau, L., Shapiro, N. M., Guilbert, J., Campillo, M., and Roux, P.: Antipodal focusing of seismic waves observed with the USArray, Geophys. J. Int., 199, 1030–1042, doi:10.1093/gji/ggu309, 2014.

Satriano, C., Kiraly, E., Bernard, P., and Vilotte, J.-P.: The 2012 Mw 8.6 Sumatra earthquake: Evidence of westward sequential seismic ruptures associated to the reactivation of a N–S ocean fabric, Geophys. Res. Lett., 39, L15302, doi:10.1029/2012GL052387, 2012.

Satriano, C., Dionicio, V., Miyake, H., Uchida, N., Vilotte, J.-P., and Bernard, P.: Structural and thermal control of seismic activity and megathrust rupture dynamics in subduction zones: Lessons from the Mw 9.0, 2011 Tohoku earthquake, Earth Planet. Sci. Lett., 403, 287–298, doi:10.1016/j.epsl.2014.06.037, 2014.

8

Morphodynamic modelling for the entire German Bight: an initial study on model sensitivity and uncertainty

A. Plüß and F. Kösters

Federal Waterways Engineering and Research Institute, Hamburg, Germany

Correspondence to: A. Plüß (andreas.pluess@baw.de)

Abstract. Morphodynamic modelling of coastal seas and estuaries for large-scale and long-term applications is strongly affected by parameter sensitivity of process-based models. Moreover, the comparison of data-based methods with numerical model results is limited by uncertainties in measurements. These drawbacks can be partly overcome by a multi-model approach (MMA). In a case study to assess long-term sediment transport and morphodynamic processes for the German Bight, the AufMod research project applies two different methods for process-based modelling: UnTRIM-SediMorph and DELFT3D. Model sensitivity is illustrated in terms of different morphological changes for diverse porosity values. As a first step, discrepancies between individual methods are shown based on resulting sediment transport patterns.

1 Introduction

The North Sea acts as a gateway for container ships connecting European harbours to the world. Within this economical important region, natural and man-made morphological changes take place which potentially affect safety and ease of maritime traffic. The prediction of morphologic changes on temporal and spatial scales is relevant for coastal engineering applications and although it is still a scientific challenge (e.g. French and Burningham, 2009). The application and interpretation of (long-term) morphodynamic model results demand for well-founded knowledge about sediment transport and local morphological conditions in the field as well as profound knowledge of the modelling techniques.

In order to improve our current understanding of these morphological changes, the multidisciplinary research project AufMod (German acronym for "Model-based analy-

sis of long-term morphodynamic processes in the German Bight") was funded by the Federal Ministry of Education and Research (BMBF) to investigate long-term sediment transport and morphodynamic processes. The project focuses on the German Bight, located in the south-eastern part of the North Sea. AufMod takes a combined data-based and process-based modelling approach to investigate long-term sediment transport. One main scientific objective of AufMod is to identify processes and effects which are relevant for the long-term sediment transport and the morphodynamic reaction of the seabed. The first step is to investigate the uncertainty of the simulation results from different numerical methods and parameterizations.

In order to simulate sediment transport, a consistent database for bathymetric and sedimentological data is indispensable. Only when these data are available, hydrodynamic transport models can be utilized to calculate large-scale sediment transport; as a consequence of this the morphological changes. In practice there is often not sufficient field data (e.g. bathymetry, sediment properties) available to calibrate and validate a morphodynamic (MD) model thoroughly enough for a reliable forecast. The calibration is further complicated by uncertainties inherent to the numerical methods (e.g. sediment transport formulations, numerical diffusion) and physical processes not or not well represented (e.g. erodibility or stability of sediments at the bed, consolidation).

In this situation, the MD multi-model approach (MMA) offers a solution to overcome the uncertainties mentioned above for sediment transport and morphodynamic studies (Plüß and Heyer, 2008). Similar to the approach in climate models, covering the range of potential changes in the climate as a response to an increase in CO_2 (e.g. Tebaldi and Knutti, 2007), the MMA is in particular helpful to determine applicability, skill and uncertainty of available modelling systems in sediment transport studies.

In this study we focus on two aspects of sensitivity and uncertainty of numerical model results. First, bed level changes calculated from MD models heavily depend on the sedimentological properties prescribed as boundary conditions for the bed. Therefore, necessity and reliability of sedimentological properties need to be assessed. If it is possible to calculate the distribution and sorting of the seabed sediment, one might be able to reduce the amount of input data for morphodynamic model simulations. Moreover, what is the impact of observed variability in key sedimentological properties such as the initial sediment grain size distribution or the porosity of surface sediments? The initial sediment distribution is generally not well known and will vary on scales smaller than those resolved by the model. Porosity depends on the sorting of sediments as well as on compaction and thus the sedimentological history, which is not commonly taken into account in present-day morphodynamic models. The second aspect, covered in this study, is to compare sediment transport patterns from different modelling systems and thus the parameter sensitivity of different empirical sediment transport formulations.

2 Methods

Two different modelling systems are employed here. Firstly, the unstructured 3-D hydrodynamic model UnTRIM (Casulli and Zanolli, 2002) coupled with the sediment transport model SediMorph (Malcherek et al., 2005) and the unstructured version of the wave model K-model (Schneggenburger, 1998) was applied. Secondly, the structured model DELFT3D (Stelling, 1984, 1986) was used in a 2-D mode, combining hydro-, sediment and morphodynamics (Lesser et al., 2004) together with the effects of waves (Booij and Holthuisen, 1996).

The MD simulations cover the time span from 1996 to 2008 and are steered using spatial and temporal varying forcing based on operational forecast model results provided by national authorities, taking into account the whole variability of tides, external surge, river run-off, wind and waves. The large-scale and long-term sediment transport models comprise coastal areas, islands, shore lines and connected estuaries. Because of the versatile effects and interactions between the individual regions (estuary–coastal region–shelf), different approaches for grid generation of the individual methods were necessary. The unstructured UnTRIM model consists out of 77 500 triangular elements with a resolution varying from 24 km in the outer North Sea down to 80 m in the estuaries (BAW, 2013). The vertical resolution changes from 50 m off the coast down to 1 m in intertidal areas. The structured DELFT3D model was set up in a vertical integrated (2-D), nested configuration: (a) North Sea model with 455×489 curvilinear grid cells with a variable resolution from 1 km

in the outer North Sea down to 1 km near the coast and (b) German Bight model with 762×1046 curvilinear grid cells with a variable resolution from 1.8 km in deep water down to 100 m in the estuaries.

At the open boundaries of both North Sea models, water levels predicated on tidal constituents out of the global model FES 2004 (Lyard et al., 2006) and external surges based on the operational model of the German Hydrographic Office (BSH) have been prescribed. UnTRIM solves the hydrodynamics using a time step of 120 s; DELFT3D uses 150 s. River run-off is taken to be variable within the inner German Bight resulting out of measurements. For the continental run-off, seasonally averaged mean values were assigned. The atmospheric forcing (wind speed) is taken from model results of the operational model of the German National Meteorological Service (DWD). The wave model has been subsequently run with pre-calculated water levels and the same atmospheric forcing as the hydrodynamic model.

For the long-time calculations annual bathymetries are computed with the utilization of the time varying digital bathymetry model year by year provided by the AufMod database (see http://projekt.mdi-de.org/services/verwandte-projekte/40-aufbau-von-integrierten-modellsystemen.html). For long-term morphodynamic simulations, it is crucial to prescribe the sediment distribution and composition of the seabed precisely. For this, the initial sediment distribution and composition are taken from the functional AufMod sediment model, which integrates more than 76 500 measurements of grain size distribution.

Sediment transport in UnTRIM is modelled as transport of individual sediment classes either transported as suspension or as bed load following the formulation of van Rijn (1993). In the DELFT3D model the standard sediment transport formula (van Rijn 1993) is chosen taking into account both suspended and bed-load transport.

The hydrodynamic numerical (HN) model results have been validated using more than 90 tide gauge stations around the North Sea yielding sufficient accuracy of modelled water levels (BAW, 2013). Also, the simulated waves show good agreement to measurements. For suspended sediment and bed load transport, no detailed validation could be carried out as measurements appear too seldom and are only locally available. Moreover, the validation of the MD modelling results has been inhibited due to only sparse measurements in time and space, with accuracies from the measurements of the order of the morphological changes within the simulation period.

As an alternative calibration approach, the local bed shear stress, based on hindcast simulations, has been shown to correlate with the measured mean bed evolution range significantly (BER denotes the difference between the lowest and highest seabed elevation in the time span considered; Kösters and Winter, 2014).

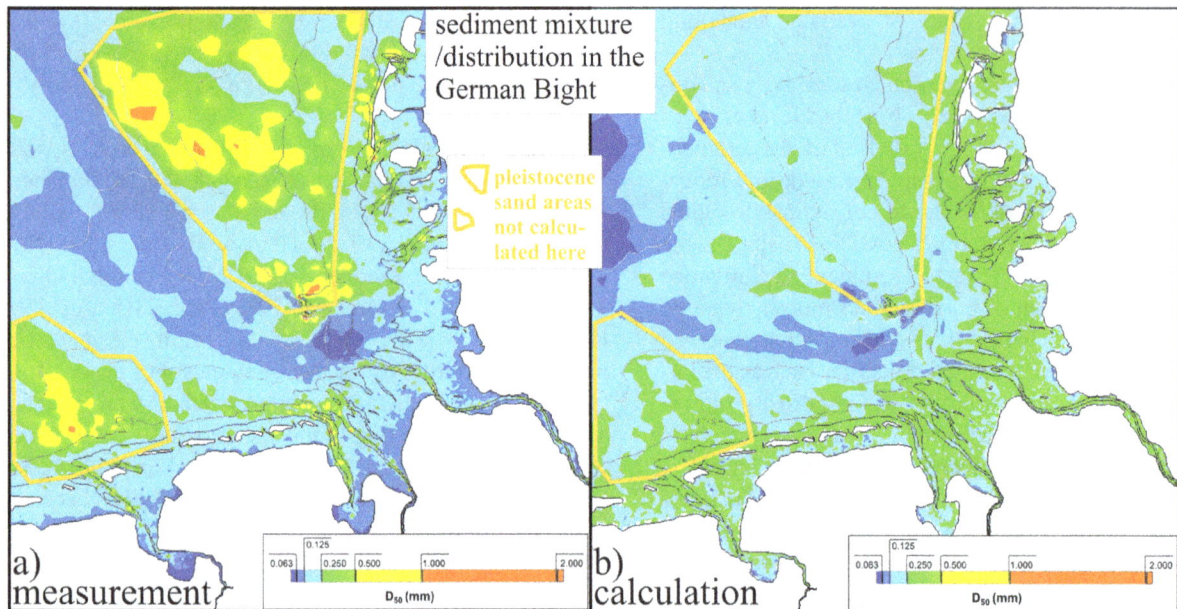

Fig. 1. Initial sediment mixture/distribution in the German Bight from **(a)** measurements and **(b)** numerical model calculation (UnTRIM).

3 Results and discussion

3.1 Reliability of sedimentological properties as model input data

To address the question if sedimentological properties of the bed can be obtained from a hydraulic sorting of sediments in order to avoid expensive measurements, a model study has been conducted. An UnTRIM simulation run is started with an initially uniform sediment distribution and compared to the measured distribution. The sediment distribution of the German Bight (Fig. 1a) was obtained as median grain size (D50) from the AufMod database. This data set updates and extends previous work (Figge, 1981) by including more than 7600 sediment samples from the German Bight. The derived D50 data have been mapped on the numerical model grid. The state of sediment distribution results from sorting of Holocene sediments due to hydrodynamic forcing and outcropping Pleistocene sediments in regions of Holocene sediment deficiency (Zeiler et al., 2008). Sediments mainly consist of well-sorted fine sand with regions of medium sand at the North Frisian coast. In the northern part of the East Frisian shelf, glacial gravel deposits can be found. Finer sediments are present in the Wadden Sea (e.g. at tidal flats, in the estuaries, in the mud deposition area close to Helgoland and in the drowned river Elbe valley). In tidal channels or in tidal inlets, coarser sediments are observed. The model calculates the sorting of bed material for a run time of 60 days taking into account the effects of tidal, wind- and wave-driven bed shear stresses. The initial sediment distribution has been set to a uniform distribution of spatially averaged conditions of the German Bight as obtained from the

AufMod data set represented by four bed load fractions of the following: very coarse sand (vcSa, grain size 1500 mum, fraction 0.3 %), coarse sand (cSa, 750 mum, 2.9 %), medium sand (mSa, 375 mum, 16.2 %), fine sand (fSa, 187.5 mum, 36.5 %) and five suspended load fractions of very fine sand (vfSa, 94 mum, 25.9 %), coarse silt (cSi, 46.5 mum, 9.6 %), medium silt (mSi, 23.5 mum, 3.7 %), fine silt (fSi, 12 mum, 3.3 %), and very fine silt (vfSi, 6 mum, 1.6 %). The resulting sediment distribution from the sorting model experiment is shown as D50 in Fig. 1b.

Although an exact reproduction of the sedimentology of the German Bight cannot be expected from the numerical model, several characteristic features are reproduced. The comparison of the observed and modelled sediment distribution shows that typical accumulation areas of finer sediments behind barrier islands, in estuaries and in extension of the estuaries are reproduced by the model. Coarser sediments in tidal channels of estuaries and tidal inlets are mostly captured but tend to be underestimated in the model results. Deposits of coarse sand on the shelf can only partly be captured by the model; notably the extensive area of coarse sediments on the North Frisian shelf cannot be reproduced. Sediment sorting in energetic tidal channels mainly reflects present-day tidal forcing conditions and can qualitatively reproduce characteristic features after only 60 days of hydrodynamic forcing. Similarly the redistribution of fines in back-barrier regions becomes apparent after this short period. However, main features of the observed sediment properties are absent such as coarse sediments at the North Frisian shelf which originate from glacial periods. Moreover, even though the sediment distribution is qualitative similar, the grain size distribution can differ quantitatively by more than an order of magnitude.

Difference of bed level changes: porosity =0.5 vs. 0.6

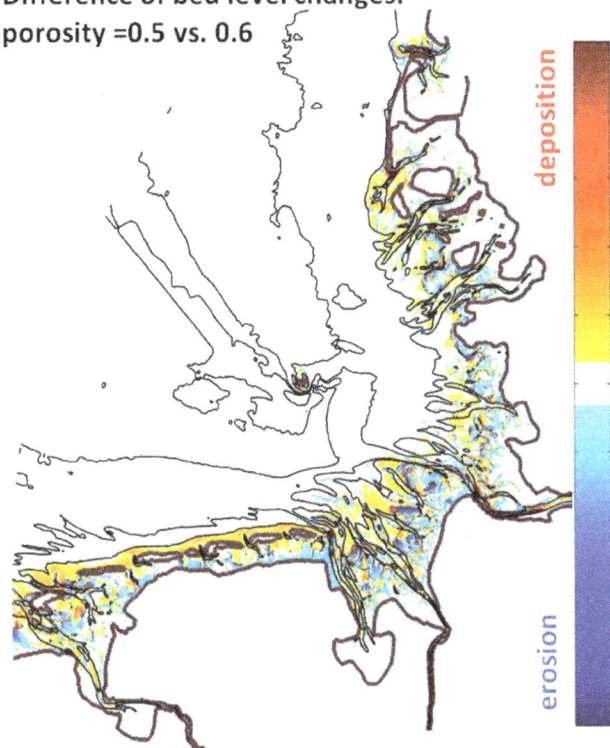

Fig. 2. Differences in bed level changes after 1 yr MD simulation with DELFT3D (2-D mode) due to a change of the porosity parameter from 0.5 to 0.6.

As the sediment transport strongly depends on the sediment grain size (as can be seen easily from sediment transport formulas, e.g. van Rijn 1993), it has to be concluded that, for study areas with inhomogeneous sediments such as the German Bight, hydrodynamically calculated sediment distributions are not reliable for large-scale sediment transport applications. This emphasizes the need for a detailed sedimentological investigation prior to numerical modelling. And it is not only the sediment grain size distribution which affects model results but also the space variable porosity of the seabed.

The porosity of surface sediments is not well known and difficult to measure. Due to this, commonly the assumption of a space and time constant porosity is made. It is expected that larger porosity yields a stronger morphological reaction because sediment movement is easier if less densely packed, and the same amount of transported sediment results in a larger volume change. In MD models such as DELFT3D, morphological changes are calculated from residual sediment mass transport and then are converted into volume changes using the user-defined porosity, and a strong influence is evident. As an example, the difference in bed evolution after 1 yr of simulation for changing the porosity from 0.5 to 0.6 is assessed as a sensitivity study. The experiment has been carried out using the DELFT3D mod-

elling system applying five sediment fractions: coarse sand (cSa, 750 mum), medium sand (mSa, 375 mum), fine sand (fSa, 187.5 mum), coarse silt (cSi, 46.5 mum) and fine silt (fSi, 12 mum) together with the default DELFT3D settings for sediment transport. The differences in bed level changes between the two porosity settings show a spatially structured response (Fig. 2). In deeper areas off the coast and islands, more deposition and less erosion occur. In contrast to this, shallow tidal areas are more erosive, and less deposition can be found here. The physical mechanism behind this is that with increased porosity sediments from the tidal flats can be more easily mobilized by, for example, wave action and then be transported by tidal currents to deeper areas.

Again, these results illustrate the importance of a precise description of the sediment properties at the seabed for reliable morphodynamic calculations.

3.2 Uncertainties and differences of sediment transport for different modelling systems

The two aforementioned modelling systems UnTRIM-SediMorph and DELFT3D have been calibrated independently for the year 2006. Both models are compared in order to investigate the effect of the different numerical methods and model parameterizations.

To demonstrate the main differences and similarities in time and space, the spatial distribution of sediment transport is shown for bed load transport (Fig. 3) and suspended load transport (Fig. 4). In addition a time series at Helgoland is presented (Fig. 5). The spatial sediment transport is shown for a stormy situation on 9 February 2006 during flood (Figs. 3, 4 left) and ebb (Figs. 3, 4 right) tide.

In both model simulations the spatial distribution of bed load transport is more pronounced near the coast and the estuary mouths for both flood and ebb tide. In contrast, the suspended load transport is more widespread over the German Bight. During flood tide the maximum is located near the coast/estuary mouths; during ebb tide suspended sediment transport is extended seaward. Overall the areal distributions between both models show qualitatively a good agreement; quantitatively the results vary due to differences in model parameterizations. Note that DELFT3D is set up as a depth-integrated 2-D model, whereas UnTRIM-SediMorph is run in 3-D with a vertical resolution between 1 m and 50 m.

The temporal distribution of sediment transport is shown in Fig. 5 as time series of depth-integrated sediment transport at Helgoland, which is located in the inner German Bight. For all three time series, the same axis dimension is used for a better comparison. The time span covers a complete spring–neap cycle (4–19 February 2006) and a storm situation (9 February 2006). Bed load transport quantities (top panel) are small compared to the suspended transport (middle panel). In the DELFT3D (2-D) run, the bed load is more dominant compared to UnTRIM (3-D). Note that the form of transport mode (suspended load, bed load) of individual sediment

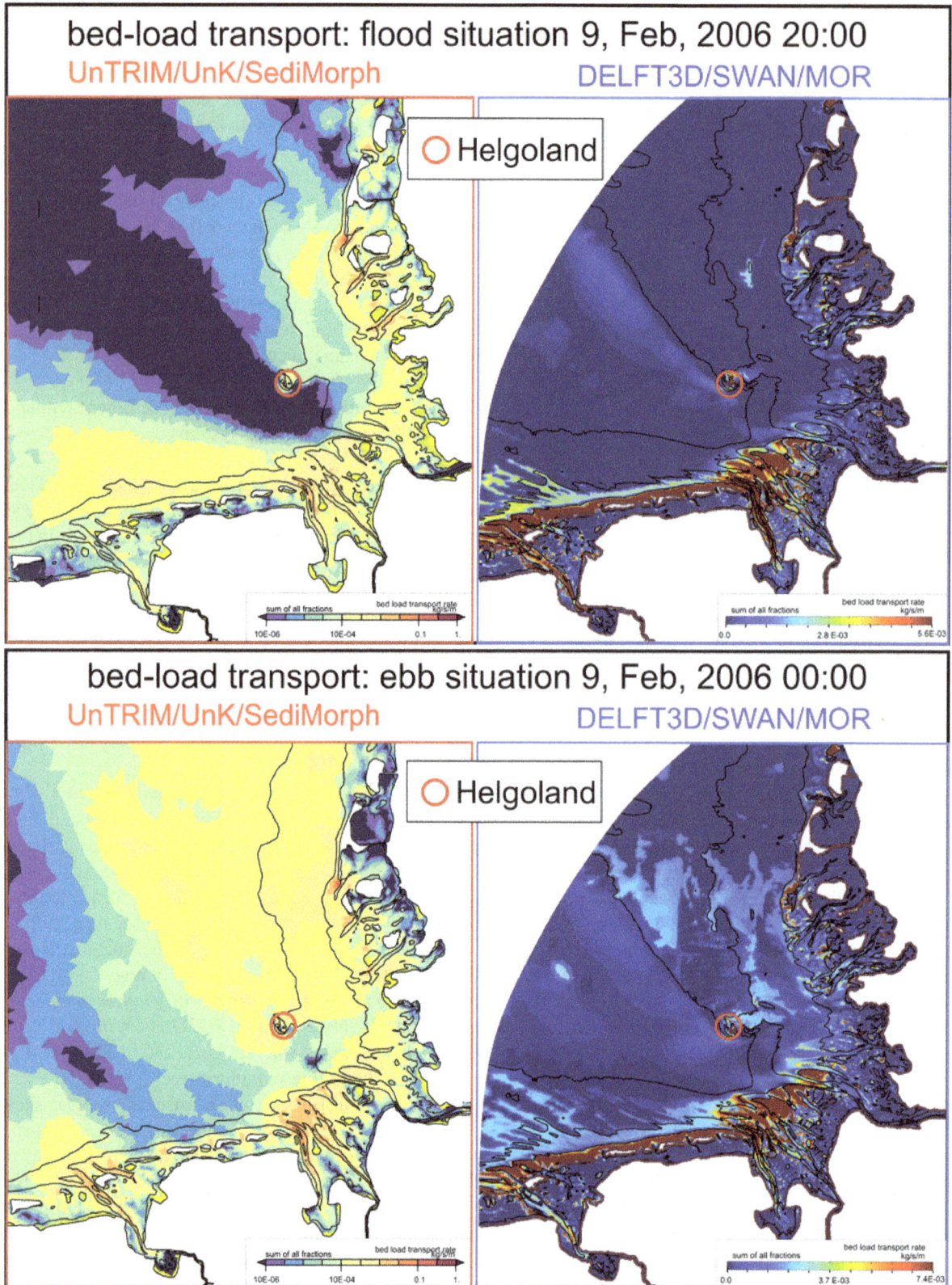

Fig. 3. Bed load transport for windy flood/ebb velocity tide in the German Bight, calculated by UnTRIM/UnK/SediMorph and DELFT3D(2-D mode)/SWAN/MOR.

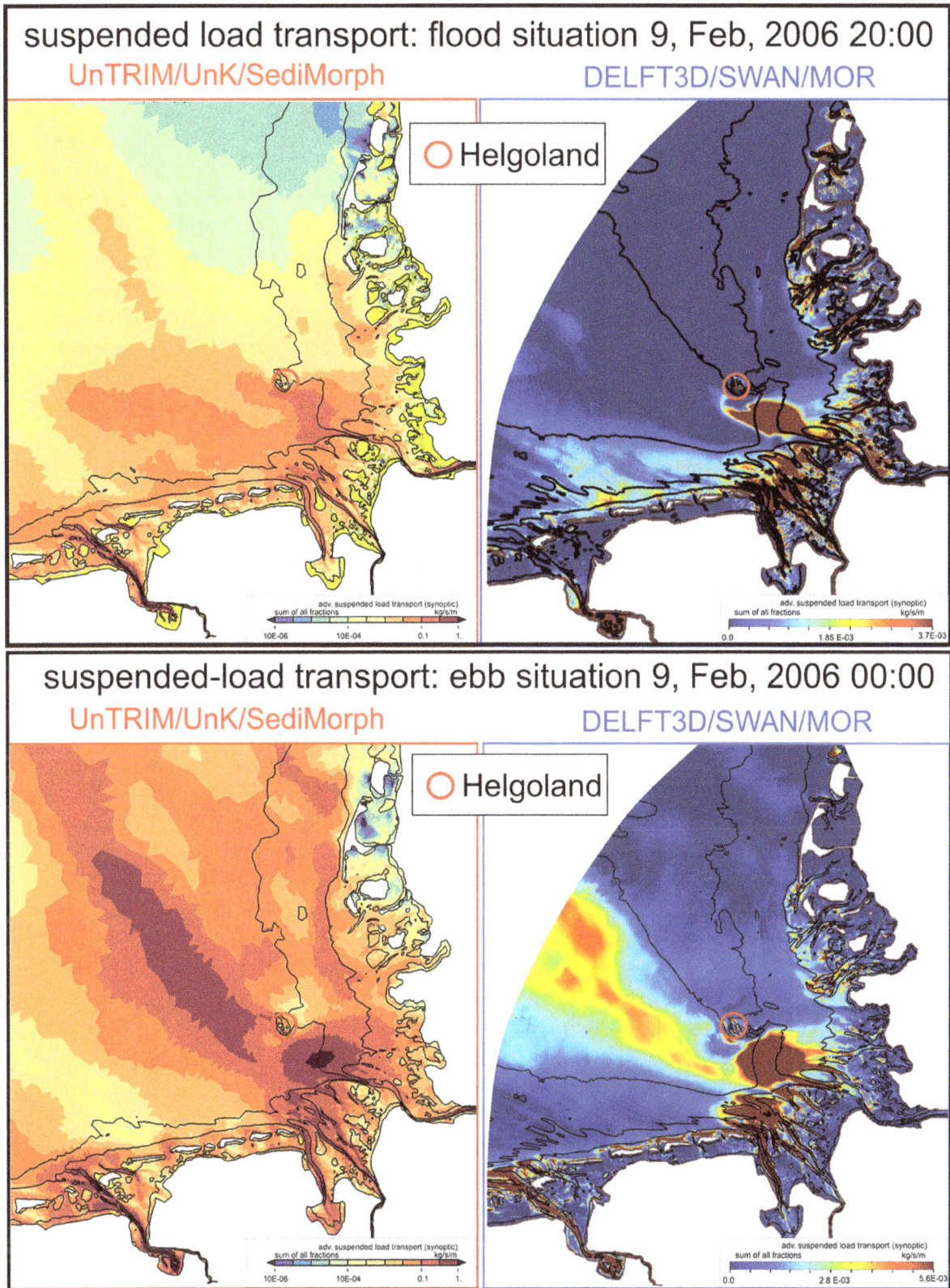

Fig. 4. Suspended load transport for windy flood/ebb velocity tide in the German Bight, calculated by UnTRIM/UnK/SediMorph and DELFT3D(2-D mode)/SWAN/MOR.

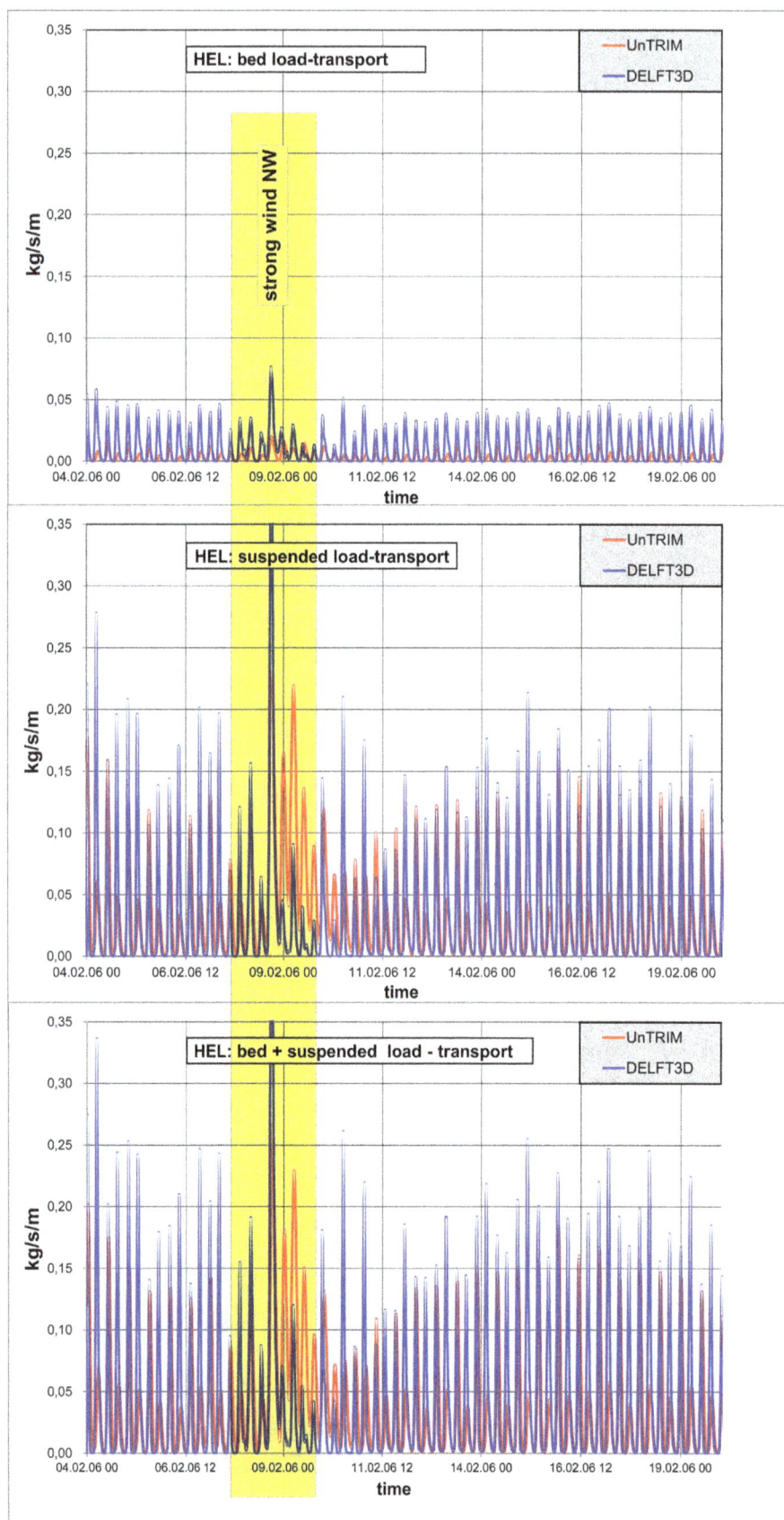

Fig. 5. Calculation of bed load, suspended load and total load transport at Helgoland (February 2006) derived from UnTRIM/UnK/SediMorph and DELFT3D(2-D mode)/SWAN/MOR.

fractions is internally determined in DELFT3D based on the flow situation, whereas it is specified by the user in UnTRIM-SediMorph. The suspended sediment transport is much larger (approx. 4 times) than bed load transport. During the storm period (7–9 February 2006), large suspended-transport rates are calculated by both UnTRIM and DELFT3D, but with slight differences in timing. Except for the stormy period, the UnTRIM and the DELFT3D results for the total sediment transport, as the sum of suspended and bed load transport, are approximately in the same range.

Considering the differences in model set-up (2-D/3-D), differences in the representation of sediment transport (transport formulas, sediment fractions and mode of transport), numerical factors (numerical diffusion/damping) and coupling hydrodynamics and waves, both models show quite comparable results. More detailed investigations are necessary to identify the influences of individual model parameterizations (grid topology/grid spacing, vertical resolution, sediment transport formulations) for a better explanation of differences in model results. However, in terms of a multi-model approach, the range of diverse model results seems quite important.

4 Conclusions

A morphodynamic modelling system for the German Bight has been set up using the DELFT3D and UnTRIM-SediMorph modelling systems. Initial sensitivity studies illustrate the need for detailed sedimentological input data sets and improved estimates for specific sedimentological properties such as porosity, which strongly influence the model results. The approach to obtain a sedimentology, based on present-day hydrodynamic forcing, fails most evidently if geological structures from the past are present. The morphodynamic multi-model approach has been illustrated by sediment transport results for the German Bight. Even though results differ in detail, the large-scale spatial and temporal results are comparable. The proposed multi-model approach, as an extension to parameter variation of individual models, might be a way forward in order to assess model uncertainty in morphodynamic modelling following the path set by climate modellers.

Acknowledgements. The authors and the AufMod-team would like to thank the Federal Ministry of Education and Research (BMBF #03KIS084) for funding of the research project.

References

BAW: Nordsee-Basismodell – Teil II: Modellsystem UnTRIM-SediMorph-Unk, a) Hydrodynamik (UnTRIM-SediMorph), BAW-report A39550270116-2a, 2013.

Booij, N., Holthuijsen, L. H., and Ris, R. C.: The "SWAN" wave model for shallow water, Coast. Eng. Proc., 1, 25, 1996.

Casulli, V. and Zanolli, P.: Semi-Implicit Numerical Modelling of Non-Hydrostatic Free-surface Flows for Environmental Problems, Math. Comput. Model., 6, 1131–1149, 2002.

Figge, K.: Begleitheft zur Karte der Sedimentverteilung in der Deutschen Bucht 1:250000 (Nr. 2900) – DHI, 1981.

French, J. R. and Burningham, H.: Coastal geomorphology: trends and challenges, Prog. Phys. Geogr., 33, 117–129, 2009.

Kösters, F. and Winter, C.: Exploring German Bight coastal morphodynamics based on modelled bed shear stress, Geo Marine Lett., 34, 21–36, doi:10.1007/s00367-013-0346-y, 2014.

Lesser, G. R., Roelvink, J., van Kester, J., and Stelling, G.: Development and validation of a three-dimensional morphological model, Coastal Eng., 51, 883–915, 2004.

Lyard, F., Lefevre, F., Letellier, T., and Francis, O.: Modelling the global ocean tides: modern insights from FES2004, Ocean Dynam., 53, 394–415, 2006.

Malcherek, A., Piechotta, F., and Knoch, D.: Mathematical Module SediMorph. Technical report. Hamburg: Bundesanstalt für Wasserbau (BAW), 2005.

Plüß, A. and Heyer, H.: Morphodynamic Multi-Model Approach for the Elbe estuary. In Dohmen-Jansen and Hulscher, editors, River, Coastal and Estuarine Morphodynamics: RCEM 2007, 1, 113–117, London, Taylor and Francis Group, 2008.

Schneggenburger, C.: Spectral Wave Modelling with Nonlinear Dissipation, Hamburg: Dissertation, Universität Hamburg, FB Geowissenschaften, 1998.

Stelling, G. W.: On the construction of computational methods for shallow water flow problems, Rijkwaterstaat, the Netherlands: Technical Report 35, 1984.

Stelling, G. W.: Practical aspects of accurate tidal computations, J. Hydr. Eng., 112, 802–817, 1986.

Tebaldi, C. and Knutti, R.: The use of the multi-model ensemble in probabilistic climate projections, Philos. Trans. Roy. Soc. a – Math. Phys. Eng. Sci., 365, 2053–2075, 2007.

Zeiler, M., Schwarzer, K., Bartholomä, A., and Ricklefs, K.: Seabed morphology and sediment dynamics, Die Küste, 74, 31–44, 2008.

9

Improvements in Data Quality, Integration and Reliability: New Developments at the IRIS DMC

T. Ahern, R. Benson, R. Casey, C. Trabant, and B. Weertman

IRIS Data Management Center, 1408 NE 45th Street, Seattle, WA 98105, USA

Correspondence to: T. Ahern (tim@iris.washington.edu)

Abstract. With the support of the US National Science Foundation (NSF) and on behalf of the international seismological community, IRIS developed a Data Management Center (DMC; Ahern, 2003) that has for decades acted as a primary resource for seismic networks wishing to make their data broadly available, as well as a significant point of access for researchers and monitoring agencies worldwide that wish to access high quality data for a variety of purposes. Recently IRIS has taken significant new steps to improve the quality of and access to the services of the IRIS DMC. This paper highlights some of the current new efforts being undertaken by IRIS. The primary topics include (1) steps to improve reliability and consistency of access to IRIS data resources, (2) a comprehensive new approach to assessing the quality of seismological and other data, (3) working with international partners to federate seismological data access services, and finally (4) extensions of the federated concept to extend data access to data from other geoscience domains.

1 Building resiliency in IRIS data services through auxiliary data centers

For several decades IRIS has relied on a single centralized data center in Seattle, Washington to provide all services to the community. In 2006 we created an Active Backup System at the PASSCAL Instrument Center in Socorro, New Mexico. This backup system held copies of all the primary waveform data, key software source and binaries, documentation and a variety of other information in case of a catastrophic event at the primary center such as fire or earthquake. In 2009 the system was relocated to the UNAVCO facility in Boulder, Colorado to take advantage of the higher bandwidth at that location. In 2013 the DMC transitioned the Active Backup concept to a fully functioning Auxiliary Data Center (ADC) where ultimately all of the services of the primary IRIS data center would be replicated and available at all times. The first auxiliary center is located at the Livermore Valley Open Campus (LVOC) at Lawrence Livermore National Laboratory (LLNL). The location of the ADC allows it to be connected to the High Performance Computing (HPC) environment at LLNL and helps in IRIS' goal to place the entire IRIS archive in the proximity of supercomputing resources. The key development that enabled IRIS to do this relies on the Service Oriented Architecture (SOA) that IRIS has developed. The replication of the DMC functionality was greatly aided by relying on web services that have been adopted by the International Federation of Digital Seismograph Networks (FDSN) as well as a small number of additional services that IRIS has developed and uses internally as well as exposing them external to the DMC.

Most of IRIS' applications use these web services and it is much simpler to deploy systems at multiple locations once the web service infrastructure is deployed. This infrastructure is fully deployed at the ADC and the ADC services function identically to the services at the primary DMC.

In the future it is IRIS' plan to leverage an external load balancing system that will seamlessly route some requests to the primary DMC and others to the ADC based on business rules such as how busy one system is over the other system, geographic proximity to one or the other services, or other business rules yet to be determined. Currently the global load balancing is not in place. It is possible to access the web services and other installed applications at the ADC if the URL is known. In fact we are currently running the MUSTANG QA system at both locations leveraging the web services internally so the proof of concept has been heavily exercised. Specifically IRIS services that operate at both the DMC and

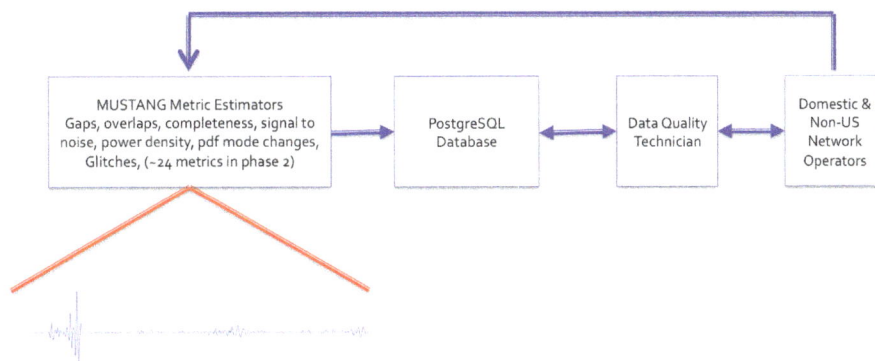

Figure 1. The new MUSTANG system not only identifies potential problems with data quality but also builds in a feedback loop between MUSTANG and network operators. This feedback will sometimes be able to help resolve the problem that caused the identified problem and later MUSTANG estimates should verify the improvement in the data quality.

the ADC include the BUD real time system, the ring server that replicates files across a computer network and so in principle data ingestion from any data source can take place at either the DMC or the ADC. However real time data ingestion for real time streams takes place primarily at the DMC and would have to be switched to the ADC manually in the event of a failure. Support for access tools such as BreqFast, WILBER3, SeismiQuery, WebRequest, and the new MUSTANG system all operate at both the DMC and the ADC. In principle it would be possible to install additional ADCs in the US or around the world if resources were available.

2   Enhanced quality assessment of time series data

At the current time a very ambitious project called MUSTANG is soon to enter operational status at the IRIS DMC and the ADC. As new time series arrive at IRIS either in real time or by delayed file transfer procedures, a system of roughly 50 metric calculators derives statistical metrics that characterize a day's worth of waveform data. Such things as gaps per day, overlaps per day, means, RMS values, medians, extreme values and other statistical measures are estimated by MUSTANG algorithms. Additional metrics that are calculated include measures of latency, power spectral density, power density functions and a variety of other time series comparisons between multiple time series (collocated sensors, comparisons to nearby stations, comparisons to synthetic tide estimates are calculated) as well as the signal to noise ratio for windows of data recorded during larger events. The entire list of metrics being calculated can be found at http://service.iris.edu/mustangbeta/metrics/1/query and the list will evolve dynamically as new ways of looking at data quality are determined with time.

Figure 1 shows the basic concept of MUSTANG. Time series enter the DMC, metric calculators are run, normally about one day after real time, and all relevant metrics are estimated from the new data. These metrics are stored in

a PostgreSQL database and are made accessible through a set of web services similar to the other data access services that are available at IRIS and some other FDSN centers. Data technicians are alerted to patterns in a single metric or a combination of multiple metrics that are indicative of a data problem. The technicians will validate the data problem, try to identify the source of the problem and then communicate with knowledgeable people at the seismic network from which the data came.

At the current time (November 2014) MUSTANG is in beta mode but appears stable and will enter a production phase early in 2015. The system is presently being used to calculate all the metrics for all the relevant data in the archive. By the end of 2014 we anticipate that all metrics will have been calculated for IRIS generated data (_GSN, _PASSCAL, _OBSIP, _US-Array) as well as significant portions of the data from networks that share their data with IRIS. A coverage service will be available that allows one to quickly assess whether or not metrics have been calculated for a specific network, station, channel for a given time period.

A sophisticated recalculation component of MUSTANG is being developed to know when to recalculate metrics. This can happen when any of the following occur: (1) a new version of the time series is received, (2) relevant metadata is updated, or (3) the algorithm itself changes. When completed the automated recalculation engine will trigger recalculation of just those metrics that need recalculation. When complete, metrics should not become stale for any reason and users will have confidence that the metrics they view are correct and represent the metadata and waveform state in the holdings of the DMC. Ultimately we intend to make use of the various metrics to enable data requestors to filter the data they receive from a request to the IRIS DMC based on the values of the MUSTANG metrics.

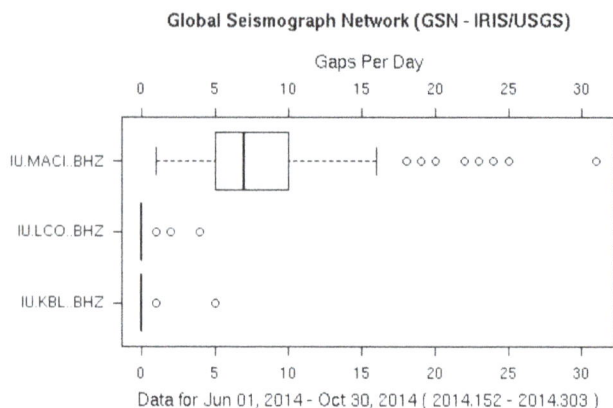

Figure 2. The MUSTANG Data Browser allows visualization of metrics. For instance Box Plots for an entire network can be displayed and quickly allow the operator of a specific seismic network to identify specific stations that have problems indicated by any specific metric. This example shows a boxplot for three stations of the IRIS/USGS network FDSN Network Code IU. The box visible for the topmost station shows the 25th, 50th (median) and 75th percentile range of the gaps per day metric. By looking at long time spans for entire networks in one display, a network analyst can quickly identify problematic stations within their network. The small circles show outliers for the given station.

3   Federation of seismological data centers

Seismological activities have been coordinated globally since the late 1980's by the FDSN (www.fdsn.org). Driven be coordinated definitions of web services and standardized XML schema (Casey and Ahern, 2011), in Working Groups II and III of the FDSN, identical services have now been deployed at 3 data centers in the United States and 6 centers in Europe. In the US, the participating data centers include (IRIS, the Northern California Earthquake Data Center (NCEDC), and the National Earthquake Information Center (NEIC) of the US Geological Survey (USGS). In Europe the participating centers include the French National Data Center (RESIF), the Swiss Seismological Service in Zurich (ETHZ), the Italian National Center for Geophysics and Volcanology (INGV), the ORFEUS Data Centre in the Netherlands, the GEOFON data center in Germany, and the International Seismological Centre in the UK.

Each of these centers has exposed FDSN standard web services that accept identical parameters as a query string in the URL as well as delivering the same FDSN approved XML document resulting from the query. These XML definitions include StationXML for metadata about seismic stations and channels and QuakeML for returning catalogs of earthquakes and other seismic events. Waveform data are returned through identical services in miniseed format defined by the FDSN or as a variety of other formats that include picture files, sound files, or ASCII files. Waveforms are not returned as XML.

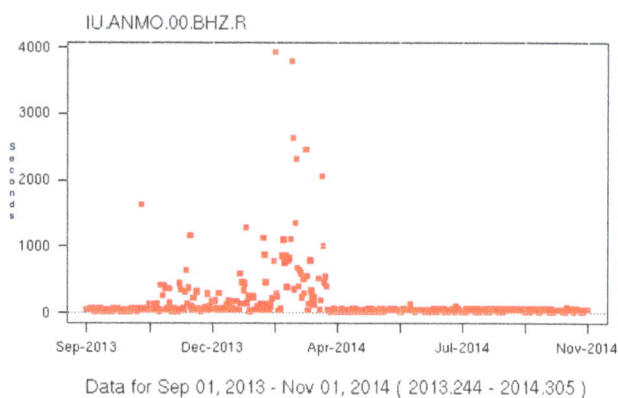

Figure 3. The MUSTANG Data Browser can also display values of metrics for arbitrary lengths of time as requested by a user. The picture above shows the latency for the vertical channel from the GSN station in Albuquerque, New Mexico, USA. The latency is shown for a time range of 14 months in this figure. In the new MUSTANG system latency is estimated once every 4 h that allows most latency problems to be detected and when possible, corrected.

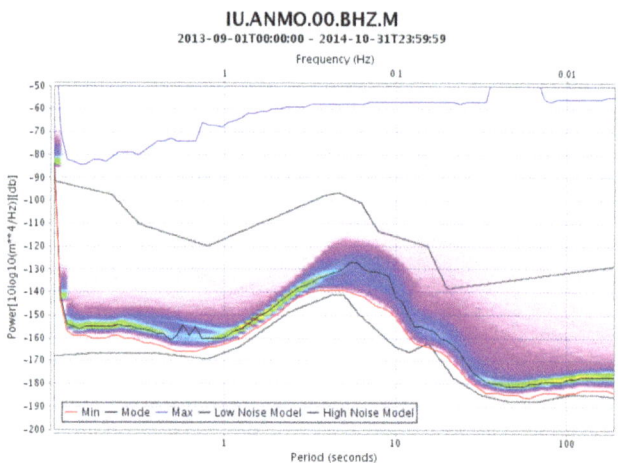

Figure 4. Probability Density Function (PDF) Plots (McNamara and Buland, 2004). The new MUSTANG Data Browser continuously generates estimates of the power spectral density functions. From these raw values, PDFs can be generated for specific network-station-channels for arbitrary time spans. PDFs are an extremely powerful tool that characterizes the noise across a broad range of frequencies for a given seismic station. When compared to the Low Noise Model (shown by the bottommost grey line in the above figure) a stations performance can easily be ascertained.

The key to the federated services is that other than the leftmost portion of the URL (that points to a specific center), the right portion of the URL is identical and the resulting XML document or miniseed data are also identical.

This simplifies the manner in which an external user can interact with all of the federated centers.

IRIS is also developing a federator. The federator works as follows. On a routine basis the IRIS federator queries the

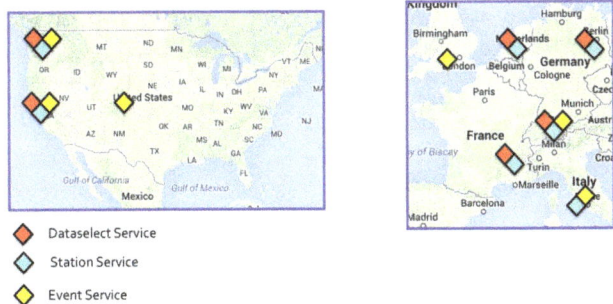

Figure 5. These maps show the various FDSN services that have been placed into operation at various data centers in the US and Europe. The red diamond indicates that time series services are in place; blue indicates that station metadata services are in operation, and yellow means that FDSN event services are in operation.

holdings of all known federated centers to determine the various time series holdings at each center. It databases this information in a PostgreSQL database at the DMC. An external client that could be written in a variety of languages can formulate a request for data in a specific region, bounding box, or within a specific distance of a point. The IRIS federator returns information from which specific URLs to retrieve the waveform and metadata directly from various federated data centers that hold the information.

4 EarthCube and cross disciplinary data integration

As part of an EarthCube Building Block project, IRIS is seeing if the simple web services concepts developed within seismology can be extended to other geoscience communities. Driven by a use case in geodynamics, and working with multiple partners in the US our goal is to ease data discovery, data access, and data usability across several fields in the earth sciences. The Geoscience Web Services (GeoWS) building block, advocates for standard approaches in the development of simple web services that include standardized parameter naming conventions, URL usages, similar documentation styles, and availability of URL builders to show how URLs to access services can be properly formed, Our funded partners include Caltech (GPLATES), Columbia University (marine geoscience data), IRIS (seismology), SDSC (hydrology), UNAVCO (geodesy), and UNIDATA (atmospheric sciences). These six groups will expose their data holdings through similar style web services. Simultaneously, IRIS is working directly with other groups in geoscience to expose their data holdings through simple web services. These include (1) superconducting gravimeter data, (2) gravity data, magnetic data, structural geology data sets, volcanological data and data from three other large facilities including the National Geophysical Data Center, Ocean Observatory Initiative (OOI), and NEON the National Ecological Observatory Network.

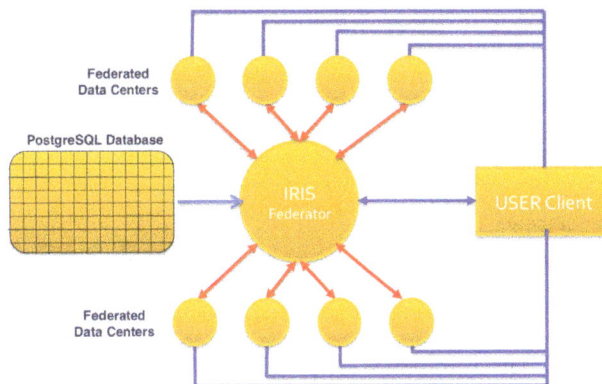

Figure 6. The IRIS Federator Catalog Web Service. The IRIS federator catalog is a web service that enables a client application to recover data of interest from the federated system. Periodically, the IRIS federator queries (red arrows) all of the known external data centers (small yellow circles) and stores the state of data holdings of the federated system in a PostgreSQL database. Using the federator catalog web service to perform client-side federation is a 2-step process: (1) a user client submits a request to the catalog service (blue arrow) to extract a list of data centers and data that match the query and (2) the user client submits the requests directly to the identified data centers. The federator can be queried in a manner whereby all instances of a seismogram can be returned or it can use a powerful set of business rules to return data from the "authoritative" center only. If this system proves effective in meeting data user's needs it will solve many aspects of accessing the data wanted when those data are managed in multiple centers around the world.

While we are certain progress will be made we are also aware that it is impossible to make interdisciplinary data seamlessly accessible across these 14 domains. For this reason we are working closely with the Global Earth Observation System of Systems (GEOSS) brokering group that will offer mediation services across a subset of these domains. If successful it will ease the task of integration of data from these 14 domains.

Acknowledgements. Much of the work involved in these projects drew heavily upon the work of the FDSN working groups especially Working Groups II and III. In addition to WG II chair Reinoud Sleeman of ORFEUS, we would like to particularly thank Marcelo Bianci at GFZ as well as Luca Trani and Alessandro Spinuso at ORFEUS for their very active involvement and timely comments. We would also like to acknowledge the support received by Steve Bohlen, Bill Walter, and Jennifer Aquillino of Lawrence Livermore National Laboratory of the US Department of Energy for helping with the details of establishing the ADC at LLNL.

The developments presented in this paper were supported by several grants from the National Science Foundation including EAR-1261681 (SAGE), ICER-1343709 (EarthCube), and ICER-1321600 (COOPEUS)

References

Ahern, T. K.: The FDSN and IRIS Data Management System: Providing easy access to terabytes of information, International Handbook of Earthquake and Engineering Seismology, 81, 1645–1655, 2003.

Casey, R. and Ahern, T.: Web Services for Seismic Data Archives, Geoinformatics: Cyberinfrastructure for the Solid Earth Sciences, Cambridge University Press, Part V, 13, 210–223, 2011.

McNamara, D. E. and Buland, R. P.: Ambient Noise Levels in the Continental United States, B. Seismol. Soc. Am., 94, 1517–1527, 2004.

10

A theoretical analysis of river bars stability under changing channel width

S. Zen, G. Zolezzi, and M. Tubino

Department of Civil, Environmental and Mechanical Engineering, via Mesiano 77, 38123, Trento, Italy

Correspondence to: S. Zen (simone.zen@unitn.it)

Abstract. In this paper we propose a new theoretical model to investigate the influence of temporal changes in channel width on river bar stability. This is achieved by performing a nonlinear stability analysis, which includes temporal width variations as a small-amplitude perturbation of the basic flow. In order to quantify width variability, channel width is related with the instantaneous discharge using existing empirical formulae proposed for channels with cohesionless banks. Therefore, width can vary (increase and/or decrease) either because it adapts to the temporally varying discharge or, if discharge is constant, through a relaxation relation describing widening of an initially overnarrow channel towards the equilibrium width. Unsteadiness related with changes in channel width is found to directly affect the instantaneous bar growth rate, depending on the conditions under which the widening process occurs. The governing mathematical system is solved by means of a two-parameters (ϵ, δ) perturbation expansion, where ϵ is related to bar amplitude and δ to the temporal width variability. In general width unsteadiness is predicted to play a destabilizing role on free bar stability, namely during the peak stage of a flood event in a laterally unconfined channel and invariably for overnarrow channels fed with steady discharge. In this latter case, width unsteadiness tends to shorten the most unstable bar wavelength compared to the case with constant width, in qualitative agreement with existing experimental observations.

1 Introduction

River bars have been extensively studied, analytically, as an instability phenomenon of an incompressible flow over a flat mobile bed in a single-thread river channels. Practical reasons of this interest lays in the need to predict bars formation in channelized rivers and the related scour and deposition processes that can affect navigation or damage engineering structures (e.g. bank protection).

Most of these bar theories have been developed under simplifying assumptions of steady discharge and fully sediment transporting cross section, therefore their application is strictly suitable mostly for single-thread channels and alternate bar patterns. Despite many simplifying assumptions, bar theories have been effective in supporting quantitative understanding of bar processes (e.g. conditions of occurrence; lenght scales) and have received quantitative support from laboratory experiments. Despite their simplifying assumptions, their application has been extended to be used as physically-based predictors of alluvial channel pattern (e.g. Parker, 1976; Crosato and Mosselman, 2009) with some degree of success.

Moreover evidence has been reported that rather simple and regular bar dynamics can take place also in complex channel morphologies, as wandering (Church and Rice, 2009) or braiding (Zolezzi et al., 2012). In this latter study the formation and downstream migration of alternate and central bar patterns has been observed in a main individual branch of a braided river during a bar-forming flood event below bankfull stage. In such case, the channel transporting sediments expands its width during the rising stage of the flood, and is afterwards contracted during the falling limb. Application of the classical bar theory using constant discharge and channel width values, averaged over the flood duration, yields several discrepancies in bar wavelength and dominant transverse modes.

In principle, however, bar theory can be applied to laterally unconfined channels, provided temporal variations of discharge and actively transporting channel width are properly accounted for.

The aim of the present work is therefore to investigate, on a modelling basis, the stability of river bars in channels where the active width changes with time, thus possibly affecting bar morphodynamics. Because the channel width-to-depth ratio plays a key role on bar development, its temporal variations associated with the widening process can be expected to affect the dynamics of contemporarily developing bars. Existing theoretical analyses (e.g. Repetto et al., 2002) have investigated the influence of spatial planform variability on bars formation showing how spatial changes in channel width, influencing river bars, may produce planform instability and a related tendency to braid. The role of width unsteadiness may become relevant especially in laterally unconfined channels with non-cohesive banks, as it has been observed in laboratory experiments on the initiation of braided and of "pseudo-meandering" streams (Ashmore, 1982, 1991; Bertoldi and Tubino, 2005; Visconti et al., 2010). Evidence of this dynamics has been provided, also, by field observation on an artificially re-shaped natural river consequentially to a series of flood events (Lewin, 1976); this highlights the mutual influence between planform and bar instability in streams where the evolution of bed and banks occur at comparable time scales. Despite the fact that unsteadiness is ubiquitous in natural river systems, only very few theoretical analyses have addressed the role of flow unsteadiness on bars formation (Tubino, 1991; Hall, 2004) while an analysis of the role of width unsteadiness is even lacking. Overall, the present work is therefore expected to contribute a novel theoretical understanding about the applicability of analytical bar theories to real river systems.

2 Methods

2.1 Conceptual approach

In order to understand the impact of width unsteadiness on bar stability in a straight channel, we first need to characterize the processes that lead to temporal changes in the active channel width. Their mathematical description will then be included in existing approaches for classical bar stability analyses, allowing to build the desired model. Four main hydromorphological configurations, based on different combinations of channel width and discharge variability, are examined (Fig. 1): (a) ideal channels with both constant width and discharge, like in most common laboratory flume experiments on river bars; (b) channelized rivers with constant width subject to streamflow variability and often developing alternate bars; (c) constant discharge flowing in an initially overnarrow channel with non cohesive banks, as typical of the initial experimental condition of physical models of braided rivers; and (d) laterally unconfined channels subject to floods, as it can be the case of the main active branches of braided rivers. Configuration (a) is representative of most of the conditions to which classical bar theories strictly ap-

ply. Bar dynamics in configuration b) has been investigated only by the theories of Tubino (1991), Hall (2004) and by the field investigations by Welford (1994). The present contribution specifically aims to investigate free bar stability referring to configurations (c) and (d).

2.2 Quantification of width unsteadiness

In order to keep the mathematical problem suitable for analytical solutions, simple relationships to express channel width variability are needed. A first, simple attempt has been made by using classical regime formulaes to relate bankfull discharge with channel width in single-thread channels. Such relationships are strictly valid under long-term, equilibrium conditions: therefore the present approach is based on the assumption of an "instantaneous" validity of regime equations. Note that this assumption should be less crude than it may appear, because we are interested here more in a simple mathematical law describing the trends and order of magnitude of width unsteadiness rather than in a predictive formnula quantitatively valid for a specific case. To this aim we use the empirical formula proposed by Ashmore (1982) to predict the width at river equilibrium stage for anabranch channels in braided rivers:

$$W^* = \alpha_W Q^{*n_W} \qquad \alpha_W = 8.1019, \ n_B = 0.4738 \qquad (1)$$

In Eq. (1) W^* represent the dimensional channel width and Q^* the related flow discharge in equilibrium conditions. Hereinafter we indicate with a star (*) dimensional quantities. According to our purposes relation (Eq. 1) can be read as the equilibrium width to which a laterally unconfined channel tends asimptotically for a given discharge. A suitable dimensionless expression for Eq. (1) has been derived to fit a dimensionless mathematical approach. Such expression is obtained by following the procedure proposed by Parker et al. (2007). In the following relation the dimensionless channel half-width B is also introduced for formal consistency with existing bar theories:

$$W = 2B = 2\alpha_n Q^{n_W} \qquad (2)$$

$$\alpha_n = 0.5\alpha_W F_0^{n_W} \bar{\beta}^{n_W-1} \bar{d}_s^{-5/2n_W+1} \qquad (3)$$

where the dimensionless parameters are defined as follows:

$$F_0 = \frac{U_0^*}{\sqrt{gD_0^*}} \qquad \bar{\beta} = \frac{B_0^*}{D_0^*} \qquad \bar{d}_s = \frac{d_s^*}{D_0^*} \qquad (4)$$

with d_s^* the median sediment grain size, B_0^* the reference half-channel width value and U_0^*, D_0^* the components of the uniform flow chosen as reference scales.

Two different time variables are considered when examining both cases. The variable τ will vary on the externally imposed time scale of width and discharge unsteadiness, while the variable t will be used to denote the relevant "intrinsic" time variable for the temporal morphodynamics of bars.

These two variables will be considered as mutually independent within the present theoretical analysis, while for applications of the theory to specific real cases the related time scales shall be quantified and possible interactions assessed. In general, the externally imposed time scale is quite "fast", i.e. much shorter than that associated with temporal width adjustments that occur in regulated rivers because of land-use changes, gravel mining, dam construction (Surian and Rinaldi, 2003).

2.2.1 Constant discharge

This is the case of an initially overnarrow channel fed by constant discharge(Fig. 1c), for which the minimum width for bars formation predicted by the bar theories is larger than the imposed initial width (Bertoldi and Tubino, 2005). To describe the widening process, we simply assume that the erosion occurs homogeneously at the banks and that the channel tends, asimptotically, to achieve the equilibrium width at a widening rate that linearly decreases with the distance from the equilibrium conditions. This simplified law is based on the consideration that the erosion process, increases the channel width and decreases the sediment transport of the flow, reducing, in turn, its ability of entering sediments from the banks and thus slowing down the widening process. This process occurs faster while the initial geometry of the channel is far from equilibrium and reduces its effect progressively approaching the final conditions. Therefore, under this assumption, the relaxation relation (Eq. 5) is proposed whereby the value of channel width ($2B_E^*$) at equilibrium with the specified discharge is evaluated through Eq. (2):

$$\begin{cases} B^*(\tau^*),_{\tau^*} = \left(\frac{U^*}{B^*}\right)_E (B_E^* - B^*(\tau^*)) \\ \\ B^*(\tau^* = 0) = B_0^* \\ B^*(\tau^* \to \infty) = B_E^*. \end{cases} \tag{5}$$

The complete solution for this case reads:

$$B^*(\tau^*) = B_E^* + (B_E^* - B_0^*)\left(-e^{-\alpha\tau^*}\right), \tag{6}$$

which, scaled by the unperturbed half channel width B_E^*, can be written in its dimensionless form as:

$$B(\tau) = \frac{B_E^*}{B_E^*} + \frac{(B_E^* - B_0^*)}{B_E^*}\left(-e^{-\alpha\tau}\right). \tag{7}$$

In Eq. (7) τ is a dimensionless time variable defined as

$$\tau = k\frac{U_E^*}{(B_E^* - B_0^*)}\tau^*, \tag{8}$$

where the parameter k accounts for the erodibility of both banks, assumed to be uniform in space.

2.2.2 Unsteady discharge

This configuration can be representative of a main active branch of a braided river that laterally expands during a flood event (Zolezzi et al., 2012). To predict the temporal variability of the active (i.e. sediment-transporting) channel width, we extend the validity of equilibrium regime formulae to instantaneous discharge values during a flood event. Despite being rather crude, this assumption can be justified by the need to use a simple, physics-based relationship in a first theoretical attempt of this type. In addition, better alternatives don't seem to be available at present, despite recent approaches to investigate planform evolution of river channel with self evolving banks (Parker et al., 2011). We therefore assume the channel width being instantaneously at equilibrium with the imposed flow hydrograph and according to Eq. (2) we write this relationship as:

$$B(\tau) = \alpha_n Q(\tau)^{n_W} \qquad \tau = \sigma_T^* \tau^* \tag{9}$$

where τ denotes dimensionless time, τ^* dimensional time and σ_T^* reciprocal of the flood event duration. In this configuration, Eq. (9) has been assumed to describe both channel narrowing and widening.

2.3 Mathematical formulation

The analysis refers to a straight channel with erodible bed and banks made of homogeneous non-cohesive sediment. The governing equations are the 2-D shallow water equation and the sediment continuity equation, which can be written in the following dimensionless form:

$$\frac{\partial U}{\partial t} + U\frac{\partial U}{\partial s} + V\frac{\partial U}{\partial n} + \frac{\partial H}{\partial s} + \bar{\beta}\frac{\tau_s}{D} = 0; \tag{10}$$

$$\frac{\partial V}{\partial t} + U\frac{\partial V}{\partial s} + V\frac{\partial V}{\partial n} + \frac{\partial H}{\partial n} + \bar{\beta}\frac{\tau_n}{D} = 0; \tag{11}$$

$$\frac{\partial D}{\partial t} + \frac{\partial UD}{\partial s} + \frac{\partial VD}{\partial n} = 0; \tag{12}$$

$$\frac{\partial}{\partial t}(F_0^2 H - D) + Q_0\left(\frac{\partial Q_s}{\partial s} + \frac{\partial Q_n}{\partial n}\right) = 0. \tag{13}$$

In Eqs. (10)–(13) classical scalings for theoretical river morphodynamics are employed: the longitudinal and transversal variables s and n are scaled by the half-channel width $\bar{B}_0^*$, chosen as reference, the averaged velocity components (U, V), the water depth D, the free surface elevation H and the shear stress are scaled using the reference state quantities $(\bar{U}_0^*, \bar{D}_0^*)$. The reference, basic state is defined through three dimensionless parameters: the mean width ratio $\bar{\beta}$, the mean relative roughness $\bar{d}_s$ and the mean Shields parameter $\bar{\theta}$, which read:

$$\bar{\beta} = \frac{\bar{B}_0^*}{\bar{D}_0^*} \qquad \bar{d}_s = \frac{d_s^*}{\bar{D}_0^*} \qquad \bar{\theta} = \frac{S}{\Delta\bar{d}_s} \tag{14}$$

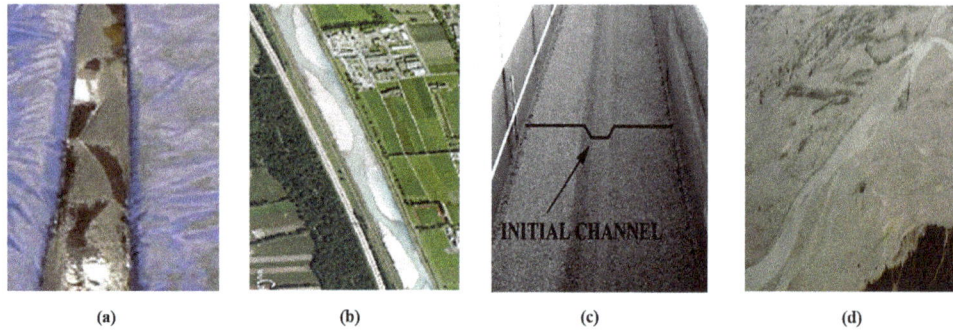

Fig. 1. Examples of channels belonging to the four reference configurations considered in the stability analysis. Each class presents different combinations of discharge and width unsteadiness. More in detail: (**a**) constant discharge and width (courtesy of Grecia A. Garcia Lugo); (**b**) variable discharge and fixed width, a reach of the Alpine Rhine in Switzerland; (**c**) erodible channel with unsteady width and constant discharge; (**d**) natural, laterally unconfined channel with unsteady width and discharge, Tagliamento River (NE Italy).

where S is the longitudinal slope of the channel, $\Delta = (\rho_s/\rho - 1)$ the submerged sediment gravity and d_s^* the sediment diameter.

Boundary conditions in the lateral direction impose vanishing water and sediment flux orthogonal to the banks. When the active channel width changes in time, either because of bank erosion (constant discharge, Fig. 1c) or because of a combination of lateral inundation with erosional dynamics (variable discharge, Fig. 1d), the banklines are laterally moving at the timescale imposed by the process of temporal width variation (described through Eq. 20). Under those conditions, the vanishing lateral flux condition is assumed to apply at the instantaneous (moving) bank line position. Overall, the adopted approach results in neglecting the effect of sediment supply to the channel associated with lateral erosion on the process of bar stability. This seems reasonable given this first theoretical attempt although investigation of the actual role played by that effect will deserve attention in the future.

2.4 Perturbation solution

We solve the governing differential problem through a nonlinear, two-parameters perturbation approach. We then investigate under which conditions the reference uniform basic flow is unstable with respect to infinitesimal and sinusoidal perturbations of the bed elevation and of the other relevant flow quantities. The following two-parameters perturbation expansion, say for the water depth D, is adopted:

$$D = 1 + \epsilon A(t)[(S_m d_{10})E_1(s,t) + c.c.] + \delta d_{01}(\tau)$$
$$+ \epsilon\delta[A(t)S_m d_{11}E_1(s,t)d_{01}(\tau) + c.c.] + O(\epsilon^2, \delta^2) \quad (15)$$

where $c.c.$ denotes the complex conjugate, ϵ and δ are small parameters related to the amplitude of free bars and to the

rate of width unsteadiness respectively. Moreover:

$$S_m = \sin\left(\frac{\pi}{2}mn\right) \quad C_m = \cos\left(\frac{\pi}{2}mn\right) \quad m = 1,2,... \quad (16)$$

$$E_1 = \exp(i\lambda s) \quad (17)$$

where λ is the dimensionless bar wavenumber.

By substituting the structure of Eq. (15) into Eqs. (10)–(13), the original differential system is transformed into a series of linear homogeneous algebraic systems, at each order of approximation comparing in Eq. (15). The key property of this two-parameters perturbation expansion is that the $O(\epsilon\delta)$ is the lowest at which the spatial pattern of free bars is reproduced, because the solution at the order $O(\delta)$ is a perturbation of the reference basic flow and thus it is spatially uniform. The details of the solution procedure at the different orders of approximations are reported below.

2.4.1 $O(\epsilon)$: classical bar stability under conditions of constant width and discharge

By substituting Eq. (15) into Eqs. (10)–(13) and collecting all terms at the leading order ϵ, the classical linear free bar stability is recovered. This requires solving the following linear system:

$$\mathbf{L}_{10}\begin{pmatrix} u_{10} \\ v_{10} \\ h_{10} \\ d_{10} \end{pmatrix} = \begin{pmatrix} 0 \\ 0 \\ 0 \\ 0 \end{pmatrix} \quad (18)$$

where the linear differential operator $\mathbf{L}_{10}$ is reported, in its extended form, in Appendix A. Solution of Eq. (18) reveals that the amplitude of bars behaves exponentially in time:

$$A(t) = \exp(\Omega_{10}t) \quad (19)$$

with $\Omega_{10} = \Omega_{10,R} + i\Omega_{10,I}$, $\Omega_{10,R}$ bars growth rate and $\Omega_{10,I}$ bars angular frequency. Solution of Eq. (18) for the unknowns u_{10}, v_{10}, h_{10} and d_{10} with the parameters $\Omega, \lambda, \bar{\beta}, \bar{\theta}$

and $\bar{d}_s$ requires a solvability condition, which allows calculation of the growth rate of bars and their angular frequency, for a given combination of bar wavenumber λ, the unperturbed width ratio $\bar{\beta}$, Shields parameter $\bar{\theta}$ and relative roughness $\bar{d}_s$.

2.4.2 $O(\delta)$: linear correction to the basic flow related to width unsteadiness

If the length of the channel is much shorter than the typical length of a flood wave, the temporal variability of channel width can be modelled as a temporal sequence of instantaneously uniform flows. To fit within the perturbation scheme, the adopted empirical relations for width unsteadiness are expanded in power of the small parameter δ, which takes slightly different meanings depending on the considered hydromorphological configuration. In both the examined configurations (c) and (d) in Fig. 1 the perturbed channel width is then written in the form:

$$B(\tau) = 1 + \delta b_{01}(\tau) \quad \delta << 1 \tag{20}$$

where b_{01} is described through different functional expressions in the two cases, as it occurs for δ. More specifically, for initially overnarrow channels subject to constant discharge and erodible banks:

$$b_{01}(\tau) = -e^{-\tau} \quad \delta = \frac{B_E^* - B_0^*}{B_E^*}, \tag{21}$$

while for the flood event over a laterally unconfined channels it holds:

$$b_{01}(\tau) = \alpha_n n_B q_{01}(\tau) \quad Q_0(\tau) = 1 + \delta q_{01}(\tau)$$
$$\delta = \frac{Q_{max}^* - Q_0^*}{Q_0^*}. \tag{22}$$

Here the unsteady discharge term q_{01} is assigned as input data and it represents the functional shape of the given flow hydrograph.

According to the adopted perturbation approach, an analytical solution expressing the correction to the reference uniform flow due to width unsteadiness is obtained for the two configurations, (c) and (d). Therefore the unsteadiness-corrected basic flow reads:

$$U_0 = 1 + \delta u_{01}(\tau) + O(\delta^2), \tag{23}$$

$$D_0 = 1 + \delta d_{01}(\tau) + O(\delta^2). \tag{24}$$

By feeding the expansion (Eq. 23) into the governing Eqs. (10–13) we find for the unsteady discharge configuration (d):

$$d_{01}(\tau) = p(1-\alpha)q_{01}(\tau), \tag{25a}$$

$$u_{01}(\tau) = (1-p)(1-\alpha)q_{01}(\tau); \tag{25b}$$

and, for steady discharge configuration (c):

$$d_{01}(\tau) = -pb_{01}(\tau), \tag{26a}$$

$$u_{01}(\tau) = -(1-p)b_{01}(\tau). \tag{26b}$$

In the above expressions the parameters p and α take the value:

$$p = \frac{2}{3 - C_D} \qquad \alpha = \alpha_n n_B. \tag{27}$$

2.4.3 $O(\epsilon\delta)$: effect of width variations on free bar stability

By considering that the solution at the order δ is not dependent on the longitudinal variable s, the smallest order at which the spatial dependence of the fundamental is reproduced is the order $\epsilon\delta$ (Eq. 15). At this order, therefore, the solution accounts for the effect of the temporal width variation on free bar instability. Using the mathematical operator $\mathbf{L}_{10}$ introduced in Eq. (18), at the order $\epsilon\delta$ the governing differential system reads:

$$\mathbf{L}_{10}\begin{pmatrix} u_{11} \\ v_{11} \\ h_{11} \\ d_{11} \end{pmatrix} = \begin{pmatrix} -a_{11}^1 u_{10} - a_{14}^1 d_{10} \\ -a_{22}^1 v_{10} \\ -a_{31}^1 u_{10} - a_{32}^1 v_{10} - a_{34}^1 d_{10} \\ i\Omega_{11}K_1 + (-i\Omega_{10})K_2 - Q_0\Phi_0 K_3 \end{pmatrix};$$
$$K_1 = (F_0^2 h_{10} - d_{10}), \quad K_2 = K_1 \Phi_T [2u_{01}(\tau) + c_D d_{01}(\tau)],$$
$$K_3 = a_{41}^1 u_{10} + a_{42}^1 v_{10} + a_{43}^1 h_{10} + a_{44}^1 d_{10}, \tag{28}$$

where $\mathbf{L}_{10}$ is the same linear algebraic operator found at the leading $O(\epsilon)$ and the coefficients a_{ij}^1, K_i are related to the nonlinear, unsteady effect arising from the interaction between the fundamental perturbation (ϵ) and the unsteady correction to the basic flow due to width variability (δ). The expression of a_{ij}^1, K_i are reported in Appendix A for the sake of brevity.

As for the leading order, since the determinant of $\mathbf{L}_{10}$ vanishes, a solvability (eigenrelation) analogous to that occurring at the leading order holds:

$$f(\Omega_{11}, \lambda; \Omega_{10}, \bar{\beta}, \bar{\theta}, \bar{d}_s) = 0, \tag{29}$$

which allows to compute the correction $\Omega_{11} = \Omega_{11,R} + i\Omega_{11,I}$ to bar growth rate and angular frequency at the order $\epsilon\delta$. As it can be easily seen from Eqs. (28) and (29), the solution of the system (Eq. 28) depends on the solution found at the previous order $O(\epsilon)$, so the problem is solved in cascade. The corrected expression of the complex number Ω accounting for width unsteadiness is therefore:

$$\Omega(\tau) = \Omega_{10} + \delta\Omega_{11}(\Omega_{10})b_{01}(\tau) \tag{30}$$

where Ω_{10} and Ω_{11} account respectively for the growth rate associated to the fundamental perturbation and for the component related to the unsteadiness due to channel width variability.

3 Results

Results obtained by performing the analysis described above are here reported for the two analyzed morphological configurations: a laterally unconfined channel during a flood event

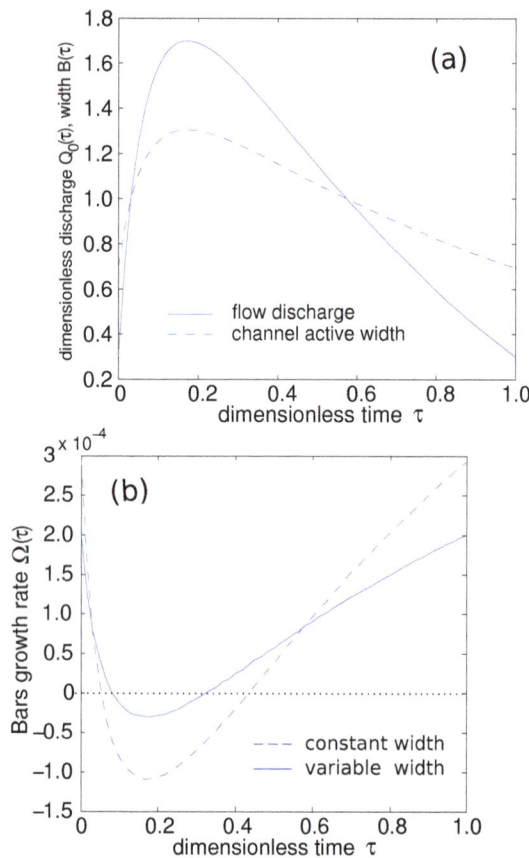

Fig. 2. (a) Temporal variability of the dimensionless discharge $Q_0(\tau)$ and associated temporal variability of the dimensionless channel width $B(\tau)$. **(b)** Values of the total bar growth rate $\Omega(\tau)$ for the flow event in **(a)** for laterally confined (dashed line) and unconfined (solid line) channels.

Fig. 3. The linear growth rate of alternate bars (real part of Ω_{11}) is plotted vs. the Shields stress $\bar{\theta}$ for different values of the roughness parameter $\bar{d}_s$ for the case of variable width with **(a)** variable, **(b)** constant discharge.

(unsteady discharge and width) and widening of an initially overnarrow channel under constant discharge.

3.1   Unsteady discharge and variable width

Figure 2 refers to the case of a flood event that occurs in a laterally unconfined channel. Figure 2 shows the dimensionless hydrograph of the flood event $Q_0(\tau)$ together with the linearized temporal variation of channel width, expressed by Eq. (20). The reference flow (basic state) has been assumed as the uniform flow with constant width occurring for the value $Q_0 = 1$.

As already pointed out by Tubino (1991) free bars under unsteady flow conditions are more stable during the rising and peak stage of the hydrograph, because the aspect ratio β is decreasing and the critical value for bar formation β_c is increasing. In a channel with constant width the decrease of β during the rising stage of the flood is related to decrease in water depth, while the increase of β_c is mainly associated with the increase in the Shields stress. On the contrary, bars

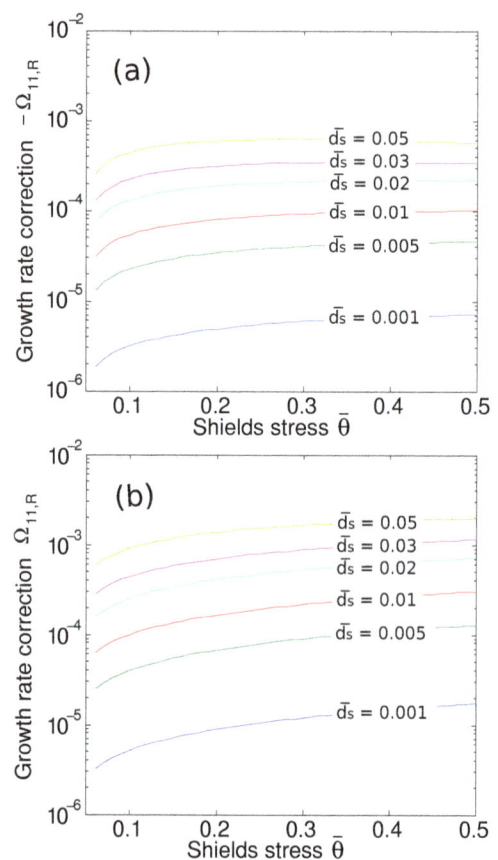

are more likely to form during the decreasing phase where the value of β increases against a decreasing β_c value.

Figure 2 shows that accounting for temporal width variability results in damping the stabilizing effect associated with flow unsteadiness in a channel where the width does not change in time. Namely, when discharge exceeds the reference flow value ($Q_0(\tau) > 1$) the reduction in the bar growth rate is less pronounced than in the case with constant width.

As the temporal behavior of the complete solution is fundamentally controlled by the structure assigned to the temporal variable q_{01} or b_{01}, it is informative to focus on the contribution given by the unsteady term $\Omega_{11,R}$, which is invariably negative, as reported in Fig. 3a.

3.2   Constant discharge and variable width

Application of the bar theory with reference flow parameters that instantaneously adapt to the evolving width value indicate that free bars tend to form during the widening process because the value of the aspect ratio β increases, while the critical value β_c is decreasing, being β_c a growing function of the Shields stress, which decreases with widening.

Moreover, results obtained by applying the present theoretical analysis to an initially overnarrow channel, characterized by erodible banks and constant discharge, reveals a net positive contribution of the width unsteadiness to bar growth rate, $\Omega_{11,R} > 0$ (Fig. 3b). Channel widening is therefore predicted to enhance bar instability compared to the configuration with constant width. Moreover, as it can be recognized from Eq. (21), this effect is stronger when the difference between the initially imposed channel width and the equilibrium width is higher (Fig. 4). Width unsteadiness may therefore determine a complete reversal of bar stability conditions compared to the constant width case. This behaviour is related to the continuity equation for the fluid phase: in channels with constant discharge, the widening process is associated with a decrease of the cross-sectional averaged velocity and depth (Eq. 26), while in channels with variable discharge the temporal width variability is driven by the flood hydrograph: therefore widening during the rising stage is related to a contemporary increase in both cross-sectional averaged velocity and depth (Eq. 25). Such continuity effect mathematically determines opposite signs of the term b_{01} in the two configurations, which are then reflected in the opposite signs of $\Omega_{11,R}$ (Fig. 3a and b).

More general results can be presented by examining the dependence of the most unstable wavelength at the initial time on the Shields stress θ and on the relative roughness d_s (Fig. 4b). This was achieved by assigning as reference state (equilibrium) a wide enough channel to guarantee barforming condition at the initial stage of the erosion process, where, the width is smaller. Shorter bars are therefore promoted, as the most unstable wavenumber is larger. Theoretical predictions of instability enhancement of shorter bars are in qualitative agreement with the experimental observations of Bertoldi and Tubino (2005).

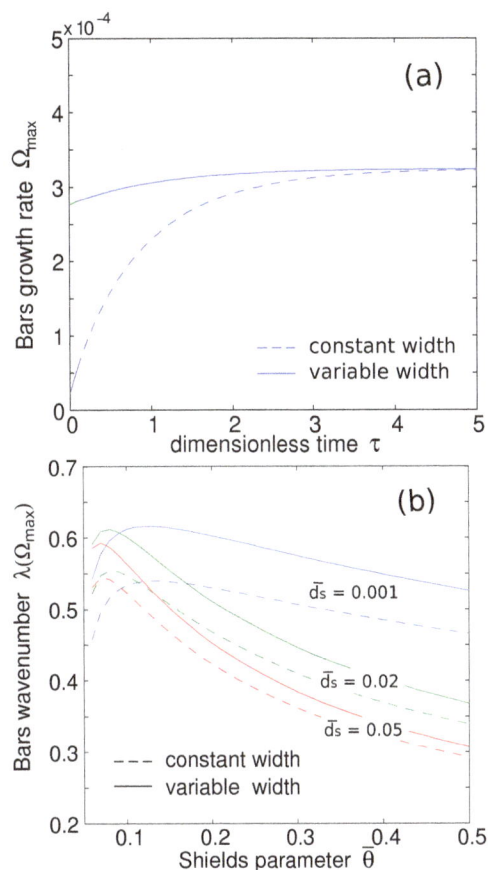

Fig. 4. Comparison between values predicted by applying the present theory (solid line) and the classical linear bar theory applied by assuming steady flow at each temporal step (dashed line). (a) shows difference in maximum bar growth rate and (b) shows the values of the most unstable bar wavenumber plotted against the Shields parameter $\bar{\theta}$ for different value of the roughness parameter $\bar{d}_s$ at the initial time of the process.

4 Discussion

A novel stability analysis for free bars in channels with temporally variable channel width has been developed. Simple relationships describing the variability of the active channel in time allow us to develop a two-parameters perturbation expansion able to quantify the role of width unsteadiness by the interaction of the fundamental bar perturbation with the correction of the basic flow associated with the widening/narrowing processes. Despite the adopted simplifying hypothesis to describe the widening process in rivers channel, the method developed allows a first quantitative investigation of bars dynamics in laterally unconfined channels. Considering that little has been done in literature to investigate the effect of temporal width adjustments (see Parker et al., 2011), another novel point of the work consists in the derivation of a physically-based relationship for width variability under both steady and unsteady discharge conditons. The suitability of the adopted approach finds qualitative sup-

port in the experiments of Bertoldi and Tubino (2005) and of Visconti et al. (2010), who specifically mentioned that in the initial stage of a run, bank erosion increases the channel width keeping the channel straight. The assumption of width variation to be small (say initial stage of laboratory experiment) seems to be reasonable in the light of this. Moreover, data provided from laboratory experiments can be used to validate the relaxation relation here proposed as well as the related time scale adopted.

Results obtained show, curiously, two formally opposite behaviours for the unsteady term $\Omega_{11,R}$ when considering channel width variability. The sign of $\Omega_{11,R}$ is controlled in a rather complex way by the variables defining the basic flow: β, θ, ds and the growth rate computed for the fundamental perturbation $\Omega_{10,R}$ as reported in Eq. (29). Namely, in the case of unsteady discharge this term is found to be negative, while in the other case it gives a positive contribution to bars growth rate. Moreover, during the rising and peak stage of a flood in a laterally unconfined channel, bar stability is increased, thus suggesting a non-trivial effect of width unsteadiness. In association to this destabilizing action, the most unstable bars present a shorter length compared to that predicted by the same theory developed for steady channel width.

Overall, theoretical predictions are in qualitative agreement with both the field observations of Zolezzi et al. (2012) and the experimental observation of (Bertoldi and Tubino, 2005). Further research is needed to apply the present theory to real cases, both in the field and in the laboratory, to assess to which extent the present theory can actually provide a relevant step towards an increased applicability of bar theories to complex channel geometries. Further research shall also concentrate to derive physically-based relationships for channel width variability.

Acknowledgements. This work has been partially carried out within the SMART Joint Doctorate (Science for the MAnagement of Rivers and their Tidal systems) supported by the Erasmus Mundus programme of the European Union. The authors are grateful to Brad Murray and the anonymous reviewer for their suggestions, which were useful in improving the paper.

References

Ashmore, P.: Laboratory modelling of gravel braided stream morphology. Earth Surf. Process. Land., 7, 201–225, 1982.

Ashmore, P.: How do gravel-bed rivers braid? Can. J. Earth Sci., 28, 326–341, 1991.

Bertoldi, W. and Tubino, M.: Bed and bank evolution of bifurcating channels, Water Resour. Res., 41, W07001, doi:10.1029/2004WR003333, 2005.

Church, M. and Rice, S. P.: Form and growth of bars in a wandering gravel-bed river, Earth Surf. Process. Land., 34, 1422–1432, 2009.

Colombini, M., Seminara, G., and Tubino, M.: Finite-amplitude alternate bars, J. Fluid Mech., 181, 213–232, 1987.

Crosato, A. and Mosselman, E.: Simple physics-based predictor for the number of river bars and the transition between meandering and braiding. Water Resour. Res., 45, W03424, doi:10.1029/2008WR007242, 2009.

Hall, P.: Alternating bar instabilities in unsteady channel flows over erodible beds, J. Fluid Mech., 499, 49–73, doi:10.1017/S0022112003006219, 2004.

Lewin, J.: Initiation of bed forms and meanders in coarse grained sediment. Geolog. Soc. Am. Bull., 87, 281–285, 1976.

Parker, G.: On the cause and characteristic scales of meandering and braiding in rivers, J. Fluid Mech., 76, 457–479, doi:10.1017/S0022112076000748, 1976.

Parker, G., Shimizu, Y., Wilkerson, G. V., Eke, E. C., Abad, J.D., Lauer, J. W., Paola, C., Dietrich, W. E., and Voller, V. R.: A new framework for modeling the migration of meandering rivers. Earth Surf. Process. Land., 36, 70–86, doi:10.1002/esp.2113, 2011.

Parker, G., Wilcock, P. R., Paola, C., Dietrich, W. E., and Pitlick, J.: Physical basis for quasi-universal relations describing bankfull hydraulic geometry of single-thread gravel bed rivers, J. Geophys. Res., 112, F04005, doi:10.1029/2006JF000549, 2007.

Repetto, R., Tubino, M., and Paola, C.: Planimetric instability of channels with variable width, J. Fluid Mech., 457, 79–109, 2002.

Surian, N. and Rinaldi, M.: Morphological response to river engineering and management in alluvial channels in italy, Geomorphology, 50, 307–326, 2003.

Tubino, M.: Growth of alternate bars in unsteady flow, Water Resour. Res., 27, 37–52, 1991.

Visconti, F., Camporeale, C., and Ridolfi, L.: Role of discharge variability on pseudomeandering channel morphodynamics: Results from laboratory experiments, J. Geophys. Res., 115, F04042, doi:10.1029/2010JF001742, 2010.

Welford, M.: A field-test of Tubino's (1991) model of alternate bar formation, Earth Surf. Process. Land., 19, 287–297, doi:10.1002/esp.3290190402, 1994.

Zolezzi, G., Bertoldi, W., and Tubino, M.: Morphodynamics of bars in gravel-bed rivers: Bridging analytical models and field observations, in Gravel-bed Rivers: Processes, Tools, Environments, edited by Church, M., Biron, P. M., and Roy, A., Chap. 6, 69–89, John Wiley & Sons, Chichester, UK, 2012.

Appendix A

Coefficients of the linear system at the order $O(\epsilon)$ and $O(\epsilon\delta)$

Collecting the terms at the order (ϵ) and $(\epsilon\delta)$ in the system

$$a_{i1}(t)u_1 + a_{i2}(t)v_1 + a_{i3}(t)h_1 + a_{i4}(t)d_1 = 0,$$
$$i = 1, 2, 3, 4 \tag{A1}$$

the coefficients $a_{i,j}(t)$ can be expanded as

$$a_{ij}(t) = a_{ij}^0 + a_{ij}^1(t) \quad i, j = 1, 2, 3, 4 \tag{A2}$$

and for the specific case they read as follow:

$$a_{11}^0 = i\lambda + 2\beta_0 C_0$$
$$\quad a_{11}^1(t) = a_{11}^0 u_{01}(t) + [2\beta_0 C_0(C_D - 1)]d_{01}(t);$$
$$a_{13}^0 = i\lambda$$
$$\quad a_{13}^1(t) = 0;$$
$$a_{14}^0 = \beta_0 C_0(C_D - 1)$$
$$\quad a_{14}^1(t) = [2\beta_0 C_0(C_D - 1)]u_{01}(t) +$$
$$\quad [\beta_0 C_0(P_{11} + C_D(C_D - 1) + 2]d_{01}(t);$$
$$a_{22}^0 = i\lambda + \beta_0 C_0$$
$$\quad a_{22}^1(t) = [i\lambda + \beta_0 C_0]u_{01}(t) + \beta_0 C_0(C_D - 1)d_{01}(t);$$
$$a_{23}^0 = \frac{\pi}{2}$$
$$\quad a_{23}^1(t) = 0;$$
$$a_{31}^0 = i\lambda$$
$$\quad a_{31}^1(t) = a_{31}^0 d_{01}(t);$$
$$a_{32}^0 = -\frac{\pi}{2}$$
$$\quad a_{32}^1(t) = a_{32}^0 d_{01}(t);$$
$$a_{34}^0 = i\lambda$$
$$\quad a_{34}^1(t) = a_{34}^0 u_{01}(t);$$
$$a_{41}^0 = 2i\lambda\Phi_T$$
$$\quad a_{41}^1(t) = [2i\lambda(P_{21F} - \Phi_T)]u_{01}(t) + 2i\lambda P_{21D}d_{01}(t);$$
$$a_{42}^0 = -\frac{\pi}{2}$$
$$\quad a_{42}^1(t) = -a_{42}^0 u_{01}(t);$$
$$a_{43}^0 = \frac{\pi^2 r F_0^2}{4\beta_0 \sqrt{\theta_0}}$$
$$\quad a_{43}^1(t) = -a_{43}^0 u_{01}(t) + \left[-\frac{C_D}{2}a_{43}^0\right]d_{01}(t);$$
$$a_{44}^0 = -\frac{\pi^2 r F_0^2}{4\beta_0 \sqrt{\theta_0}} + i\lambda\Phi_T C_D$$
$$\quad a_{44}^1(t) = \left[\frac{a_{43}^0}{F_0^2} + i\lambda C_D P_{21F}\right]u_{01}(t)$$
$$+ \left[\frac{C_D a_{43}^0}{2F_0^2} + i\lambda(\Phi_T C_D P_{11} + C_D P_{21D})\right]d_{01}(t),$$
$$\tag{A3}$$

having defined

$$P_{11} = C_D\left(\frac{C_D}{2} - 1\right), \tag{A4a}$$

$$P_{21F} = -2\theta_c \Phi_T, \tag{A4b}$$

$$P_{21D} = -C_D\theta_c \Phi_T \tag{A4c}$$

and

$$C_D = \frac{C_{f,D}|_0}{C_{f0}}, \tag{A5a}$$

$$\Phi_T = \frac{\theta_0}{\Phi_0}\Phi_{,\theta}|_{\theta_0}, \tag{A5b}$$

where C_f denotes the friction coefficient, Φ the intensity of bedload transport and the subscript 0 refers to the reference state.

The system 18 for the order $O(\epsilon)$ is therefore written in the form

$$\begin{pmatrix} a_{11}^0 & 0 & a_{13}^0 & a_{14}^0 \\ 0 & a_{22}^0 & a_{23}^0 & 0 \\ a_{31}^0 & a_{32}^0 & 0 & a_{34}^0 \\ a_{41}^0 Q_o \Phi_o & a_{42}^0 Q_o \Phi_o & a_{43}^0 Q_o \Phi_o F_0^2(-i\Omega_{10}) & a_{44}^0 Q_o \Phi_o F_0^2(i\Omega_{10}) \end{pmatrix} \begin{pmatrix} u_{10} \\ v_{10} \\ h_{10} \\ d_{10} \end{pmatrix} = \begin{pmatrix} 0 \\ 0 \\ 0 \\ 0 \end{pmatrix}. \tag{A6}$$

The matrix of the system A6 has been represented in the text through the linear differential operator $\mathbf{L}_{10}$.

11

The 2013 Earthquake Series in the Southern Vienna Basin: location

M.-T. Apoloner[1], G. Bokelmann[1], I. Bianchi[1], E. Brückl[2], H. Hausmann[3], S. Mertl[4], and R. Meurers[3]

[1]Department of Meteorology and Geophysics, University of Vienna, Vienna, Austria
[2]Department of Geodesy and Geoinformation, Vienna University of Technology, Vienna, Austria
[3]Zentralanstalt für Meteorologie und Geodynamik, Vienna, Austria
[4]Mertl Research GmbH, Vienna, Austria

Correspondence to: M.-T. Apoloner (maria-theresia.apoloner@univie.ac.at)

Abstract. Eastern Austria is a region of low to moderate seismicity, and hence the seismological network coverage is relatively sparse. Nevertheless accurate earthquake location is very important, as the area is one of the most densely populated and most developed areas in Austria.

In 2013 a series of earthquakes with magnitudes up to 4.2 was recorded in the Southern Vienna Basin. With portable broadband, semi-permanent, and permanent installed seismic sensors from different institutions it was possible to record the main- and aftershocks with an unusual multitude of close-by seismic stations.

In this study we combine records from all available stations up to 240 km distance in one dataset. First, we stabilize the location with three stations deployed in the epicentral area. The higher network density moves the location of smaller magnitude events closer to the main shocks, with respect to preliminary locations achieved by permanent and semi-permanent networks. Then we locate with NonLinLoc using consistent picks, a 3-D velocity model and apply station corrections. This second approach results in stable epicenters, for limited and even changing station availability.

This dataset can then be inspected more closely for the presence of regional phases, which then can be used for more accurate localizations and especially depth estimation. Further research will address directivity effects and the asymmetry in earthquake intensity observed throughout the area, using double differences and cross-correlations.

1 Introduction

The study area is situated at the transition of the Eastern Alps to the Pannonian Basin and the Western Carpathians. The Vienna Basin is, due to the vicinity to Vienna, one of the most densely populated and developed areas in the region.

Instrumentally recorded seismicity in the area is moderate, with a maximum registered magnitude of around 5. The Vienna Basin Fault System occasionally shows earthquakes with magnitudes larger than 4, for example in 1938 close to Ebreichsdorf one event with a magnitude of 5.0. Historical records (e.g., Gutdeutsch et al., 1987) and paleoseismicity (e.g., Hintersberger et al., 2010) indicate that even stronger earthquakes occur, more infrequently.

The Austrian seismological network is built of very high quality stations. However, due to large inter-station distances the allocation of an events to a fault is not always definite. Earthquake location and depth estimation accuracy can be increased with different approaches: additional seismic stations, particularly close to the epicenter, are the easiest way of improvement. As most location techniques are strongly dependent on the velocity model, the use of a regionally adapted model has a significant impact. The location technique itself has a big influence as well, in particular as seismic data and the velocity model enter in different ways.

In this article, we show an earthquake series south of Vienna in fall 2013, and which steps can be taken to improve and stabilize location. We deployed additional stations and collected all available data up to 240 km distance to form a comprehensive dataset. We use two different location techniques with two distinct velocity models.

This article illustrates the change in location and tries to allocate the earthquakes to known faults.

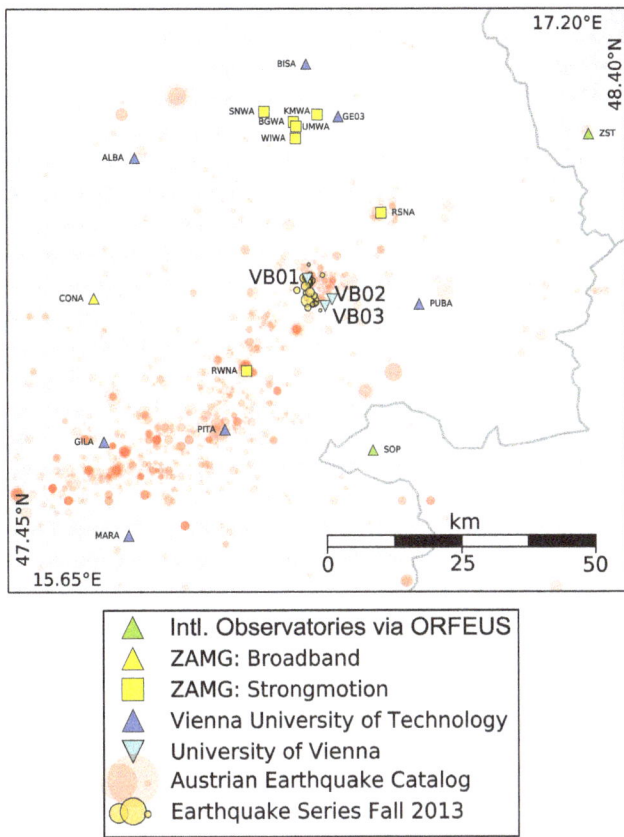

Figure 1. Local station networks and earthquake distribution.

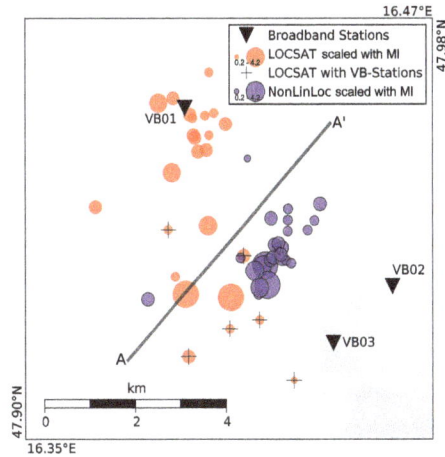

Figure 2. Comparison of epicentral locations calculated with different approaches. Profile A parallel to fault direction by Peresson and Decker (1997).

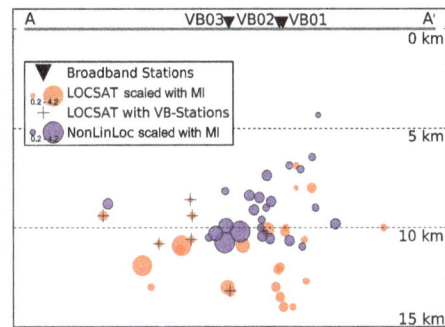

Figure 3. Earthquake depth distribution along profile A (see Fig. 2 for details).

2 Dataset

2.1 Seismicity

In fall 2013 an earthquake series was recorded in the Southern Vienna Basin close to Ebreichsdorf. Two main shocks with a local magnitude of 4.2 as well as about 30 aftershocks with magnitudes below 3.0 were observed. Although in 2000 a similar earthquake swarm was registered, with an epicentral distance to the closest permanent station of 20 km, hypocentral parameters can be only constrained within limits. This time the events were recorded by a multitude of networks as shown in Fig. 1.

2.2 Velocity models

We used two velocity models for this investigation. The global IASP91 (Kennett, 1991) 1-D velocity model, which is used for event location in the ZAMG bulletin. The 3-D P- and S-velocity models by Behm et al. (2007a, b), which are available for Eastern Austria, were also used.

2.3 Seismic stations

The Zentralanstalt für Meteorologie und Geodynamik (ZAMG) operates a seismic network of permanent broadband and strong motion stations in Austria. As part of the ALPAACT Project (Mertl and Brückl, 2010), by the Technical University of Vienna (TU) 10 temporary seismic stations record seismic data in and around the Vienna Basin. Also, seismic records of the national networks surrounding Austria have been obtained via ORFEUS web service. Supplementary data from the nuclear power plant monitoring stations in Hungary was made available by GeoRisk Earthquake Engineering. After the second main shock, the University of Vienna deployed three seismic broadband stations (VB01, VB02 and VB03) as close as possible to the epicenters of the magnitude 4.2 events, as given in the ZAMG bulletin. We assembled all available data from the stations in above-mentioned networks to a comprehensive dataset for this study. We used stations up to 240 km distance, as the 3-D model has limited coverage.

Table 1. Hypocentral parameters with LOCSAT and NonLinLoc, events located with LOCSAT using VB01, VB02 and VB03 are marked with ∗ at the beginning of the line.

Origin Time (CEST) DD.MM.YY HH:MM	Ml	LOCSAT			NonLinLoc		
		Lon	Lat	Depth	Lon	Lat	Depth
4.9.13 11:00	1.9	16.4088	47.9599	8.0	16.4369	47.9447	9.8
20.9.13 02:06	4.2	16.4103	47.9270	10.9	16.4206	47.9331	10.2
20.9.13 02:42	1.7	16.4032	47.9550	14.0	16.4257	47.9364	10.6
20.9.13 02:44	1.0	16.4028	47.9615	6.9	16.4274	47.9395	6.9
20.9.13 03:17	1.3	16.4040	47.9578	14.0	16.4242	47.9364	9.0
20.9.13 23:24	1.2	16.4054	47.9620	10.6	16.4353	47.9416	9.0
24.9.13 13:53	2.7	16.3892	47.9639	10.1	16.4198	47.9331	9.3
25.9.13 10:08	1.7	16.3936	47.9648	12.1	16.4222	47.9341	8.4
28.9.13 17:19	0.6	16.3942	47.9309	13.0	16.4185	47.9276	10.5
1.10.13 23:54	1.4	16.3992	47.9610	10.0	16.4275	47.9438	6.5
2.10.13 04:09	2.0	16.3985	47.9617	10.2	16.4230	47.9346	9.1
2.10.13 05:12	1.3	16.4040	47.9697	12.7	16.4275	47.9417	7.1
2.10.13 05:26	2.1	16.4008	47.9547	13.0	16.4257	47.9343	8.5
2.10.13 05:33	1.8	16.3999	47.9573	12.0	16.4234	47.9370	7.4
2.10.13 17:17	4.2	16.3972	47.9276	11.9	16.4212	47.9294	10.8
2.10.13 19:38	1.6	16.3995	47.9578	13.5	16.4246	47.9372	8.7
2.10.13 19:42	2.8	16.3932	47.9507	10.9	16.4174	47.9320	9.9
3.10.13 00:11	0.2	16.4008	47.9856	8.0	16.4152	47.9534	4.3
3.10.13 00:18	1.9	16.3706	47.9441	11.1	16.3859	47.9266	8.8
3.10.13 00:21	-.-	16.3590	47.9386	10.0	16.3964	47.9815	9.0
*5.10.13 01:27	1.0	16.3921	47.9398	9.4	16.4333	47.9397	10.9
*7.10.13 18:22	1.5	16.3981	47.9157	9.4	16.4224	47.9420	10.7
*7.10.13 19:47	0.3	16.4292	47.9112	8.6	16.4249	47.9359	10.2
*13.10.13 23:26	0.5	16.4188	47.9227	10.6	16.4269	47.9340	10.0
*14.10.13 02:34	1.9	16.4141	47.9349	13.2	16.4253	47.9351	10.4
*16.10.13 02:19	1.4	16.4100	47.9210	10.8	16.4284	47.9335	9.6
23.10.13 19:34	2.6	16.4037	47.9406	13.0	16.4189	47.9289	10.3
15.11.13 16:31	0.9	16.4348	47.9685	10.0	16.4132	47.9345	8.2

3 Hypocenter location

3.1 Routine processing with 3 local stations

The locations given in the bulletin are calculated using the LOCSAT (Nagy, 1996) algorithm with the 1-D velocity model. P- and S-arrivals of the Austrian station network and international stations close to the border are automatically picked and manually reviewed. Besides, the stations PUBA, MARA, SITA and GUWA of the ALPAACT network are included in routine analysis. Six aftershocks were also recorded with the close-by deployed stations VB01, VB02 and VB03. P- and S-picks from those stations were added to the routine processing. The locations using the local stations are marked with ∗ in Table 1. The mean inter-event distance of the newly located earthquakes decreases from 2.0 to 0.9 km with inclusion of the VB-stations. The use of these stations also moves the smaller aftershocks closer to the main events.

3.2 Advanced processing routine with comprehensive consistent dataset

We picked all station data available to us to compile a consistent arrival dataset. With the program NonLinLoc (Lomax et al., 2000) it was possible to use the 3-D velocity model. However, it does not include small-scale inhomogeneities like the underground beneath a station. This leads to shifts in event location, depending on station availability. This influence was reduced by calculating station corrections in the first location run and applying them in the final run. The improvement in location can be deduced from the mean inter-event distance, which is reduced from 1.8 to 1.1 km after applying station corrections. Especially smaller events with less picks move closer to the rest of the swarm. Events with magnitudes bigger than 2.0 do change their location only negligibly. Hence, it was possible to get stable epicenters even for sparse station configurations. According to Geller (1976) the source radius of the two 4.2 earthquakes was estimated at less than 200 m, based on their Mb of 3.6, which was taken from the ZAMG Bulletin.

The final epicenter locations are mostly within an area of 1×2 km as shown in Fig. 2. The mean hypocentral depth is around 9 km, which is typical for this region as described, e.g. in Lenhardt et al. (2007). This means that the hypocenters are beneath the principal displacement zone of the flower structure which is assumed for this area (Beidinger and Decker, 2011). Furthermore, the epicenters show a southwest to north-east pattern which maps them to the Vienna Basin Fault System. Final locations are listed in Table 1 and are also available in QuakeML format in the digital Supplement related to this article.

4 Conclusions

In this study we used a multitude of data available for post-processing. As expected, a higher network density around the epicenter improves the stability of earthquake locations, even if a global 1-D velocity model is used. With NonLinLoc, 3-D velocity model and station corrections it is possible to get stable epicenters even for changing and sparse station configurations.

Although the events can be associated with a nearby fault, no space-time pattern of the main shocks and their aftershocks can be seen with the methods applied and the data used. Therefore, we will investigate this earthquake swarm further using cross-correlation between events.

With this dataset it might be possible to identify additional regional phases like sP, PmP or sPmP. Those could then be used in further research to improve estimation of hypocentral parameters for other events, where less data is available.

Acknowledgements. We would like to thank ZAMG, TU, ORFEUS and GeoRisk Earthquake Engineering for making available seismic data for this study. Topographic data used in maps was taken from SRTM (Jarvis et al., 2008) and historic seismicity from the Austrian Earthquake Catalog (ZAMG, 2014) before 2013. The ZAMG-Bulletin is available online via autodrm@zamg.ac.at. For picking the software Seismon by Mertl (2010) was used. Plots were created with ObsPy (Beyreuther et al., 2010).

References

Behm, M., Brückl, E., Chwatal, W., and Thybo, H.: Application of stacking and inversion techniques to three-dimensional wide-angle reflection and refraction seismic data of the Eastern Alps, Geophys. J. Int., 170, 275–298, 2007a.

Behm, M., Brückl, E., and Mitterbauer, U.: A new seismic model of the Eastern Alps and its relevance for geodesy and geodynamics, VGI Österreichische Zeitschrift für Vermessung & Geoinformation, 2, 121–133, 2007b.

Beidinger, A. and Decker, K.: 3D geometry and kinematics of the Lassee flower structure: Implications for segmentation ans seismotectonics of the Vienna Basin strike-slip fault, Austria, Tectonophysics, 499, 22–40, 2011.

Beyreuther, M., Barsch, R., Krischer, L., Megies, T., Behr, Y., and Wassermann, J.: ObsPy: A python toolbox for seismology, Seismol. Res. Lett., 81, 530–533, 2010.

Geller, R. J.: Scaling relations for earthquake source parameters and magnitudes, B. Seismol. Soc. Am., 66, 1501–1523, 1976.

Gutdeutsch, R., Hammerl, C., Mayer, I., and Vocelka, K.: Erdbeben als historisches Ereignis – Die Rekonstruktion des niederösterreichischen Erdbebens von 1590, Springer Verlag Wien, Heidelberg, New York, 1987.

Hintersberger, E., Decker, K., and Lomax, J.: Largest earthquake north of the Alps excavated within the Vienna Basin, Austria, in: European Seismological Commission (ESC) 32nd General Assembly, Montpellier, France, Abstract T/Sd2/TU/05., 2010.

Jarvis, A., Reuter, H., Nelson, A., and Guevara, E.: Hole-filled seamless SRTM data V4, available at: http://srtm.csi.cgiar.org (last access: 24 April 2014), 2008.

Kennett, B. L. N.: IASPEI 1991 seismological tables, Research School of Earth Sciences, Australian National University, 1991.

Lenhardt, W., Freudenthaler, C., Lippitsch, R., and Fiegweil, E.: Focal-depth distributions in the Austrian Eastern Alps based on macroseismic data, Austrian Journal of Earth Sciences, 100, 66–79, 2007.

Lomax, A., Virieux, J., Volant, P., and Berge, C.: Advances in seismic event locations, chap. Probabilistic earthquake location in 3D and layered models: Introduction of a Metropolis-Gibbs method and comparison with linear locations, 101–134, Kluwer, Amsterdam, 2000.

Mertl, S.: Seismon, available at: http://www.stefanmertl.com/ (24 April 2014), 2010.

Mertl, S. and Brückl, E.: ALPAACT seismological and geodetic monitoring of ALpine – PAnnonian ACtive Tectonics Annual Report – Research Year 2009, Tech. rep., Austrian Academy of Sciences, 2010.

Nagy, W.: New region-dependent travel-time handling facilities at the IDC; Functionality, testing and implementation details, Tech. Rep. 96/1179, SAIC, 1996.

Peresson, H. and Decker, K.: Far-field effects of Late Miocene subduction in the Eastern Carpathians: E–W compression and inversion of structures in the Alpine-Carpathian-Pannonian region, Tectonics, 16, 38–56, 1997.

ZAMG: Austrian earthquake catalogue of earthquakes from 1200 to 2013 A.D. (Austria), Computer file of earthquakes with Ml and defining picks >= 6, E, Zentralanstalt für Meteorologie und Geodynamik, 2014.

12

How to create a very-low-cost, very-low-power, credit-card-sized and real-time-ready datalogger

M. Bès de Berc[1]**, M. Grunberg**[2]**, and F. Engels**[2]

[1]Institut de Physique du Globe, UMR7516, Université de Strasbourg/EOST, CNRS, 5 rue René Descartes, 67084 Strasbourg, France

[2]Réseau National de Surveillance Sismique, UMS830, Université de Strasbourg/EOST, CNRS, 5 rue René Descartes, 67084 Strasbourg, France

Correspondence to: M. Bès de Berc (mbesdeberc@unistra.fr)

Abstract. In order to improve an existing network, a field seismologist would have to add some extra sensors to a remote station. However, additional ADCs (analogue-to-digital converters) are not always implemented on commercial dataloggers, or, if they are, they may already be used. Installing additional ADCs often implies an expensive development, or the purchase of a new datalogger. We present here a simple method to take advantage of the ADCs of an embedded computer in order to create data in a seismological standard format and integrate them within the real-time data stream from the station.

Our first goal is to plug temperature and pressure sensors on the ADCs, read data and record them in *mini-seed* format (*seed* stands for Standard for the Exchange of the Earthquake Data), and eventually transfer them to a central server together with the seismic data, by using *seedlink*, since *mini-seed* and *seedlink* are standard for seismology.

1 Motivations

Inside and outside vault temperatures, outside vault atmospheric pressure (Beauduin et al., 1996), wind speed, or, as shown recently, geomagnetic pulsations for high-latitude stations (Kozlovskaya and Kozlovsky, 2012), have a strong influence on broadband seismic signals. Therefore, environmental measurements represent valuable data as well as the seismic signals which should be stored, ideally by the seismic datalogger, in order to take advantage of the same transmission protocol. This implies that the number of auxiliary ADCs (analogue-to-digital converters) needed by the datalogger increases since main ADCs (24/26 bits) are designed for sampling seismometer analogue outputs. This motivated our study and we specifically work around the auxiliary channels. Several commercial dataloggers offer additional channels, but often not enough or at a prohibitive price. Table 1 presents the availability of auxiliary channels and the sampling environmental parameters for a series of commercial dataloggers that are currently used in French seismic networks.

2 Datalogger

According to our criteria, a reliable datalogger for environmental data includes the following features:

- at least four ADCs, 100 dB dynamic range at a 1 Hz sampling rate;

- a real-time transmission protocol, with an adjustable buffer (1 to 24 h);

- a local data archive, recorded in *mini-seed*, keeping the last 3 months;

- a robust and efficient delayed-time transmission protocol;

- synchronisation with GPS;

- sampling of all ADCs synchronised together;

- weak power consumption; and

- robust hardware.

Table 1. Comparison of common dataloggers used at EOST. ADCs mentioned here are the auxiliary ADCs, available for sampling environmental parameters. SE means single-ended and DI means differential. Some dataloggers offer extra options like Q330/Q330HR, allowing a total configuration of type and dynamic of its auxiliary ADCs, or Staneo offering solutions through RS232 probes.

Manufacturer	Model	Available auxiliary ADCs	Type	Bits	Dyn (V)	Notes
Quanterra	Q330HR	8	SE	16	0–50 V	Can be used in differential mode, dynamic settable
Quanterra	Q330	8	SE	16	0–50 V	Can be used in differential mode, dynamic settable
Quanterra	Q330S/S+	0				
Reftek	RT-130	0				
Nanometrics	Taurus	4	DI	12	±2.5 V	
Nanometrics	Centaur	0				
Staneo	D6BB	3	SE	12	0–12 V	Offers solutions through RS232 sensors

Sampling those data at a frequency higher than 1 Hz may not be necessary, since we are only interested in improving the long-period noise. In order to estimate the needed dynamic range, we used temperature and pressure data from the Geoscope network during the year 2012. We first calculated the power spectrum density (PSD) of all signals, and then extracted the maximum and minimum PSD shown in Figs. 1 and 2. The Geoscope network includes broadband stations installed over the world, covering a large range of climate conditions. According to Figs. 1 and 2, a dynamic range of about 100 dB is the best performance at a frequency range of 1e-5 to 5e-1 Hz. Those data have to be both sent through the same transmission protocol and locally recorded in the same format as the seismological data, i.e. using *seedlink* and *miniseed*, commonly used in seismology. A robust delayed-time protocol has to be implemented in case of loss of transmission, allowing the recovery of the missing data, only, with an optimised algorithm.

3 Choice of software and hardware

We specifically use standard and robust toolkits: a Linux distribution, running *gpsd* and *ntpd* for synchronisation with an Usb GPS receiver (XBU-353 from UsGlobalSat), *rsync* for delayed-time transmission protocol and *seiscomp3* for real-time acquisition and transmission through its *seedlink* protocol (Gempa GmbH, 2014), and finally local archive capacity through its *slarchive* plugin. In order to study the possibilities of using a commercial embedded computer, more often used in robotics, we used a BeagleBone Black (BBB) Rev C.1. Its main features are an ARM Cortex A8 (32 bit) processor at 1 Ghz, 512 MB of DDR3 Ram, 4 GB onboard flash, eight ADCs of 12 bits and a dynamic of 0–1.8 V, and 1 W of power consumption. Other industrial embedded computers are probably better suited for our purpose, but we selected this one on the criteria of price and ease of use, but also because the BeagleBone Black is provided with a Debian 7 operating system, and an active community maintains

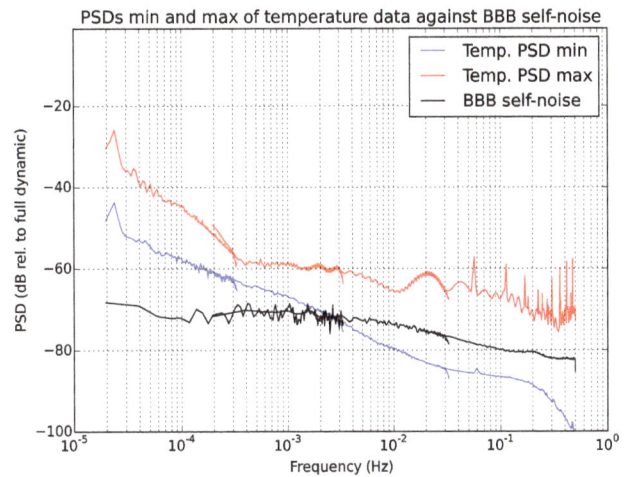

Figure 1. Maximum and minimum values of power spectral densities relative to the full dynamic of temperature, calculated from 10 stations of the Geoscope network. Those values are compared to the self-noise of a BBB's ADC at the optimal sampling frequency.

the needed documentation and libraries. The main first issue was the determination of the exact dynamics of its ADCs, because only few manufacturers communicate the real dynamic range, since it is not a crucial parameter in robotics. Therefore, we looked at the three-channel coherence analysis (Sleeman et al., 2006). Figures 1 and 2 show the actual dynamic range of ADCs by calculating their self-noise compared to both minimum and maximum PSD values of temperature and pressure, respectively. Compared to the Geoscope stations, a loss of performance occurs between 0.003 and 0.5 Hz (up to 15 dB at 0.4 Hz) for temperature and between 0.08 and 0.5 Hz (up to 5 dB) for pressure, due to the lower quality of BBB's ADC against Geoscope's auxiliary channels.

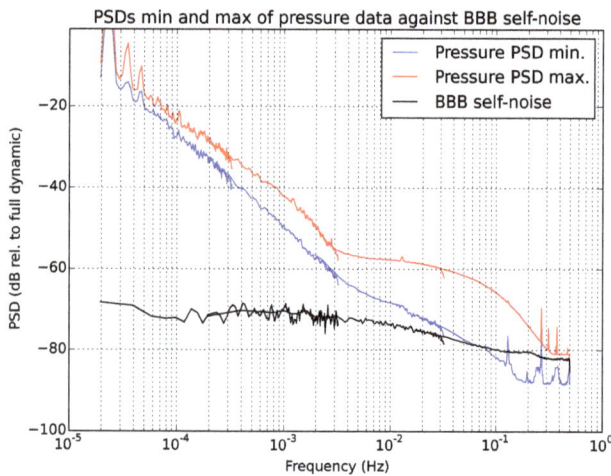

Figure 2. Maximum and minimum values of power spectral densities relative to the full dynamic of pressure, calculated from five stations of the Geoscope network. Those values are compared to the self-noise of a BBB's ADC at the optimal sampling frequency.

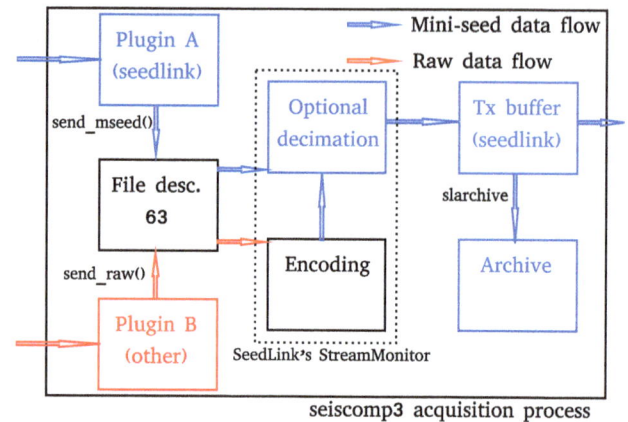

Figure 3. Seiscomp3 acquisition process scheme. If StreamMonitor is not enabled (by default), the mini-seed data are sent from file descriptor 63 directly to the transmission buffer, and raw data are ignored. Running bbb_plugin implies activating this module within seiscomp3.

4 Setting up BBB

4.1 Loading Linux and softwares on board

The board is provided with an already installed Debian 7 system and beagleboard.org provides the latest ready-to-load system image. The methods for loading a Linux operating system are well documented (BeagleBoard.org Foundation, 2014a). We choose to keep this Debian distribution, mainly because it is a well-known robust operating system and because *rsync*, *gpsd* and *ntpd* are packaged, making them easy to install through Debian's utility *aptitude*.

4.2 Compiling and configuring seiscomp3 on the BBB

The sources of the packages are available on the *seiscomp3* download site (http://www.seiscomp3.org/downloader). The compilation is long, and several dependencies need to be installed. We installed the gui system in addition to the base package, because BBB includes a graphic processor and therefore the configuration of *seiscomp3* is possible through its graphic menu *scconfig*. For a non-negligible gain of time, a BBB system can be emulated on a work station with *qemu*, and *seiscomp3* can be compiled on it.

5 Sampling ADCs through seiscomp3: bbb_plugin

In the standard distribution, the ADCs are readable in a simple system file. However, standard drivers are not able to handle a continuous analogue application at a high sample rate (1 kHz maximum). Moreover, one of the advantages of the BeagleBone Black is a sub-module of its processor (AM335x from TI) called PRU, supporting that kind of "fast" analogue

application. We then wrote a *seedlink* acquisition plugin filling those roles:

1. drive the PRU to launch sampling;

2. copy data in a buffer; and

3. send buffer to *seiscomp* server.

Each library necessary to write this program is available on the Internet. We used the *libpruio* library, specifically written to manage the PRU on BBB (BeagleBoard.org Foundation, 2014b), and the development file provided with the *seiscomp3* package. We called this simple plugin *bbb_plugin*. Since *seiscomp3* is able to manage the *mini-seed* conversion and the numerical decimation/filtering (see Sect. 7) through its *StreamMonitor* (Gempa GmbH, 2014), we used the *send_raw_depoch*() function in *bbb_plugin* in order to avoid the coding of the *mini-seed* conversion. Our code is available at https://github.com/mbdb/bbb_plugin. Figure 3 shows a diagram of the *seedlink* acquisition.

6 Analogue considerations

Before sampling any data, an analogue anti-aliasing filter per ADC has been designed in order to respect the Shannon–Nyquist theorem. Furthermore, our sensors have an output voltage range non-compatible with the input range of the BBB (0–1.8): we wanted to use a hybrid temperature/humidity sensor (EE08 from EplusE, output 0–10 V) and a pressure sensor (HD9408T Baro from DeltaOhm, output 0–5 V), commonly used in our networks. We applied the common strategy to use a simple first-order RC filter/voltage divider. We designed it to cut frequencies larger than the upper usable frequency (0.5 Hz) of decimated data, but much

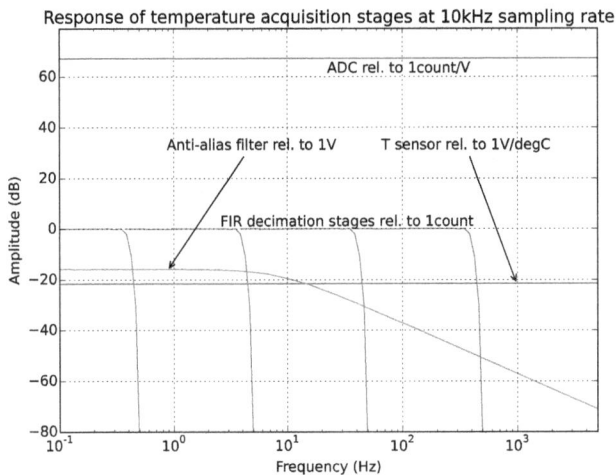

Figure 4. Response over frequency of every stage at a 10 kHz sampling rate for a EE08 (from epluse) temperature sensor. The overall response in dB is the sum of all stages; therefore, for data recorded at 1 Hz, the gain is equal to the addition of the T sensor gain, the anti-alias filter's optimal gain, and ADC's gain. Only frequencies between 0.4 and 0.5 Hz (i.e. just before the Nyquist frequency) could be attenuated by the last FIR (Finite Impulse Response) filter stage.

lower than the Nyquist frequency (5 kHz). This analogue stage uses only two resistors and one capacitor per ADC. Figure 4 shows simulations of acquisition stages, and particularly the response of that anti-alias filter in the case study of temperature acquisition. This specific filter gives an attenuation of ~ 70 dB at the Nyquist frequency.

7   Digital considerations

A basic concept of analogue–digital conversion is related to the data oversampled at the highest possible frequency, and then digitally decimated to the usable frequencies (Asch, 2003). The quantisation noise of a datalogger is uniformly distributed between DC and the Nyquist frequency and is independent of the sampling rate. Therefore the effective noise power density due to the quantisation at the band of interest is lower at the highest frequency rates.

Since it is difficult to estimate the noise reduction (a new filter design would be required at each sampling rate and a large number of data needed), we used the decimation option within the *seiscomp3 StreamMonitor* module and measured the highest frequency acceptable in terms of resource consumption. We therefore chose a sampling rate at 10 kHz. Figure 4 shows the response of the FIR decimation stages in the case study of temperature acquisition.

8   Conclusions and discussions

Our first tests were conclusive for using a commercial embedded computer (BeagleBone Black) to sample environmental data. Its characteristics give a loss of performance in the range 0.001–0.5 Hz for temperature and 0.1–0.5 Hz for pressure, compared to the instrumentation used in the Geoscope network. Except for its standard box, which is not sealed and not suitable for harsh environments, this kind of instrumentation seems to be appropriate for field operations. It is compelling to note that our prototype instrument has been in the field for 3 months, and did not show any signs of deficiency. Other more ambitious projects exist for using a BeagleBone Black and its optional capes, as a real seismological station (Mertl and Lettenbichler, 2014), with associated constraints (135–140 dB dynamic range, good synchronisation, etc.). These tests suggest as well that this instrumentation could fulfill a central role in a station. Up to now, the BBB has only been considered as a secondary source of data at a station, which presents a problem, because the data centre receives two distinct sources of data, and has to associate different streams for each station. To overcome this limitation, another strategy has been tested: (1) we loaded a BBB with a *seiscomp3* server and set up the acquisition of the seismological datalogger on it; and (2) we also installed our own software (*bbb_plugin*) for the acquisition of its ADCs, and set up the server to mix the streams. This latter configuration has advantages:

– the data centre communicates with a unique source of data per station; and

– the unique model of the source permits a total abstraction of the local instrumentation, allowing a standardisation of transmission protocols and local archive structure for every station.

Acknowledgements. The authors would to thank Cécile Doubre and the anonymous reviewers for their suggestions, which were useful in improving the paper.

References

Asch, G. D. (Ed.): Acquisition de données, du capteur à l'ordinateur, 2nd Edn., Dunod, Saint-Jean-de-Braye, France, Chap. 16, 278–279, 2003.

BeagleBoard.org Foundation: Getting started with BeagleBone & BeagleBone Black, available at: http://beagleboard.org/getting-started, last access 22 September 2014a.

BeagleBoard.org Foundation: libpruio (fast and easy D/A -I/O), available at: http://beagleboard.org/project/libpruio, last access: 1 October 2014b.

Beauduin, R., Lognonné, P., Montagner, J. P., Cacho, S., Karczewski, J. F., and Morand, M..: The effects of the atmospheric pressure on seismic signals or How to improve the quality of a station, B. Seismol. Soc. Am., 86, 1760–1769, 1996.

Gempa GmbH: Seedlink, A realtime waveform server implementing the Seedlink protocol, available at: http://www.seiscomp3. org/wiki/doc/applications/seedlink, last access: 22 September 2014.

Kozlovskaya, E. and Kozlovsky, A.: Influence of high-latitude geomagnetic pulsations, Geoscientific Instrumentation, Methods Data Syst., 1, 85–101, 2012.

Mertl, S. and Lettenbichler, A.: Towards a community environmental observation network, available at: http://www.mertl-research.at/ceon/2014/05/ poster-presentation-at-the-egu-general-assembly-2014, last access: 30 March 2015.

Sleeman, R., van Wettum, A., and Trampert, J.: Three-channel correlation analysis: A new technique to measure instrumental noise of digitizers and seismic sensors, B. Seismol. Soc. Am., 96, 258–271, 2006.

13

Displacement-based error metrics for morphodynamic models

J. Bosboom[1] **and A. J. H. M. Reniers**[2,1]

[1]Faculty of Civil Engineering and Geosciences, Delft University of Technology, the Netherlands
[2]Applied Marine Physics, Rosenstiel School of Marine & Atmospheric Science, University of Miami, USA

Correspondence to: J. Bosboom (j.bosboom@tudelft.nl)

Abstract. The accuracy of morphological predictions is generally measured by an overall point-wise metric, such as the mean-squared difference between pairs of predicted and observed bed levels. Unfortunately, point-wise accuracy metrics tend to favour featureless predictions over predictions whose features are (slightly) misplaced. From the perspective of a coastal morphologist, this may lead to wrong decisions as to which of two predictions is better. In order to overcome this inherent limitation of point-wise metrics, we propose a new diagnostic tool for 2-D morphological predictions, which explicitly takes (dis)agreement in spatial patterns into account. Our approach is to formulate errors based on a smooth displacement field between predictions and observations that minimizes the point-wise error. We illustrate the advantages of this approach using a variety of morphological fields, generated with Delft3D, for an idealized case of a tidal inlet developing from an initially very schematized geometry. The quantification of model performance by the new diagnostic tool is found to better reflect the qualitative judgement of experts than traditional point-wise metrics do.

1 Introduction

Quantitative validation methods for morphodynamic models are often grid-point based; they compare observations and predictions per grid-point and compute various metrics for the entire set or subset of grid-points (e.g., Sutherland et al., 2004). Unfortunately, point-wise accuracy metrics, such as the commonly used MSE (Mean-Squared Error) and RMSE (Root-Mean-Squared Error), tend to penalize, rather than reward, the model's capability to provide information on features of interest, such as scour holes, accumulation zones and migrating tidal channels. For instance, a prediction of a morphological feature that is correct in terms of timing and size, but is misplaced in space, may not outperform even a flat bed, which is inconsistent with the common judgement of morphologists (Fig. 1). This "double penalty effect" (Bougeault, 2003), which applies in full when a feature is misplaced over a distance equal or larger than its size, makes it difficult to demonstrate the quality of a high variability prediction (Anthes, 1983; Mass et al., 2002). Clearly, a high quality validation process requires alternative validation techniques that account for the spatial structure of 2-D morphological fields.

For the verification of weather variables (e.g. precipitation), methods are being actively developed to quantify forecast performance based on spatial structure; see for instance Casati et al. (2008) and Gilleland et al. (2009) for an overview. One of the approaches in meteorology, now also pioneered in other fields (e.g., Haben et al., 2014; Ziegeler et al., 2012), is to find an optimal deformation of the predictions that minimizes the misfit with observations. This optimal deformation can be obtained by employing one of many existing image matching methods, of which optical flow techniques, designed to estimate motion, are probably most well-known in the coastal community. The result of the image matching or warping is a vector field of displacements, which can be regarded as a displacement error field. In addition, an intensity or amplitude error field may be defined as the difference between the deformed prediction and the observations (e.g., Marzban and Sandgathe, 2010), which can be seen as the point-wise error if no penalty applies for misplacements.

Existing verification methods, based on field deformation of meteorological fields, not only differ in the applied image matching method, but also in the approach to the subsequent extraction of map-mean errors. Keil and Craig (2009) determine RMS (root-mean-squared) intensity and mean displacement errors within the boundaries of precipitation features, which they then combine into a single error metric.

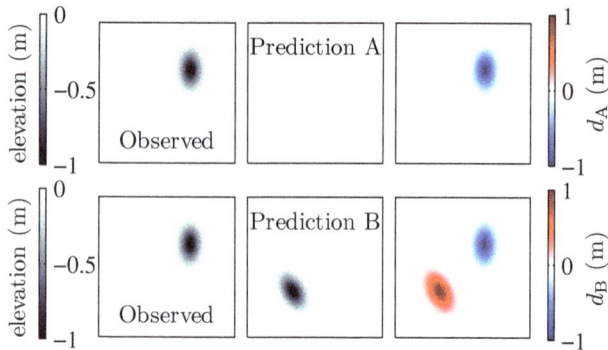

Fig. 1. The "double penalty effect". Top panels: the featureless prediction A has a non-zero difference d_A between predicted and observed depth values at the location of the observed feature only. Lower panels: prediction B, which reproduces the feature at the wrong location, is penalized twice (d_B is non-zero both where the predicted feature is and where it should be) and is thus diagnosed with a twice as large (R)MSE as prediction A.

The latter requires the normalization of the two errors to put each term on equal footing, which introduces two parameters to the formulation. In contrast, Gilleland et al. (2010) propose a combined error metric that besides the post-warp RMS intensity error and the mean displacement error also takes the original RMS intensity error into account, enabling a more fair comparison of forecast performance. Their metric, however, is not easily applicable since it requires three user-chosen weights that are dependent on the error terms themselves.

The goal of this paper is to quantify morphodynamic model performance, while taking the spatial characteristics of 2-D morphology into account. Using a field deformation technique, we have developed and tested a new diagnostic tool for the validation of 2-D morphological predictions. It includes a location (displacement) error metric and a robust and physically intuitive combined error metric that incorporates both location and intensity error. The combined metric rewards predictions to the degree that a larger error reduction can be obtained with smaller displacements. As a reference, we use the subjective but very powerful method of visual inspection of morphological patterns by experts.

Our method is outlined in Sect. 2, along with a brief description of the image warping method that we have adopted to calculate the optimal deformation. Next, in Sect. 3, we put the new diagnostic tool to the test, using morphological fields generated with Delft3D for an idealized case of a tidal inlet developing from an initially very schematized geometry. Section 4 concludes with a summary of our findings and the implications for morphodynamic model validation.

2 Method

This section outlines our two-step approach to quantify the (dis)agreement between 2-D morphological patterns. Section 2.1 describes the first step of deforming (or warping) the predicted morphology to minimize the point-wise error with observations. Next, Sect. 2.2 formulates two new error metrics, a mean location error that is distilled from the displacement vector fields and a single-number error metric that measures both the correspondence with respect to location and intensity (i.e. depth-values).

2.1 Warping method

The measure of closeness between images or spatial fields is encountered in many fields from radiography to meteorology. This has led to the development of a multitude of image matching methods that, depending on the scientific field, are also named registration or warping methods. The goal of such methods is to find the optimal transformation that maps each point of a static image to a corresponding point (with the same intensity) in the moving image. Within the context of morphodynamic model validation, the static image represents the observed depth field o and the moving image the predicted depth field p.

Of all the available techniques, the class of optical flow techniques, designed to estimate small displacements in temporal image sequences, is probably the most well-known in our field. The basic assumption of optical flow is that the intensity of a moving object does not change appreciably in the considered time interval. We employ the efficient, non-rigid (i.e. allowing for free-form deformations) registration technique named Demon's registration (Thirion, 1998), which bears similarities to optical flow, in an implementation by Kroon and Slump (2009). The Demon's approach can be considered as similar to a minimization of the sum of square image intensities between the deformed predictions and observations (Pennec et al., 1999). It is therefore consistent with our quest to find the optimal deformation of the predictions that minimizes the point-wise (R)MSE.

The estimated backward pixel displacements $\boldsymbol{B}^* = (B_x^*, B_y^*)$ that are required for a given point in a static image (the observations in our validation context) to match the corresponding point in a moving image (the predictions) is given by Thirion (1998):

$$\boldsymbol{B}^* = \frac{(I_p - I_o)\nabla I_o}{|\nabla I_o|^2 + \alpha^2(I_p - I_o)^2} \tag{1}$$

in which α is a normalization factor that is equal to 1 in the original method and I_o and I_p are the intensities of the static and moving image, respectively. The latter are taken as the observed and predicted depth fields, normalized by scaling between 0 and 1. Since Eq. (1) is based on local information, it is solved iteratively while including Gaussian smoothing as a regularization criterion. This ensures that a realistic, smooth displacement field is found instead of an irregular field that nonetheless minimizes the sum of squares. The normalization factor is chosen as $\alpha = 2.5$ in line with Kroon and Slump (2009) and the standard deviation of the Gaussian smoothing window as $\sigma = 4$. These parameters are kept constant for all registrations presented in Sect. 3. The forward displacements $\boldsymbol{F}^* = (F_x^*, F_y^*)$ from the moving to the static image can be determined from $\boldsymbol{B}^*$ after the registration. Note that when in the following the subscript $*$ is dropped, we refer to the displacement fields transferred to a physical distance.

For the purpose of model validation, we interpret $d_0 = p_0 - o$, with p_0 the prediction prior to warp, as the total pointwise error and $d_1 = p_1 - o$, with p_1 the deformed prediction as follows from the registration, as the point-wise error if no penalty is imposed for location disagreement. Next, we use this perspective in the formulation of map-mean errors.

2.2 Formulation of new error metrics

From the Demon's registration (Sect. 2.1), we obtain the optimal displacement vector field between predictions and observations as well as the optimal deformation of the predictions. "Optimal" in this context means that the sum of squares between the deformed predictions and observations is minimized, such that $0 \leq \text{RMSE}_1 \leq \text{RMSE}_0$, where RMSE_0 and RMSE_1 are the root-mean-squared errors before and after the warp, respectively. Note that we have preferred the RMSE over the MSE, since the first is measured in the same units as the data. Out of two predictions that have the same RMSE_0, a prediction that has similar morphological features as the measurements, albeit displaced, may receive a lower RMSE_1 than a prediction that is not able to reproduce the observed morphological features at all. Thus, the RMSE_1 is expected to diagnose the agreement between morphological fields if a zero penalty is imposed for misplacements of features. However, which of the two predictions is valued the better prediction by morphologists not only depends on RMSE_0 and RMSE_1, but also on the magnitude of the displacements required to obtain the error reduction. Therefore, we expect that the similarity in both location and intensity between morphological patterns can be fully assessed using three error metrics in concert: RMSE_0, RMSE_1 and a mean location error $\bar{D}$ that we will formulate next from the displacement vector fields.

It is tempting to define $\bar{D}$ as the arithmetic mean of $D = \sqrt{(B_x{}^2 + B_y{}^2)}$, the field of displacement magnitudes. However, it should be realised that the optical flow problem is underconstrained; for a single grid-point, we only have information on the displacements normal to the contour lines, whereas along the contour lines the displacements are ambiguous (the so-called aperture problem). In the Demon's approach, the Gaussian smoothing acts as the necessary additional constraint, requiring that nearby grid-points have similar displacements. As a consequence, non-zero displacements may be found along depth contours in morphologically inactive regions (see Sect. 3), whereas these displacements do not improve the match between the deformed prediction and the observations. Therefore, we propose a weighted mean location error that weights the local backward displacement magnitudes D with their effect on the reduction of the local squared error. In this way, displacements are only taken into account to the extent that they contribute to the minimization of the sum of squares. This yields:

$$\bar{D} = \frac{\sum_{i=1}^{n} w_i D_i}{\sum_{i=1}^{n} w_i}; \quad w_i = \frac{\text{SE}_{0,i} - \text{SE}_{1,i}}{\sum_{i=1}^{n} (\text{SE}_{0,i} - \text{SE}_{1,i})} \qquad (2)$$

Here $\text{SE}_0 = (p_0 - o)^2$ and $\text{SE}_1 = (p_1 - o)^2$ are the local squared errors before and after the warp, respectively, n is the number of equidistant points in the spatial domain and $\sum_{i=1}^{n} w_i = 1$. Note that $\text{RMSE}_j = \sqrt{n^{-1} \sum_{i=1}^{n} \text{SE}_{j,i}}$, with $j = [0, 1]$.

Whereas model performance is usually diagnosed based on RMSE_0 only, we now have two additional metrics RMSE_1 and $\bar{D}$. In Sect. 3, it is demonstrated that considering these three metrics in concert allows a full assessment of model quality, avoiding the double penalty effect for misplaced features. In practice, guidance may be required on how to weight these three metrics. Besides, the morphologist may sometimes desire a single-number summary of model performance, especially if automated calibration routines are used. To serve these needs, we propose an adjusted RMS error measure, RMSE_w, that is computed from a field of weighted squared errors SE_w. The latter are determined by locally weighting SE_0 and SE_1. The purpose of the weighting procedure is to locally relax the requirement of an exact match to an extent determined by the local displacement magnitude. Figure 2 illuminates the weighting procedure for the ith gridpoint; an error reduction is awarded that is a fraction $1 - \delta_i$ of the full error reduction potential ($\text{SE}_{0,i} - \text{SE}_{1,i}$). Here, $\delta_i = D_i / D_\text{max}$ and D_max is a maximum displacement length above which no relaxation is allowed. A larger fraction $1 - \delta_i$ is allowed for smaller displacement magnitudes D_i, with a maximum of $1 - \delta_i = 1$ and thus $\text{SE}_{\text{w},i} = \text{SE}_{1,i}$ for $D_i = 0$ m. For $D_i \geq D_\text{max}$, we have $1 - \delta_i = 0$ and thus $\text{SE}_{\text{w},i} = \text{SE}_{0,i}$. Note that D_max is a user-defined, physically intuitive parameter that is dependent on the prediction situation and the goal of the simulation. It can be seen as the maximum distance over which morphological features may be displaced for the prediction to still get (some) credit for predicting these features. We now have for RMSE_w:

$$\text{RMSE}_\text{w} = \sqrt{\frac{\sum_{i=1}^{n} \text{SE}_{\text{w},i}}{n}} \qquad (3)$$

where

$$\text{SE}_\text{w} = \text{SE}_1 + \delta(\text{SE}_0 - \text{SE}_1) \qquad (4)$$

$$\delta_i = \frac{D_i}{D_\text{max}} \text{ for } D_i \leq D_\text{max}; \quad \delta_i = 1 \text{ for } D_i > D_\text{max} \qquad (5)$$

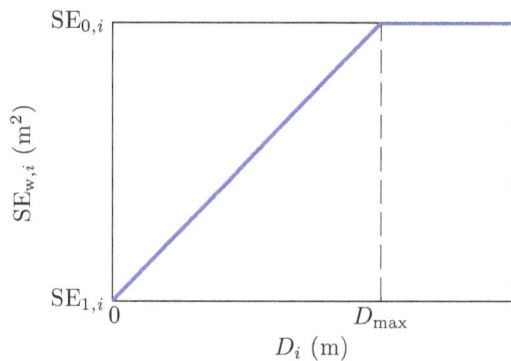

Fig. 2. Weighted squared error for the ith gridpoint $SE_{w,i}$, which is the sum of the local squared error after the warp $SE_{1,i}$ and a penalty for misplacements $\delta(SE_{0,i} - SE_{1,i})$ with $\delta = D_i/D_{max}$. The penalty ranges from 0 for $D_i \to 0$ to $(SE_{0,i} - SE_{1,i})$ for $D_i = D_{max}$, a user-defined maximum displacement length. For $D_i \geq D_{max}$ the full point-wise error applies and $SE_{w,i} = SE_{0,i}$.

In conclusion, $RMSE_w$ as an error metric rewards forecasts to the degree that a larger error reduction can be obtained by smaller displacements. By definition, $RMSE_1 \leq RMSE_w \leq RMSE_0$. If the error reduction due to the image deformation is negligible or can only be obtained with displacements equal to or larger than D_{max}, the diagnosed error is equal to the original error prior to the deformation $RMSE_0$. If, on the other hand, the displacements required to minimize the point-wise error are very small relative to D_{max}, we have $RMSE_w \approx RMSE_1$. The justification for this approach lies in the tendency of coastal morphologists to credit a prediction for the reproduction of features, albeit displaced, while imposing a relatively small penalty for misplacement. The intuitive weighting of these two aspects is mimicked by the user-defined parameter D_{max}.

3 Application

Below, the new error metrics are used to diagnose the correspondence between model-generated pairs of morphological patterns for an idealized tidal inlet as well as the relative ranking between the pairs. The fields have been generated for the idealized case of a tidal inlet developing from an initially very schematized geometry (Roelvink, 2006). First, Sect. 3.1 demonstrates that the location error $\bar{D}$ is able to capture the overall misplacement of the morphological patterns. Next, in Sect. 3.2, the combined error metric $RMSE_w$ is put to the test. Two examples are shown where the $RMSE_w$ makes the right the decision as to which of two predictions is the better prediction while the conventional, purely point-wise $RMSE_0$ fails to do so.

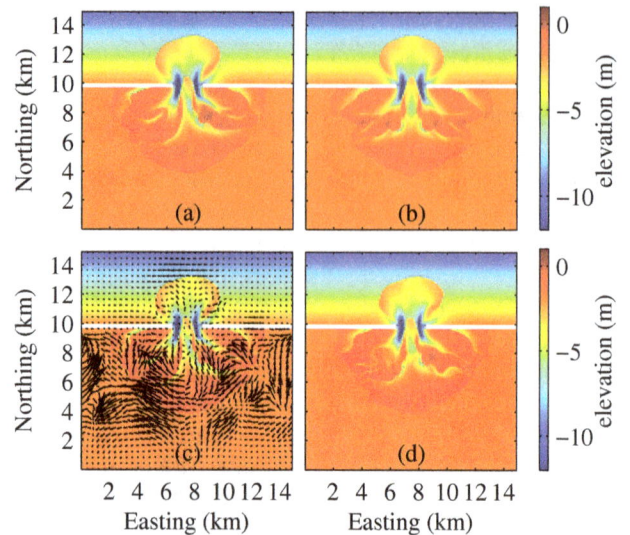

Fig. 3. Example of the image warp: **(a)** the "observations", calculated using Delft3D with Coriolis at 53° N; **(b)** the predictions, calculated at 0°; **(c)** the backward displacement vector field B of the observations towards the predictions, shown on top of the observations; and **(d)** the predictions deformed to more closely match the observations.

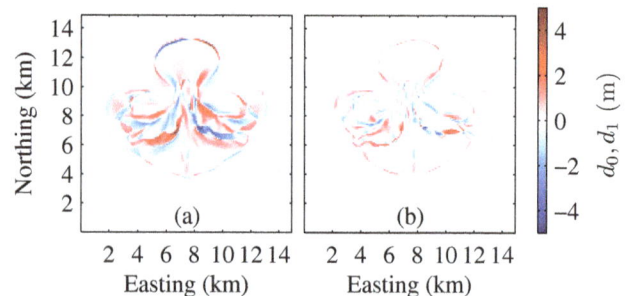

Fig. 4. Point-wise error fields for the predicted depth field at 0°: **(a)** the total error $d_0 = p_0 - o$ before the warp; **(b)** the error $d_1 = p_1 - o$ after the warp, to be regarded as the remaining point-wise error if no penalty applies for location disagreement.

3.1 Location error

In this subsection, we consider a subset of the model-generated depth fields which only differ with respect to the latitude, and hence Coriolis parameter, used in the model. Of four depth fields, we label the field generated at 53° N as the "observations" (Fig. 3a) and consider the other fields, for latitudes 90° N, 0° and 90° S, as three competing predictions. Even though the predictions are not shown here, it will not come as a surprise that the point-wise error $RMSE_0$ is smallest for 90° N and largest for 90° S (Table 1).

In order to determine $RMSE_1$ and $\bar{D}$, the image warping method is applied, following the procedure outlined in Sect. 2, and illustrated here for the prediction at 0° (Fig. 3b). The deformed prediction that matches the observations most

Table 1. Errors for competing predictions that differ with respect to the latitude, and thus the Coriolis parameter, used in Delft3D. The model results for 53° N are regarded as the "observations".

Latitude	RMSE_0 (m)	RMSE_1 (m)	$\bar{D}$ (m)
90° N	0.29	0.12	180
0°	0.52	0.26	350
90° S	0.73	0.35	710

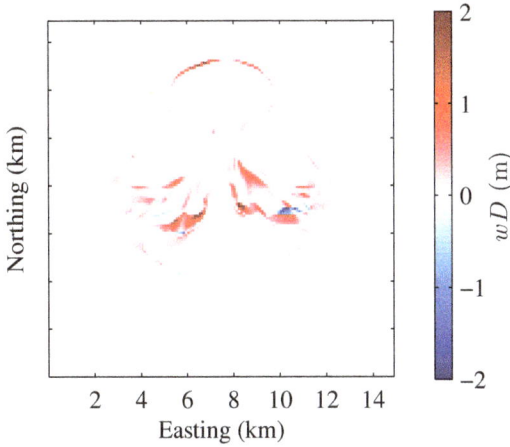

Fig. 5. Weighted displacements wD for the prediction at 0°. Here $D = \sqrt{(B_x^2 + B_y^2)}$ is the field of displacement magnitudes computed from the backward displacement vector field $\boldsymbol{B}$ (see Eq. 1), and w is determined according to Eq. (2).

closely is shown in Fig. 3d and the corresponding backward vector displacement field $\boldsymbol{B}$ in Fig. 3c. As explained in Sect. 2.1, in the inactive outer regions, physically unrealistic displacements are found along depth contours, since no penalty is imposed in the minimization for displacements along depth contours. As will be illustrated next, this is solved for in the formulation of $\bar{D}$ (Eq. 2).

The difference d_0 between the predictions prior to the warp and the observations is shown in Fig. 4a, whereas Fig. 4b shows the difference d_1 after the warp. Note that taking the root-mean-square of d_0 and d_1 yields RMSE_0 and RMSE_1, respectively. From d_0, the double penalty problem is clearly observed; for instance at the edges of the ebb-tidal delta, an error is diagnosed both where the delta is present in the observations but absent from the predictions and vice versa. After the warp, both errors have practically disappeared, such that they will not count towards RMSE_1, demonstrating again that RMSE_1 should be regarded as the point-wise error if no penalty for misplacement is taken into account. For the prediction at 0°, $\text{RMSE}_1/\text{RMSE}_0 = 0.5$, and slightly smaller ratios are found for the other two predictions (Table 1).

The weighted dispacements wD, with $D = \sqrt{(B_x^2 + B_y^2)}$ and w according to Eq. (2), are shown in Fig. 5. Inherent to the use of the squared error to determine w is that larger error reductions are heavily weighted. Here, we have never-

Table 2. Subjective ranking (with 1 being the best prediction) and errors for competing predictions, generated with Delft3D for various boundary conditions. The "observations" are taken as the model outcome at 0° (cf. Sect. 3.1). The values for RMSE_w hold for $D_{\max} = 3000$ m.

Prediction	Ranking	RMSE_0 (m)	RMSE_1 (m)	$\bar{D}$ (m)	RMSE_w (m)
A	1	0.78	0.38	610	0.49
B	2	0.77	0.53	770	0.60
C	3	1.16	0.56	860	0.78
D	4	0.96	0.77	1230	0.84

theless chosen this weighting since squared errors are consistent with the minimization as performed by the registration method as well as with the use of the (R)MSE as the point-wise metric, which is common in morphodynamic model validation. Note that for the computation of $\bar{D}$ (Eq. 2), we require the backward (from the observations to the predictions) rather than the forward displacements; for each point in the observational domain, these provide the distance at which the point in the predictions is located that is shifted to the considered location in the observations. Summing wD for the entire domain yields a location error $\bar{D} = 350$ m at 0° (Table 1).

The values for $\bar{D}$ for the three predictions demonstrate a qualitative behaviour consistent with the error in latitude and hence Coriolis effect in the various predictions (Table 1). In fact, all three error metrics, RMSE_0, RMSE_1 and $\bar{D}$ diagnose the predictions for 90° N and 90° S as the best and worst predictions, respectively. Next, we will consider situations in which a ranking consistent with expert judgement is only obtained by considering these three metrics in concert, using an appropriate weighting, or from RMSE_w.

3.2 Ranking according to the combined error metric

In this subsection, we present an example, again using depth fields generated with the Delft3D model of the schematized tidal inlet, that demonstrates that RMSE_w outperforms the traditional score RMSE_0. Now, the model results at a latitude of 0° (see Sect. 3.1) are assumed to be the "truth". Four competing predictions are considered that are generated at 0° with various changes to the model boundary conditions (w.r.t. tidal amplitude and flow direction). Figure 6 shows the four predictions, the "observations" and the deformed predictions that minimize the point-wise error.

We have labelled the predictions according to a subjective ranking based on visual inspection, with A the prediction with the closest match with the observations and D, the worst prediction. We have a slight preference for prediction B over C, but it is possible that other morphologists would tend to regard C as the better prediction. Not surprisingly, the relative ranking as diagnosed by RMSE_0 deviates from the expert ranking (Table 2); based on RMSE_0 one would wrongfully

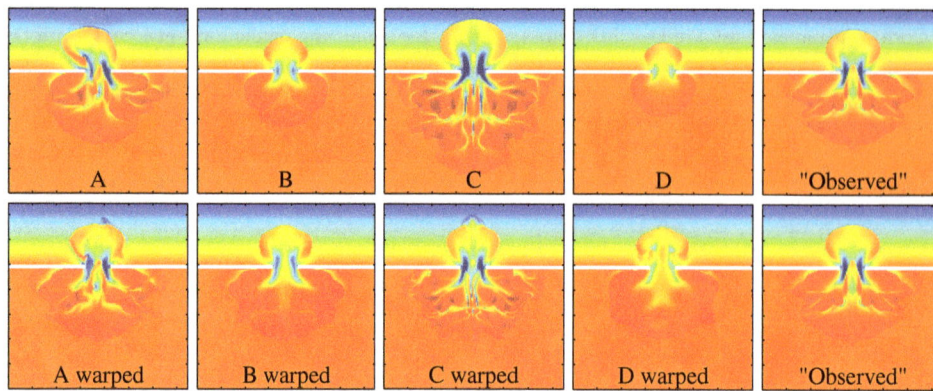

Fig. 6. Predictions A, B, C and D, the "observations" (taken as the model results for 0°) and the corresponding deformed predictions that minimize the point-wise mismatch between predictions and observations. The labels are chosen such that the lower the label in the alphabet, the higher the quality that the prediction is probably diagnosed with upon visual inspection. The axes are as in Fig. 3.

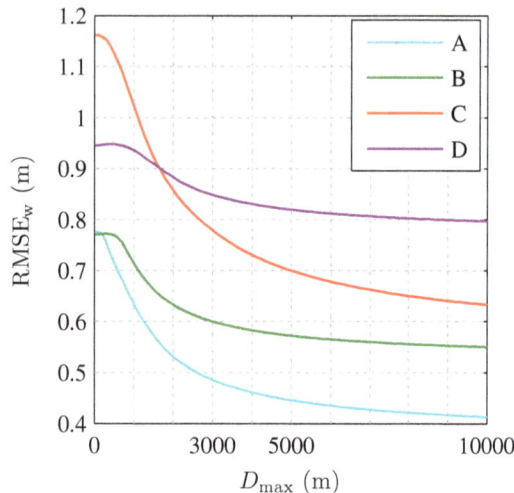

Fig. 7. The combined error metric $RMSE_w$ as a function of D_{max} for predictions A, B, C and D. The larger D_{max}, the more the requirement of an exact match is relaxed; for $D_{max} \to 0$, we have $RMSE_w \to RMSE_0$ and for $D_{max} \to \infty$, we have $RMSE_w \to RMSE_1$.

conclude that predictions A and B perform equally well and that prediction D outperforms prediction C.

The values of $RMSE_1$, $\bar{D}$ and $RMSE_w$ for the respective predictions provide the necessary additional information on model performance (Table 2). The smaller $RMSE_1$ for prediction A than for prediction B shows that if no penalty is imposed for misplacements, prediction A receives a lower error than B. Moreover, a smaller average displacement $\bar{D}$ is required to minimize the point-wise error. Thus, even though no distinction can be made based on $RMSE_0$, we can conclude that pattern A more closely corresponds to the observations than pattern B. Clearly, considering the values of $RMSE_0$, $RMSE_1$ and $\bar{D}$ in concert leads to a diagnosis of

relative model performance of A and B in line with visual inspection.

To determine $RMSE_w$, a value for D_{max} must be chosen. A defendable choice would be to limit D_{max} to the scale of the morphological features of interest. For this particular case, $D_{max} = 3000\,m$ is considered appropriate, being in the order of magnitude of the seaward extent of the ebb-tidal delta. In general, of course, D_{max} must be chosen in accordance with the goal of the simulation.

Figure 7 shows that with $D_{max} = 3000\,m$, $RMSE_w$ reports a higher quality for prediction A than for prediction B, regardless of the exact choice for D_{max}. Only if one decides to not allow any relaxation of the requirement of an exact match ($D_{max} = 0\,m$), $RMSE_w$ is identical to the full point-wise error $RMSE_0$ and no distinction can be made between A and B. If one wishes to allow the full error reduction potential ($D_{max} \to \infty$), we have $RMSE_w = RMSE_1$.

Table 2 illuminates that prediction C, the prediction with the largest $RMSE_0$, has a much larger potential for error reduction by warping than prediction D; notwithstanding the larger $RMSE_0$, $RMSE_1$ is smaller for prediction C than for D and at a smaller mean displacement $\bar{D}$. The relatively small error reduction potential for D is a result of the fact that features not present in the predictions remain absent after the warping procedure, as evident in the deformed predictions in Fig. 6. As a result, $RMSE_1$ remains relatively high for D, rightfully penalizing the prediction for the absence of the observed features. A conclusive answer as to whether C or D is the better prediction now requires a (subjective) weighting of $RMSE_0$, $RMSE_1$ and $\bar{D}$. Conveniently, the weighting between location errors, pre-warp and post-warp intensity errors is already provided by the formulation of $RMSE_w$, allowing a quantitative single-number comparison between predictions C and D. For $D_{max} = 3000\,m$, the values for $RMSE_w$ indicate that prediction C outperforms D (Fig. 7), consistent with the ranking based on visual inspection. Naturally, the occurence of this ranking reversal, as compared to the ranking based on $RMSE_0$, depends on the chosen value of D_{max}.

4 Conclusions

We have developed a new diagnostic tool for morphodynamic model validation. It employs an image warping method that finds the smooth displacement field between predictions and observations that minimizes the point-wise error. Two new metrics are proposed: (1) a location error $\bar{D}$ that is determined as a weighted mean distance between morphological fields; and (2) a combined error metric $RMSE_w$ that takes both location and intensity errors into account.

A full appreciation of the quality of a prediction can be obtained when considering $\bar{D}$ in concert with both the original point-wise error $RMSE_0$ and the point-wise error of the deformed predictions, $RMSE_1$. In order to quantify the relative performance between predictions, a (subjective) weighting of these three metrics must be carried out. Alternatively, the weighting is already provided by $RMSE_w$ that combines all relevant information on location errors and pre- and post-warp intensity errors.

The combined error metric credits predictions to the degree that a larger error reduction can be obtained with smaller displacements. It reduces to $RMSE_0$ if all displacements are larger than a user-defined D_{max} and to $RMSE_1$ for displacements that are negligible relative to D_{max}. The latter can be seen as the maximum distance over which morphological features may be displaced for the prediction to still get (some) credit for predicting these features. The appropriate choice for D_{max} depends on the prediction situation and the goal of the simulation. Since it only requires a single, physically intuitive parameter, $RMSE_w$ provides a robust basis for comparison.

An example of a schematized tidal inlet has demonstrated that $RMSE_w$ outperforms the conventional validation approach based on a strictly point-wise metric such as $RMSE_0$. In situations where morphological features are misplaced, point-wise accuracy metrics tend to favour predictions that underestimate variability. For the schematized tidal inlet, it was shown that, as opposed to $RMSE_0$, the new combined error metric $RMSE_w$ makes choices as to which of two predictions is better, which are consistent with visual validation by experts.

Acknowledgements. The authors wish to thank both reviewers for their constructive comments. Ian Townend (HR Wallingford) is thanked for stimulating discussions and helpful comments.

References

Anthes, R. A.: Regional models of the atmosphere in middle latitudes, Monthly weather review, 111, 1306–1335, doi:10.1175/1520-0493(1983)111<1306:RMOTAI>2.0.CO;2, 1983.

Bougeault, P.: The WGNE survey of verification methods for numerical prediction of weather elements and severe weather events, CAS/JSC WGNE Report, 18, WMO/TD-NO.1173, Appendix C, 1–11, http://www.wcrp-climate.org/documents/wgne18rpt.pdf, 2003.

Casati, B., Wilson, L., Stephenson, D., Nurmi, P., Ghelli, A., Pocernich, M., Damrath, U., Ebert, E., Brown, B., and Mason, S.: Forecast verification: current status and future directions, Meteorol. Appl., 15, 3–18, doi:10.1002/met.52, 2008.

Gilleland, E., Ahijevych, D., Brown, B. G., Casati, B., and Ebert, E. E.: Intercomparison of spatial forecast verification methods, Weather Forecast., 24, 1416–1430, doi:10.1175/2009WAF2222269.1, 2009.

Gilleland, E., Lindström, J., and Lindgren, F.: Analyzing the image warp forecast verification method on precipitation fields from the ICP, Weather Forecast., 25, 1249–1262, doi:10.1175/2010WAF2222365.1, 2010.

Haben, S., Ward, J. A., Vukadinovic Greetham, D., Singleton, C., and Grindrod, P.: A new error measure for forecasts of household-level, high resolution electrical energy consumption, Int. J. Forecast., 30, 246–256, doi:10.1016/j.ijforecast.2013.08.002, 2014.

Keil, C. and Craig, G. C.: A displacement and amplitude score employing an optical flow technique, Weather Forecast., 24, 1297–1308, doi:10.1175/2009WAF2222247.1, 2009.

Kroon, D.-J. and Slump, C. H.: MRI modality transformation in demon registration, in: From Nano to Macro, 2009, ISBI'09, IEEE International Symposium on Biomedical Imaging, 963–966, IEEE, doi:10.1109/ISBI.2009.5193214, 2009.

Marzban, C. and Sandgathe, S.: Optical flow for verification, Weather Forecast., 25, 1479–1494, doi:10.1175/2010WAF2222351.1, 2010.

Mass, C. F., Ovens, D., Westrick, K., and Colle, B. A.: Does increasing horizontal resolution produce more skillful forecasts?, B. Am. Meteorol. Soc., 83, 407–430, doi:10.1175/1520-0477(2002)083<0407:DIHRPM>2.3.CO;2, 2002.

Pennec, X., Cachier, P., and Ayache, N.: Understanding the "demons algorithm": 3D non-rigid registration by gradient descent, in: Medical Image Computing and Computer-Assisted Intervention–MICCAI'99, 597–605, Springer, doi:10.1007/10704282_64, 1999.

Roelvink, J. A.: Coastal morphodynamic evolution techniques, Coastal Engineering, 53, 277–287, doi:10.1016/j.coastaleng.2005.10.015, 2006.

Sutherland, J., Peet, A., and Soulsby, R.: Evaluating the performance of morphological models, Coastal Engineering, 51, 917–939, doi:10.1016/j.coastaleng.2004.07.015, 2004.

Thirion, J.-P.: Image matching as a diffusion process: an analogy with Maxwell's demons, Medical Image Analysis, 2, 243–260, doi:10.1016/S1361-8415(98)80022-4, 1998.

Ziegeler, S. B., Dykes, J. D., and Shriver, J. F.: Spatial error metrics for oceanographic model verification, J. Atmos. Ocean. Tech., 29, 260–266, doi:10.1175/JTECH-D-11-00109.1, 2012.

14

Acoustic measurements of a liquefied cohesive sediment bed under waves

R. Mosquera, V. Groposo, and F. Pedocchi

Instituto de Mecánica de los Fluidos e Ingeniería Ambiental (IMFIA), Facultad de Ingeniería, Universidad de la República, Montevideo, Uruguay

Correspondence to: R. Mosquera (rmosquer@fing.edu.uy)

Abstract. In this article the response of a cohesive sediment deposit under the action of water waves is studied with the help of laboratory experiments and an analytical model. Under the same regular wave condition three different bed responses were observed depending on the degree of consolidation of the deposit: no bed motion, bed motion of the upper layer after the action of the first waves, and massive bed motion after several waves. The kinematic of the upper 3 cm of the deposit were measured with an ultrasound acoustic profiler, while the pore-water pressure inside the bed was simultaneously measured using several pore pressure sensors. A poro-elastic model was developed to interpret the experimental observations. The model showed that the amplitude of the shear stress increased down into the bed. Then it is possible that the lower layers of the deposit experience plastic deformations, while the upper layers present just elastic deformations. Since plastic deformations in the lower layers are necessary for pore pressure build-up, the analytical model was used to interpret the experimental results and to state that liquefaction of a self consolidated cohesive sediment bed would only occur if the bed yield stress falls within the range defined by the amplitude of the shear stress inside the bed.

1 Introduction

The erosion of cohesive sediment deposits presents significant differences with the erosion of non-cohesive deposits. One of the main differences is the strong dependence of the cohesive bed erosion on the previous consolidation process. Mehta (1991) characterized the erosion mechanisms of cohesive sediment beds into three types: *surface erosion*, when the flocs or aggregates are entrained one by one as a result of the hydrodynamic lift and drag; *Mass erosion*, when a slip surface is generated inside the deposit and all the material above this surface is mobilized; and *Destabilization of the sediment-water interface*, when processes within the bed induce the formation of fluid mud.

Two mechanism for the formation of fluid mud have been proposed in the literature, sudden failure due to the large shear stresses imposed by "large" waves on a "soft bed" and progressive pore pressure build-up under the successive action of "small" waves on a "partially consolidated bed". This article explores the necessary conditions that would lead to the occurrence of each of these mechanisms. The second of these two mechanisms is technically described as liquefaction of the bed due to pore pressure build-up. Terzaghi et al. (1996) established that the stress at any location inside a sediment-water mixture has two components: one component is a hydrostatic stress state that acts with equal intensity in every direction, which is associated to the pore-water pressure. The other component, called effective stress, is associated with the stress supported by the solid phase. The solid phase is considered the skeleton of the mixture and provides its shearing resistance.

Liquefaction of a sediment-water mixture is related to the increase of pore-water pressure and the corresponding decrease of the effective stress. When liquefaction occurs, the water-sediment mixture loses its shearing resistance and behaves as a dense fluid. During liquefaction, the mechanical characteristics of the sediment bed change so dramatically that marine structures fail, buried pipelines emerge, and navigation channels get silted in a matter of hours. Under the action of waves, liquefaction is a major concern in sediment beds with low permeability such as silt, clay or very fine sand (De Groot et al., 2006; Jeng, 2003).

Silt and clay water mixtures are usually referred as mud. For analytical purposes, mud can be considered to be a visco-elastic material, having an elastic response for small deformations and shear stresses, and a viscous response for large deformations (for an extense review see de Wit et al., 1994). These large deformations are associated with shear stresses τ overpassing a certain threshold, called yield stress τ_Y. Under oscillatory flow, the shear stresses could overpass the yield stress during part of the wave cycle producing non-elastic deformations in some regions of the bed. In high porosity and low permeability deposits, this non-elastic deformations may induce the progressive reorganization of sediment particles and the reduction of pore volume. This reduction may progressively lead to the increase (build-up) of the pore pressure and the associated reduction of the effective stresses, eventually triggering liquefaction (see for example the work of Sumer and Fredsøe, 2002, for non-cohesive sediments).

In this article we explore the liquefaction mechanism in a cohesive-sediment bed under the action of regular waves. For this aim we ran experiments in a laboratory wave flume, and simultaneously registered the pore pressure and the mud bed velocity. To our knowledge this is the first time these simultaneous measurements are performed, making it possible to clearly differentiate the two mechanisms that can generate fluid mud under the action of waves: sudden shear failure in soft deposits, and progressive pore pressure build-up in partially consolidated ones. For this second mechanism, a theoretical model that only considers elastic deformations of the bed is used to estimate the deformations within the bed, and to predict regions where non-elastic deformations of the bed and pore pressure build-up may occur.

2 Theoretical model

In order to have non-elastic deformations and pore pressure build-up, shear stress τ must be larger than the yield stress τ_Y in some regions of the bed during part of the wave cycle. If the sediment grains are loosely packed, the shear stresses generated by the successive waves will gradually rearrange the grains, reducing of the pore volume, and if the permeability of the soil is low, increasing the pore-water pressure (Sumer and Fredsøe, 2002). To study this necessary condition for liquefaction, the shear stresses inside the bed are estimated assuming an elastic response of the bed. If τ remains smaller than τ_Y during the whole cycle, only elastic deformations would occur, the pore volume would not change after a complete wave cycle, and pore pressure build-up would not take place. However, if τ becomes larger than τ_Y at some point during the cycle, a permanent deformation of the bed will occur, the pore-water pressure may build-up and liquefaction would be eventually observed. Once liquefaction has occurred, the elastic model is not able to predict the actual stress anymore. However, the elastic model can be used to qualitatively explore the necessary conditions for liquefaction and interpret the experimental observations.

Yamamoto et al. (1978) studied the response of a poro-elastic semi-infinite bed under regular water waves using a quasi-static model. Here the same set of equations is used, but for studying the response of a finite thickness poro-elastic bed. Furthermore, the equations are expressed in dimensionless form, defining a set of dimensionless parameters, which facilitates the psychical interpretation of the different terms in the equations, and their relevance for the occurrence of liquefaction. The case of a finite thickness poro-elastic bed was studied before by Spierenburg (1987). However, Spierenburg solutions were approximate, while the harmonic solutions presented here are exact. The detailed deduction of these exact harmonic solutions exceeds the scope of this article and can be found in Mosquera (2013).

Based on the model developed by Biot (1941) and assuming that the pore-water flow follows Darcy's law, the following equation for the conservation of pore-water can be written

$$\frac{k}{\gamma}\Delta p = \frac{n}{K'}\frac{\partial p}{\partial t} + \frac{\partial \epsilon}{\partial t}, \tag{1}$$

where k is the coefficient of permeability of the soil, γ is the unit weight of the pore-water, p is the pore-water pressure, n is the porosity, K' is the apparent bulk modulus of pore-water, t is the time, and ϵ is the volume strain of the porous medium. Considering that the mud behaves in an elastic way, the equations of equilibrium may be expressed as

$$G\Delta u + \frac{G}{1-2v}\frac{\partial \epsilon}{\partial x} = \frac{\partial p}{\partial x}, \tag{2}$$

$$G\Delta w + \frac{G}{1-2v}\frac{\partial \epsilon}{\partial z} = \frac{\partial p}{\partial z}, \tag{3}$$

where u and w are the horizontal and vertical components of the mud displacement, respectively; v is the Poisson's ratio of the mud and G is its shear modulus. Remembering the definition of the volume strain

$$\epsilon = \frac{\partial u}{\partial x} + \frac{\partial w}{\partial z}, \tag{4}$$

now we have four equations from Eqs. (1) to (4), one for each unknown variable (p, u, w and ϵ). Additionally, effective stresses within the bed can be related to the bed strains using Hooke's law

$$\sigma'_x = 2G\left(\frac{\partial u}{\partial x} + \frac{v}{1-2v}\epsilon\right), \tag{5}$$

$$\sigma'_z = 2G\left(\frac{\partial w}{\partial z} + \frac{v}{1-2v}\epsilon\right), \tag{6}$$

$$\tau_{xz} = G\left(\frac{\partial u}{\partial z} + \frac{\partial w}{\partial x}\right), \tag{7}$$

where σ'_x and σ'_z are the effective normal stress in horizontal and vertical directions respectively and τ_{xz} is the shear stress on a vertical or horizontal plane.

In order to solve the mud motion and the pore-water pressure field, it is necessary to know the boundary conditions of the problem. A dynamic condition is used for the water-mud interface. Since cohesive sediment beds can be considered smooth, the shear stress friction factor is relatively low. For the experimental conditions considered in this article, it can be shown that tangential stresses are at least one order of magnitude smaller than τ_Y, and as a first approximation waves can be considered to just impose normal stresses on the bed.

Assuming that the pressure varies continuously from the water column into the top pores of the bed, the boundary conditions at the mud-water interface are obtained

$$\sigma'_z = 0, \tag{8}$$

$$\tau_{xz} = 0, \tag{9}$$

$$p = P_0 \cos(\lambda x - \omega t), \tag{10}$$

where λ is the wave number and ω the angular frequency of the surface waves, and P_0 is the pressure amplitude imposed by the waves on the bed surface. P_0 is calculated using the Airy wave theory as

$$P_0 = \frac{\gamma H}{2 \cosh(\lambda h)}. \tag{11}$$

The lower boundary of the mud deposit is considered rigid and impermeable, allowing for no vertical displacement and no vertical pore-water flux. Additionally, two possible conditions for this boundary are considered: complete adherence at the rigid boundary and a perfectly slipping boundary.

For the complete adherence case

$$u = 0, \tag{12}$$

and for the perfect-slip plane case

$$\tau_{xz} = 0. \tag{13}$$

The first two boundary conditions at the lower boundary are

$$w = 0, \tag{14}$$

$$\frac{\partial p}{\partial z} = 0. \tag{15}$$

Taking scales for the different variables, it could be shown that, apart from the Poisson's ratio ν, the problem is defined by three dimensionless numbers

$$d\lambda, \tag{16}$$

$$\kappa = \frac{1 - \nu}{1 - 2\nu} \frac{2nG}{K'}, \tag{17}$$

$$\delta = \frac{1 - \nu}{1 - 2\nu} \frac{2kG}{\gamma d^2 \omega (\kappa + 1)}. \tag{18}$$

$d\lambda$ is the geometric ratio between the bed thickness and the wave length, κ is the ratio between the water compressibility and sediment bed compressibility, and δ is the ratio between the ability of the water to flow within the bed and the pore pressure variation along the wave cycle. For a cohesive sediment bed saturated with water, both permeability k and shear modulus G are small, and therefore both κ and δ can be considered small quantities.

Length scales for u and w can be conveniently defined as

$$U_0 = \frac{1 - 2\nu}{1 - \nu} \frac{P_0}{2G\lambda}, \tag{19}$$

and

$$W_0 = \frac{1 - 2\nu}{1 - \nu} \frac{d P_0}{2G}, \tag{20}$$

respectively.

Finally a system of equations and boundary conditions in dimensionless form can be solved for a given set of the four dimensionless numbers ($\nu = 0.3$, $d\lambda = 0.512$, $\kappa = 2.65 \cdot 10^{-4}$, $\delta = 1.80 \cdot 10^{-3}$). These numbers were obtained from Eqs. (16) to (18) using the following characteristics for the mud bed: $k = 10^{-6}$ ms^{-1}, $n = 0.3$, $\nu = 0.3$, $G = 4.8 \cdot 10^5$ Nm^{-2}, $\gamma = 9800$ Nm^{-3}, $K' = 1.9 \cdot 10^8$ Nm^{-2}, and $d = 0.15$ m; under the following wave forcing: $h = 0.176$ m, $\omega = 4.24$ rads^{-1}, $H = 0.10$ m, giving $\lambda = 3.41$ m^{-1}, $P_0 = 413$ Pa, $U_0 = 7.20 \cdot 10^{-5}$ m and $W_0 = 3.69 \cdot 10^{-5}$ m. All the characteristics are considered uniform within the bed. The numerical values were taken from the literature and are considered representative of the experimental conditions discussed here.

Under these conditions it was possible to determine the profiles of p, u, w and the effective stress state (σ'_x, σ'_z and τ_{xz}) for the two possible bottom boundary conditions. Figure 1 shows the dimensionless amplitude profiles of these magnitudes. Once the stress state is known, the amplitude of the maximum shear stress at a point for any plane direction τ can be computed. For both bottom boundary conditions the τ profile is found to be non-zero at the top layer of the deposit, where $\tau = \tau_0$. For the complete adherence case, the τ profile has a local maximum and a local minimum, finally increasing toward the bottom of the deposit, where $\tau = \tau_d$. For the perfect slip case, the τ profile monotonically increases toward the bottom of the deposit. It is interesting to note that τ reaches higher values for the perfect-slip case that for the complete adherence case showing the relevance of the phase shift among the different components of the shear stress tensor.

Although the τ profile was found under the hypothesis of pure elastic motion, some relevant observations can be made regarding the possible occurrence of liquefaction in regions where the maximum shear stress exceeds the mud yield stress τ_Y, and non-elastic deformations may occur. Comparing the mud yield stress τ_Y with the shear stress profile τ, three scenarios may be defined:

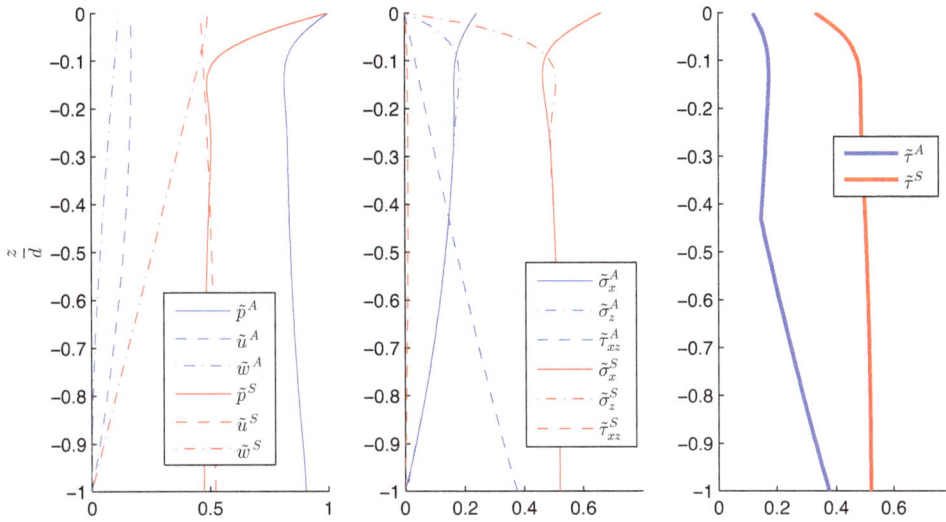

Fig. 1. Non dimensional profiles of $\tilde{p} = \frac{|p|}{P_0}$, $\tilde{u} = \frac{|u|}{U_0}$, $\tilde{w} = \frac{|w|}{W_0}$, $\tilde{\sigma}_x = \frac{|\sigma'_x|}{P_0}$, $\tilde{\sigma}_z = \frac{|\sigma'_z|}{P_0}$, $\tilde{\tau}_{xz} = \frac{|\tau_{xz}|}{P_0}$ and maximum shear stress amplitude for both bottom boundary conditions, complete adherence (A) and perfect slip (S).

- If $\tau_0 > \tau_Y$ the top layers of the deposit will flow during part of the wave cycle. This behavior will be observed immediately after the first wave acts on the deposit. Since the top layers are flowing, it is easier for pore-water to liberate its pressure, reducing the chances of pressure build-up.

- If $\tau_d > \tau_Y > \tau_0$, plastic deformations are observed in the lower layers where $\tau > \tau_Y$. In this region the permanent deformations will lead to the reorganization of the soil skeleton and, depending on the mud permeability, pore pressure build-up would take place.

- If $\tau < \tau_Y$ over the whole deposit, only elastic deformations will be expected and no pore pressure build-up can occur.

3 Methods

Previous laboratory work regarding liquefaction focused on measuring pore-water pressure, due to the importance of this variable on the phenomena. For the experiments shown here a laboratory acoustic velocity meter was used to measure the mud flow simultaneously with the pore pressure. These velocity meter have high spatial and temporal resolution, and are practically non-intrusive. Most acoustic instruments are designed to measure in water, but it is possible to use them to perform velocity measurements in concentrated sediment mixtures (Gratiot et al., 2000; Salehi and Strom, 2011). For the present experiments, velocity measurements through the water column and inside the upper layers of the deposit (first 3 cm) were performed using an Ultrasonic Velocity Profiler (UVP) produced by MetFlow (Pedocchi and García, 2011) equipped with a 2 MHz transducer. The trans-

ducer was placed 12 cm above the mud deposit at a 55° angle with the horizontal. It should be pointed out that the UVP measures the projection of the velocity vector on the direction of the sensor. However, in shallow waters the water motion can be considered horizontal near the bed and the projection of the velocity measured by the UVP can be considered a good proxy for the actual velocity.

The experiments presented here were performed in a wave flume located at the Instituto de Mecánica de los Fluidos e Ingeniería Ambiental (IMFIA). The flume is 0.51 m wide, 0.76 m deep, and 16 m long. The flume bottom was modified by placing a false bottom of 15 cm high. The false bottom covered the entire flume, with the exception of a 1.8 m gap in the middle of the flume (Fig. 2). At the front of the wavemaker, a 2.8 m long ramp allowed a smooth transition from the flume bottom towards the false bottom. At the opposite end of the flume a permeable beach absorbed the incident waves. The beach reflection in terms of the wave energy was close to 1 % (Mosquera and Pedocchi, 2013). All five experiments described next were performed under the same hydrodynamics conditions: regular waves generated by a piston-type wavemaker oscillating with a period $T = 1.48$ s, and producing $H = 10$ cm hight waves in a water depth $h = 17.6$ cm.

Seven wet-wet pore pressure transducers, model 26PCAFA6D produced by Honeywell, were mounted at different locations inside the mud bed as shown in Fig. 2. PVC tubing, 3/16 inch (4.8 mm) internal diameter and 9/32 inch (7.1 mm) external diameter, was used to connect the measurement points P# with the pressure transducers. The measuring range of the pressure transducers was ± 0.70 mH$_2$O (± 6.9 KPa). Iron pieces were attached to the flume bottom to strongly hold the piping and avoid their motion. The air from inside the tubing was carefully drained

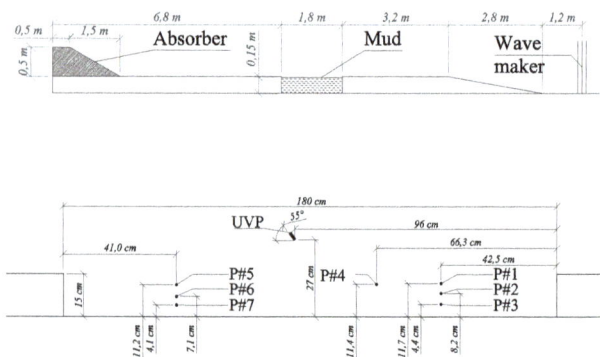

Fig. 2. Scheme of the wave flume and the experimental set-up, showing the location of the UVP transducer and the seven pore pressure sensors.

in order to obtain high quality measurements. At the end of each tubing a filter was placed to prevent sediments from flowing into the measuring system and care was taken to avoid the filters from getting clogged with sediment.

Artificial kaolinite clay was used for the mud deposits. The sediment density was $2635 \, \text{kg m}^{-3}$. Its granulometry was determined by two different techniques, with ASTM (2007a) standard method and with a laser diffraction technique using a Mastersizer 2000 produced by Malvern. Results were in good agreement, giving a mean diameter of $7.1 \, \mu\text{m}$ and a standard deviation of $0.2 \, \text{mm}$. Also ASTM (2007b) and ASTM (2007c) standards where applied for the determination of a liquid limit of 24 and a plastic limit of 18, an inorganic clay of low plasticity according to Casagrande (1932) classification. The density of the mud was determined before the experiments by extracting a sample of the upper layers of the deposit, and applying the ASTM (2010) and ASTM (2009) standards, with the help of a LJ16 Infrared Dryer produced by Mettler Toledo.

In order to produce beds that would develop different responses under the same wave forcing several mud deposits were prepared as follows. After the flume was filled with water, the sediments were mixed with the water in the water column. Two partition walls were used to prevent the sediment mixture from flowing away from the bottom depression area during the mixing, sedimentation, and consolidation of the bed. Different consolidation times, ranging from days to weeks, were used to generate deposits with different densities and yield stresses. Experiment # 1 had a bed density of $1548 \, \text{kg m}^{-3}$, #2 $1599 \, \text{kg m}^{-3}$, #3 $1738 \, \text{kg m}^{-3}$, #4 $1608 \, \text{kg m}^{-3}$ and #5 $1660 \, \text{kg m}^{-3}$. Experiment #3 was performed over the bed left by Experiment #2.

4 Results and discussion

Figure 3 summarizes the results of the experiments. The top colored charts in this figure show the velocity profile series measured with the UVP during sixty waves. The black an

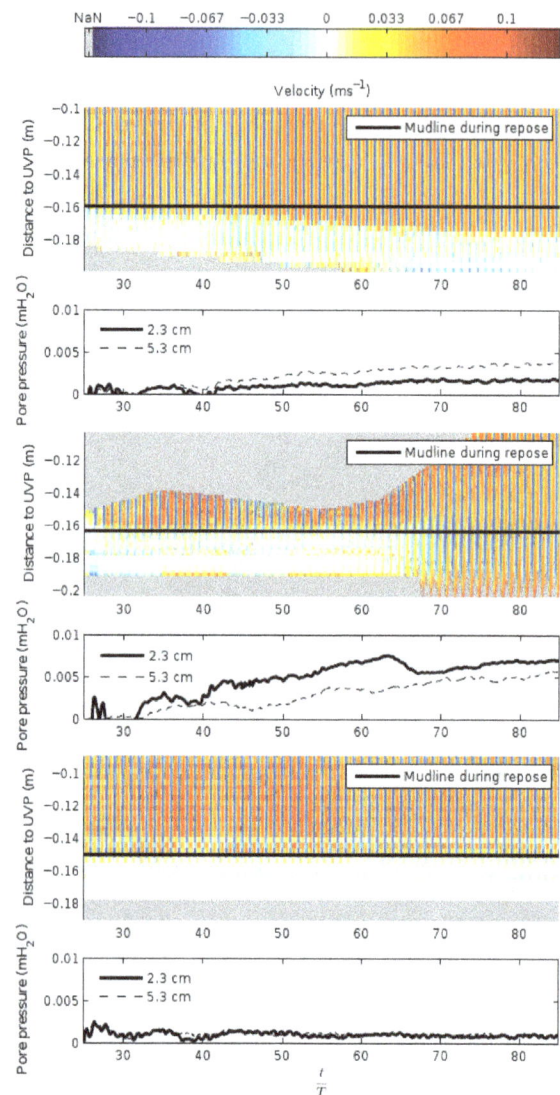

Fig. 3. From top to bottom: Experiment #1 ($\rho_b = 1548 \, \text{kg m}^{-3}$), progressive motion of the bed starting from the upper layers, corresponding to the $\tau_Y < \tau_0$ condition. Experiment #4 ($\rho_b = 1608 \, \text{kg m}^{-3}$), liquefaction of the bed due to pore pressure build-up, corresponding to the $\tau_d > \tau_Y > \tau_0$ condition. Experiment #3 ($\rho_b = 1738 \, \text{kg m}^{-3}$), no mobilization of the bed, corresponding to the $\tau_Y > \tau_d$ condition. The colored charts show the velocity profile series measured with the UVP at each experiment during sixty waves, the black line indicates the location of the top of the bed at the beginning of the experiment, cold colours indicate velocities towards the sensor and warm colours away from the sensor, grey zones indicate low quality data due to low acoustic backscatter intensity. The black an white charts shows the period-averaged pore-water pressure for two different heights (2.3 cm and 5.3 cm below the initial mud-water interface).

white charts show the period-averaged pore-water pressure for two different heights (2.3 cm and 5.3 cm below the initial mud-water interface). For the present experiments the maximum shear stress at the upper layer was of the order of $\tau_0 = 50$ Pa. Three types of bed response were observed in the experiments: motion of the upper layers of the deposit, starting with the first waves; liquefaction of the bed, after several waves had pass over the deposit; and "no motion".

The first response (motion of the upper layers) was observed during Experiment #1 ($\rho_b = 1548$ kg m^{-3}), which had the lowest bed density. The bed motion started on the upper layers with the first wave and slowly progressed down into the deposit. Figure 1 illustrates this response, which corresponds to the $\tau_0 > \tau_Y$ case presented at the end of Sect. 2. For this case the shear stress overpasses the yield stress and the top layer fails as the first wave travels over the bed. As the first layer fails, the elastic model does not apply to that layer anymore. However, it may still be applied to the next layer down, which will also fail. This process can therefore progress, slowly mobilizing successive layers of the deposit as it was observed in this experiment. A slight build-up of pore-water pressure was observed during Experiment #1 on both sensors but it was not large enough to produce liquefaction.

The second response (liquefaction of the bed) was observed in Experiments #2 and #4 ($\rho_b = 1599$ kg m^{-3} and $\rho_b = 1608$ kg m^{-3} respectively). A gradual build-up of pore pressure was measured by the pressure sensors and no motion on the top layers was measured by the UVP. These experiments can be considered to fall in the $\tau_Y > \tau_0$ class. The shear stresses near the surface were only able to produce elastic deformations. However, as the shear stress increases with depth, the lower layers suffered plastic deformations. Then, the successive action of waves slowly reduced the voids volume in these layers increasing the pore-water pressure. After tens of waves had passed over the bed, the entire bed abruptly started to move and the pore pressure at 2.3 cm sensor descended to a value near the recorded value by the 5.3 cm depth sensor. The difference between Experiments #2 and #4 was the number of waves that induced liquefaction, in Experiment #2 (not shown here) only ten waves were needed. It is interesting to note that the 2.3 cm sensor presented a faster pore pressure build-up. This can be explained in part because the yield stress of the bed probably increases with depth, instead of being constant as supposed here. Additionally, if complete adherence at the bottom of the deposit is considered, Fig. 1 shows that this sensor was located close to τ's local maximum, where the largest plastic deformations should be expected.

Finally, during Experiments #3 and #5 ($\rho_b = 1738$ kg m^{-3} and $\rho_b = 1660$ kg m^{-3} respectively) neither bed motion nor significant pore-water pressure built up were observed, even after hundreds of waves had passed over the bed. These two experiments are considered to had fallen in the $\tau_d < \tau_Y$ case and therefore neither bed plastic deformation nor pore pressure build-up were possible.

5 Conclusions

For some time, researchers have suggested two mechanisms to explain the generation of fluid mud under waves: progressive pore pressure build-up in partially consolidated deposits, and sudden shear failure in soft deposits. Both mechanisms were observed in the experiments presented here. And an explanation for each of these responses was discussed depending on the ratio between the bed yield stress and the shear stress profile imposed inside the bed by the action of waves. Experiments #2 and #4 are particularly relevant since they showed that a partially consolidated bed under moderate waves can suddenly get mobilized, even though the individual waves were not able to mobilize the deposit. This type of fatigue failure is particularly dangerous since the usual maximum wave hight criteria would fail to predict it.

The simultaneous measurements of the bed velocity field with the UVP and the pore-water pressure with the pressure transducers, made possible to clearly identify the failure mechanisms. The poro-elastic solutions for the bed deformation showed that the shear stress increases down into a finite thickness bed, and plastic deformations may occur in the lower layers of the deposit while only elastic deformations are possible in the top layers. If the permeability of the deposit is low enough, these plastic deformations of the lower layers would induce the build-up of pore pressure and lead to the liquefaction of the mud bed.

Acknowledgements. We would like to thank the Agencia Nacional de Investigación e Innovación (ANII), Uruguay, for the financial support given to the first author under its scholarships program "Posgrados Nacionales" (BE_POS_2010_1_2578, 2010). Additionally part of the work presented here was done under the grant of PR_FMV_2009_1_1890 also from the ANII, this support is greatly appreciated. The authors would like to thank the IMFIA machine shop staff (D. Barboza and R. Zouko) for their help with the experimental set-up. We would also like to thank G. Sanchez and A. Bologna from the Instituto de Ingeniería Química of Facultad de Ingeniería, Uruguay; and Susana Vinzón and Marcos Gallo from the Area de Engenharia Costeira of the Universidad Federal de Río de Janeiro (UFRJ), Brasil for their help with the characterization of the sediment samples. The collaboration with the UFRJ was possible thanks to the CAPES-UdelaR academic exchange program (026/2010) of the CAPES Foundation, Brazil, and Universidad de la República, Uruguay.

References

ASTM: D422-63 Standard Test Method for Particle-Size Analysis of Soils, American Society for Testing and Materials, West Conshohocken, PA, 2007a.

ASTM: D423-66 Method of Test for Liquid Limit of Soils, American Society for Testing and Materials, West Conshohocken, PA, 2007b.

ASTM: D424-54 Standard Method of Test for Plastic Limit, American Society for Testing and Materials, West Conshohocken, PA, 2007c.

ASTM: D7263-09 Standard Test Methods for Laboratory Determination of Density (Unit Weight) of Soil Specimens, American Society for Testing and Materials, West Conshohocken, PA, 2009.

ASTM: D2216-10 Standard Test Methods for Laboratory Determination of Water (Moisture) Content of Soil and Rock by Mass, American Society for Testing and Materials, West Conshohocken, PA, 2010.

Biot, M.: General theory of three-dimensional consolidation, J. Appl. Phys., 12, 155–164, 1941.

Casagrande, A.: The structure of clay and its importance in foundation engineering, Journal of Boston Society of Civil Engineers, 419, 168–209, 1932.

De Groot, M., Bolton, M., Foray, P., Meijers, P., Palmer, A., Sandven, R., Sawicki, A., and Teh, T.: Physics of liquefaction phenomena around marine structures, J. Waterw. Port C.-ASCE, 132, 227–243, 2006.

de Wit, P., Kranenburg, C., and Winterwerp, J.: Liquefaction and erosion of mud due to waves and current: Experiments on China Clay, Internal report, 128 pp., Delft University of Technology, the Netherlands, 1994.

Gratiot, N., Mory, M., and Auchère, D.: An acoustic Doppler velocimeter (ADV) for the characterisation of turbulence in concentrated fluid mud, Cont. Shelf Res., 20, 1551–1567, 2000.

Jeng, D. S.: Wave-induced sea floor dynamics, Appl. Mech. Rev., 56, 407–429, 2003.

Mehta, A.: Review notes on cohesive sediment erosion, Coastal Sediments 1991, 40–53, 1991.

Mosquera, R.: Generalized response of a poro-elastic bed to water waves, Instituto de Mecánica de los Fluidos, Internal Report, 2013.

Mosquera, R. and Pedocchi, F.: Decomposition of incident and reflected surface waves using an Ultrasonic Velocity Profiler, Coast. Eng., 71, 52–59, 2013.

Pedocchi, F. and García, M. H.: Acoustic measurement of suspended sediment concentration profiles in an oscillatory boundary layer, Cont. Shelf Res., 46, 87–95, doi:10.1016/j.csr.2011.05.013, 2011.

Salehi, M. and Strom, K.: Using velocimeter signal to noise ratio as a surrogate measure of suspended mud concentration, Cont. Shelf Res., 31, 1020–1032, 2011.

Spierenburg, S. E. J.: Seabed Response to Water Waves, Ph.d.dissertation, Delft University of Technology, the Netherlands, 1987.

Sumer, B. and Fredsøe, J.: The Mechanics of Scour in the Marine Environment, World Scientific, 2002.

Terzaghi, K., Peck, R., and Mesri, G.: Soil mechanics in engineering practice, Vol. 3rd Edn., Wiley, 1996.

Yamamoto, T., Koning, H., Sellmeijer, H., and van Hijum, E.: On the response of a poro-elastic bed to water waves, J. Fluid Mech., 87, 193–206, 1978.

15

Retrieval of Ocean Bottom and Downhole Seismic sensors orientation using integrated MEMS gyroscope and direct rotation measurements

A. D'Alessandro and G. D'Anna

Istituto Nazionale di Geofisica e Vulcanologia, Centro Nazionale Terremoti, Italy

Correspondence to: A. D'Alessandro (antonino.dalessandro@ingv.it)

Abstract. The absolute orientation of the horizontal components of ocean bottom or downhole seismic sensors are generally unknown. Almost all the methods proposed to overcome this issue are based on the post-processing of the acquired signals and so the results are strongly dependent on the nature, quantity and quality of the acquired data. We have carried out several test to evaluate the ability of retrieve sensor orientation using integrated low cost MEMS gyroscope. Our tests have shown that the tested MEMS gyroscope (the model 1044_0–3/3/3 Phidget Spatial Precision High Resolution) can be used to measure angular displacement and therefore to retrieve the absolute orientation of the horizontal components of a sensor that has been subjected to rotation in the horizontal plane. A correct processing of the acquired signals permit to retrieve, for rotation at angular rate between 0 and $180° \text{ s}^{-1}$, angular displacement with error less $2°$.

1 Introduction

Ocean Bottom Seismometers (OBS) are fundamental tools for the monitoring and study of seismogenetic offshore areas (D'Alessandro et al., 2009, 2012a, 2013a, 2014a; D'Alessandro, 2014a). An OBS is a stand-alone seismic station that can be deployed in open sea until a depth of several kilometres (Mangano et al., 2011). During the OBS descent phase along the seawater column, the seismic sensor can be undergo to random rotations on the horizontal plane. How and how much the sensor has rotated before reaching the ocean bottom is unknown and so it is not possible to retrieve the correct orientation of the its horizontal components. Similar problem can occur in the installation of seismic sensor in

deep borehole. Indeed, to reduce the background noise level, seismic sensors are often installed on downhole, at depths of several tens of meters. The orientation of the sensor is usually determined with a gyroscopic tool that is lowered into the hole and into the keyed hole-lock (Holcomb, 2002). However, the gyro tools are relatively fragile and very expensive (Holcomb, 2002). For these reasons, several authors have proposed alternative cheaper methods for orienting borehole sensors.

Almost all the methods proposed to retrieve the correct horizontal components orientation, are based on the post-processing of the acquired signals (Anderson et al., 1987; Duennebier et al., 1987; Nakamura et al., 1987; Chiu et al., 1994; Michaels, 2001; Baker and Stevens, 2004; Oye and Ellsworth, 2005; Zheng and McMechan, 2006; Ekström and Busby, 2008; D'Alessandro et al., 2009, 2013a; Grigoli et al., 2012; Stachnik et al., 2012; Ringler et al., 2012, 2013; Zha et al., 2013). These techniques employ different approaches (polarization analysis, cross-correlation measurements, synthetic seismograms fitting), different datasets (shots, earthquakes, seismic noise) using different part of the seismic wave-field (P or S wave arrival times, Rayleigh waves, full waveforms), but are all based on the post-processing of the acquired data. Anderson et al. (1987), Duennebier et al. (1987) and Nakamura et al. (1987) used the amplitude ratio of water waves and first arrivals generated by air gun shots from known locations to orientate OBS. Chiu et al. (1994) used the late arrivals times (mainly S wave) to orientate strong motion sensors installed on borehole. Michaels (2001) proposed a method based on principal component analysis to determine the sensors orientation, relative to source polarization direction using SH wave. Holcomb (2002) test several techniques

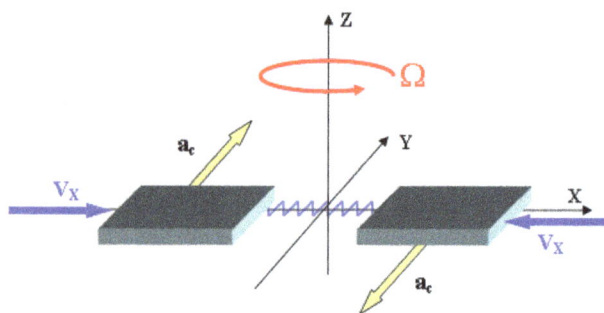

Figure 1. Sketch of the principle of operation of a MEMS gyroscope.

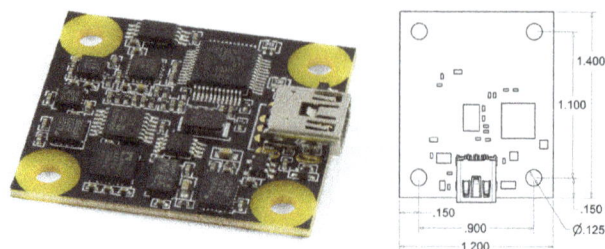

Figure 2. MEMS model 1044_0 (3/3/3 PhidgetSpatial Precision High Resolution) produced by the Canadian company Phidget Inc.; the reported measures are in inches.

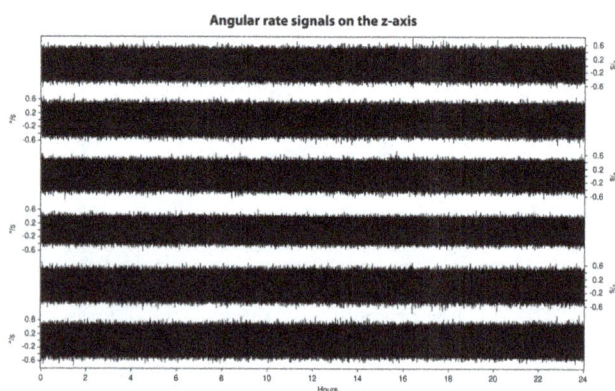

Figure 3. 24 h angular rate signals (z axis) recorded by the six MEMS gyroscope here tested.

based on the comparison of the sensor signal outputs with that of a reference-oriented sensor. The methods proposed by Baker and Stevens (2004) and also adopted by Stachnik et al. (2012), is based on the polarization of observed Rayleigh-waves. Oye and Ellsworth (2005) determined sensors absolute orientations comparing azimuths obtained from P wave polarization analysis and theoretical azimuths estimated from ray-tracing. Zheng and McMechan (2006) used traces cross-correlation to infer relative angles between adjacent geophone pairs in borehole arrays. Ekström and Busby (2008) determine sensor orientation by examining correlations between observed and synthetic surface-wave over a range of orientations. D'Alessandro et al. (2009, 2013a) employed a method based on noise unbiased polarization analysis of the first P-wave arrival time of well-located events. Grigoli et al. (2012) proposed a complex linear least-squares method applicable on full waveform records. Ringler et al. (2012, 2013) developed a technique for estimating relative sensor azimuths by inverting for the orientation with the maximum correlation to a reference instrument, using a nonlinear parameter estimation routine. Zha et al. (2013) proposed a method based on polarization analysis of Rayleigh waves retrieved from ambient noise cross-correlation.

However, all these methods are not error-free and not always applicable: methods based on active sources are not applicable in seismic passive monitoring campaigns; methods based on synthetic waveforms are strong dependent on accuracy of the source parameters estimations and are generally computationally intensive; methods based on polarization analysis are clearly strong dependent on the quality of the data in term of number of seismic events recorded, azimuthal coverage and signal to noise ratio; methods base on events or noise cross-correlation can be applicable only if an array of sensor is deployed, but are not applicable to individual sensors or sensor-very far from each other.

For all the above reasons it would be desirable a not expensive direct method for the determination of the absolute orientation of the sensor horizontal components, not dependent on the nature, quantity and quality of the data acquired. The simplest solution to this problem would be the installation,

together with the sensors, of an electronic compass able to measure the real horizontal component orientation respect to the magnetic North. However, as well known, all the seismic sensors currently used in earthquakes monitoring produces moderate to strong electromagnetic fields, which make the data recorded by an electronic compass, placed in their proximity, unusable for the described purpose.

For this reason, in this work we propose an alternative method for the estimation of the absolute orientation of horizontal components of a sensor, based on the use of MEMS (Micro Electro-Mechanical Systems) gyroscope.

In the following sections, after describing in detail the technical characteristics of the MEMS gyroscope, we will describe the tests carried out to verify the suitability of this device in determining the actual orientation of the horizontal components of a seismic sensor placed on downhole or on ocean bottom.

2 MEMS gyroscopes

MEMS devices have been recognized as one of the most promising technologies of the XXI century, able to revolutionize both the industrial world and that of the consumer products. In the 90s, MEMS sensors revolutionized the automotive airbag system and are today widely used in lap-

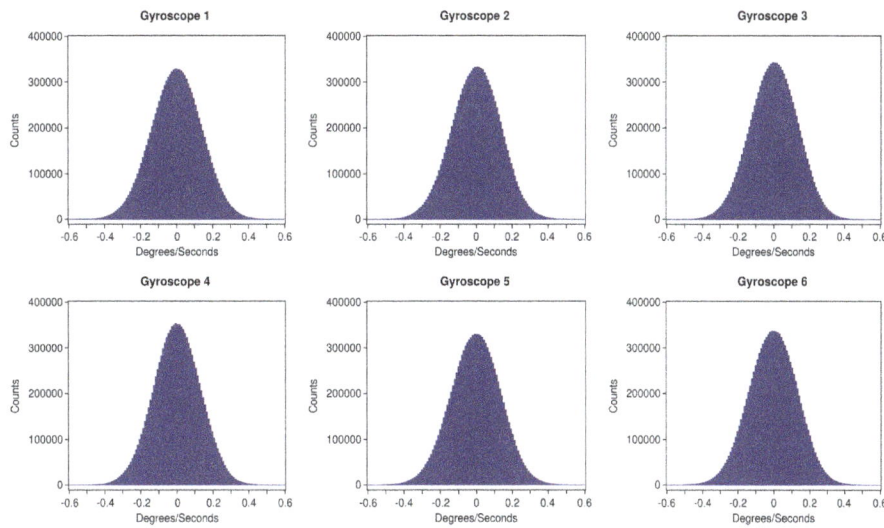

Figure 4. Angular rate histograms of the signals of Fig. 3.

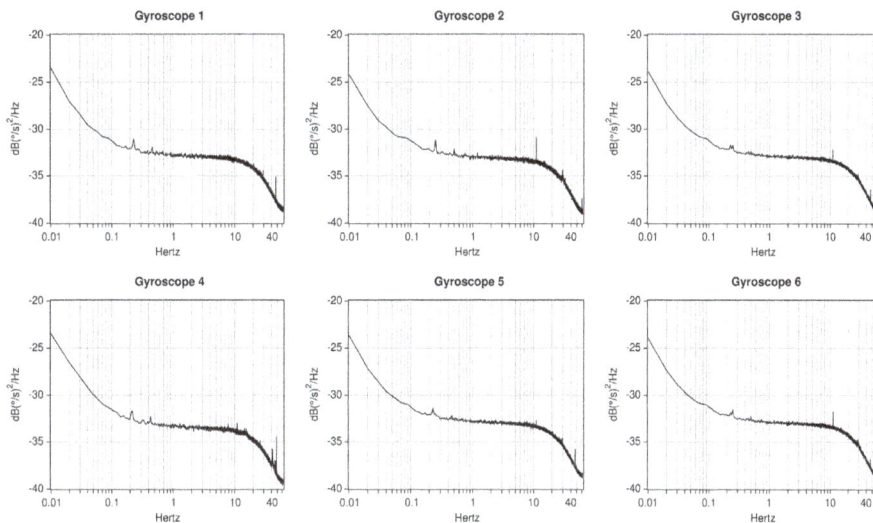

Figure 5. Angular rate PSD of the signals of Fig. 3.

tops, games controllers and mobile phones. Thanks to the great commercial success, the research and development of MEMS technology actively continues all over the world. A MEMS device is a system capable of receiving information from the environment by translating the physical quantities into electrical impulses. The MEMS sensors can measure phenomena of various kinds: mechanical (sound, acceleration and pressure), thermal (temperature and heat flux), biological (cell potential), chemical (pH), optical (intensity of light radiation, spectroscopy), magnetic (intensity of flow). Due to their versatility, MEMS sensors are increasingly being used in a wide field of science, including the physical, engineering, medical and earthquakes (D'Alessandro and D'Anna, 2013; D'Alessandro et al., 2014b; D'Alessandro, 2014b) one.

A gyroscope is a rotating physical device that, due to the law of angular momentum conservation, tends to maintain its rotation axis oriented in a fixed direction. A homogeneous mass, which quickly turns around its principal axis of inertia, develops centrifugal forces that appear to be perfectly in balance among them. In this way, no internal force can disturb the rotational motion and generate vibrations or alterations of the same rotational motion. A gyroscope is therefore an instrument that can be used to measure the angular velocity of a body around a generic axis, respect to an inertial reference system.

The angular velocity can be derived through different physical principles. The MEMS gyroscopes generally exploit the inertial accelerations that arise due to the motion of the sensor with respect to a non-inertial reference system, which

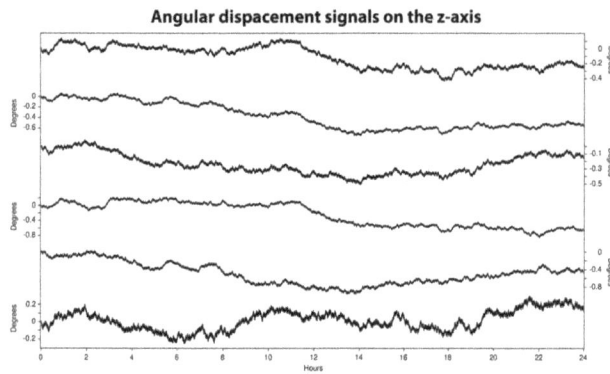

Figure 6. Angular displacement determined from the signals of Fig. 3.

is the Coriolis acceleration. Figure 1 schematised the principle of operation of a MEMS gyroscope. The system consists of two masses joined by means of an elastic element. Subjecting the system to the action of a forcing in the x direction and to the rotation imposed by an angular velocity along the z axis (perpendicular to the xy plane), it manifests the appearance of an acceleration directed orthogonally to the trajectory (in y direction), that is precisely the Coriolis acceleration.

The MEMS device here tested is the model 1044_0 (3/3/3 Phidget Spatial Precision High Resolution) produced by the Canadian company Phidget Inc (Fig. 2). This device integrates a three-axis gyroscope but is also equipped with an accelerometer and a magnetometer, both tri-axial. The sensor was chosen from among hundreds of MEMS gyroscope on the market based on his performance, cost and in relation to the purpose of this paper. The circuit of transduction is internal to the device and is of the digital type, for which the outputs are already in a digital format and proportional to the measured quantity.

The integrated gyroscope features three sensing elements oriented along three mutually orthogonal axes. In precision mode, the gyroscope have resolution of 0.02 and $0.013° \text{s}^{-1}$ on horizontals and vertical axis, respectively. The self-noise standard deviation is $0.095° \text{s}^{-1}$ and the maximum measurable angular rate is 300 and $400° \text{s}^{-1}$ for the horizontals and vertical axis, respectively.

Moreover, the MEMS device has a low current consumption, estimated at a maximum current consumption of 55 mA. Power can be supplied via a USB connection with a potential difference between 4.4 and 5.3 V. The sampling step can be set between a minimum of 4 ms sample^{-1} to a maximum of 1 s sample^{-1}. The analogical to digital conversion is done automatically by means of a 16-bit digital converter. Furthermore, the temperature range of operation is extremely large (-40–$85 °C$) ensuring the operation of the device even in the most extreme temperature conditions. All specifications above described, together with the small size of the device ($3 \times 3.5 \times 0.4$ cm) and its low cost, make this MEMS device

suitable for co-installation with seismic sensor in borehole or in OBS.

3 Performance evaluation

In order to evaluate the performance of the gyroscope included in the MEMS model 1044_0, we have tested six different devices. Since the main aim of this work is to check the suitability of this equipment to retrieve the horizontal components rotation, in that follows, we consider only the signals relative to the vertical (z) axis of the gyroscope. In each of the tests described below the sampling frequency has been set equal to 125 Hz.

The first test was performed in order to verify the parameters declared by the manufacturer on the gyroscope self-noise. The six sensors have been co-installed in a quiet place and signals acquired for a week. Figure 3 show 24 h of signals recoded on the z axis by the six MEMS gyroscopes tested. The recorded signals show the typical features of the self-noise generated by electronic systems. Such self-noise shows good temporal stationarity throughout the whole observation period for all the six sensors tested. Their amplitude histograms show a typical Gaussian distribution (Fig. 4); all the signals have zero mean ($< 10^{-9} ° \text{s}^{-1}$) and standard deviation between $0.130 \text{ e } 0.138° \text{s}^{-1}$, slightly greater than that declared by the manufacturer. The Skewness ranges between -0.026 and -0.042 while the Kurtosis ranges between -0.018 and -0.050.

For our purposes, that is the determination of the sensor horizontal rotation we are interested to the angular displacement. Angular displacement can be determined from angular rate by means of time integration. However, time integration without any data pre-processing can lead to a large drift in the resulting signals. This drift is generated by the very low frequency component of the original signals that, after time integration, as suggest by the theorem of convolution, are largely amplified (each spectral coefficient result divided by the angular frequency).

Figure 5 shows the angular rate Power Spectra Density (PSD) of the signals of Fig. 3. The PSD was calculated using the periodogram method. To calculate the average PSD, each signal was split into half-overlapped time windows 100s long and a Hanning window was applied to each time window to reduce spectrum distortion due to the signals truncation.

We can observe as at very low frequency (< 0.1 Hz) the self-noise generated by the MEMS gyroscopes have high power. To reduce the drift, before time integration to derive angular displacement, we applied a high-pass filter with cutoff frequency of 0.2 Hz.

Figure 6 show the same signals of Fig. 3 after time integration using Simpson's rule. We can observe as the resulting angular displacement signals have little offset (between -0.491 and $0.016°$) and standard deviation between 0.115 and $0.320°$. These values can be considered respectively as

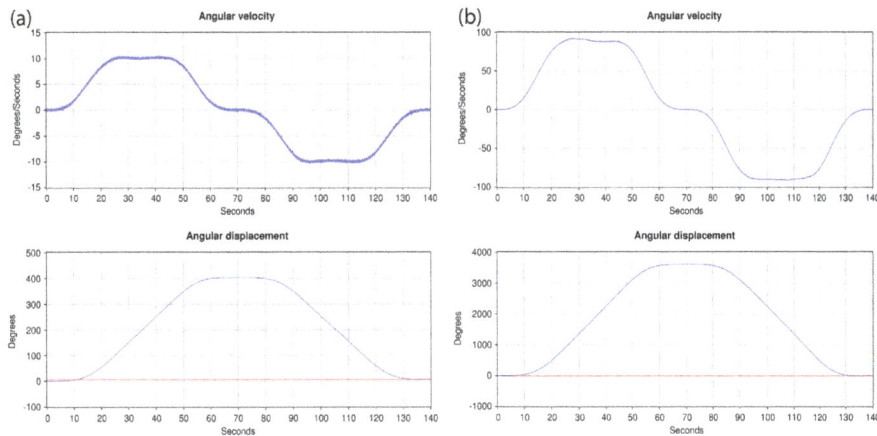

Figure 7. Signals recorded by a MEMS sensor during the rotation test at angular rate of (**a**) 10 and (**b**) 90 ° s^{-1}. The dashed red line indicated the real stop position of the rotating table.

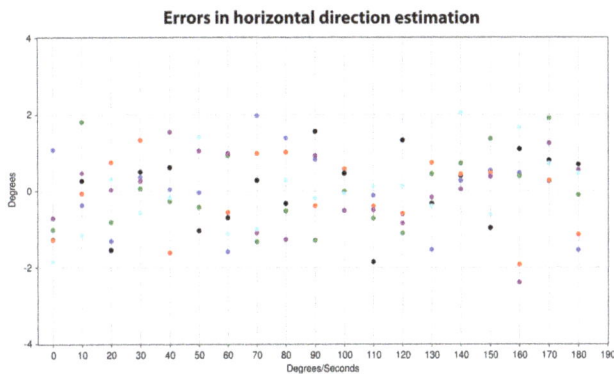

Figure 8. Differences between measured and determined angular displacements as function of the angular rate: black = gyroscope 1, blue = gyroscope 2, red = gyroscope 3, green = gyroscope 4, violet = gyroscope 5, cyan = gyroscope 6.

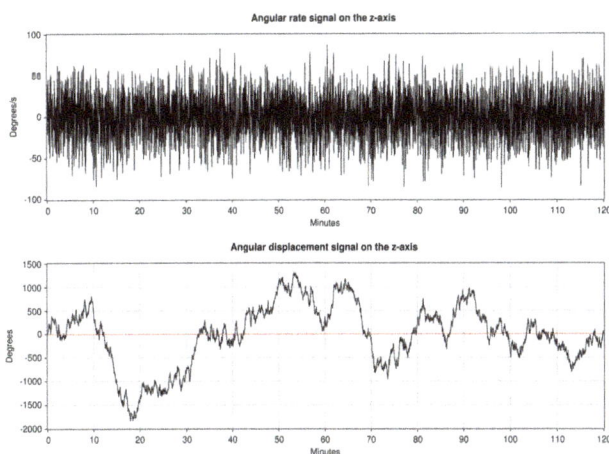

Figure 9. Signals recorded by a MEMS sensor during a random horizontal rotation for a period of 2 h.

a measure of the best accuracy and precision achievable by this instrument when used to derive angular displacements.

The second test was performed in order to verify the ability of the MEMS gyroscope to retrieve angular displacement when it undergoes to a rotation at a given angular velocity. The tests were conducted using a rotating table capable of generating rotation at about constant angular rate. The gyroscopes were subjected, for several tens of seconds, to a rotation on z-axis at constant angular rate, first clockwise and then counter clockwise, to go back to its starting position. The tests were performed generating rotation at angular rate between 0 and 180 ° s^{-1} with step of 10 ° s^{-1}.

Figure 7 show the angular rate and the derived angular displacement signals recorded on the z axis of a MEMS gyroscope subjected to a rotation at angular rate of 10 and 90 ° s^{-1}. Following the angular rate signals of Fig. 7, we can see as from a standing start, the sensor is subjected to a clockwise rotation with increasing angular rate to reach the maximum angular rate (0–20 s), that in the two examples reported in Fig. 7 are 10 and 90 ° s^{-1}. Following the gyroscope is subjected to a clockwise rotation at constant angular rate (20–50 s) and subsequently the angular rate is decreased up to stop the rotation (50–70 s) and the process is symmetrical repeated in counter clockwise (70–140 s).

In this test, the theoretical final angular displacement should be zero, but in reality our rotating table lead the sensor to a position that does not coincide exactly with the initial one (in the figure the final angular displacement is indicated with a red dashed line). For this reason, in each test, the final angular displacement was then measured with a high precision goniometer with angular resolution of 0.1°. For each test, we determined the difference between the final angular displacement measured and that determined from gyroscope signals.

Figure 8 show, for the six tested gyroscope, the differences between the measured and determined angular displacements as function of the angular rate. We can observe as only in 2 cases out of 114 (2 %) the error is greater than 2°, with mean standard deviation equal to 0.93°.

We can therefore say that, for rotations at angular rate between 0 and $180°\,\mathrm{s}^{-1}$, the MEMS gyroscope here tested, can used to determine the final angular displacement with error less than 2° in the 98 % of cases.

However, the OBS falling time is generally greater than 140s. From our experience, the OBS falling velocity is between $35\text{--}55\,\mathrm{m\,min}^{-1}$ and so for deployment in deep water the fall can last more than an hour. For this reason, in order to simulate the descent phase in the sea of an OBS, we have carried out an additional test. In this last test, we applied to each MEMS sensor a random horizontal rotation for a period of 2 h (Fig. 9). The obtained result is in agreement (error less than 2°) with those of the tests previously carried out. The error does not increase, as it is not linked to the accuracy of the measurements, but only to the presence of the instrument self-noise, which can be considered random and stationary.

4 Conclusions

As well known, the magnitude detection threshold and location performance of a seismic network are mainly related to the noise level and geometry of the stations that make up the network (D'Alessandro et al., 2011a, b, 2012b, 2013b, c, d; D'Alessandro and Stickney, 2012; D'Alessandro and Ruppert, 2012; D'Alessandro and D'Anna, 2013). OBS are increasingly used to monitoring offshore areas while downhole sensors always have been used to reduce natural noise level. However, due to sensor rotation during installation, the absolute orientation of the horizontal components are generally unknown. This can be a major problem that can limit data analysis and interpretation. Indeed, the absolute orientation of the horizontal components is critical for many seismic analysis techniques such as receiver functions, body- and surface-wave polarization, anisotropy and surface wave dispersion analysis.

In this paper, we have proposed a simple method to retrieve ocean bottom and downhole seismic sensors horizontal components orientation using a low cost MEMS gyroscope. We have tested the gyroscope included in the MEMS model 1044_0-3/3/3 produced by the Canadian company Phidget Inc, by means of a rotation table. Our test have showed that this MEMS gyroscope can be used to retrieve angular displacement, by means of simple time integration of the angular rate signals. However, to improve the accuracy is necessary to apply an appropriate high-pass filter to remove the low frequency components generated by the device self-noise. A correct processing of the signals permit to retrieve, for rotation at angular rate between 0 and $180°\,\mathrm{s}^{-1}$, the final angular displacement with error less than 2° in the 98 % of cases, also for long duration rotations (2 h).

It is clear that co-installing the gyroscope with a seismic sensor, this can be used to retrieve the absolute sensor horizontal axis orientation. This technology/method is promising, but clearly will require additional work to make operational.

The next step will be the creation of a stand-alone module equipped with single board computer and a battery pack able to record the signals generated by the gyroscope for several days. The module would be down with the sensor and recovered after the installation for downhole and at the end of the monitor campaign for OBS.

Acknowledgements. We are grateful to the anonymous reviewer, Adam Ringler and the Editor Damiano Pesaresi for their constructive comments and suggestions. Special thanks to Chester Fitchett and the Phidgets Inc. (Canada) which provided free of charge the tested devices.

References

Anderson, P. N., Duennebier, F. K., and Cessaro, R. K.: Ocean borehole horizontal seismic sensor orientation determined from explosive charges, J. Geophys. Res.-Solid Earth, 92, 3573–3579, doi:10.1029/JB092iB05p03573, 1987.

Baker, G. E. and Stevens, J. L.: Backazimuth estimation reliability using surface wave polarization, Geophys. Res. Lett., 31, 1–4, doi:10.1029/2004GL019510, 2004.

Chiu, H. C., Huang, H. C., Leu, C. L., and Ni, S. D.: Application of polarization analysis in correcting the orientation error of a downhole seismometer, Earthquake Eng. Struct. Dynam., 23, 1069–1078, 1994.

D'Alessandro, A., D'Anna, G., Luzio, D., and Mangano, G.: The INGV's new OBS/H: analysis of the signals recorded at the Marsili submarine volcano, J. Volcanol. Geotherm. Res., 183, 17–29, doi:10.1016/j.jvolgeores.2009.02.008, 2009.

D'Alessandro, A., Luzio, D., D'Anna, G., and Mangano, G.: Seismic Network Evaluation through Simulation: An Application to the Italian National Seismic Network, B. Seismol. Soc. Am., 101, 1213–1232, doi:10.1785/0120100066, 2011a.

D'Alessandro, A., Papanastassiou, D., and Baskoutas, I.: Hellenic Unified Seismological Network: an evaluation of its performance through SNES method, Geophys. J. Int., 185, 1417–1430, doi:10.1111/j.1365-246X.2011.05018.x, 2011b.

D'Alessandro, A. and Ruppert, N.: Evaluation of Location Performance and Magnitude of Completeness of Alaska Regional Seismic Network by SNES Method, B. Seismol. Soc. Am., 102, 2098–2115, doi:10.1785/0120110199, 2012.

D'Alessandro, A. and Stickney, M.: Montana Seismic Network Performance: an evaluation through the SNES method, B. Seismol. Soc. Am., 102, 73–87, doi:10.1785/0120100234, 2012.

D'Alessandro, A., Mangano, G., and D'Anna, G.: Evidence of persistent seismo-volcanic activity at Marsili seamount, Ann. Geophys., 55, 213–214, doi:10.4401/ag-5515, 2012a.

D'Alessandro, A., Danet, A., and Grecu, B.: Location Performance and Detection Magnitude Threshold of the Romanian National Seismic Network, Pure Appl. Geophys., 169, 2149–2164, doi:10.1007/s00024-012-0475-7, 2012b.

D'Alessandro, A. and D'Anna, G.: Suitability of low cost 3 axes MEMS accelerometer in strong motion seismology: tests on the LIS331DLH (iPhone) accelerometer, B. Seismol. Soc. Am., 103, 2906–2913, doi:10.1785/0120120287, 2013.

D'Alessandro, A., Mangano, G., D'Anna, G., and Luzio, D.: Waveforms clustering and single-station location of microearthquake multiplets recorded in the northern Sicilian offshore region, Geophys. J. Int., 194, 1789–1809, doi:10.1093/gji/ggt192, 2013a.

D'Alessandro, A., Badal, J., D'Anna, G., Papanastassiou, D., Baskoutas, I., and Özel. M. M.: Location Performance and Detection Threshold of the Spanish National Seismic Network, 1859–1880, Pure Appl. Geophys., 170, 1420–9136, doi:10.1007/s00024-012-0625-y, 2013b.

D'Alessandro, A., Gervasi, A., and Guerra, I.: Evolution and strengthening of the Calabrian Regional Seismic Network, Adv. Geosci., 36, 11–16, doi:10.5194/adgeo-36-11-2013, 2013c.

D'Alessandro, A., Scarfì, L., Scaltrito, A., Di Prima, S., and Rapisarda, S.: Planning the improvement of a seismic network for monitoring active volcanic areas: the experience on Mt. Etna, Adv. Geosci., 36, 39–47, doi:10.5194/adgeo-36-39-2013, 2013d.

D'Alessandro, A.: The Marsili Seamount, the biggest European volcano, could be still active!, Current Sci., 106, p. 1339, 2014a.

D'Alessandro, A.: Monitoring of earthquakes using MEMS sensors, Current Sci., 107, 733–734, 2014b.

D'Alessandro, A., Guerra, I., D'Anna, G., Gervasi, A., Harabaglia, P., Luzio, D., and Stellato, G.: Integration of onshore and offshore seismic arrays to study the seismicity of the Calabrian Region: a two steps automatic procedure for the identification of the best stations geometry, Adv. Geosci., 36, 69–75, doi:10.5194/adgeo-36-69-2014, 2014a.

D'Alessandro, A., Luzio, D., and D'Anna, G.: Urban MEMS based seismic network for post-earthquakes rapid disaster assessment, Adv. Geosci., 40, 1–9, doi:10.5194/adgeo-40-1-2014, 2014b.

Duennebier, F., Anderson, P., and Fryer, G.: Azimuth determination of and from horizontal ocean bottom seismic sensors, J. Geophys. Res., 92, 3567–3572, doi:10.1029/JB092iB05p03567, 1987.

Ekström, G. and Busby, R. W.: Measurements of Seismometer Orientation at USArray Transportable Array and Backbone Stations, Seismol. Res. Lett., 79, 554–561, doi:10.1785/gssrl.79.4.554, 2008.

Grigoli, F., Cesca, S., Dahm, T., and Krieger, L.: A complex linear least-squares method to derive relative and absolute orientations of seismic sensors, Geophys. J. Int., 188, 1243–1254, doi:10.1111/j.1365-246X.2011.05316.x, 2012.

Holcomb, G. L.: Experiments in seismometer azimuth determination by comparing the sensor signal outputs with the signal output of an oriented sensor, USGS Open-File Report: 2002-183, pp. 205, 2002.

Mangano, G., D'Alessandro, A., and D'Anna, G.: Long-term underwater monitoring of seismic areas: design of an Ocean Bottom Seismometer with Hydrophone and its performance evaluation, OCEANS 2011 IEEE Conference, 6–9 June, Santander, Spain, in: OCEANS 2011 IEEE Conference Proceeding, pp. 9, doi:10.1109/Oceans-Spain.2011.6003609, 2011.

Michaels, P.: Use of principal component analysis to determine downhole tool orientation and enhance SH-waves, J. Environ. Eng. Geophys., 6, 175–183, 2001.

Nakamura, Y., Donoho, P. L., Roper, P. H., and McPherson, P. M.: Large offset seismic surveying using ocean-bottom seismographs and air gun: instrumentation and field technique, Geophysics, 52, 1601–1611, 1987.

Oye, V. and Ellsworth, W. L.: Orientation of three-component geophones in the San Andreas fault observatory at depth pilot hole, Parkfield, California, B. Seismol. Soc. Am., 95, 751–758, 2005.

Ringler, A. T., Edwards, J. D., Hutt, C. R., and Shelly, F.: Relative azimuth inversion by way of damped maximum correlation estimates, Comput. Geosci., 43, 1–6, 2012.

Ringler, A. T., Hutt, C. R., Persefield, K., and Gee, L. S.: Seismic Station Installation Orientation Errors at ANSS and IRIS/USGS Stations, Seismol. Res. Lett., 84, 926–931, doi:10.1785/0220130072, 2013.

Stachnik, J. C., Sheehan, A. F., Zietlow, D. W., Yang, Z., Collins, J., and Ferris, A.: Determination of New Zealand Ocean Bottom Seismometer Orientation via Rayleigh-Wave Polarization, Seismol. Res. Lett., 83, 704–713, doi:10.1785/0220110128, 2012.

Zha, Y., Webb, S. C., and Menke, W.: Determining the orientations of ocean bottom seismometers using ambient noise correlation, Geophys. Res. Lett., 40, 1–6, doi:10.1002/grl.50698, 2013.

Zheng, X. and McMechan, G. A.: Two methods for determining geophone orientations from VSP data, Geophysics, 71, 87–97, 2006.

16

Repeated erosion of cohesive sediments with biofilms

K. Valentine[1,*], G. Mariotti[1,**], and S. Fagherazzi[1]

[1]Department of Earth and Environment, Boston University, 22015 Boston, USA
[*]now at: Boston College, 02467 Chestnut Hill, USA
[**]now at: Massachusetts Institute of Technology, 02139 Cambridge, USA

Correspondence to: K. Valentine (kendallv@bu.edu)

Abstract. This study aims to explore the interplay between biofilms and erodability of cohesive sediments. Erosion experiments were run in four laboratory annular flumes with natural sediments. After each erosion the sediment was allowed to settle, mimicking intermittent physical processes like tidal currents and waves. The time between consecutive erosion events ranged from 1 to 12 days. Turbidity of the water column caused by sediment resuspension was used to determine the erodability of the sediments with respect to small and moderate shear stresses. Erodability was also compared on the basis of the presence of benthic biofilms, which were quantified using a Pulse-Amplitude Modulation (PAM) Underwater Fluorometer. We found that frequent erosion lead to the establishment of a weak biofilm, which reduced sediment erosion at small shear stresses (around 0.1 Pa). If prolonged periods without erosion were present, the biofilm fully established, resulting in lower erosion at moderate shear stresses (around 0.4 Pa). We conclude that an unstructured extracellular polymeric substances (EPS) matrix always affect sediment erodability at low shear stresses, while only a fully developed biofilm mat can reduce sediment erodability at moderate shear stresses.

1 Introduction

Muddy coastlines are common in macrotidal environments and near large rivers and deltas, hosting highly productive ecosystems (Woodroffe, 2002). The cohesive sediments forming mudflats are constantly reworked by waves and currents giving rise to an ever-changing landscape (deJonge and van Beusekom, 1995; Mariotti and Fagherazzi, 2013; Fagherazzi and Mariotti, 2012). A full understanding of the mechanisms responsible for the erosion of cohesive sediments in mudflats is therefore important for the preservation of these delicate environments.

The purpose of this study is to examine the effects of repeated erosion events on the same cohesive sediment surface under similar conditions, elucidating how biotic and abiotic time-dependent processes affect erodability on a time scale on the order of weeks.

Historically, the main approaches used to study cohesive sediments are laboratory flumes (van Leussen and Winterwerp, 1990; Schieber et al., 2007; Young and Southard, 1978), in situ flumes (Amos et al., 1992; Young and Southard, 1978), field studies (DeVries, 1992), and modeling (Sanford and Maa, 2001). The primary difficulty in studying erosion of cohesive sediments is the vast number of variables that influence erosion, ranging from grain size to ionic charge and sediment composition (Mehta et al., 1989; Schieber et al., 2007). Due to these difficulties, there are many unanswered questions about the elementary processes involved in the erosion of cohesive sediments.

Our experiments have some similarities with those of Winterwerp et al. (1993), in which the same sediments were subjected to several "tidal" cycles of erosion and deposition. In Winterwerp et al. (1993) experiments the bed was deposited from the water column, was highly erodible, suspended sediment concentration reached values up to $10\,\mathrm{g\,L^{-1}}$, and fluid mud formed in the slack water phase. On the contrary, our experiment had placed beds, which are known to be well consolidated and are likely to experience Type II erosion (Winterwerp et al., 2012). As a result, the suspended sediment concentration measured in our experiments never exceeded $1\,\mathrm{g\,L^{-1}}$, reflecting conditions commonly found in sheltered mudflats (Mariotti and Fagherazzi, 2011) and salt marshes (Christiansen et al., 2000).

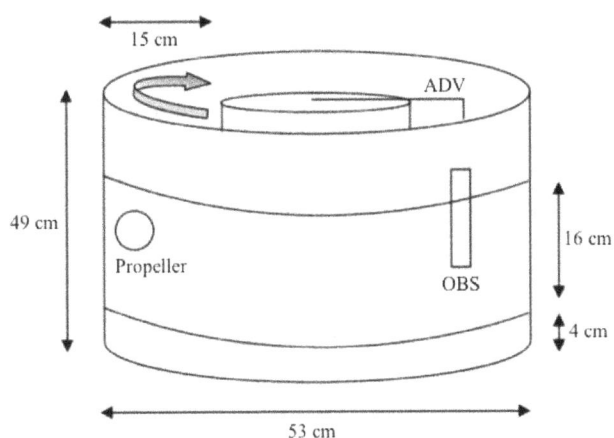

Fig. 1. Annular flume used for erosion experiments. Schematic of the flume apparatus. The flow in the flume is clockwise. The outer walls and the bottom of each flume were durable opaque plastic and the inside walls were glass.

Of increasing interest to the study of erosion of cohesive sediments is the role that biofilms play in physical processes (Paterson et al., 1998; Tolhurst et al., 2003, 2008; Underwood and Smith, 1998). A biofilm of microphytobenthos (MPBs), as used in this study, is a mixed community of microorganisms and their secretions, called extracellular polymeric substances (EPS) that surround sediment particles (Decho, 2000). It has been shown that microbial EPS contributes to grain-grain adhesion in fine sediments (Dade et al., 1990; Tolhurst et al., 1999; Yallop et al., 1994; Paterson, 1995; Taylor et al., 1999). De Brouwer et al. (2002) found no increase in sediment stability in the presence of extracted EPS, while Dade et al. (1990) saw an increase in sediment stability in the presence of extracted bacterial EPS, indicating that even in eroded material, residual EPS could cause grain-grain adhesion. Additionally, studies that examined well developed biofilms reported an increase in erosion threshold and sediment stability (De Brouwer et al., 2000; Tolhurst et al., 2008; Paterson et al., 1998). Maximum erosion thresholds in tidal flats with diatom mats has been measured immediately preceding the maximum biomass of the mat (Stal, 2010), suggesting that the maturity of the diatom mat, in addition to the diatom biomass, can affect the erosion threshold and sediment stability. An experimental study showed that sediments with biofilm had a critical shear stress 5 to 10 times larger than abiotic sediments, while no significant differences were found between a weak and a fully developed biofilm (Neumeier et al., 2006).

To sum up, biofilms potentially stabilize cohesive sediments, but the magnitude of this effect is largely unknown. In particular, it is not clear how the type, structure, physiology, and development stage of the biofilm quantitatively affect erodability (Yallop et al., 1994). Three mechanisms of biogenic stabilization have been identified: network formation by the filamentous cyanobacteria, formation of amorphous organic linkages between non-cohesive sediment grains and accumulation of an EPS matrix (Yallop et al., 1994). While the amount of EPS is likely to be conserved during repeated erosion and deposition events, the biofilm network is destroyed when the biofilm is eroded away, strongly linking biostabilization to hydrodynamic disturbances experienced by the sediments.

Because of the finite time needed for a biofilm to develop, Mariotti and Fagherazzi (2012) suggested both the intensity and frequency of intermittent disturbances, such as tidal currents and wind waves, determine whether the biofilm can approach a fully developed state. This has prompted questions regarding the exposure of cohesive sediments with and without biofilms to repeated stresses, mimicking tidal conditions. Our experiments further explore this dichotomy to see if the sediment itself tends to reach a steady state of erosion after repeated exposure to flow.

This complex interplay between physical and biological processes can provide valuable information as to how mudflats and their ecosystems function in terms of cohesive sediment dynamics.

2 Methods

Annular flumes were designed and constructed based on modifying the design of previous flume studies (Fig. 1, see also Amos et al., 1992; Thompson et al., 2003). A propeller was placed in the flume at the top of the water column, which was 16 cm deep. An Acoustic Doppler Velocimeter (ADV) and an Optical Backscatter Sensor (OBS) were positioned in the flume, on the opposite side of the propeller. The ADV measured the three-dimensional velocity at 32 Hz, sampling a volume 4 cm above the bed, centered in the flume cross section. An Optical Backscatter Sensor was positioned 5 cm below the water surface, and it was calibrated using samples from the water column.

Sediment was collected from a tidal flat in the Rowley River, Massachusetts, USA. Collection took place on days with average temperatures and no precipitation. At low tide, approximately the top 1–2 cm of sediment was removed from the mudflat. The sediment was sieved with a 2 mm mesh (#10) to remove macrofauna. The sediment was 60–65 % clay/silt and 35–40 % sand (d_{50} of coarse fraction $= 313.9\,\mu m$) and can be classified as a sandy cohesive mud. Estuarine water was collected during ebb tide and was later decanted to remove remaining suspended sediment. The sediment was frozen ($-18\,^\circ C$) for 12 h to eliminate excess biota (Ford et al., 1999; Tolhurst et al., 2008). The sediments were then thawed in a warm water bath. Once thawed, the sediment was mixed to make a homogenous slurry. An even layer of sediment (~ 4 cm) was placed at the bottom of each flume. The tanks were then filled with salt water to a height

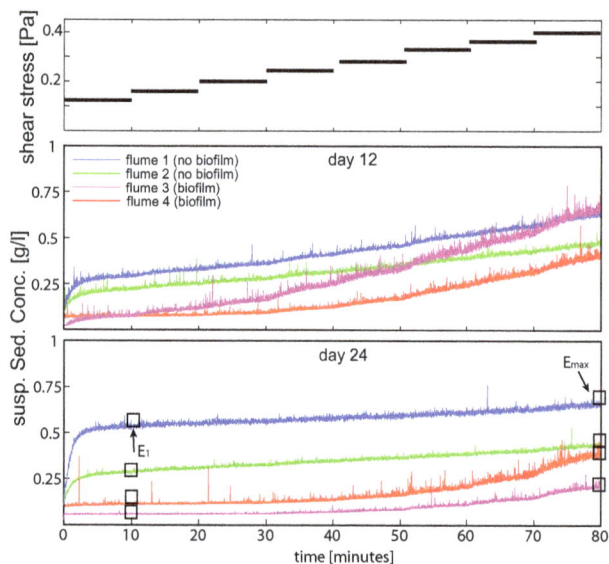

Fig. 2. Example of an erosion experiment. **(A)** Applied shear stresses. **(B, C)** Suspended sediment concentration as a function of time, in two different days. The values of E_1 and E_{max} are indicated by arrows.

of 16 cm. In two control flumes bleach was regularly added to create an abiotic environment (Kim et al., 2008).

Erosion events were simulated at intervals varying between 1 and 12 days. The erosion frequency of 12 days mimics natural spring-neap tidal cycles. Additionally, 12 days is a realistic time interval between wind events that are able to create wave-induced resuspension (D'Alpaos et al., 2013). During each erosion event, the velocity of the propeller was increased every 10 min for 80 min until reaching a maximum value, generating currents ranging from $0.1\,\mathrm{m\,s}^{-1}$ to $0.2\,\mathrm{m\,s}^{-1}$. These velocities relate to shear stresses of approximately 0.1 to 0.4 Pa (Fig. 2a). Then the propeller was removed from the flume and the sediment was allowed to settle until the next erosion event. Hence, our experiments mimic an environment in which sediments are eroded and deposited in the same area, such as a mudflat that is spatially uniform and isolated from external sediment sinks or sources.

Erosion was measured by using the turbidity of the water column, to determine the amount of suspended material. Bed shear stresses were computed with the quadratic stress law, $\tau = C_D \rho U^2$, where U is the total velocity magnitude, ρ is the water density, set equal to $1035\,\mathrm{g\,L}^{-1}$, and C_D is the drag coefficient, set equal to 0.004 (see also Thompson et al., 2003). This method was found to be more reliable than using the Reynolds shear stresses, which were affected by large velocity fluctuations.

In the flumes without bleach, biofilms were cultured on the sediment surface. Fluorescent lamps (Hydrofarm FLT22 2 tube T5, 6400 K) provided a 12 h light/dark cycle for photosynthetic growth of the biofilm that was already present in

the natural sediments. Air temperature was kept within the range of 20–25 °C. Nutrients were regularly added according to the Redfield Ratio to maintain biofilm health. The water in the flumes was filtered using a filter pump to remove algae growing in the water column (Fluval 105 Filter Pump). Filtering was done at a slow rate so that the underlying sediment was not disturbed and after the sediment had settled out of the water column for at least 20 h, so that no sediment nor suspended biofilm fragments were removed.

Biofilm density was determined using a Pulse Amplitude Modulator (PAM) sensor (Diving-PAM Underwater Fluorometer, Heinz Walz, Germany), which is a non-invasive method to determine biofilm growth. The minimum fluorescence yield was used as a proxy for chl a concentration (Schreiber, 2004), which can be linearly related to the EPS concentrations (Buchsbaum et al., 2008). PAM measurements were taken every 24 h, before the erosion events, under constant light conditions and keeping the probe approximately 4 mm above the sediment surface. Each measurement had 30 points, approximately equally distributed over the whole sediment surface, and averaged together. PAM measurements were calibrated to chl a sampling. To measure chl a, samples of approximately 1 g were taken weekly from each flume using a shallow scoop. The samples were then frozen at -18 °C. The chl a from the samples was extracted using acetone and it was measured using a Turner Design Flourometer at 485 nm wavelength (Dalsgaard et al., 2000).

3 Results

Each erosion experiment consists of a time series of suspended sediment concentration. To synthetically describe each erosion experiment we used two parameters: E_{max} and E_1 (Fig. 2). E_{max} is the turbidity output from the final stage in the erosion experiment, and it is interpreted as the sediment erodability at moderate disturbances (0.4 Pa), which are often associated with tidal currents or light storms on mudflats (Mariotti and Fagherazzi, 2013). E_1 is the turbidity output after exposure to the lowest velocity, and it is interpreted as the erodability at low disturbances (0.1 Pa), such as those associated with daily breezes or limited tidal velocities. This value is likely associated with the erosion of the unconsolidated layer of the sediment deposited over a more consolidated bed.

In the bleached flumes (#1 and 2), the chl a concentration remained at a constant low value, indicating that they were indeed abiotic (Fig. 3a, b). In the flume with sporadic erosion events (#3), the biofilm chl a increased during periods without erosion, and decreased significantly after the two erosion events at day 12 and 22 (Fig. 3c). Visual and microscopic analysis revealed the presence of a diatom biofilm in flume 3, and an absence of it in flume 1 and 2. In the flume eroded every other day (#4), there was a gradual increase in

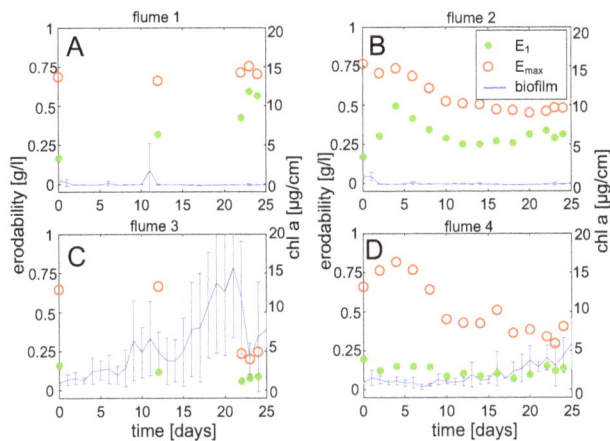

Fig. 3. Erodability (E_{max} and E_1) and chl a during the flume experiments. (**A**) Flume 1, eroded every 12 days. (**B**) Flume 2, eroded every other day. (**C**) Flume 3, eroded every 12 days. (**D**) Flume 4, eroded every other day. Flumes 1 and 2 were bleached.

biomass (Fig. 3d), which however remained much lower than that of flume 3 throughout the entire experiment.

In flume 1, E_{max} was constant for the entire duration of the experiment and E_1 increased with time up to 200 % of the initial value (Fig. 3a). During the erosion at day 12, the difference in E_{max} between flume 1 and 3 was small, but E_1 in the biotic flume was 50 % lower than in the bleached flume. During the erosion at day 22 both E_{max} and E_1 in the biotic experiment were about 20 % of those in the bleached one.

In both flume 2 and 4, E_{max} decreased by about 30 % and reached a similar value after about 10 days (Fig. 3b, d). In flume 4 E_1 remained constant, while in flume 2 E_1 increased by about 100 % in the first 4 days (Fig. 3b, d).

4   Discussion

Sediment consolidation decreased the total erodability (E_{max}) in flume 2. Similarly, consolidation caused the decrease in total erodability in flume 4, since the decrease occurred before the slight increase in biofilm biomass after day 15. The presence of a small amount of biofilm in flume 4 is the reason why E_1 remained relatively low during the experiment. A small amount of biofilm stabilized the freshly deposited material over the more consolidated bed. The presence of a large amount of biofilm in flume 3 promoted a similar stabilization (Fig. 3c), while the presence of bleach removed this effect (Fig. 3a, b). As a result, in the bleached flumes, sediments were highly erodible, even when subjected to extremely low shear stresses. Sediments deposited from the previous erosion event were unable to stick to the underlying bed without biofilm, causing a progressive increase in E_1.

During the erosion at day 12 in flume 3, the presence of biofilm was unable to affect the total erosion E_{max}, which

resulted in a value almost identical to the bleached flume. Hence, the biofilm allowed stabilization with respect to very small disturbances (E_1), but was unable to sustain moderate shear stresses (E_{max}). This observation is in accordance with the large decrease in biofilm biomass after the erosion events, likely caused by detachment of the biofilm. On the other hand, biofilms likely caused the sharp decrease in E_{max} in flume 4 at day 24. In this case the biofilm had enough time to grow and increase its resistance to erosion.

To summarize, biofilm biomass is strongly controlled by erosive events. Frequent disturbances prevent the establishment of a resistant biofilm. Nonetheless, a very weak biofilm is always present, increasing grain-grain cohesion, and promoting stabilization at very small shear stresses (E_1). It is possible that this effect is promoted by the EPS present between the sediment grains, but without a coherent network which uniformly covers the bed and shelter it from the overlying flow. This type of biofilm persists even if the sediments are eroded, since the system is closed and all the eroded material is allowed to settle.

These findings are in agreement with previous research. The grain-to-grain adhesion observed at low values of shear stress agrees with the slight stabilization by non-structured biofilms found by Dade et al. (1990). Additionally, we confirmed that a mature biofilm can largely increase the stability of cohesive sediments (De Brouwer et al., 2000; Tolhurst et al., 2008; Paterson et al., 1998). Our findings support those previously discussed by Yallop et al. (1994) in that the structural state and maturity of the biofilm is critical in determining the magnitude of sediment stabilization.

In order to limit erosion at large shear stresses, the biofilm requires a prolonged period without disturbances to allow its growth, as suggested by previous modeling results (Mariotti and Fagherazzi, 2012). Once the biofilm reaches a critical state, it can significantly reduce the erodability at moderate disturbances (the decrease in E_{max} at day 24 in flume 3). To provide such stabilization, the biofilm needs to form a coherent network and cover the sediment surface.

5   Conclusions

Our experiments demonstrate that biofilm growth is strongly controlled by erosive events. If the biofilm is frequently disturbed, its growth is inhibited. In this case, it is able to offer stabilization only to very small disturbances, possibly stabilizing a thin layer of freshly deposited sediment. On the other hand, if long periods without erosion occur, the biofilm can fully develop and is able to reduce erodability from large shear stresses.

Even if no biofilm is visible on the sediment surface, small-magnitude biofilm stabilization may still occur. Biofilm EPS might periodically erode, deposit, and mix with the sediments, increasing grain-to-grain cohesion. This result emphasizes the need to quantify not only the biofilm biomass, but also its structure in order to understand biostabilization.

Finally, biofilm biostablization may occur at the same time as abiotic time-dependent stabilization, for instance in the consolidation of freshly deposited material, which occurs on time scales of a few weeks. This could possibly complicate the study of biostabilization in the field.

Acknowledgements. We would like to thank W. Fulweiler, A. Vieillard, and T. Viggato for their help. This research was supported by NSF awards OCE-0924287, DEB-0621014 (VCR-LTER program), and OCE-1238212 (PIE-LTER program). We also thank the Boston University UROP program for its generous support. We also thank J. Malarkey, G. Coco, and an anonymous reviewer for their contributions.

References

Amos, C. L., Grant, J., Daborn, G. R., and Black, K.: Sea Carousel – A benthic, annular flume, Estuar. Coast. Shelf S., 34, 557–577, 1992.

Buchsbaum, R. N., Deegan, L. A., Horowitz, J., Garritt, R. H., Giblin, A. E., Ludlam, J. P., and Shull, D. H.: Effects of regular salt marsh haying on marsh plants, algae, invertebrates and birds at Plum Island Sound, Massachusetts, Wetl. Ecol. Manag., 17, 469–487, 2008.

Christiansen, T., Wiberg, P. L., and Milligan, T. G.: Flow and sediment transport on a tidal salt marsh surface, Estuar. Coast. Shelf S., 50, 315–331, 2000.

Dade, W. B., Davis, J. D., Nichols, P. D., Nowell, A. R. M., Thistle, D., Trexler, M. B., and White, D. C.: Effects of bacterial exopolymer adhesion on the entrapment of sand, Geomicrobiol. J., 8, 1–16, doi:10.1080/01490459009377874, 1990.

D'Alpaos, A., Carniello, L., and Rinaldo, A.: Statistical mechanics of wind wave-induced erosion in shallow tidal basins: Inferences from the Venice Lagoon, Geophys. Res. Lett., 40, 3402–3407, doi:10.1002/grl.50666, 2013.

Dalsgaard, T. (Ed.), Nielsen, L. P., Brotas, V., Viaroli, P., Underwood, G., Nedwell, D., Sundback, K., Rysgaard, S., Miles, A., Bartoli, M., Dong, L., Thornton, D. C. O., Ottosen, L. D. M., Castaldelli, G., and Risgaard-Petersen, N.: Protocol Handbook for NICE – Nitrogen Cycling in Estuaries: A Project under the EU Research Programme: Marine Science and Technology (MAST III), National Environmental Research Institute, Silkeborg, Denmark, 2000.

De Brouwer, J. F. C., Bjelic, S., de Deckere, E. M. G. T., and Stal, L. J.: Interaction between biology and sedimentology in a mudflat (Biezelingse Ham Westerschelde the Netherlands), Cont. Shelf Res., 20, 1159–1178, 2000.

De Brouwer, J. F. C., Ruddy, G. K., Jones, T. E. R., and Stal, L. J.: Sorption of EPS to sediment particles and the effect on the rheology of sediment slurries, Biogeochemistry, 61, 57–71, 2002.

Decho, A. W.: Microbial biofilms in intertidal systems: an overview, Cont. Shelf Res., 20, 1257–1273, 2000.

De Jonge, V. N. and van Beusekom, J. E. E.: Wind- and tide-induced resuspension of sediment and microphytobethos from tidal flats in the Ems estuary, Limnol. Oceanogr., 40, 766–778, 1995.

DeVries, J. W.: Field Measurements of the erosion of cohesive sediments, J. Coast. Res., 8, 312–318, 1992.

Fagherazzi, S. and Mariotti, G.: Mudflat runnels: Evidence and importance of very shallow flows in intertidal morphodynamics, Geophys. Res. Lett., 39, L14402, doi:10.1029/2012GL052542, 2012.

Ford, R. B., Thrush, S. F., and Probert, P. K.: Macrobenthic colonisation of disturbances on an intertidal sandflat: the influence of season and buried algae, Mar. Ecol.-Prog. Ser., 191, 163–174, 1999.

Kim, J., Pitts, B., Stewart, P. S., Camper, A., and Yoon, J.: Comparison of the antimicrobial effects of chlorine, silver Ion, and tobramycin on biofilm, Antimicrob. Agents Ch., 52, 1446–1453, 2008.

Mariotti, G. and Fagherazzi, S.: Asymmetric fluxes of water and sediments in a mesotidal mudflat channel, Cont. Shelf Res., 31, 23–36, 2011.

Mariotti, G. and Fagherazzi, S.: Modeling the effect of tides and waves on benthic biofilms, J. Geophys. Res.-Biogeo., 117, G04010, doi:10.1002/jgrf.20134, 2012.

Mariotti, G. and Fagherazzi, S.: Wind waves on a mudflat: The influence of fetch and depth on bed shear stresses, Cont. Shelf Res., 60, S99–S110, doi:10.1016/j.csr.2012.03.001, 2013.

Mehta, A., Hayter, E., Parker, W., Krone, R., and Teeter, A.: Cohesive sediment transport. I: Process description, J. Hydraul. Eng., 115, 1076–1093, 1989.

Neumeier, U., Lucas, C., and Collins, M.: Erodibility and erosion patterns of mudflat sediments investigated using an annular flume, Aquat. Ecol., 40, 543–554, 2006.

Paterson, D. M.: The biogenic structure of early sediment fabric visualized by low-temperature scanning electron microscopy, J. Geol. Soc., 152, 131–140, 1995.

Paterson, D. M., Wiltshire, K. H., Miles, A., Blackburn, J., Davidson, I., Yates, M. G., McGrorty, S., and Eastwood, J. A.: Microbiological mediation of spectral reflectance from intertidal cohesive sediments, Limnol. Oceanogr., 43, 1207–1221, 1998.

Sanford, L. P. and Maa, J. P.-Y.: A unified erosion formulation for fine sediments, Mar. Geol., 179, 9–23, 2001.

Schieber, J., Southard, J., and Thaisen, K.: Accretion of mudstone beds from migrating floccule ripples, Science, 318, 1760–1763, 2007.

Schreiber, U.: Pulse-Amplitude-Modulation (PAM) Fluorometry and Saturation Pulse Method: An Overview, in: Chlorophyll a Fluorescence, edited by: Papageorgiou, G. C. and Govindjee, P., Springer, the Netherlands, 279–319, 2004.

Stal, L. J.: Microphytobenthos as biogeomorphological force in intertidal sediment stabilization, Ecol. Eng., 36, 236–245, 2010.

Taylor, I. S., Paterson, D. M. Y., and Mehlert, A.: The quantitative variability and monosaccharide composition of sediment carbohydrates associated with intertidal diatom assemblages, Biogeochemistry, 45, 303–327, 1999.

Thompson, C. E. L., Amos, C. L., Jones, T. E. R., and Chaplin, J.: The manifestation of fluid-transmitted bed shear stress in a smooth annular flume-a comparison of methods, J. Coast. Res., 19, 1094–1103, 2003.

Tolhurst, T. J., Black, K. S., Shayler, S. A., Mather, S., Black, I., Baker, K., and Paterson, D. M.: Measuring the in situ shear stress of intertidal sediments with the Cohesive Strength Meter (CSM), Estuar. Coast. Shelf S., 49, 281–294, 1999.

Tolhurst, T. J., Jesus, B., Brotas, V., and Paterson, D. M.: Diatom migration and sediment armouring – an example from the Tagus Estuary, Portugal, Hydrobiologia, 503, 183–193, 2003.

Tolhurst, T. J., Consalvey, M., and Paterson, D. M.: Changes in cohesive sediment properties associated with the growth of a diatom biofilm, Hydrobiologia, 596, 225–239, 2008.

Underwood, G. J. C. and Smith, D. J.: Predicting epipelic diatom exopolymer concentrations in intertidal sediments from sediment chlorophyll a, Microb. Ecol., 35, 116–125, 1998.

van Leussen, W. and Winterwerp, J. C.: Laboratory experiments on sedimentation of fine-grained sediments: A state-of-the-art review in the light of Experiments with the Delft Tidal Flume, in: Residual Currents and Long-term Transport, edited by: Cheng, R. T., Springer, New York, USA, 241–259, 1990.

Winterwerp, J. C., Cornelisse, J. M., and Kuijper, C.: A laboratory study on the behavior of mud from the Western Scheldt under tidal conditions, in: Nearshore and Estuarine Cohesive Sediment Transport, edited by: Mehta, A. J., American Geophysical Union, Washington D.C., USA, 295–313, 1993.

Winterwerp, J. C., van Kesteren, W. G. M., van Prooijen, B., and Jacobs, W.: A conceptual framework for shear flow–induced erosion of soft cohesive sediment beds, J. Geophys. Res.-Oceans, 117, 2156–2202, doi:10.1029/2012JC008072, 2012.

Woodroffe, C. D.: Coasts: form, process and evolution, Cambridge University Press, Cambridge, United Kingdom, 2002.

Yallop, M. L., de Winder, B., Paterson, D. M., and Stal, L. J.: Comparative structure, primary production and biogenic stabilization of cohesive and non-cohesive marine sediments inhabited by microphytobenthos, Estuar. Coast. Shelf S., 39, 565–582., 1994.

Young, R. N. and Southard, J. B.: Erosion of fine-grained marine sediments: sea-floor and laboratory experiments, Geol. Soc. Am. Bull., 89, 663–672, 1978.

17

Significant technical advances in broadband seismic stations in the Lesser Antilles

A. Anglade[1], A. Lemarchand[3], J.-M. Saurel[2], V. Clouard[2], M.-P. Bouin[1,3], J.-B. De Chabalier[1,3], S. Tait[3], C. Brunet[3], A. Nercessian[3], F. Beauducel[3], R. Robertson[4], L. Lynch[4], M. Higgins[4], and J. Latchman[4]

[1]Observatoire Volcanologique et Sismologique de Guadeloupe (OVSG/IPGP), Le Houëlmont
97113 Gourbeyre, Guadeloupe, French West Indies
[2]Observatoire Volcanologique et Sismologique de Martinique (OVSM/IPGP), Morne des Cadets,
97250 Fonds Saint Denis, Martinique, French West Indies
[3]Institut de Physique du Globe de Paris (IPGP), Paris, France
[4]Seismic Research Centre (SRC/UWI), St. Augustine, Trinidad and Tobago, West Indies

Correspondence to: A. Lemarchand (arnaudl@ipgp.fr)

Abstract. In the last few years, French West Indies observatories from the Institut de Physique du Globe de Paris (IPGP), in collaboration with The UWI Seismic Research Centre (SRC, University of West Indies), have modernized the Lesser Antilles Arc seismic and deformation monitoring network. 15 new, permanent stations have been installed that strengthen and expand its detection capabilities. The global network of the IPGP-SRC consortium is now composed of 20 modernized stations, all equipped with broadband seismometers, strong motion sensors, Global Positioning System (GPS) sensors and satellite communication for real-time data transfer. To enhance the sensitivity and reduce ambient noise, special efforts were made to improve the design of the seismic vault and the original Stuttgart shielding of the broadband seismometers (240 and 120s corner period). Tests were conducted for several months, involving different types of countermeasures, to achieve the highest performance level of the seismometers. GPS data, realtime and validated seismic data (only broadband) are now available from the IPGP data centre (http://centrededonnees.ipgp.fr/index.php?&lang=EN). This upgraded network feeds the Caribbean Tsunami Warning System supported by UNESCO and establishes a monitoring tool that produces high quality data for studying subduction and volcanic processes in the Lesser Antilles arc.

1 Introduction

Following the submarine earthquake of Sumatra on 26 December 2004 and the subsequent devastating tsunami, UNESCO has been orchestrating activities and immediate action to establish a tsunami and other coastal hazards Early Warning System (EWS) in the Caribbean and Adjacent Regions. The immediate response included the establishment of an Interim Tsunami Advisory Information Service to the Caribbean Sea and Adjacent Regions through the Pacific Tsunami Warning Center in Hawaii. Meanwhile, since 2005, the Intergovernmental Oceanographic Commission of UNESCO (IOCUNESCO) has coordinated Intergovernmental Coordination Group (ICG) discussions and working groups to address all kinds of issues with a view to establish a fully functional Tsunami Warning System in the Caribbean Region (ICG/CARIBE EWS). In this context, existing seismic network operators of the Caribbean countries have been meeting in regular technical workshops for many years to agree the specifications of seismic monitoring stations that contribute to the Tsunami Warning System.

Two major thrust earthquakes killed several hundred people and destroyed cities during historical times: the 1843 M 8.5 earthquake and the 1839 M 8.0 earthquake (e.g., Bernard and Lambert, 1988; Shepherd, 1992; Feuillet et al., 2011; Hough, 2013). In 1843, evidence of tsunamis was reported in Antigua and Nevis, and the oldest earthquake known also triggered a tsunami in 1690 as the sea withdrew over a distance of 200 m at Charleston in Nevis (Lander et al., 1997). Nowadays, the high level of coastal development/tourist infrastructure is critical to the economies of the Caribbean islands states which are vulnerable to several major telluric hazards such as landslides, volcanic eruptions, and especially earthquakes and tsunamis. Observing, understanding and monitoring these hazards can help mitigate their impact. Indeed, comparison of damage caused by the recent massive earthquakes and tsunamis that shook the subduction zones of Sumatra ($M = 9.3$, 2004; e.g. Singh et al., 2011) and Japan ($M = 9.1$, 2011; e.g. Mori et al., 2011), reveals a difference of approximately one order of magnitude in the scale of the loss of life, mostly attributable to disparate observational capacity, levels of scientific understanding and preparedness in the two regions.

The Lesser Antilles subduction zone spans over 1000 km from Trinidad to the Virgin Island and the regional seismicity has been recorded and processed since the mid-1950's by the SRC (http://www.uwiseismic.com), based in Trinidad and Tobago and since 1980, by the two French IPGP West Indies Volcanological and Seismological Observatories (OVS), located in Martinique and Guadeloupe. Those centres were originally dedicated to volcano monitoring, and used to operate local networks producing their own earthquake solutions for events in the vicinity of French islands, but even though they have been contributing with arrivals from regional events to the SRC since 1953, recording the seismicity of the Lesser Antilles arc was not the main goal. Most of the stations in operation since the beginning of the seismologic observations in the Lesser Antilles have been short-period, vertical component seismic sensors, with analog telemetry.

In 2006 and 2007, KNMI (Koninklijk Nederlands Meteorologicsh Institut), USGS (United States Geological Survey) and SRC committed themselves to update their seismic networks with digital broadband stations (Fig. 1) and real-time communications to meet the tsunami warning requirements. However, these modernized stations, while greatly improving regional monitoring capability, did not cover the whole Antilles arc (French West Indies and Grenada Islands). Hence around 2008, IPGP began modernization of Guadeloupe and Martinique seismic networks, in collaboration with the Seismic Research Centre (University of West Indies-Trinidad) and initiated data sharing in real-time via satellite communications (VSAT). In 2010, IPGP and SRC had the opportunity to further fill gaps in the geometry of the seismic network of the Antilles arc with four additional stations installed in Cariaccou (Grenada), Saint-Lucia, Dominica and Antigua, but also to strengthen the network by adding several radomes, notably providing protection from severe weather. Moreover IPGP and SRC paved the way to a new strategy regarding the installation of regional seismic broadband stations to achieve high scientific standards regarding data quality. Thus best practice in observational seismology (Trnkoczy et al., 2011; Forbriger, 2012) and GPS networks (Sakic, 2013; UNAVCO) has been implemented to achieve performance goals and special care was taken in the construction of the seismic vault

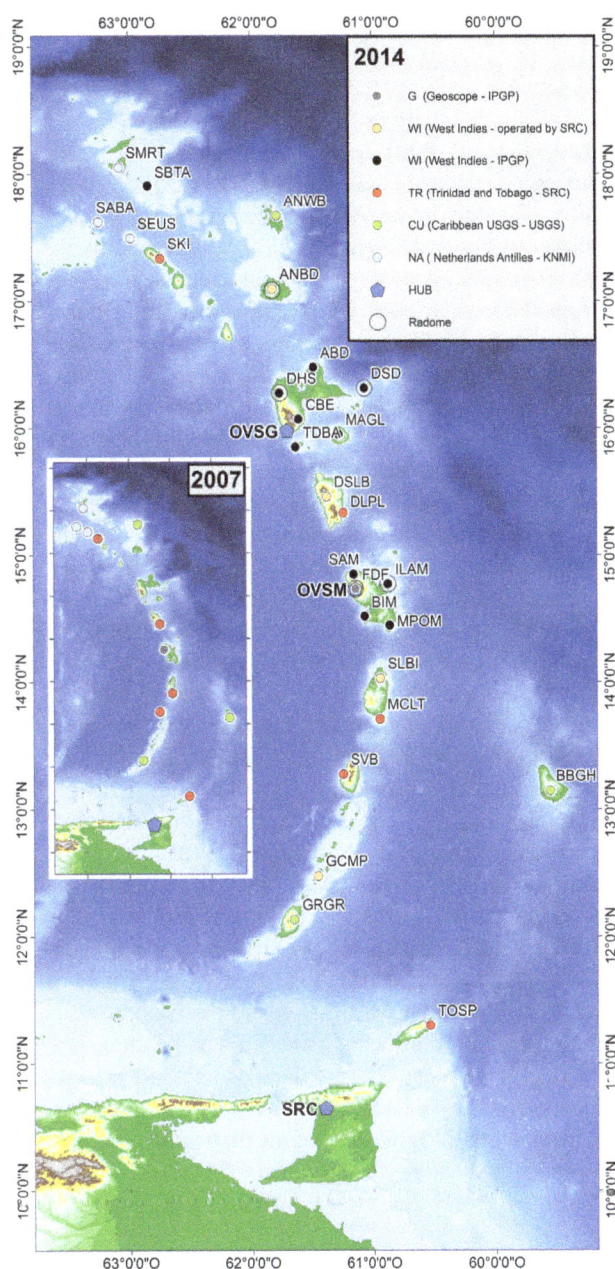

Figure 1. Digital broadband seismic stations in the Lesser Antilles in 2007 and 2014, from IPGP (France, WI and G), KNMI (Netherlands, NA), USGS (United States of America, CU) and SRC (Trinidad and Tobago, TR and WI).

and sensor shielding with systematic tests carried out. Finally, SRC and the two French West Indies observatories designed, installed and now jointly operate an arc scale seismic network in Lesser Antilles. This paper, describes: (i) the station distribution within the context of existing regional networks and the design strategy of the network, (ii) the global characteristics of stations, (iii) strategies used to improve data quality, (iv) results achieved with this network.

2 The WI network

The WI network was installed by SRC and IPGP between 2008 and 2014. It is composed of stations geographically distributed all along the Lesser Antilles subduction zone, from Grenadine islands in the south, to St Barthélémy in the north (Fig. 1), and from Eastern Guadeloupe in the west to La Désirade in the east. Together with the other networks (CU, NA, G and TR), the WI stations complete the coverage of the subduction zone. The Caribbean is subject to hurricanes, earthquakes and volcanic events, therefore, measures were employed to promote system robustness and reliability; damage to any one component does not lead to complete network failure. Data are collected in real-time by three VSAT hubs located in Guadeloupe, Martinique and Trinidad, at the OVSG/IPGP, the OVSM/IPGP and at the SRC respectively. The VSAT technology allows us to share the data without the use of any terrestrial link or Internet access, both of which are vulnerable in case of natural disasters. The system is designed to send data from any one station to the three hubs. Such redundancy promotes data recovery, in the event of failure of up to two of the three hubs. We chose to cover four stations and one hub with a radome that can resist winds of up to 300 kph, thus protecting a minimal network for early warning and strong earthquake detection with winds up to 210 kph. Those stations are: (a) ANBD in Antigua in the northern part of the arc, where most of the hurricanes go, (b) DSD in La Désirade and (c) DHS in Deshaies, which form the largest possible east-west line in the arc, (d) and ILAM in la Caravelle which is the closest land to the subduction zone. There is now at least one broadband station in each major island of the arc and along an east-west line where the islands arc is the broadest, making the subduction zone well instrumented and allowing large-scale studies of the processes at work and the output of enhanced data products (e.g. focal mechanisms, source characterization). Because of the particular elongated shape of the islands along the arc, this network can also be viewed as a large seismometer antennae for global studies.

3 Station design

The stations all share the following features: (1) very Small Aperture Terminal (VSAT) satellite telemetry, (2) solar power and 10-days battery autonomy to ensure high reliability, (3) multi sensors stations with broadband seismometer, accelerometer and continuous GPS, (4) vault and installation design to minimize environmental effects on the instruments (Fig. 2). They share with the TR network the same VSAT satellite telemetry and they have the same kind of solar power source.

In order to sample the whole range of expected tectonic movements in a subduction region: from very-long term slow movements to strong high frequency shaking, each station is equipped with three sensors. GPS sensors measure long-term

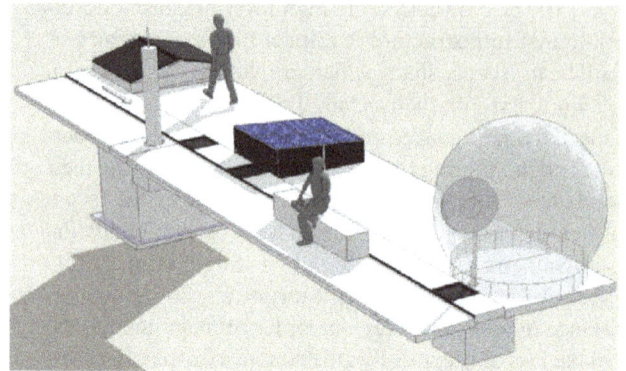

Figure 2. Perspective view of a VSAT station with, from top to bottom; the seismic vault, the GPS monument, the solar power station and the VSAT antenna under a radome (by courtesy of Atelier David Besson-Girard Paysagiste, Paris).

displacement; broadband seismometers (from 50 Hz up to 120s or 240s in the flat band) record earthquakes from magnitude 3 and force feedback, ±2 G accelerometers record, without being saturated, strong earthquakes, including any that might occur close to the station. Stations are deployed in quiet environment as far from anthropogenic noise as possible.

The solar system that powers the station is designed to provide 60 Watts continuously for at least 10 days without sunlight. This robust solar system is comprised of two distinct sub-systems, each with the capacity to deliver a minimum power of 30 Watts in case one element fails. Four stations have their VSAT antenna protected with a radome, which can withstand hurricane winds up to 300 kph, thus maintaining a minimal network with location capability for moderate to large magnitude earthquakes, under such conditions. The geodetic GPS antenna is located on top of a two-meter pedestal with a 1 cubic meter concrete foundation to promote long-term stability. Care has been taken in the design of the top of the pedestal in order to remove multi-path reflections by employing a dome shape. The geodetic sensor at some stations is shared with the COCONet network (UNAVCO project, EAR-1042906/9, http://coconet.unavco.org/coconet.html) and those were installed by UNAVCO in accordance with their standards. Both seismometer and accelerometer are installed in a 2 m deep vault, on a common seismic pier. One station is installed in a 4 m borehole drilled in hard rock.

4 Vault design

The vault design (Fig. 3) addresses several needs: cost, environmental insulation, ground coupling, instrument security, available land (Saurel et al., 2012). A drain pipe was installed in each vault to reduce the risk of the vault being flooded. Vaults have been installed, insofar as possible, at the opposing side of the site from the VSAT antennae, in order to min-

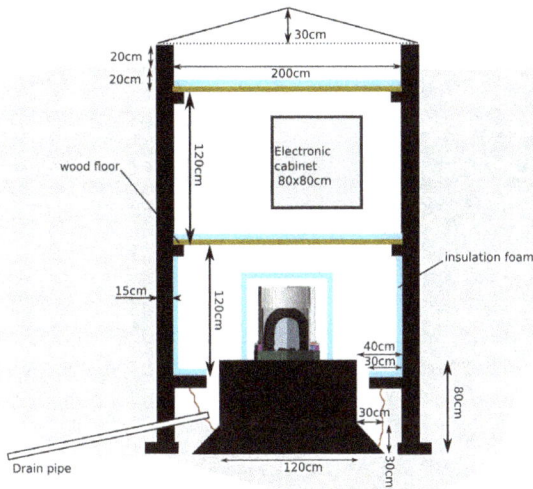

Figure 3. Seismic vault cross section, dimensions are in centimeters. The solid black represents the concrete structure, the solid blue parts are insulation foam and the brown lines represents the wooden floors.

imize any transmission of vibrations from the antennae mast to the instruments. The seismic pier, which can accommodate a broadband seismometer, with its insulation, and an accelerometer, is mechanically isolated from the vault walls. Thermal insulation and ground coupling are both achieved by excavating 2 m of soil. When possible, the seismic pier is directly grouted on to the rock basement. When rock was not available, one cubic meter of concrete was poured before the seismic pier was erected. The sensors are separated from the electronics in order to avoid any electric, magnetic and thermal mutual influence. This also allows maintenance of the electronics with minimal disturbance to the sensors' thermal environment. The vault is further insulated with 6 cm thick Styrofoam panels covering every floor. Additionally a box, in the same material, covers the seismic sensor.

5 CASSIS shielding

Changes in barometric pressure contribute significantly to seismometer noise level (Zürn and Widmer-Schnidrig, 1995; Beauduin et al., 1996). While there is little to be done to compensate for the local Earth induced tilt, the buoyancy effect on seismometer masses can be attenuated by proper shielding. The original Stuttgart shielding (Widmer-Schnidrig and Kurrle, 2006) has been proven to decrease the vertical noise level of broadband sensors by several dBs, without degrading performance of the horizontal axes. IPGP engineers from Geoscope (Stutzmann et al., 2000) and the volcanic observatory team added some innovative features to this initial design. This is the CASSIS shielding (CASserole SIS-mologique – seismic cooking pot) which is still based on the granite slab and "cooking pot" pair (Fig. 4). The granite base

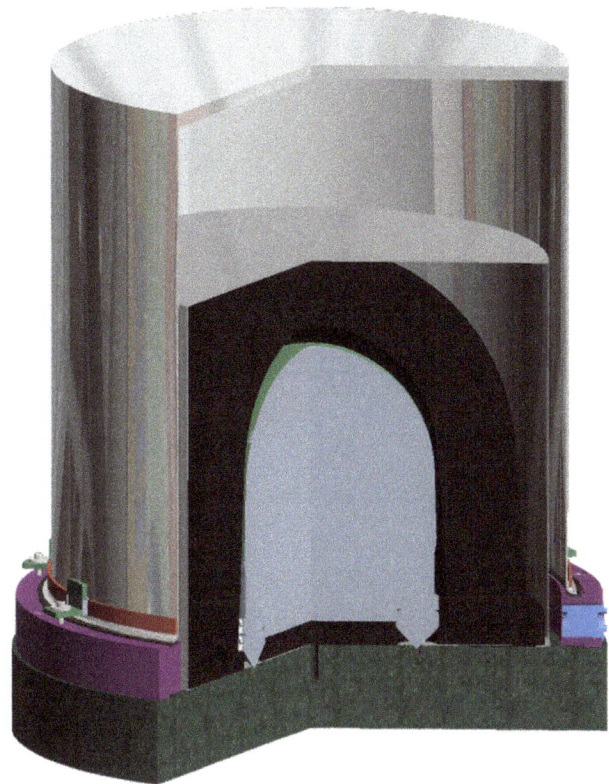

Figure 4. CASSIS seismometer insulation with, from the inside to the outside; the Trillium seismometer, the foam cover, the mu-metal cover, the stainless steel cover bolted to the granite base plate. The cable goes out through the blue joint on the right hand side of the schematic.

plate is engraved with a North-South indication, along with the positions for three different sensors: Streckeisen STS2 family, Nanometrics Trillium120PA and Nanometrics Trillium240. The design may be adjusted to host other seismometers once their footprint is known. A metal ring (either aluminium or stainless steel), with a neoprene gasket, is bolted to the granite plate. This allows the use of the original sensor cables through a hole, which is sealed afterwards with a compressed rubber gasket. This decreases the number of connectors on the electrical signal path, which enhances performance and is always a good measure for long-term installations. A foam cover fits the sensor and provides thermal insulation via a thin air gap that also suppresses thermal convection currents around the sensor. Finally, a stainless steel "cooking pot" is bolted and sealed to the metal ring, providing proper barometric pressure shielding. This barometric shielding is now in use in the Geoscope network and replaces the original Stuttgart design.

Extensive tests were conducted on this shielding in Martinique Geoscope vault (FDF station), which hosts an STS2 shielded with the original Stuttgart design. The Trillium120PA and Trillium240 sensors were tested. It was

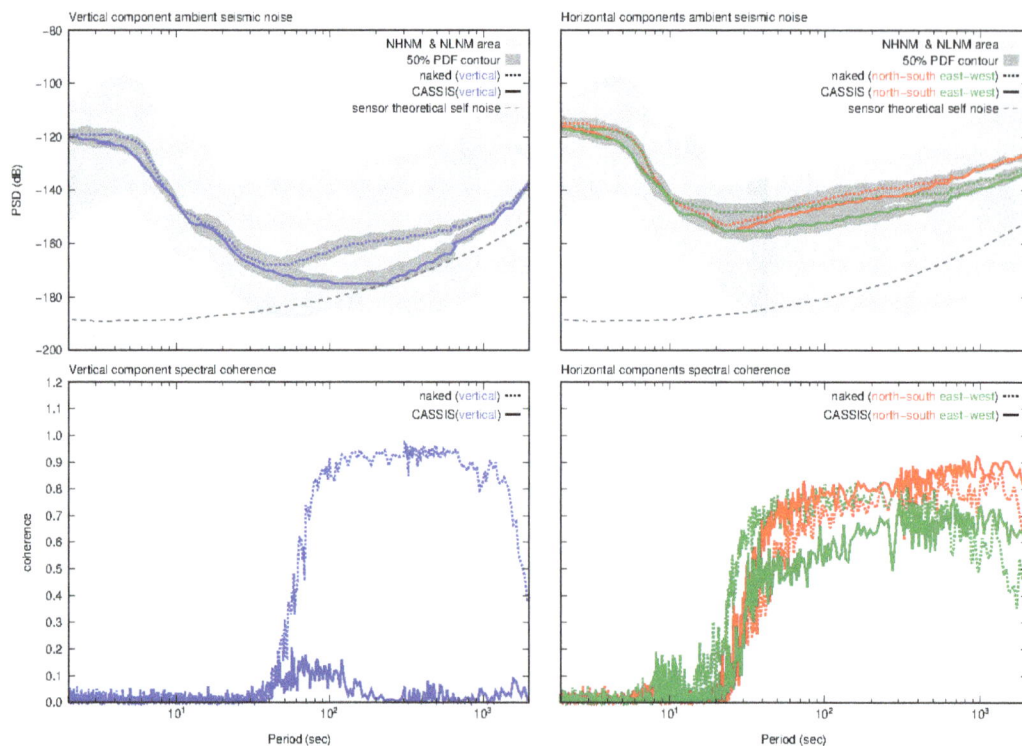

Figure 5. Influence of the CASSIS barometric shielding on Trillium120PA noise level. CASIS has a significant impact on the vertical component noise level (blue). The spectral coherence plots confirms the barometric pressure origin of the noise. Top-left figure compares vertical noise level of the shielded (solid lines) and naked (dotted lines) seismometers, with the 50 % PDF contour (dark shaded area). Idem for horizontal components in top-right figure. Bottom-left figure compares vertical spectral coherence between velocity and barometric pressure of the shielded (solid lines) and naked (dotted lines) seismometers. Idem for horizontal components in bottom-right figure. Barometric and seismic signals were down-sampled to 1 sps, and high pass filtered at 0,2 mHz before the spectral coherence computation. Ambient noise plots combines raw 1 and 20 sps data. Thick dotted gray line represents the theoretical self noise level of the seismometer, according to Nanometrics (P. Devanney, personal communication, 2012) and the light shaded area the NHNM and NLNM boundaries (Peterson, 1993).

found that the barometric shielding for the Trillium240 did not yield significant changes in noise levels. It is considered that a very rigid sensor housing may reduce sensitivity to barometric pressure, which would explain the results obtained on this sensor. The shielding did not alter instrument performance, which was maintained on all axes. The noise level on the Trillium120PA was improved by the CASSIS shielding, up to 10 dBs in the 100s range (Fig. 5, upper plots). We calculated spectral coherence between the seismic sensor signals converted to velocity and barometric pressure as recorded by the Geoscope microbarometer (Fig. 5, bottom plots). The results clearly show that, within the 40s to 1000s range, the coherence between the Trillium120PA signals and the barometric pressure is very high for all the axes. When installing the sensor under the CASSIS shielding, the same plot shows almost no coherence in that same period range for the vertical component. The remaining coherence on the horizontal component might be due to tilt and not to a direct pressure effect on the masses. Those results are consistent with the TANK shielding used in Observatoire de Grenoble

(OSUG) seismic network, which was inspired by CASSIS design (Langlais et al., 2013).

Finally, efforts were made to reduce another type of external influence, observed in 2007 at BFO (Black Forest Observatory) (Forbriger, 2007; Forbriger et al., 2010) on very broadband sensors, including an early design of Trillium240. Wielandt (2007) stated in a Nanometrics Trillium240 test report, conducted in 2007, that this new sensor was more sensitive to terrestrial magnetic fields than the reference STS2. But when removing this influence using active coils around the seismometer, he was able to demonstrate a noise level at least on a par with that of the STS2. Therefore, a mu-metal cover was designed to damp magnetic field variations by 100 dB between 50s and 600s, which is the range of the strongest natural variations during magnetic storms. During the tests, significant solar eruptions occurred that generated perturbations of the Earth magnetic field (Kp index of 80 for the whole day). The records from the closest Intermagnet station in Puerto Rico were compared with the signals recorded by the three very broadband sensors in use on the Geoscope pier at the time : the reference Geoscope STS2,

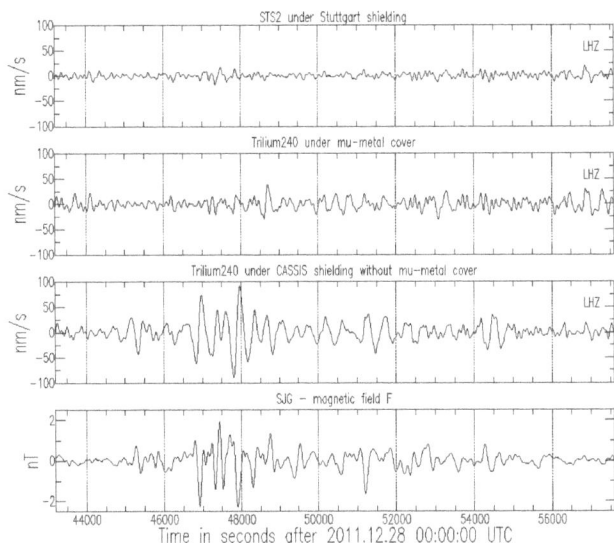

Figure 6. From top to bottom, Geoscope STS2 with Stuttgart shielding which has a low sensitivity to magnetic fields, Trillium240 with mu-metal cover which proves to reduce the magnetic field influence, Trillium240 with CASSIS shielding without mu-metal cover showing the influence of Earth magnetic field and San Juan (Puerto Rico) filtered value of Earth magnetic field. All signals were filtered between 1 and 10 mHz, and seismic recording were deconvolved to velocity.

a Trillium240 under the CASSIS shielding only and a Trillium240 under the mu-metal cover only. Figure 6 shows a 4-hour record of the four signals. Because the closest magnetic field record we could use is from San Juan (Puerto Rico) station, it may include some local effects, either due to a different geology or local sources. Similarly, local disturbances in Martinique might have produced some effects on the seismometer records. This explains why some parts of the magnetic signal differ with those on the seismic signal. Vertical seismic components are deconvolved to velocity. All signals are filtered between 100s and 1000s. We can clearly see the impact of the mu-metal cover: near complete removal of the magnetic influence on the Trillium240 was achieved. As a result, this shielding has been installed on all Trillium240 sensors in the network.

6 Network performance

These measures have led to a good level of performance across the network, homogeneous at the arc scale. Figure 7 shows the noise performance of 6 Trillium240 and 6 Trillium120PA stations installed in Guadeloupe, Martinique, Carriacou, Dominica, Antigua and Saint-Lucia, during November 2014. The stations PSDs were calculated and merged with PQLX software (McNamara and Boaz, 2005). Then, PDFs median noise and their 50, 80 and 90 % contours were calculated. Despite very different siting condi-

tions (rock, clay, natural soil, volcanic deposits), the vertical PDF median noise level above 10s is close to the NLNM (New Low Noise Model, Peterson, 1993). It is also important to highlight that the stations are installed on islands and are never more than 50 km from the shore and the breaking waves of the Atlantic Ocean. The small 50 and 80 % contours, for most of the period range, show that the care taken in station site selection, vault infrastructure and sensor shielding design insulate the sensors from environmental influences and was, therefore, worthwhile. The Trillium120PA plot, shows a similar noise curve to that of the test in Fig. 5, and demonstrates that the CASSIS shielding works as expected in the field . At shortest periods, the difference between the stations is more apparent with larger PDF contours. This is normal since high frequency noise is mainly caused by local sources and thus more site specific. Lastly, both Trillium120PA and Trillium240 subsets of stations show an overall PDF median noise close to their respective sensor theoretical self noise level.

7 Conclusions

This redundant and robust regional network now allows homogeneous earthquake location in the region and a minimal level of monitoring and early warning in the event of major natural events without relying on terrestrial data transmission links. The completion of this network makes high quality broadband seismic data from the Lesser Antilles subduction zone available. It should lead to improved understanding of the tectonic processes at work in this region and allow for improved hazard assessments. This network meets its main three objectives, allowing accurate location of earthquakes in this zone, providing real-time data for the CARIBE-EWS with a high level of availability and offering quality data sets for researchers to understand Lesser Antilles subduction. This network deployment shows how important it is to invest in station infrastructures to achieve high quality data and a reliable and resilient network.

Acknowledgements. We would like to thank everyone, from Trinidad and Tobago, Paris, Martinique and Guadeloupe, involved in the different projects, spanning many years, that led to the completion of our network. In particular: SRC team, who pioneered the installation of VSAT broadband stations in the region, board of direction of IPGP and Sara Bazin, who fully supported the Lesser Antilles network project, Christian Anténor-Habazac and his colleagues from Guadeloupe observatory who installed the first VSAT station in a French territory, Tanguy Maury and his colleagues from Martinique observatory for their efforts during the VSAT station installations in Martinique. We are also grateful to all the funding French agencies without which this project would not have been possible: VSAT stations in Martinique (2), Saint Lucia (1), Carriacou (1), Dominica (1), Antigua (1) were co-fund by the European INTERREG Tsuareg project, lead in Martinique. The hub of Martinique was fund by the Research Ministry, VSAT stations

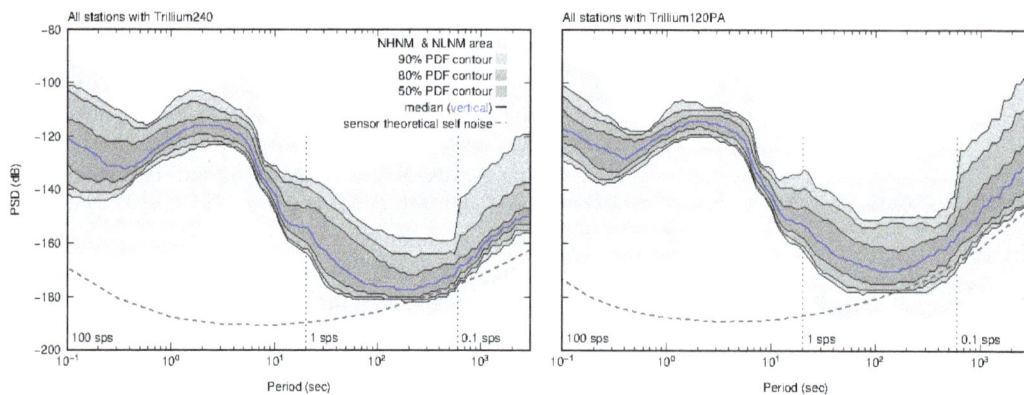

Figure 7. PDF plots of 6 Trillium240 and 6 Trillium120PA stations in the Lesser Antilles during the same time period (from the 27 October to the 26 November), showing the PDF median line, 90, 80 and 50 % PDF contours for the vertical ambient noise. CASSIS shielding improvement, as shown in Fig. 5, can be seen on the whole network, for both seismometers. Thick dotted gray line represents the theoretical self noise levels of the seismometer, according to Nanometrics (P. Devanney, personal communication, 2012). The light shaded area represents the NHNM and NLNM boundaries (Peterson, 1993). At long periods, the contours are affected by the small number of time windows used for the PDF estimation, reducing the precision of the PSD. Vertical dashed lines highlight the three sample rates used for PSD computation in three period domains. Despites varied local conditions, we can note how close the median noise and its lower contours are to the sensor theoretical self noise between 30s and 1000s. There is more variations at short periods, as the noise is mainly affected by local sources (wind, breaking waves, human activities) which are hardly attenuated by any insulation.

in Martinique (2) by the French Environment Ministry, the VSAT stations and the hub of Guadeloupe (7) by CPER PO Guadeloupe funds, co-funded by the French Ministry of Research, the French Ministry of Environment, the Guadeloupe Regional Council and IPGP. We thank all the local communities and authorities for their support, the private owners or the local authorities that made land available for the station installation in their territories. The results presented in this paper rely on the data collected at the San Juan geomagnetic observatory, Puerto Rico. We thank USGS, for supporting its operation and INTERMAGNET (Kerridge, 2001) for promoting high standards of magnetic observatory practice (www.intermagnet.org).

References

Beauduin, R., Lognonné, P., Montagner, J. P., Cacho, S., Karczewski, J. F., and Morand, M.: The effects of atmospheric pressure changes on seismic signals or how to improve the quality of a station, B. Seismol. Soc. Am., 86, 1760–1769, 1996.

Bernard, P. and Lambert, J.: Subduction and seismic hazard in the northern Lesser Antilles: Revision of the historical seismicity, B. Seismol. Soc. Am., 78, 1965–1983, 1988.

Feuillet, N., Beauducel, F., and Tapponnier, P.: Tectonic context of moderate to large historical earthquakes in the Lesser Antilles and mechanical coupling with volcanoes, J. Geophys. Res., 116, B10308, doi:10.1029/2011JB008443, 2011.

Forbriger, T.: Reducing magnetic field induced noise in broad-band seismic recordings, Geophys. J. Int. 169, 240–258, 2007.

Forbriger, T., Widmer-Schnidrig, R., Wielandt, E., Hayman, M., and Ackerley, N.: Magnetic field background variations can limit the resolution of seismic broad-band sensors, Geophys. J. Int., 183, 303–312, 2010.

Forbriger, T.: Recommendations for seismometer deployment and shielding. Information Sheet IS 5,4, in: New Manual of Seismological Observatory Pratice (NMSOP-2), edited by: Bormann, P., 2nd Edn., IASPEI, GFZ german Research Center for Geoscience, Potsdam, 2012.

Hough, S. E.: Missing great earthquake, J. Geophys. Res., 118, 1098–1108, 2013.

Kerridge, D.: INTERMAGNET: WorldWide Near-Real-Time Geomagnetic Observatory Data, available at: http://www.intermagnet.org/index-eng.php, 2001.

Lander, F., Whiteside, L. S., and Lockridge, P. A.: Caribbean tsunami, an Initial History, Natural Hazards and Hazards Management in the Greater Caribbean and Latin America, 3, 1–18, 1997.

Langlais, M., Vial, B., and Coutant, O.: Improvement of broadband seismic station installations at the Observatoire de Grenoble (OSUG) seismic network, Adv. Geosci., 34, 9–14, doi:10.5194/adgeo-34-9-2013, 2013.

McNamara, D. E. and Boaz, R. I.: Seismic Noise Analysis System Using Power Spectral Density Probability Density Fonctions: A Stand-Alone Software Package, USGS Open-File Report 2005–1438, 2005.

Mori, N., Takahashi, T., Yasuda, T., and Yanagisawa, H.: Survey of 2011 Tohoku earthquake tsunami inundation and run-up, Geophys. Res. Lett., 38, L00G14, doi:10.1029/2011GL049210, 2011.

Peterson, J.: Observations and Modeling of Seismic Background Noise, USGS, Open-File Report 93-322, 1993.

Sakic, P.: Spéfications de monumentation utilisés dans le cadre du projet GPSJURA, Observatoire des sciences de l'univers de Besançon, Réseau national GPS (RENAG), 2013.

Saurel, J.-M., Anglade, A., Maury, T., Lemarchand, A., Lynch, L., Higgins, M., Clouard, V., Bouin, M.-P., Robertson, R., de Chabalier, J.-B., Tait, S., Nercessian, A., and Latchamn, J. L.: Ecar virtual broadband seismic network: vault construction techniques and seismometer insulation, Caribbean Wave II conference, 2012.

Shepherd, J. B.: Comment on "Subduction and seismic hazard in the Lesser Antilles" by Pascal Bernard and Jérôme Lambert, B. Seismol. Soc. Am., 82, 1534–1543, 1992.

Singh, S. C., Hananto, N., Mukti, M., Robinson, D. P. , Das, S., Chauhan, A., Carton, H., Gratacos, B., Midnet, S., Djajadihardja, Y., and Harjono, H.: Aseismic zone and earthquake segmentation associated with a deepsubducted seamount in Sumatra, Nat. Geosci., 4, 308–311, doi:10.1038/NGEO1119, 2011.

Stutzmann, E., Roult, G., and Astiz, L.: Geoscope Station Noise Levels, B. Seismol. Soc. Am., 90, 690–701, 2000.

Trnkoczy, A., Bormann, P., Hanka, W., Holcomb, L. G., Nigbor, R. L., Shinaohara, M., Shiobara, H., and Suyehiro, K.: Chapter 7, Site Selection, Preparation and Installation of Seismic Stations, in: New Manual of Seismological Observatory Pratice (NMSOP-2), edited by: Bormann, P., 2nd Edn., IASPEI, GFZ german Research Center for Geoscience, Potsdam, 2011.

UNAVCO Resources: GNSS Station Monumentation, available at: http://facility.unavco.org/kb/questions/104, 2010.

Widmer-Schnidrig, R. and Kurrle, D.: Evaluation of Installation Methods for Streckeisen STS-2 Seismometers, Deutschen Geophysikalischen Gesellschaft (DGG), 2006.

Wielandt, E.: Two tests of the Trillium240 very-broad-band seismometer, personal report, November 2007.

Zürn, W. and Widmer-Schnidrig, R.: On noise reduction in vertical seismic records below 2 mHz using local barometric pressure, Geophys. Res. Lett., 22, 3537–3540, 1995.

The Hellenic Seismological Network of Crete (HSNC): validation and results of the 2013 aftershock sequences

G. Chatzopoulos, I. Papadopoulos, and F. Vallianatos

Laboratory of Geophysics and Seismology, Technological Educational Institute of Crete, Chania, Greece

Correspondence to: G. Chatzopoulos (ggh1983@hotmail.com)

Abstract. The last century, the global urbanization has leaded the majority of population to move into big, metropolitan areas. Small areas on the Earth's surface are being built with tall buildings in areas close to seismogenic zones. Such an area of great importance is the Hellenic arc in Greece. Among the regions with high seismicity is Crete, located on the subduction zone of the Eastern Mediterranean plate underneath the Aegean plate. The Hellenic Seismological Network of Crete (HSNC) has been built to cover the need on continuous monitoring of the regional seismicity in the vicinity of the South Aegean Sea and Crete Island. In the present work, with the use of Z-map software the spatial variability of Magnitude of Completeness (Mc) is calculated from HSNC's manual analysis catalogue of events from the beginning of 2008 till the end of September 2015, supporting the good coverage of HSNC in the area surrounding Crete Island. Furthermore, we discuss the 2013 seismicity when two large earthquakes occurred in the vicinity of Crete Island. The two main shocks and their aftershock sequences have been relocated with the use of HYPOINVERSE earthquake location software. Finally, the quality of seismological stations is addressed using the standard PQLX software.

1 Introduction

The number and quality of seismological networks in Europe has increased in the past decades. Nevertheless, the need for localized networks monitoring in areas of great seismic and scientific interest is constant. A very active seismic area is the Hellenic arc, located at the eastern part of Mediterranean Sea, in companion with a volcanic arc created by the subduction of the tectonic plates (McKenzie, 1972). The convergence of the Eastern Mediterranean plate and the Aegean plate at

$4\,\mathrm{cm\,yr^{-1}}$ rate creates a complex tectonic environment (Le Pichon and Angelier, 1979). In this area, more than 60 % of the total seismic energy in Europe is being released with magnitudes up to 8.3 (Papazachos, 1990). The subduction of the Mediterranean plate under the Aegean plate creates the Benioff zone of intermediate and depth earthquakes, as it has been estimated by seismological studies (McKenzie, 1972; Le Pichon and Angelier, 1979) and it has been revealed by tomographic results (Spakman, 1988; Papazachos and Nolet, 1997).

The Hellenic Seismological Network of Crete (HSNC) is a local seismological network covering and supporting the continuous need for monitoring of the front of Hellenic Arc, with its official operation started in 2004. Within almost a decade of operation, the existence of HSNC supports the application and test of modern techniques as that of earthquake early warning (Hloupis and Vallianatos, 2013, 2015). In 2013, two large earthquakes occurred in the vicinity of Crete Island. The HSNC identified more than 510 and 360 aftershocks respectively followed after the main event. The two main shocks and their aftershock sequences have been located with a GUI program that use the HYPO earthquake location software (Lee and Lahr, 1972) and relocated with HYPOINVERSE (Klein, 2002) and a modified crust model. In the present work in order to present the everyday results and to give evidence of the contribution of HSNC in the effort of creation reliable data sets, we show the preliminary results from the analysis of two aftershock sequences. In addition, the magnitude of completeness of the HSNC covered area (i.e. in the front oh Hellenic Arc) and the distribution of Power Spectral Density of HSNC's seismological stations, provide elements related with the operation of HSNC in one of the most scientifically important areas in Europe as the front of the Hellenic Arc is.

2 The development of the Hellenic Seismological Network of Crete

The HSNC started its first operation in 2004 with 4 Guralp CMG-40T 1 s sensors with flat response to velocity from 1 to 100 Hz installed in each major city of Crete. Within a period less than a decade (2004–2012) it expands to 12 online stations. Some of the initial locations of the first 6 stations were changed to more suitable locations. At the same time few of Guralp CMG-40T 1 s sensors were installed have been substituted with Guralp CMG ESPC 60 and 120 s (Hloupis et al., 2013), which are flat to velocity from 100 Hz to 0.016 Hz (60 s) and 0.0083 Hz (120 s). The next two years, three more short period sensors were upgraded with broadband. At the same time two new stations added to the seismic network to provide better coverage area. In middle of 2015 two more short period seismometers have been changed with broadband, leaving only 1 short period station in network, which will be uprated in the near future. All stations, since the network's first operation, were equipped with Reftek 130-01 digitizers (www.trimble.com/). In the beginning the digitizers continuously record data at 125 Hz, this number reduced to 100 Hz in the end of 2013 due to large amount of data stored after the network's expansion. Communication with the central station is established by using private wired ADSL MPLS VPN lines, satellite lines, and RF. The HSNC has its own private satellite hub allowing single-hop communication between central station and VSATs (Hloupis et al., 2013). Data collection is achieved using commercial RTP software (www.reftek.com). All stations are registered to International Seismological Centre (ISC) and the network is listed by International Federation of Digital Seismograph Networks (FDSN) with the assigned permanent network code HC.

The HSNC operates continuously collecting data from its 14 stations (see Fig. 1 and Table 1 for details) and 10 more stations from neighboring networks. By the of the end of 2013 the HSNC has bilateral agreements with national Observatory of Athens (NOA) (http://www.gein.noa.gr/en/), Aristotle University of Thessaloniki (AUTH) (http://geophysics.geo.auth.gr/the_seisnet/WEBSITE_2005/station_index_en.html) and Kandili Observatory and Earthquake Research Institution (KOERI) (http://www.koeri.boun.edu.tr/) ensuring a constant exchange of data. Data from neighboring networks are fed to RTP using sl2rtpd protocol transforming mini-SEED data to raw PASSCAL format. For automatic processing of seismic signal, we employ three Earthquake Monitoring Systems (EMS): Seismic Network Data Protocol (Synapse Seismic Center, Russia), SeisComp3 (Deutsches GeoForschungsZentrum, Germany) and Earthworm (USGS, USA) that run independently and in parallel. Each of the first 2 EMS is assigned for specific area monitoring and the third one is used supplementary to increase the sensitivity of the network. The Earthworm monitoring system is calibrated to work better in identify-

Table 1. Code names (as registered at ISC) and coordinates for the HSNC seismological stations.

Code	Location	Longitude	Latitude	Altitude (m)
KNDR	Koundoura	23.6248	35.2348	13.5
FRMA	Ferma	25.8555	35.0187	21.5
CHAN	Chania	24.0429	35.5193	36.0
KSTL	Kasteli	25.3374	35.2092	335.0
HRKL	Herakleio	25.1015	35.2115	81.0
PRNS	Prines	24.4260	35.3450	325.0
KTHR	Kythira	22.9938	36.2447	270.0
TMBK	Tymbaki	24.7662	35.0724	12.0
STIA	Siteia	26.0909	35.2021	93.0
KOSK	Kos	26.9785	36.7516	10.0
KLMT	Kalamata	22.0597	37.0613	6.0
THT2	Santorini	25.4218	36.4351	24.0
RODP	Rodopos	23.7577	35.5604	308.0
GVDS	Gavdos	24.0585	34.8389	348.0

ing events close to Crete region while SeisComp3 is best used for locating earthquakes in the broader area. All EMS create automated bulletins which are being distributed to registered users and authorities. In addition manual analysis of seismic events is being conducted and event messages are sent to registered users and authorities. A dedicated webpage is being updated after every event, presenting the spatial distribution of the 30 last manual located events (http://gaia.chania.teicrete.gr/uk/?cat=11). The complete catalogue of seismic events is available under request.

3 Coverage of HSNC – Quality of Stations

The spatial variability of Magnitude of Completeness (Mc) is calculated from HSNC's manual analysis catalogue of events with the use of Z-map software (http://www.seismo.ethz.ch/prod/software/zmap/index_EN) for the period of 2008 until the end of September, 2015. As indicated by Gutenberg–Richter distribution (Gutenberg and Richter, 1936) Magnitude of Completeness denotes the minimum magnitude of an earthquake, which a network can detect within the boundaries of a given region. Calculation of Mc is done using the "best combination" option of Z-map which compares the Maximum Curvature Technique (Wyss et al., 1999) and the Goodness-of-Fit Test (Wiemer and Wyss, 2000). The Mc estimation methods are based on the validity of Gutenberg–Richter Law (Mignan and Woessner, 2012). Figure 2 illustrates a map with the distribution of Mc for HSNC. The Mc value at the close vicinity of Crete Island is down to 2.0 at the western part, between 2 to 2.7 at the central and southern part of the island, and down to 3.0 at the eastern part of the coverage area. The present results regarding the HSNC magnitude of completeness map are comparable with that presented for the NOA-HUSN catalogue using both the Bayesian magni-

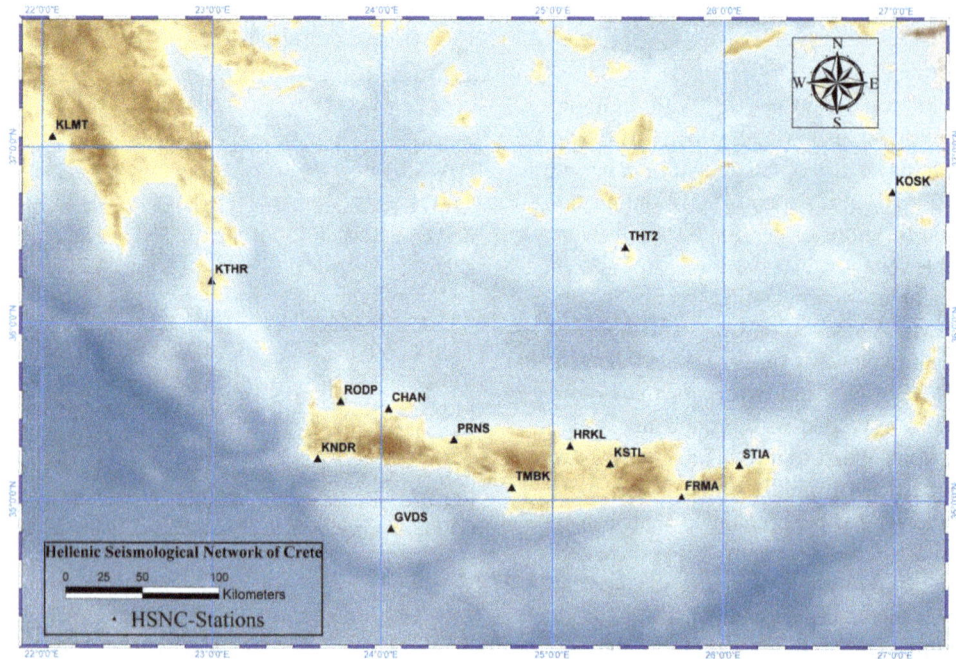

Figure 1. The HSNC seismological stations location on the front of Hellenic Arc (Crete and South Aegean region).

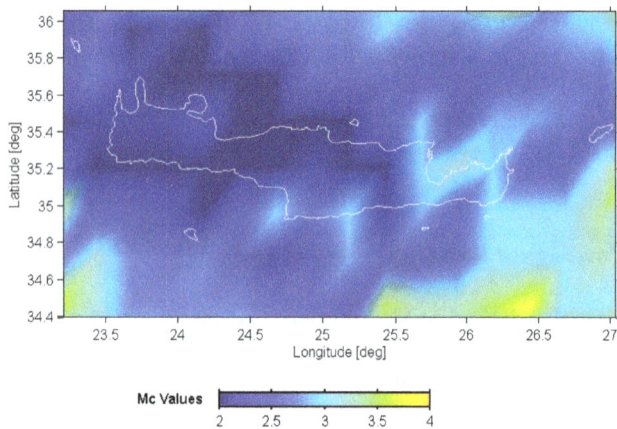

Figure 2. Magnitude of Completeness for HSNC station distribution, using the manual catalogue of the HSNC from 2008 until end of October 2015 (for details see text).

tude of completeness and the frequency -magnitude distribution (Mignan and Chouliaras, 2014).

The HSNC operation time is relative small compared to other networks in Greece. Therefore an effort to evaluate the operation results is done by comparing the magnitudes of 1500 events located by the 2 largest networks of Greece the NOA and AUTH. The common identified events, in origin time and location are from manual catalogs from 2014 to middle of 2015 for the broader area of Crete. The results illustrated on Fig. 3 suggest that the HSNC reporting magnitudes are slightly larger than the other two networks report.

To present the quality of stations deployed from the HSNC network, we use the PQLX software (McNamara and Boaz, 2005) to analyze the time series collected. PQLX calculates the distribution of Power Spectral Density (PSD) of seismological waveform with the use of a probability density function (PDF), and is considered as the standard way of presenting the quality of seismological stations. In this work, we are presenting three stations of the HSNC network, FRMA, STIA and CHAN which are located at the, eastern and western part of Crete Island (Fig. 4). The noise level at long periods is caused due to the fact that the stations of HSNC are located near urban zones. In the future the noisy stations are going to move in quieter locations. PQLX uses the whole recordings (noise and events) to calculate the PDF. System transients (cultural noise) map into a low-level background probability while ambient noise conditions reveal themselves as high probability occurrences. Cultural noise propagates mainly as high-frequency surface waves (> 1–10 Hz) that attenuate within several kilometers in distance and depth. Cultural noise shows very strong diurnal variations and has characteristic frequencies depending on the source of the disturbance. On the other hand, large teleseismic earthquakes can produce powers above ambient noise levels across the entire spectrum and are dominated by surface waves > 10 s, while small events dominate the short period, < 1 s. This presentation allows for the overall estimation of station quality and a baseline level of earth noise at each site (McNamara and Boaz, 2005).

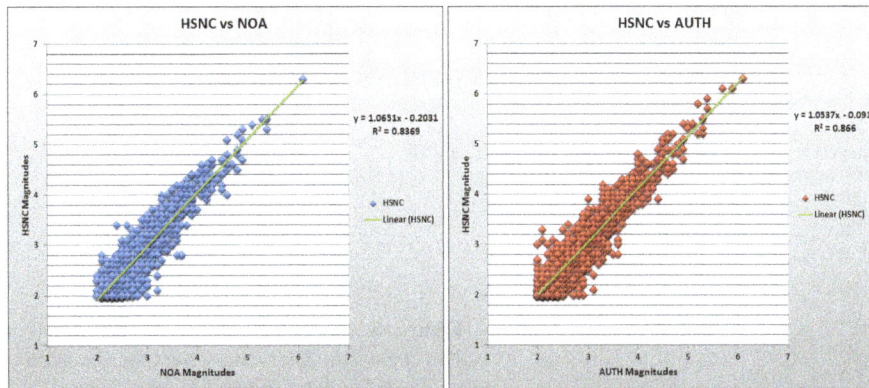

Figure 3. Comparison of the magnitudes reported by HSNC and that of NOA and AUTH. The events are from manual catalogs from 2014 to middle of 2015. The slope of the fitting show that HSNC is calculating the event magnitudes slightly higher.

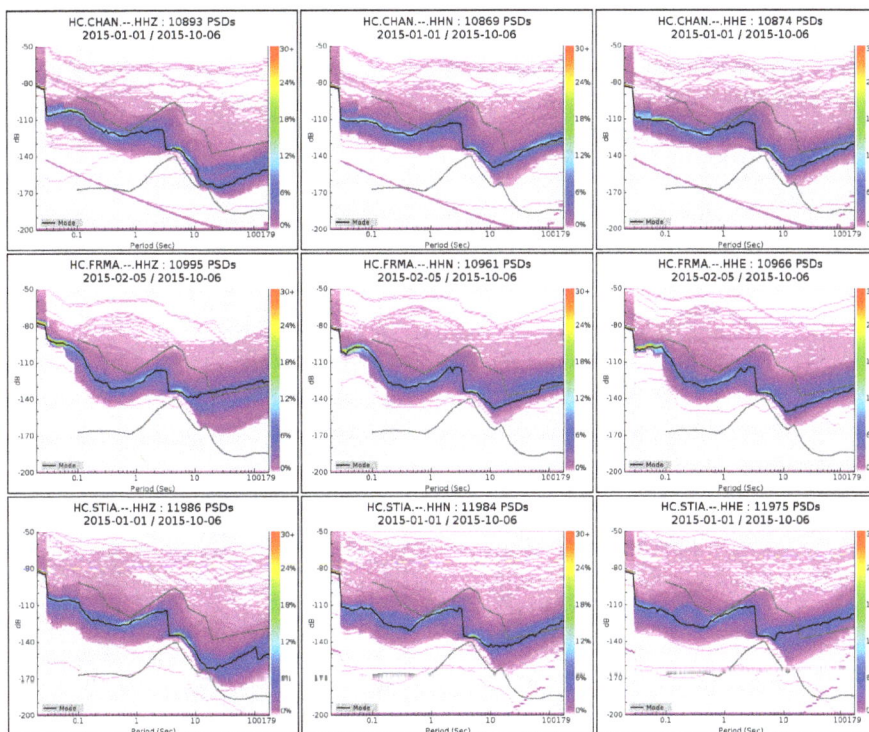

Figure 4. Distribution of Power Spectral Density for stations FRMA, KSTL and STIA of the HSNC network using Power Density Function, created using PQLX software.

4 The aftershock sequences

Within 2013, one of the seismically active periods in the front of Hellenic Arc two strong earthquakes with magnitude greater or equal to 6.0 occurred in the vicinity of Crete Island. Both mainshocks were felt in the whole Crete region. The initial almost real time location procedure has been done with the use of a dedicated software which is been adopted by the early stages of HSNC operation. The software is user friendly for picking phases and runs HYPO software but its main disadvantage is that it uses binary files with a generic crust model which is not suitable for the area. The Hellenic front arc is the initial part from the Eurasian – African subduction zone. The amphitheatric shape, created by the lithospheric plates convergence, is has been characterized by different geological environments along the arc. Therefore a generic crust model cannot be used in such a complicated structure. In order to properly relocate the aftershock sequences, we modified the Meier et al. (2004) crust models (Table 2) for both sequences (V. Karakostas, personal communication, 2014). The first results of this effort are presented in the vertical to fault plane cross sections in Fig. 5 for the 15 June 2013

Table 2. The crust models used for 15 June 2013 and 12 October 2013 relocations.

Depth (km)	Velocity P (km s^{-1})	Depth (km)	Velocity P (km s^{-1})
0.0	4.60	0.0	4.20
1.0	5.40	1.0	5.70
2.5	5.70	3.0	6.30
5.0	6.00	8.0	6.40
8.0	6.10	12.0	6.45
11.0	6.20	20.0	6.50
15.0	6.30	25.0	6.80
20.0	6.60	30.0	7.30
25.0	7.00	33.0	7.90
30.0	7.50	–	–
33.0	8.00	–	–

Table 3. The 5 strongest event errors in the determinations of the epicenters in the horizontal and vertical dimensions for the 15 June 2013 and 12 October 2013 sequnces.

Day	Hour	Magnitude	Residual	ERH	ERZ
15/06/2013	16:11:01	6	0.13	2.8	1.4
13/06/2013	21:39:04	5.8	0.22	2.9	1.5
13/06/2013	21:43:16	4.9	0.32	0.9	0.9
15/06/2013	17:22:06	4.8	0.26	0.7	0.8
19/06/2013	19:05:09	4.8	0.2	0.7	0.5
12/10/2013	13:11:53	6.2	0.16	0.8	1.3
13/10/2013	17:43:52	4.1	0.16	0.7	1
12/10/2013	13:17:00	4	0.16	0.7	1.1
12/10/2013	14:05:50	4	0.16	0.7	1
19/10/2013	2:19:21	3.9	0.15	0.7	1.1

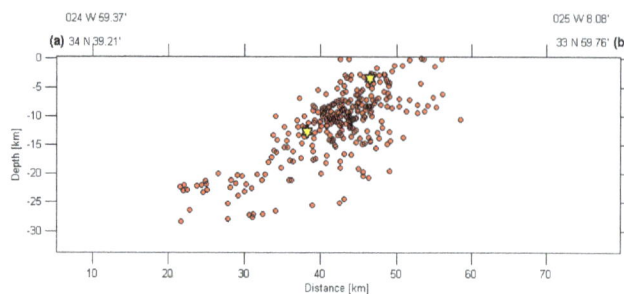

Figure 5. Cross section of the aftershock seismicity of the 15 June event. The yellow stars are the strong and the moderate events and with red circles are the aftershocks.

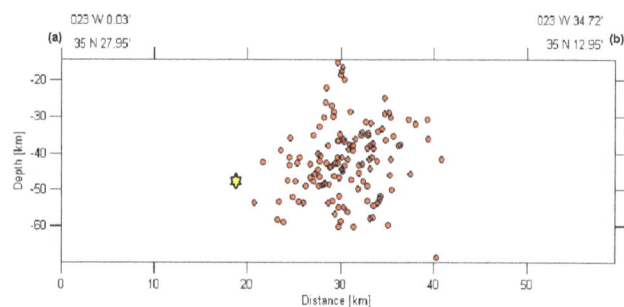

Figure 6. Cross section of the aftershock seismicity of the 12 October 2013 event. The yellow star is the strong event and with red circles are the aftershocks.

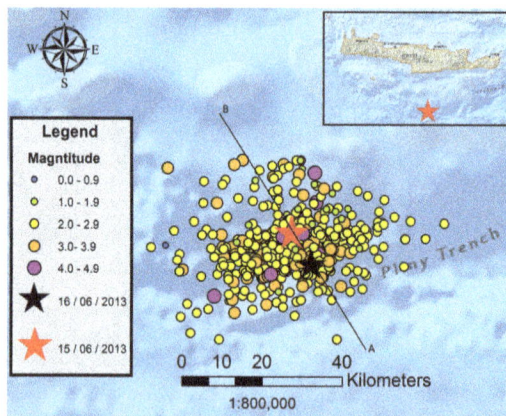

Figure 7. The 15 June 2013 Mainshock and the distribution of the aftershocks after the relocation process (see text). The two strong shocks are denoted with red and black star. The aftershocks' magnitude is represented with circles with different color and size.

event and in Fig. 6 for the 12 October 2013 event. Additionally the process errors, in horizontal and vertical dimension for five strongest events for each sequence are illustrated in Table 3. For the station time error calibration file, which is needed by HYPOINVERSE software, only large aftershocks recordings with many P and S phases (more than 21) have been used. Other important criteria for selecting the aftershocks were to be recorded in most of available stations in epicentral distances approximately less than 200 km, marked phases having very small errors in time, depth and azimuth. Parameters such as distance and RMS residual weighting also were considered. The most of the aftershock events of the 15th June sequence were recorded by 11 stations with code names: TMBK, STIA, KSTL, FRMA, ZKR, PRNS, RODP, KNDR, GVDS, HRKL and CHAN. Likewise the 12 October sequence were recorded on the following 10 stations: KTHR, VLI, CHAN, KNDR, RODP, PRNS, MHLO, HRKL, KRND and GVDS. For clarity we note that the data received from the stations VLI, MHLO and ZKR belongs to NOA and the KRND belongs to AUTH. These data contributed in the relocation procedure by providing more seismic wave phases. The results of the strong earthquakes and their aftershock are presented as an evidence of the monitoring capability of Hellenic Arc Front. Therefore the information that is provided and ploted is limited and related to seismic network operation. All the identified aftershock with minimum of 15 or more phases were used in relocation process. The available events near the fault area of the mains event considered as aftershocks. The end time for each se-

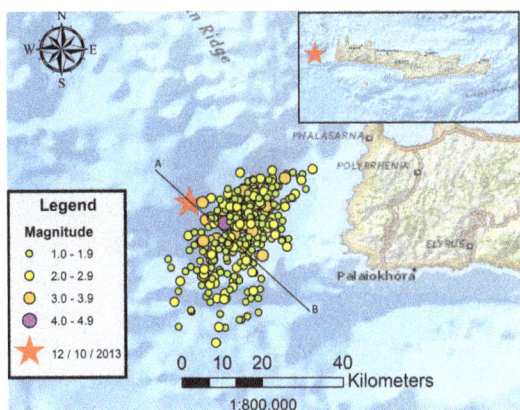

Figure 8. The 12 October 2013 Mainshock and the distribution of the aftershock after the relocation process (see text). The strong shock is denoted with red star. The aftershocks' magnitude is represented with circles with different color and size.

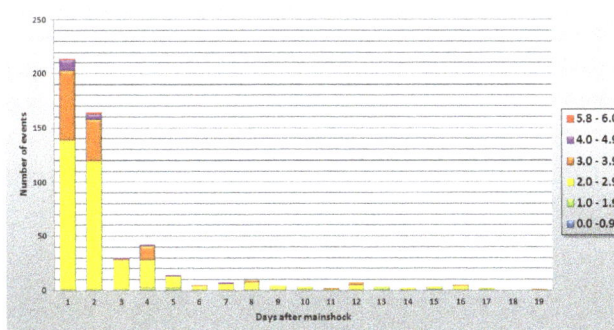

Figure 10. The daily rate of aftershocks and its magnitudes after the strong shock of the 15 June 2013.

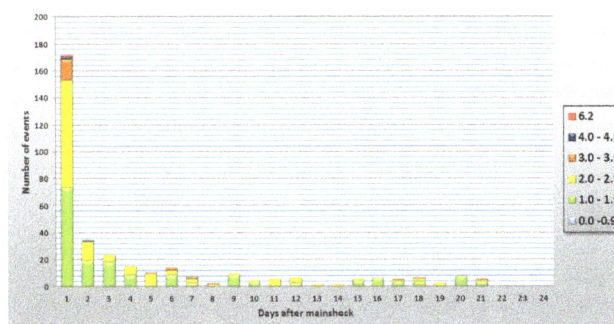

Figure 11. The daily rate of aftershocks and its magnitudes after the strong shock of the 12 October 2013.

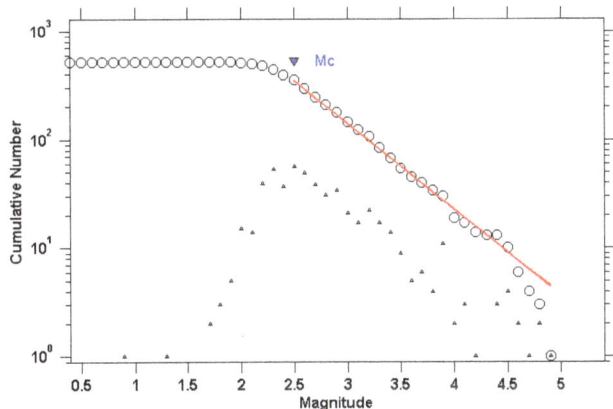

Maximum Likelihood Solution
b-value = 0.789 +/- 0.04, a value = 4.52, a value (annual) = 5.75
Magnitude of Completeness = 2.5

Figure 9. The Mmin, a and b values for the 15 June 2013 aftershock sequence (see text for details).

quence was considered a time period of 48 h or more with no earthquakes above the Mmin in the fault area. The results of the two relocations are illustrated for 15 June in Fig. 7 and for 12 October Fig. 8 respectively.

The first strong event took place on 15 June 2013 at 16:11 UTC in the front of the Hellenic Arc, south from central Crete (Latitude = 34.31, Longitude = 25.04). The magnitude and depth of the mainshock are $M_{\mathrm{w}} = 6.0$ and 15 km respectively. A second moderate activation occurred after approximately 29 h with magnitude $M_{\mathrm{w}} = 5.8$ and depth 6.5 km. The use "best combination" option of Z-map for the whole aftershock sequence provided the Gutenberg–Richter distribution, calculated the b values 0.789, the a value 4.52 and the Mmin 2.5 (Fig. 9). The moment magnitude for the strong and moderate events was calculated with Seisan (ftp:// ftp.geo.uib.no/pub/seismo/SOFTWARE/SEISAN/) software

by analyzing the spectral displacement, the stress drop and other parameters of seismic wave spectrum. The mainshock was followed for more than 510 recorded aftershocks span in a time period about 17 days. In Fig. 10 the daily rate of events after the mainshock is presented as well as the magnitude of these events. The time period between the strong and the moderate event more than 230 aftershocks occurred with the magnitude range to vary from micro (1.9–2.9) to light (4.1–4.8) with a considerable amount of events in minor range (3.0–3.9). Few hours after the moderate $M_{\mathrm{w}} = 5.8$ activation the rate of aftershocks decrease drastically, since about 280 aftershocks followed this second event in a time period about 15 days. As Fig. 10 shows there was a high seismic energy release within the first two days with most of the events having minor to light magnitudes. The rest period there is a low seismic release rate observed which leads to the conclusion that the aftershock sequence decay rate was fast.

The second strong event presented on 12 October 2013 at the western part of Crete (Latitude = 35.39, Longitude = 23.26). We note that this event has attracted the interest due to the number of different foreshocks observed (Vallianatos et al., 2014; Contadakis et al., 2015). The magnitude of the mainshock are $M_{\mathrm{w}} = 6.2$ in a depth of 47 km. The mainshock was followed by more than 360 recorded af-

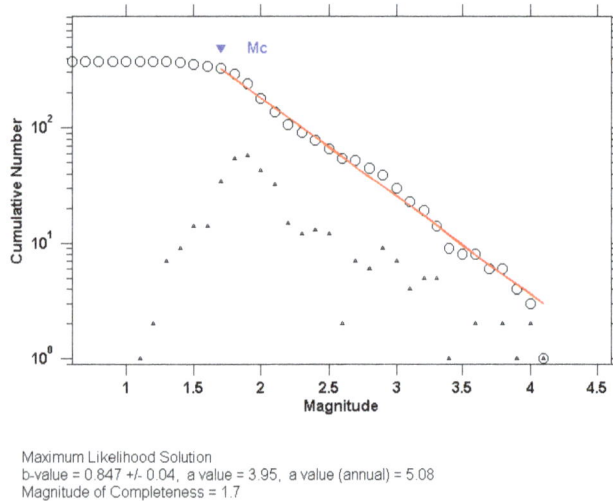

Maximum Likelihood Solution
b-value = 0.847 +/- 0.04, a value = 3.95, a value (annual) = 5.08
Magnitude of Completeness = 1.7

Figure 12. The Mmin, a and b values for the 12 October 2013 aftershock sequence (see text).

tershocks in a time window of 23 days. In Fig. 11 the daily rate of events after the mainshock is presented as well as the magnitude of these events. With the use "best combination" option of Z-map the b values calculated 0.847, as well as the a value 3.95 and the Mmin 1.7 (Fig. 12). The majority of the aftershocks were identified the first day with most of them with magnitudes range between micro and minor (range 1.1 to 3.9) and only three light events (range 4.0 to 4.1). The depth of the mainshock was probably the reason for the limited number of aftershock events and the low magnitudes observed, compared to 15 June 2013 event. It is noticeable that only the first day there was a considerable amount of seismic energy released and after that the aftershock activity is pretty low.

5 Conclusions

In order to cover the need for a reliable and modern monitoring of the seismic activity around Crete Island, Greece, the HSNC started its operations in 2004. The goal was to provide a continuous coverage at the front of the Hellenic arc and to permit the study of new methods as that of earthquake early warning (Hloupis and Vallianatos, 2013, 2015). Currently having 13 online broadband stations and one short period, the HSNC is being in constant explanations acts, as a complete observation unit able to continuous monitor the seismic activity within a very sensitive area in the East Mediterranean as that in the South of the Hellenic Arc. Using three different and independent EMS ensures a triple check on the data, enabling different triggering algorithms to be applied. As an example the earthquake events and the results of the analyses of the 2 large earthquake sequences in 2013 indicates the importance of the HSNC network and its capability to con-

tribute in the detailed monitoring of the seismic activity over the subduction zone in the Hellenic Arc.

Acknowledgements. Special thanks of the authors to V. Karakostas (AUTH) for his valuable insight and help on the methods used at hypoinverse software and providing the crust models. Also to the two anonymous referees that help us with their comments.

This work was implemented through the project IMPACT-ARC in the framework of action "ARCHIMEDES III–Support of Research Teams at TEI of Crete" (MIS380353) of the Operational Program "Education and Lifelong Learning" and is co-financed by the European Union (European Social Fund) and Greek national funds.

References

Contadakis, M. E., Arabelos, D. N., Vergos, G., Spatalas, S. D., and Skordilis, E. M.: TEC variations over the Mediterranean before and during the strong earthquake (M = 6.5) of 12th October 2013 in Crete, Greece, Journal Physics Chemistry of the Earth, doi:10.1016/j.pce.2015.03.010, 2015.

Gutenberg, B. and Richter, C. F: Magnitude and Energy of Earthquakes, Science, 21, 183–185, 1936.

Hloupis, G. and Vallianatos, F.: Wavelet-based rapid estimation of earthquake magnitude oriented to early warning, IEEE Geosci. Remote S., 10, 43–47, 2013.

Hloupis, G. and Vallianatos, F.: Wavelet-Based Methods for Rapid Calculations of Magnitude and Epicentral Distance: An Application to Earthquake Early Warning System, Pure Appl. Geophys., 172, 2371–2386, 2015.

Hloupis, G., Papadopoulos, I., Makris, J. P., and Vallianatos, F.: The South Aegean seismological network – HSNC, Adv. Geosci., 34, 15–21, doi:10.5194/adgeo-34-15-2013, 2013.

Klein, F. W.: User's Guide to HYPOINVERSE-2000, a Fortran program to solve for earthquake locations and magnitudes, Open File Report 02-171, US Geological Survey, 1–123, 2002.

Lee, W. H. K. and Lahr, J. C.: HYPO71: A computer program for determining hypocenter, magnitude, and first motion pattern of local earthquakes, US Geological Survey Open-File Report, 1–104, 1972.

Le Pichon, X. and Angelier, J.: The Hellenic Are and Trench System: a key to the neotectonic evolution of the eastern Mediterranean area, Tectonophysics, 60, 1–42, 1979.

McKenzie, D.: Active Tectonics of the Mediterranean Region, The Geophysical Journal of the Royal Astronomical Society, 30, 109–185, 1972.

McNamara, D. E. and Boaz, R. I.: Seismic Noise Analysis System, Power Spectral Density Probability Density Function: Stand-Alone Software Package, United States Geological Survey Open File Report, NO. 2005–1438, p. 30, 2005.

Meier, T., Rische, M., Endrun, B., Vafidis, A., and Harjes, H.-P.: Seismicity of the Hellenic subduction zone in the area of western and central Crete observed by temporary local seismic networks, Tectonophysics, 383, 149–169, 2004.

Mignan, A. and Chouliaras, G.: Fifty Years of Seismic Network Performance in Greece (1964–2013): Spatiotemporal Evolution of the Completeness Magnitude, Phys. Rev. Lett., 85, 657–667, doi:10.1785/0220130209, 2014.

Mignan, A. and Woessner, J.: Estimating the magnitude of completeness for earthquake catalogs, Community Online Resource for Statistical Seismicity Analysis, Version: 1.0, 1–45, doi:10.5078/corssa-00180805, 2012.

Papazachos, B. C.: Seismicity of the Aegean and surrounding area, Tectonophysics, 178, 287–308, 1990.

Papazachos, C. B. and Nolet, G. P.: P and S deep velocity structure of the Hellenic area obtained by robust nonlinear inversion of travel times, J. Geophys. Res., 102, 8349–8367, 1997.

Spakman, W.: Upper mantle delay time tomography with an application to the collision zone of the Eurasian, African and Arabian plates, PhD Thesis, Univ. of Utrecht, 53, 200 pp, 1988.

Vallianatos, F., Michas, G., and Papadakis, G.: Non-extensive and natural time analysis of seismicity before the Mw6.4, October 12, 2013 earthquake in the South West segment of the Hellenic Arc, Physica A-Statistical Mechanics and its Applications, 414, 163–173, doi:10.1016/j.physa.2014.07.038, 2014.

Wiemer, S. and Wyss, M.: Minimum magnitude of complete reporting in earthquake catalogs: Examples from Alaska, the Western US, and Japan, Bull. Seismol. Soc. Am., 90, 859–869, 2000.

Wyss, M., Hasegawa, A., Wiemer, S., and Umino, N.: Quantitative mapping of precursory seismic quiescence before the 1989, m7.1 O-sanriku earthquake, Japan, Annali di Geofisica, 42, 851–869, 1999.

19

Modeling and detection of regional depth phases at the GERES array

M.-T. Apoloner and G. Bokelmann

Department of Meteorology and Geophysics, University of Vienna, Vienna, Austria

Correspondence to: M.-T. Apoloner (maria-theresia.apoloner@univie.ac.at)

Abstract. The Vienna Basin in Eastern Austria is a region of low to moderate seismicity, and hence the seismological network coverage is relatively sparse. Nevertheless, the area is one of the most densely populated and most developed areas in Austria, so accurate earthquake location, including depth estimation and relation to faults is not only important for understanding tectonic processes, but also for estimating seismic hazard. Particularly depth estimation needs a dense seismic network around the anticipated epicenter. If the station coverage is not sufficient, the depth can only be estimated roughly. Regional Depth Phases (RDP) like sPg, sPmP and sPn have been already used successfully for calculating depth even if only observable from one station. However, especially in regions with sedimentary basins these phases prove difficult or impossible to recover from the seismic records.

For this study we use seismic array data from GERES. It is 220 km to the North West of the Vienna Basin, which – according to literature – is a suitable distance to recover PmP and sPmP phases. We use array processing on recent earthquake data from the Vienna Basin with local magnitudes from 2.1 to 4.2 to reduce the SNR and to search for RDP. At the same time, we do similar processing on synthetic data specially modeled for this application. We compare real and synthetic results to assess which phases can be identified and to what extent depth estimation can be improved. Additionally, we calculate a map of lateral propagation behavior of RDP for a typical strike-slip earthquake in our region of interest up to 400 km distance.

For our study case RDP propagation is strongly azimuthally dependent. Also, distance ranges differ from literature sources. Comparing with synthetic seismograms we identify PmP and PbP phases with array processing as strongest arrivals. Although the associated depth phases cannot be identified at this distance and azimuth, identification of the PbP phases limits possible depth to less than 20 km. Polarization analysis adds information on the first arriving Pn wave for local magnitudes above 2.5.

1 Introduction

Earthquake locations are fundamental for assessing seismic hazard. To provide these, areas with high seismicity rates and large magnitudes are instrumented with seismic stations. Particularly depth estimation requires a dense seismic network around the suspected epicenter. In contrast our study area, the Vienna Basin in Austria, a region of low to moderate seismicity with a largest instrumentally recorded magnitude around 5, is only covered sparsely with seismic stations. Nevertheless, the area is one of the most densely populated and most developed areas in Austria. In areas like these, estimation of seismic hazard has to be based on location of earthquakes with small magnitudes.

Regional Depth Phases (RDP) like sPg, sPmP and sPn are P phases converted to S at the surface. They develop at different regional distance ranges (Ma and Atkinson, 2006) and the time difference between direct and reflected phase is sensitive to epicentral depth. This property already has been successfully used for calculating depth even if an RDP is observable at least at one station. However, especially in regions with sedimentary basins, these phases prove difficult or impossible to recover from the seismic records. On the other hand, seismic arrays together with appropriate processing can be used to lower signal to noise ratio of seismic recordings and so help detect and identify phases as e.g., Rost and Thomas (2002) describes.

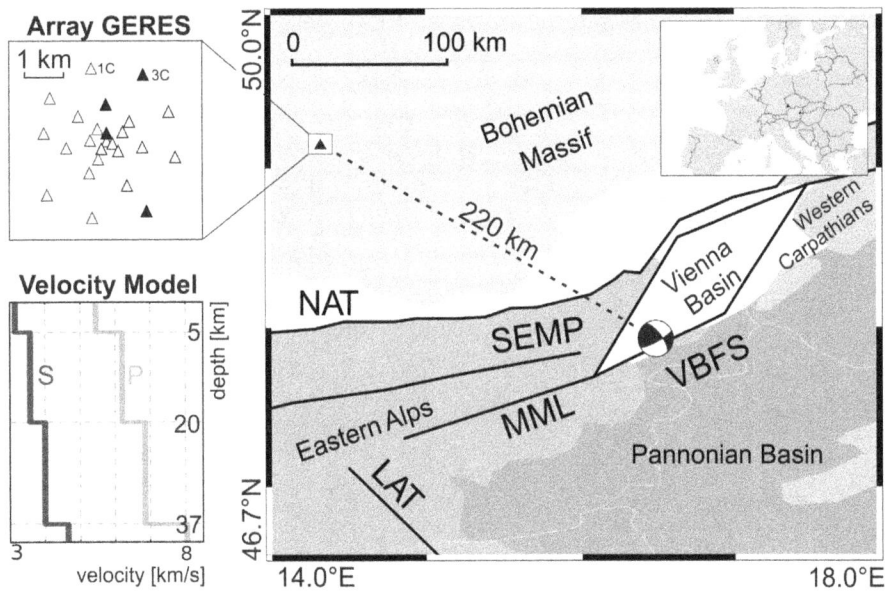

Figure 1. Right panel: schematic map of the Vienna Basin with surrounding tectonic units and main faults shown together with the focal mechanism of the Ml 4.2 earthquake from 20 Septmeber 2013 in Ebreichsdorf by Hausmann et al. (2014). Top left: layout of seismic array GERES with 1-component stations (white triangles) and 3-component stations (black triangles). Bottom left: *P* and *S* velocity model for the area by Hausmann et al. (2010)

The purpose of this study is to investigate the possibility of using a seismic array for identifying RDP in our region of interest.

In our study area we previously investigated a series of strike-slip earthquakes and relocated them (Apoloner and Bokelmann, 2015) using local stations. From this dataset we select all earthquakes with local magnitudes from 2.0 to 4.2 and analyse the records at the 220 km distant seismological array GERES (Harjes, 1990). According to the literature (Ma and Atkinson, 2006) those events should have a suitable magnitude and distance to recover PmP and sPmP phases. We use array processing on the whole array and polarization analysis on the four 3-component stations. At the same time, we perform similar processing of synthetic data specially modeled for this application. We compare real and synthetic results to assert which phases can be identified.

2 Tectonic setting

Ongoing convergence between the Bohemian Massif on the European Plate in the north and the Adriatic Plate in the south lead to lateral extrusion of crustal blocks to the east (e.g. Brückl et al., 2010), into the Pannonian Basin as is shown in Fig. 1. Two main sinistral strike-slip faults show this process: the Salzach-Enns-Mariazell-Puchberg (SEMP) fault and the seismically active Mur-Mürz-Lineament (MML). In the northwestern extension of those faults a pull-apart basin, the Vienna Basin started forming and now is filled with sedimentary layers of a few kilometers. Beneath this basin the MML branches into the Vienna Basin Fault System (VBFS)

which produces moderate seismic events with local magnitudes around 4.0. The Bohemian Massif extends beneath the Northern Alpine Transition (NAT) and forms the crystalline basement beneath the Vienna Basin (Wessely, 1983).

To approximate the underground between Ebreichsdorf and GERES we used the model proposed by Hausmann et al. (2010), which is a 4-layer simplification of the 3-D model by Behm et al. (2007).

3 Seismic data

For our research, we used the data from the Ebreichsdorf 2013 earthquake series, which was located by Apoloner and Bokelmann (2015) with all available data within 230 km and the 3-D velocity model by Behm et al. (2007). For modeling the wave propagation we used the largest event with a local magnitude of 4.2 and a depth of 10.5 km. Also, this is one of the earthquakes in the Vienna Basin where a focal mechanism was published. From this dataset we selected the six earthquakes with local magnitudes above 2.0, which are listed in Table 1. Most important for this study the events were also recorded at the seismic array GERES in Germany . The array consists of 21 1-component and four 3-component seismic stations depicted in Fig. 1 which have been continuously recording since 1991 (Harjes et al., 1993). The array is at a distance of approximately 220 km from our area of interest and has a backazimuth of 115° to the selected events.

Table 1. Hypocentral parameters of earthquakes with local magnitude above 2.0 from Apoloner and Bokelmann (2015) sorted by focal depth

ID	Origin Time (UTC)	Ml (ZAMG)	Longitude [deg]	Latitude [deg]	Depth [km]
A	20 Sep 2013 02:06	4.2	16.4230	47.9318	10.5
B	2 Oct 2013 17:17	4.2	16.4229	47.9315	10.5
C	2 Oct 2013 19:42	2.8	16.4210	47.9324	10.0
D	24 Sep 2013 13:53	2.7	16.4207	47.9322	9.8
E	23 Oct 2013 19:34	2.6	16.4202	47.9321	9.6
F	2 Oct 2013 05:26	2.1	16.4237	47.9343	8.9

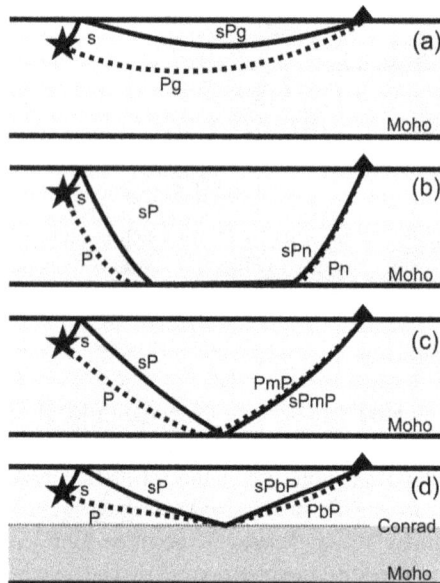

Figure 2. Sketch figure for different RDP wavepaths adapted from Ma and Eaton (2011): **(a)** Pg and sPg, **(b)** Pn and sPn, **(c)** Moho reflection PmP and sPmP, augmented by **(d)** Mid crust reflection PbP and sPbP.

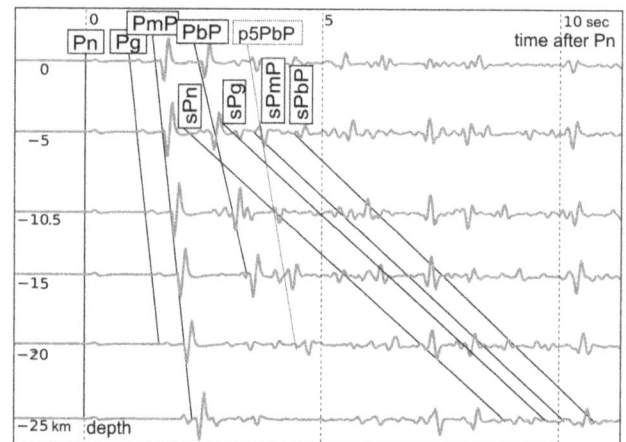

Figure 3. Synthetic seismograms of the vertical component for the central array station GEA0 for depths ranging from 0 to 25 km. Phases are arrivals annotated ackording to wave type. e.g. sPn seems to be visible, but is strongly overlapped by other phases. In addition to the RDPs, the p5PbP is recognizable.

4   Regional Depth Phase (RDP) propagation

Regional depth phases such as sPg, sPmP and sPn in combination with their reference phases Pg, PmP and Pn can be used to estimate focal depths of regional earthquakes, if they can be identified. Figure 2 sketches their wavepaths. In principle, a single station with one phase pair may be sufficient for accurately determining earthquake depth from the difference in their arrival times. Different studies, e.g., Ma and Eaton (2011) and Ma (2012) mention that regional depth phases depend mainly on epicentral distance: Between 200 and 300 km waveforms should be quite simple: a weak Pn is followed by a strong PmP and sPmP using a simple 2-layer model and a thrust type mechanism.

4.1   Waveform modeling

In the first part of this study we model RDP propagation for the tectonic setting and seismic data in our area of interest. Using the velocity model by Hausmann et al. (2010) and the focal mechanism by Hausmann et al. (2014), the source time function was estimated for both Ml 4.2 earthquakes with empirical Green functions and is approximated by a 0.2 s parabolic pulse. With those parameters synthetic seismograms were calculated using the wavenumber integration implemented by Herrmann (2013) for all stations of the array and for depths ranging from 0 to 25 km.

To gain an overview on lateral propagation behavior of RDP, we additionally calculated synthetic seismograms in a regular spaced grid around Ebreichsdorf up to 400 km distance. However, Fig. 3 shows the results for the closest grid point at an azimuth of 295° and 220 km distance, to facilitate comparison of the results to the next processing step.

To give an overview RDP propagation we processed the synthetic data as follows: we measure the maximum amplitude of arriving RDPs on the horizontal component velocity

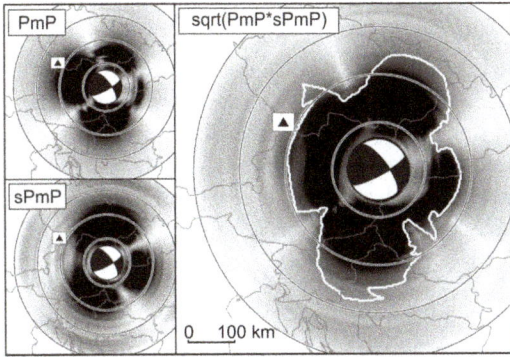

Figure 4. Lateral propagation of PmP (top left panel) and sPmP (bottom left panel) for a strike-slip earthquake in Ebreichsdorf at 10.5 km depth from synthetic traces: right panel shows combined strength of phases (grey shades) and area of minimal visible amplitude for both phases (white line).

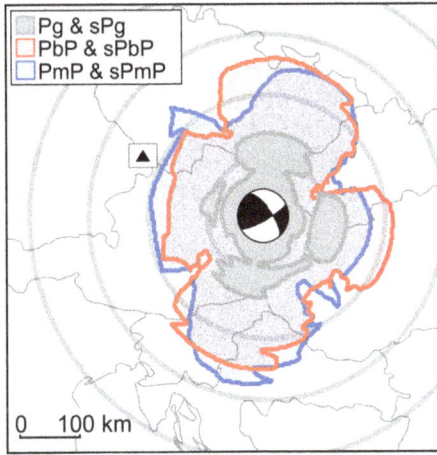

Figure 5. Areas of visibility for both phases of a depth phase pair derived from synthetic seismograms

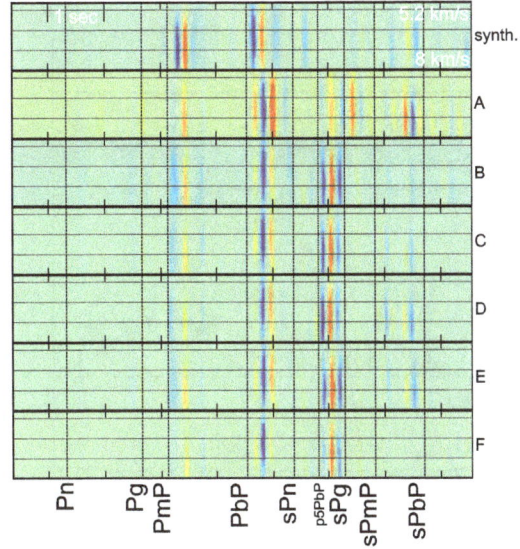

Figure 6. Vespagrams for synthetic data for the first earthquake at GERES and real data for all selected earthquakes from Table 1. Vespagrams are aligned on PmP arrival and phases were calculated with TauP of Crotwell and Owens (2011).

5 Array analysis with vespagrams

The synthetic traces in Fig. 3 as well as the spreading maps in Fig. 4 and 5 show, that PmP and PbP should be visible clearly because of their high amplitudes at GERES. However, S to P conversions like sPmP and sPbP are not visible. As mentioned before array processing can be used to improve SNR of time series by, e.g. creating vespagrams. Figure 6 shows the results for all earthquakes in our dataset in color and for the synthetics of the first earthquake A.

6 Polarization analysis of 3-component stations

Most elements of GERES are only recording the vertical movement. This is sufficient to identify backazimuth and slowness by array processing. However, GERES does also have four 3-component sensors, which are shown in the top left panel in Fig. 1. In the next step we try to use those stations to identify the phases visible in the vespagrams, using their polarization. For polarization analysis we used the method introduced by Vidale (1986). Figure 7 shows the results for earthquakes A to E in grey and for the synthetics of the first event in black. Earthquake F was not used because of low SNR due to the small magnitude.

7 Results

7.1 Regional depth phase spreading

The different propagation patterns for PmP and sPmP are shown in Fig. 4, together with the combination of both. Wave propagation for our dataset with a four layer model and a

record envelope for each grid point. Afterwards, we multiply those values for each phase pair at each grid point and normalize their power by extracting the square root. The interpolated results are shown in Fig. 4 superimposed on our area of interest. Dark coloring indicates high amplitudes for at least one phase of a depth

However, for depth estimation it is advantageous, though not necessary, to identify both. For this reason, we extract the area where both phases of a pair should have a significant amplitude. Based on the synthetic traces and real data from GERES, we know that we can identify PmP and PbP. Therefore, we assume that half the amplitude of PmP should be still visible. We draw a contour line around the area where each phase is above this minimum value and intersect the contours for each phase pair. The result for PmP and sPmP is shown in Fig. 4 in the right panel. Figure 5 puts together the contour lines for different depth phase pairs.

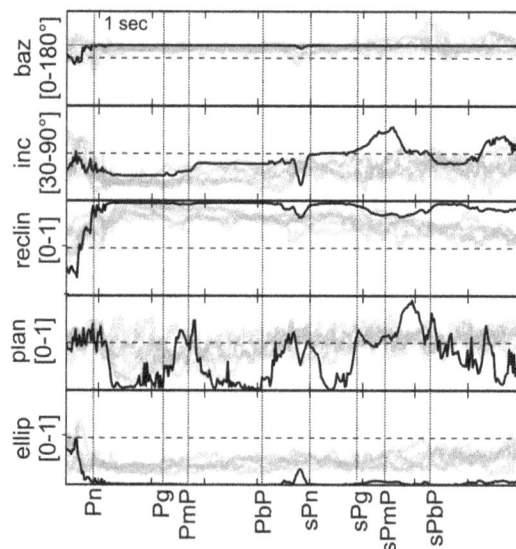

Figure 7. Polarization analysis for synthetic data (black) and earthquakes (grey). From top to bottom: backazimuth, incidence angle, rectilinearity, planarity and ellipticity.

strike-slip mechanism is more complex than anticipated from literature, which, among other factors assumes that the mechanism is not relevant. The PmP phase has propagation pattern like a P wave with four lobes with strong amplitudes between 100 and 200 km. On the other hand, the sPmP propagates like a S wave but is stronger at up to 90° azimuth to the PmP and a similar distance range. The areas where both phases are likely to be identified are depicted in the right panel in Fig. 4. Depending on the azimuth distances between 50 and 250 km can have strong amplitudes.

Figure 5 shows the outlines of three different depth phase combinations for our area of interest. Interestingly also the PbP and sPbP phase pair, which has not been used for depth estimations, shows a similar pattern to PmP and sPmP phases. Pg and sPg also show strong azimuth dependence.

7.2   Array analysis

The vespagrams for our data in Fig. 6 show a clear pattern for all earthquakes, which is also visible in the synthetics: A strong PmP with an apparent velocity of 7 km s^{-1} is followed by an even stronger PbP at 6.2 km s^{-1} even for our smallest used magnitudes. PbP arrives later than calculated, which indicated either a deeper Conrad discontinuity or higher velocities above it.

However, another strong phase follows with a high amplitude in the real data, shortly before the sPg. In the synthetics this phase is also visible and the most likely wavepath is an upward going P wave which reflects at the 5 km interface in the model and then propagates as PbP (p5PbP). The high amplitude indicates a depth below this interface and above the Conrad discontinuity. Also, the downwards reflection happens very close to the earthquakes, which would relate to a downward reflection from the bottom of the Vienna Basin.

Converted depth phases like sPg, sPmP and sPbP are not visible. Although we could not identify depth phases, depth is restricted since we know from modelling that a strong PbP is only possible for sources above the Conrad.

Furthermore, it is important to notice that PmP is the by far strongest phase arriving at GERES. This can lead to errors in location if it is mistaken for the Pn phase, which is the first arrival, but has much less energy.

The figures above show the results of the vespagrams of an event at GERES. The top figure shows the result for the recorded data, the bottom one for the synthetics. Although the Pn onset is clearly visible after processing, other phases cannot be identified because they occur very close to each other.

7.3   Polarization analysis

In the last step we analysed the polarization of data and compared it to the synthetic results (see Fig. 7). For the beginning of the P coda analysed by us the backazimuth stays stable at the estimated 115.5°. Inclination changes with the different phases arriving, in clear steps for the synthetic data and slowly for the real data. Planarity and ellipticity do not show a clear signal.

The Pn onset is clearly visible in backazimuth, incidence and rectilinearity down to a local magnitude of 2.6. Earthquake F with a Ml of 2.1 does not even show this feature and therefore was excluded.

8   Conclusions

Lateral analysis of synthetic data shows that regional depth phase propagation is strongly dependent not only on depth but also on the focal mechanism. Therefore, the knowledge of focal mechanism is important and a typical mechanism for the area needs to be used. Also, additional layers in the underground can produce strong reflections not reported by literature like PbP or p5PbP. Although no RDP pair could be identified, the visible regional phases restrict depth of the events in our dataset between a layer above and below the earthquakes, which relate to the bottom of the Vienna Basin and the Conrad discontinuity.

From comparison of the afore mentioned results to our dataset, we conclude that earthquakes from the Vienna Basin develop clear PmP and PbP arrivals at GERES. However, the converted depth phases sPn, sPg, sPmP and sPbP are not visible, not even with the improved SNR of the array. Yet, vespagram analysis can be used to identify phases in the P coda by their slowness even down to magnitudes of 2.1.

Further research will analyse RDP propagation around the Vienna Basin in more detail. One important feature that needs to be addressed is the low-velocity sediment layer, which is reported to weaken converted phases we are looking for. Also, we will investigate data from our dataset recorded at stations, which are in areas where the synthetics indicate high amplitudes for RDPs. Since time difference between RDP pairs is mainly affected by depth, as stated in Ma (2010), it should be possible to use phase readings of single phases at varying distances to estimate depth.

With newly deployed dense seismic network like the AlpArray presented in Fuchs et al. (2015) or profiles like EASI (see Plomerova et al. (2015) wave propagation could be monitored and maybe even tracked across the region. This extended knowledge of RDP behaviour can then be used to locate even small earthquakes more accurately as more information than first picks is available.

Acknowledgements. We thank the Zentralanstalt für Meteorologie und Geodynamik (ZAMG) and the Bundesanstalt für Geowissenschaften und Rohstoffe (BGR) for making available seismic data for this study. Plots were created with ObsPy (Beyreuther et al., 2010) and maps with QGIS (QGIS, 2009). Synthetics were generated with Computer Programs in Seismology by Herrmann (2013). Travel times were calculated with TauP (Crotwell and Owens, 2011).

References

Apoloner, M.-T. and Bokelmann, G.: Ebreichsdorf 2013 Earthquake Series: Relative Location, in: EGU General Assembly, 12–17 April 2015, Vienna, Austria, 2015.

Behm, M., Brückl, E., and Mitterbauer, U.: A New Seismic Model of the Eastern Alps and its Relevance for Geodesy and Geodynamics, VGI Österreichische Zeitschrift für Vermessung & Geoinformation, 2, 121–133, 2007.

Beyreuther, M., Barsch, R., Krischer, L., Megies, T., Behr, Y., and Wassermann, J.: ObsPy: A Python Toolbox for Seismology, Seismol. Res. Lett., 81, 530–533, 2010.

Brückl, E., Behm, M., Decker, K., Grad, M., Guterch, A., Keller, G. R., and Thybo, H.: Crustal structure and active tectonics in the Eastern Alps, Tectonics, 29, doi:10.1029/2009TC002491, 2010.

Crotwell, H. P. and Owens, T. J.: The TauP Toolkit: Flexible Seismic Travel-Time and Raypath Utilities Documentation Version 2.0, 2011.

Fuchs, F., Bokelmann, G., Bianchi, I., Apoloner, M.-T., and Group, A. W.: AlpArray Austria – Illuminating the subsurface of Austria and understanding of Alpine geodynamics, in: EGU General Assembly, 12–17 April 2015, Vienna, Austria, 2015.

Harjes, H.-P.: Design and siting of a new regional array in Central Europe, B. Seismol. Soc. Am., 80, 1801–1817, 1990.

Harjes, H.-P., Jost, M. L., Schweitzer, J., and Gestermann, N.: Automatic Seismogram Analysis at GERESS, Comput. Geosci., 19, 157–166, doi:10.1016/0098-3004(93)90113-J, 1993.

Hausmann, H., Hoyer, S., Schurr, B., Brückl, E., Houseman, G., and Stuart, G.: New Seismic Data improve earthquake location in the Vienna Basin Area, Austria, Austrian Journal of Earth Sciences, 103, 2–14, 2010.

Hausmann, H., Meurers, R., and Horn, N.: The 2013 Earthquakes in the Vienna Basin: Results from strong-motion and macroseismic data, in: Second European Conference of Earthquake Engineering and Seismology, Istanbul, Turkey, 2014.

Herrmann, R. B.: Computer programs in seismology: An evolving tool for instruction and research, Seismol. Res. Lett., 84, 1081–1088, doi:10.1785/0220110096, 2013.

Ma, S.: Focal Depth Determination for Moderate and Small Earthquakes by Modeling Regional Depth Phases sPg, sPmP, and sPn, B. Seismol. Soc. Am., 100, 1073–1088, doi:10.1785/0120090103, 2010.

Ma, S.: Earthquake Research and Analysis - Seismology, Seismotectonic and Earthquake Geology, chap. Focal Depth Determination for Moderate and Small Earthquakes by Modeling Regional Depth Phases sPg, sPmP, and sPn, InTech, 143–166, doi:10.5772/27240, 2012.

Ma, S. and Atkinson, G. M.: Focal Depths for Small to Moderate Earthquakes, B. Seismol. Soc. Am., 96, 609–623, doi:10.1785/0120040192, 2006.

Ma, S. and Eaton, D. W.: Combining Double-Difference Relocation with Regional Depth-Phase Modelling to Improve Hypocentre Accuracy, Geophys. J. Int., 185, 871–889, doi:10.1111/j.1365-246X.2011.04972.x, 2011.

Plomerova, J., Bianchi, I., Hetényi, G., Munzarová, H., Bokelmann, G., Kissling, E., AlpArray-EASI-Working-Group, and AlpArray-EASI-Field-Team: The Eastern Alpine Seismic Investigation (EASI) project, in: EGU General Assembly, 12–17 April 2015, Vienna, Austria, 2015.

QGIS, D. T.: QGIS Geographic Information System, Open Source Geospatial Foundation, available at: http://qgis.osgeo.org (last access: 31. August 2015), 2009.

Rost, S. and Thomas, C.: Array Seismology. Methods and Applications, Rev. Geophysics, 40, 2-1–2-27, doi:10.1029/2000RG000100, 1008, 2002.

Vidale, J. E.: Complex polarization analysis of particle motion, B. Seismol. Soc. Am., 76, 1393–1405, 1986.

Wessely, G.: Zur Geologie und Hydrodynamik im südlichen Wiener Becken und seiner Randzone, Mitteilungen der österreichischen geologischen Gesellschaft, 76, 27–68, 1983.

20

The installation campaign of 9 seismic stations around the KTB site to test anisotropy detection by the Receiver Function Technique

I. Bianchi[1], M. Anselmi[2], M. T. Apoloner[1], E. Qorbani[1], K. Gribovski[1,3], and G. Bokelmann[1]

[1]Institut für Meteorologie und Geophysik, Universität Wien, 1090 Wien, Austria
[2]Sezione Sismologia e Tettonofisica, Istituto Nazionale di Geofisica e Vulcanologia, 00143 Roma, Italy
[3]MTA CSFK Geodéziai és Geofizikai Intézet, 9400, Sopron, Csatkai E. u. 6–8, Hungary

Correspondence to: I. Bianchi (irene.bianchi@univie.ac.at)

Abstract. The project at hand is a field test around the KTB (Kontinentale Tiefbohrung) site in the Oberpfalz, Southeastern Germany, at the northwestern edge of the Bohemian Massif. The region has been extensively studied through the analysis of several seismic reflection lines deployed around the drilling site. The deep borehole had been placed into gneiss rocks of the Zone Erbendorf-Vohenstrauss. Drilling activity lasted from 1987 to 1994, and it descended down to a depth of 9101 m.

In our experiment, we aim to recover structural information as well as anisotropy of the upper crust using the receiver function technique. This retrieved information is the basis for comparing the out-coming anisotropy amount and orientation with information of rock samples from up to 9 km depth, and with high-frequency seismic experiments around the drill site.

For that purpose, we installed 9 seismic stations, and recorded seismicity continuously for two years from June 2012 to July 2014.

1 Tectonic setting of the study area

The region hosting the deployed seismic stations has been focus of several geophysical investigations during the nineties (DEKORP Research Group, 1987, 1988; Eisbacher et al., 1989; Lüschen et al., 1996; Harjes et al., 1997; DEKORP and Orogenic processes Working Groups, 1999; Muller et al., 1999). Figure 1a shows surface geological units as well as seismic reflection lines around the drilling site. The deep borehole was placed into the gneiss rocks of the Zone Erbendorf-Vohenstrauss. An alternating sequence of paragneisses and amphibolites was found, with metamorphism of upper amphibolite facies conditions. Ductile deformation produced a strong foliation of the rocks. Unexpected though was the steep inclination of that pervasive foliation, which did not correspond to previous interpretations of flat seismic reflections and mapped surface geology (Harjes et al., 1997). Previous tectonic interpretations had to be strongly modified, to explain the in-situ information from the borehole (Emmernann and Lauterjung, 1997; O'Brien et al., 1997). Principal results from the drilling and accompanying geophysical experiments are described in a special JGR volume (Haak and Jones, 1997; Emmermann and Wohlenberg, 1989).

Further geophysical experiments have been performed, especially a 3D seismic experiment around KTB, the ISO89 experiment (Fig. 1) (DEKORP Research Group, 1988; DEKORP and Orogenic processes Working Groups, 1999), as well as a Moving-Source Profiling (MSP) experiment that constrained the effective anisotropy of the upper eight kilometers (Okaya et al., 2004). This anisotropy is apparently not controlled by the faults, but instead by foliation and/or the stress-induced cracks. At mid-crustal depth, a high-conductivity zone had been suggested with maximum conductivity in North-South direction, which is perhaps consistent with the presence of a basal detachment horizon.

This metamorphic (anisotropy bearing) body was drilled till the whole length of the KTB drill. The expectations to reach the Saxothurignan-Moldanubian boundary below few kilometers of metamorphic rocks were betrayed at that time. The Saxothurignan (in the NW of the area) and Moldanubian (in the SE of the area) crustal terrains came in contact during the Variscan orogenesis. Their suture line should lie in the study area.

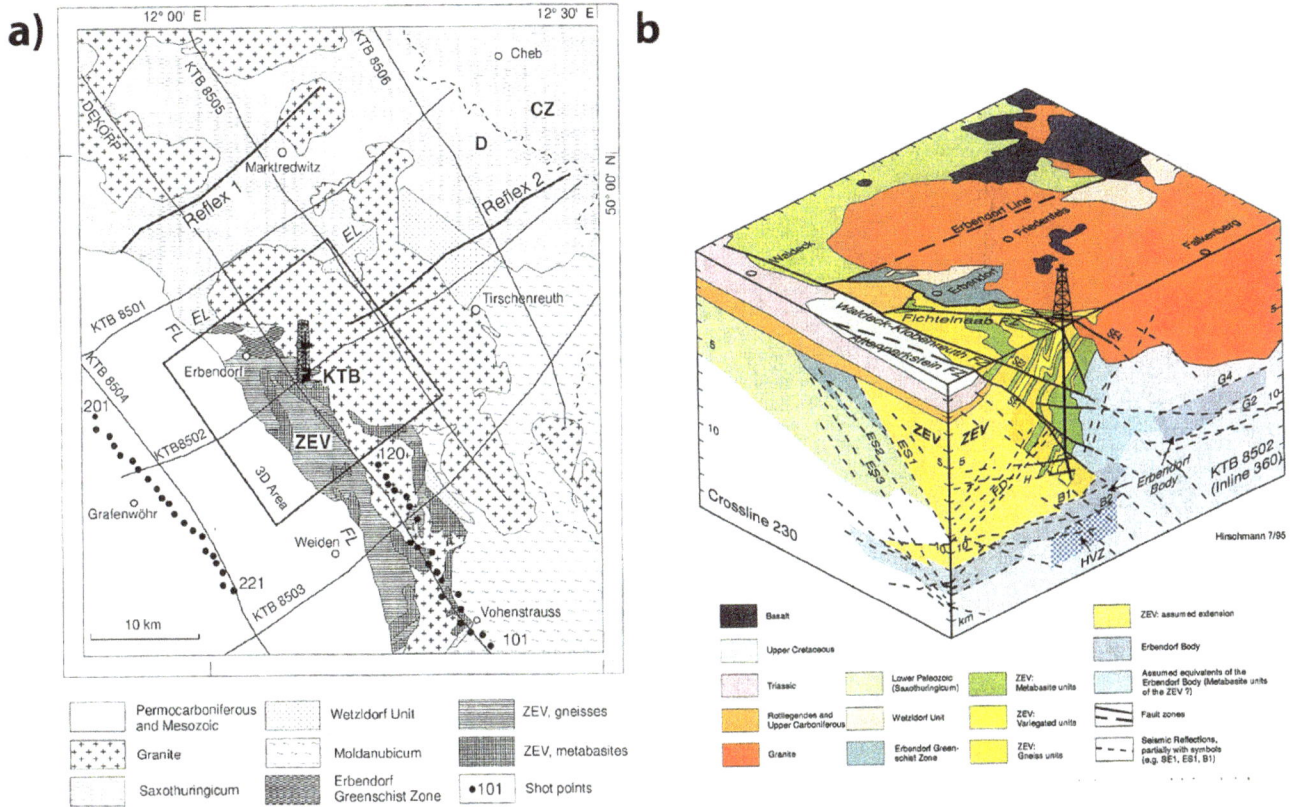

Figure 1. (a) Geology in the Oberpfalz region as well as seismic reflection lines (DEKORP Research Group, 1988; DEKORP and Orogenic processes Working Groups, 1999), and (b) a block diagram of the KTB area (Harjes et al., 1997).

Table 1. Station location.

STAT	Locality	LON	LAT	ALT
KW01	Nottersdorf	12.1192	49.8059	484
KW02	Arnoldsreuth	12.0077	49.9239	616
KW03	Kölerhof	12.2913	49.8195	558
KW04	Parkstein	12.0846	49.7246	458
KW05	Altendorf	11.9834	49.752	502
KW06	Lenkermühle	12.1184	49.7762	499
KW07	Moosbach	12.4533	49.5859	588
KW08	Öberholl	12.225	49.668	545
KW09	Siegritz	12.0682	49.8595	561

2   Scientific goal of the project

The project performs a field test around the KTB (Kontinental Tiefbohrung) site in the Oberpfalz in Southeastern Germany. The primary purpose of this experiment is to test the receiver function technique, using both the radial and transverse components, and to compare it with previous results from deep drilling, and high-frequency seismic experiments around the drill site. Beside that primary goal, we also hope to shed light on the transition between the different crustal terrains in that area, i.e. the transition be-

tween Saxothuringian and Moldanubian crustal terrains, and to study the deeper parts of that transition zone. If we can observe this transition especially detecting anisotropy by receiver functions, then there are many very promising applications of this technique.

The selected area is a perfect test laboratory due to the presence of the KTB drilling site. Crustal rocks were extracted till 9 km of depth, kept and classified few meters from the drill site; they are accessible to the public. Moreover the area was investigated by active seismic profiles, rendering accessible the knowledge of the shallow crustal structures. The final aim is to compare the anisotropy retrieved by receiver functions with anisotropy detected by previous studies; in this comparison the different frequency content of the measurements and the type of anisotropy must be considered. The wide-angle shots recorded by 3-component geophones in the KTB pilot hole provided convincing evidence for S-wave splitting in the upper crust SE of the KTB location. This anisotropy effect seemed to be related to the overall strike and dip of rock foliation (Gebrande et al., 1991).

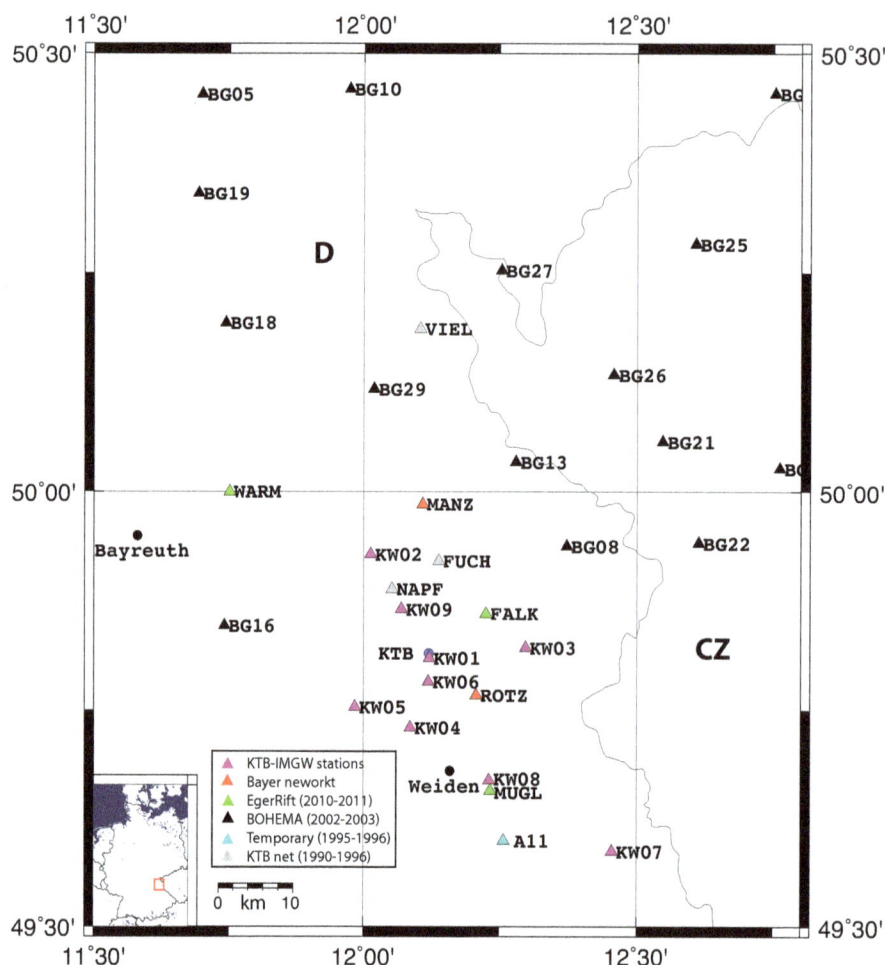

Figure 2. Map of the study are with seismic networks deployed in the study area since 1990. The blue circle represents the location of the KTB drilling site. In the inset a map of Germany with a red box displaying the location of the study area.

3   Permanent and temporary networks in the area

Before selecting the locations for the seismic station installations, we explored the literature to unravel the presence of seismic networks that ran or are still running in the area.

The permanent seismic Network in Bavaria, installed and maintained by the Geophysikalisches Observatorium der Ludwig-Maximilians Universität Ludwigshöhe in Fürstenfeldbruck is running in the study area. Stations of particular interest for this experiment are ROTZ and MANZ (Fig. 2). Recently (2010 to 2011) a broadband seismic network was installed in the area during the EgerRift experiment (e.g., Babuska and Plomerova, 2013) (Fig. 2). Records of the original KTB broadband network are available from 1990 to 1996 (Dahlheim et al., 1997). As part of a seismic broadband line the station A11 (Fig. 2) was deployed during 1995 to 1996 (Plenefisch et al., 2001). Between 2002 and 2003 broadband stations (Fig. 2) were installed over a broader area during the BOHEMA (BOhemian Massif Anisotropy and HEterogeneity) experiment (Plomerová et al., 2007). According to the

recordings available of those previously installed stations, we installed 9 more stations (Table 1, Fig. 2) in order to re-trace the geometry of the DEKORP (Deutsches Kontinentales Reflexionsseismisches Programm) and KTB8502 (Fig. 1).

4   The seismic deployment

Network geometry has to consider the shape of the metamorphic Erbendorf-Vohenstrauss body, the location of the main active seismic profiles, and the location of previously installed seismic stations. Nine stations have been deployed to sample the metamorphic body and to highlight the structural differences in the shallow crust.

Each recording site is equipped with a Reftek 130-01 digitizer, and a sensor Reftek 151B-60 (Fig. 3). Before the installation of the temporary network a site search campaign was performed. The ideal site for the location of a seismic station is characterized by a low background seismic noise. This happens far away from cities, infrastructures, quarries, indus-

Instrument	Type	Picture	Technical characteristics	
REFTEK **130-1** [REF TEK]	Digitizer		Dynamics	> 135 db
			Resolution	24 bit - 20VPP
REFTEK **151B-60**	Sensor		Frequency response	0.0166 Hz (60 sec) - 50 Hz
GARMIN	GPS antenna			
	GSM antenna			
DIGI	Modem	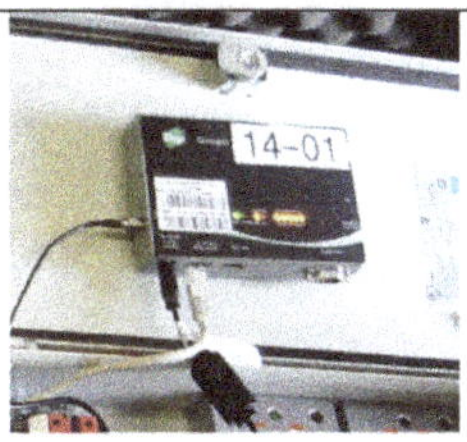		

Figure 3. From top to bottom: Reftek digitizer 130-01; seismic sensor Reftek 151-60; GPS Garmin antenna; GSM antenna; modem. All of them were used during the acquisition campaign.

trial areas, woods, rivers, lakes, and the sea. These conditions cannot always be met; moreover the final choice of the location was done on the basis of considerations that brought a compromise between these ideals locations, and the possibility of installing the stations inside buildings, in order to preserve their safety.

The stations where installed since July 2012 to October 2012 as standalone, station KW08 only was used as a test for the telemetric system. Since October 2012 all stations transmitted data through the telemetric system. The station configuration was set to a gain of 32 and sampling rates of 1 and 100 Hz.

5 Site characterization

The estimate of the background noise of a determined site and its frequency content analysis is one of the most important criteria for establishing a good performance of a seismic station.

An estimate of the noise level of the sites was performed using 1 h of recorded noise (during daytime) for each installed station. The registered noise was analyzed by its spectral characteristics (Fig. 4).

The power-spectral-density (PSD) gives the power of the recording as a function of frequency, expressed in decibel (dB) (e.g. Marzorati and Bindi, 2006) and is used to verify the amount of noise for the selected sites. Beside this we

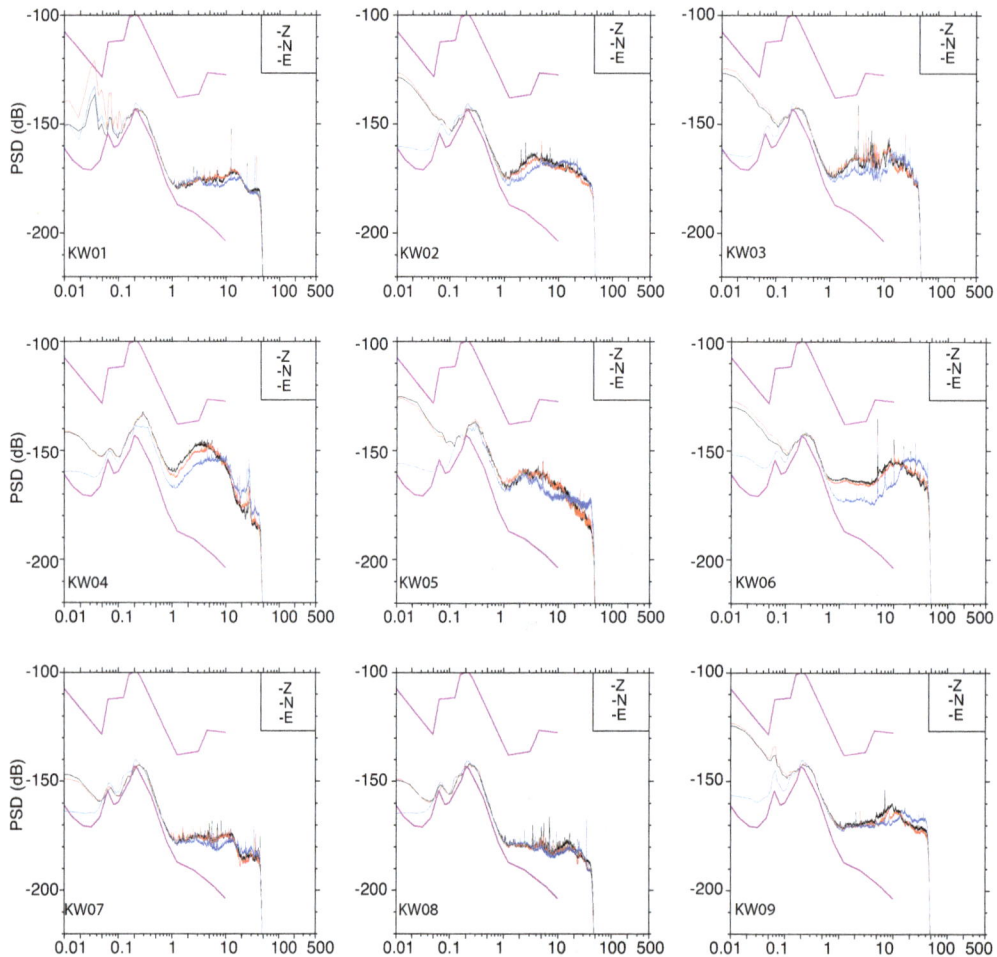

Figure 4. Power-spectral-density (PSD) in dB of the noise recorded at the stations. All sensors are 60 s, sampling 100 Hz. The top and bottom magenta lines represent the Peterson high- and low-noise curves (Peterson, 1993). Real spectra are located between these two curves. The x axis shows frequency in Hz.

analyzed the spectral ratio between the two horizontal components (N and E) with respect to the vertical one (Z), in order to estimate the soil amplification characteristics as a frequency function for each horizontal component (N and E) separately (Fig. 5). This analysis is slightly modified after Nakamura (1989) that analyzes the H/V ratio, where H is the vector sum of the horizontal vectors N and E.

To evaluate these results, we take into account the geological characteristics of the sites. The sites have been selected considering the geometrical needs of the network itself, and according to past seismic experiments data accessibility, in order to get the maximum rays coverage in the area. Most of the stations are located on the bedrock either of metamorphic or magmatic origin. The possibility to build a seismic station on such rocks ensures the background noise to be low. Two of the stations (namely KW04 and KW05) are located on a sedimentary layer; their background noise level is on average higher with respect to the other stations.

The PSD in Fig. 4, for low frequencies (less than 1 Hz) span from −180 to −130 dB, close to the low noise Peterson's reference curve (Peterson, 1993). The worse performances belong to KW04 and KW05 reaching −130 and −135 dB for low frequencies. High frequency peaks appear particularly at stations KW06 and KW03. These are located at ground floor levels of inhabited farms.

In the range of frequency needed for our analysis (0.5–4 Hz), PSD values are close to the low noise level, and therefore we regard the station deployment as satisfying.

6 Discussion

The final geometry of the installed seismic network resulted from a good balance between site availability and ideal station location.

The seismic experiment started in June 2012 and finished in July 2014, in order to get full backazimuthal coverage of the teleseismic rays at each station, necessary to investigate

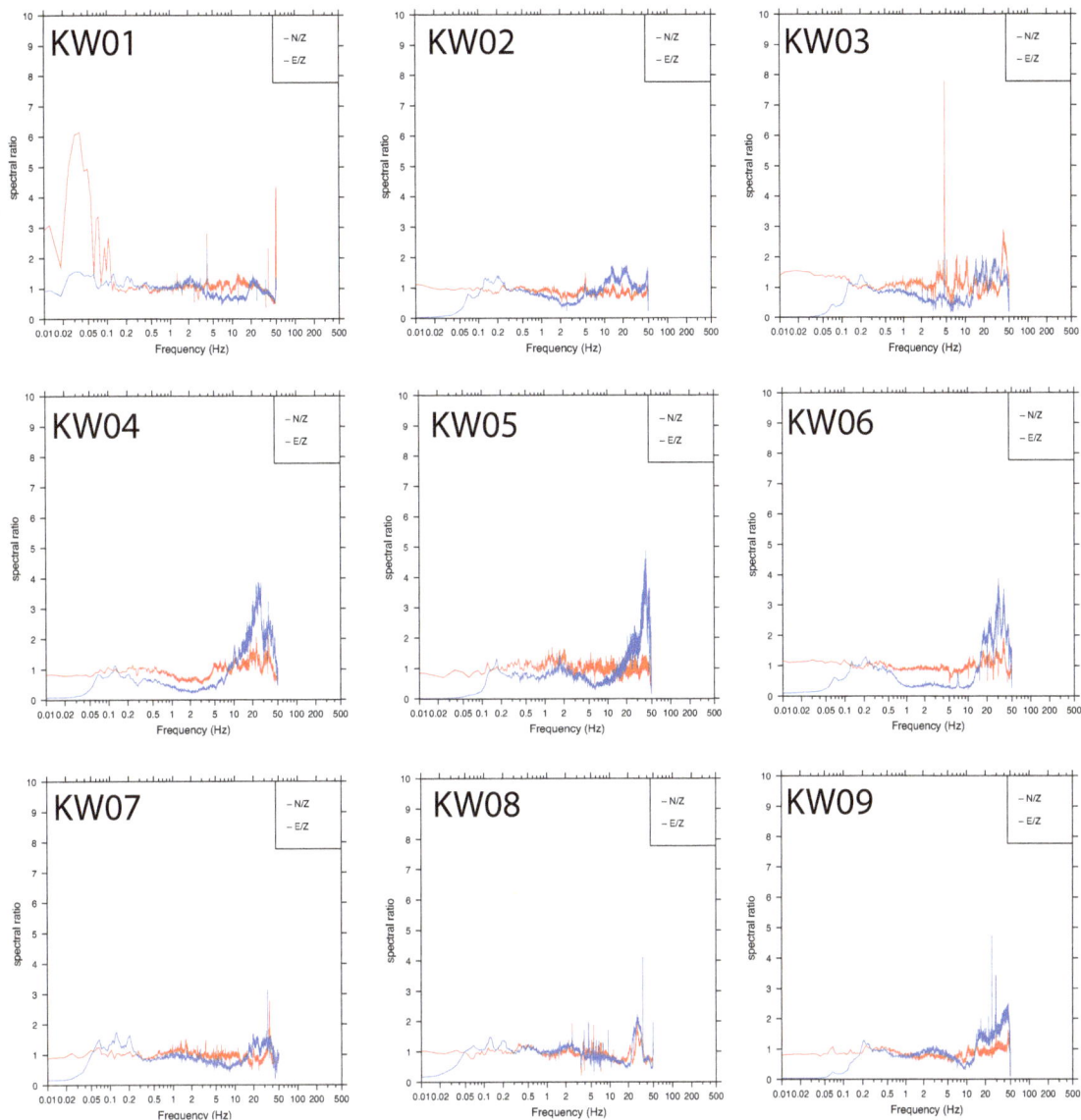

Figure 5. Spectral amplitude ratio of the two horizontal components (E and N) vs the vertical component (Z), modified after Nakamura (1989).

both the shallow and the deep crust, to unravel the seismic characteristics of the metamorphic Erbendorf body, and to explore the Saxoturingian-Moldanubian boundary zone.

Network potential resolution

In order to understand the potential resolution of the network with respect to the geometries of the structures below the surface, we show the hit surfaces of the teleseismic rays at 10 and 30 km depth, in Fig. 6a and b, respectively.

As deduced from Fig. 6a, 3 stations (KW01, KW06 and KW08) will sample the area above the Erbendorf body, entirely from the surface to its depth (at least 9 km). Data recorded at these stations have the potential of revealing the anisotropy of the metamorphic body. The other stations sam-

ple the surrounding area; teleseismic rays arriving at these stations are expected to sample the Erbendorf body at greater depth, if present.

The full ray coverage of the area at 30 km depth (Fig. 6b) will provide information on the buried boundary between the Saxoturingian and Moldanubian terrain, which nature, extent and location is still unclear, despite the intense seismic studies performed in the past in this area (Gebrande et al., 1991).

7   Conclusions

The passive seismic experiment at KTB started on June 2012 and finished in July 2014, during which a continuous seismic dataset at 9 sites has been acquired. Combining this dataset

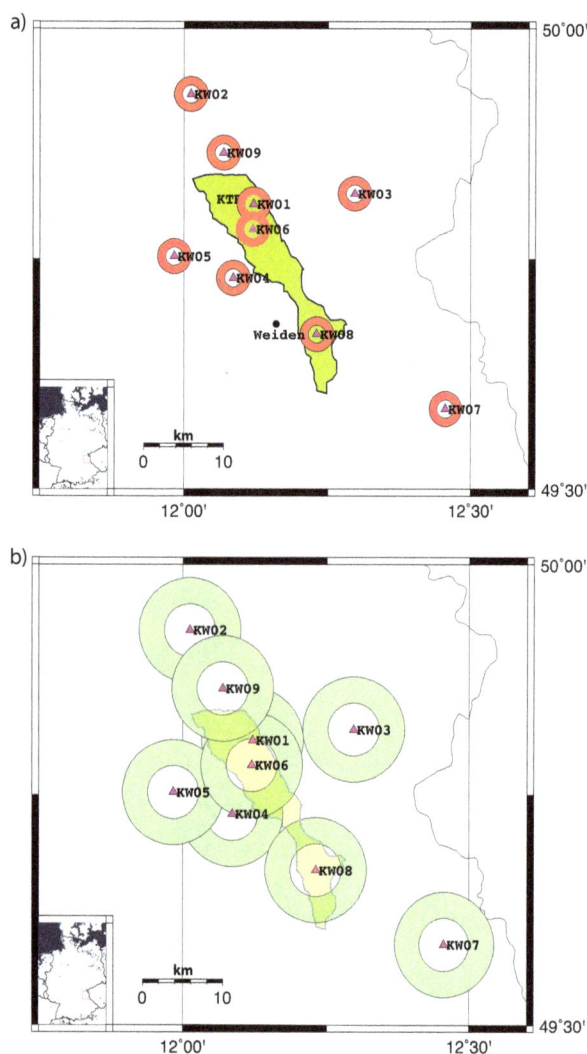

Figure 6. Map of the study area, with station location, the location of the metamorphic Erbendorf body at surface (green), and the areal coverage of teleseismic rays projected at surface for 10 km (**a**) and 30 km (**b**), considering epicentral distances of the incoming rays between 30 an 100.

with the records at the two permanent stations (Bayern Network) in the area, and collecting data from past passive seismic experiments provided a large number of seismic records which rays sample the subsurface structures from all directions. The main goal is to determine the anisotropic characteristics of the Erbendorf body and to compare them with the direct information from rock samples extracted till 9 km depth, in order to understand which of the intrinsic characteristics of the rocks cause the anisotropy effect often observed in seismic records.

The receiver function data set is a useful tool to unravel the presence of discontinuities at depth, allowing us to explore the whole crust, with the aim of deciphering the nature and location of the hidden Saxothuringian-Moldanubian boundary in the crust.

Appendix A

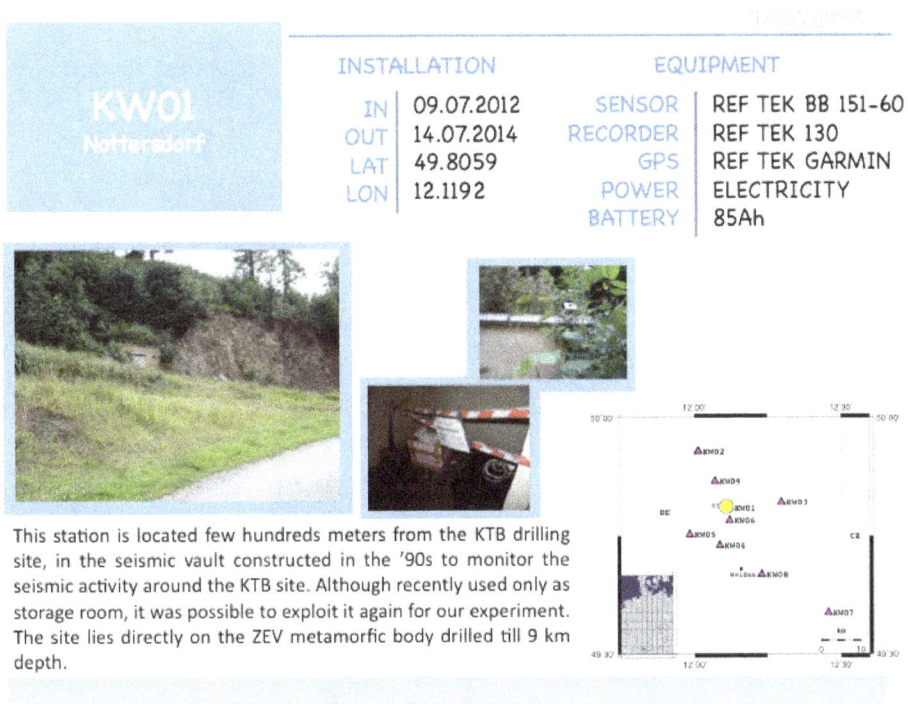

KW01 Notteradorf	INSTALLATION		EQUIPMENT	
	IN	09.07.2012	SENSOR	REF TEK BB 151-60
	OUT	14.07.2014	RECORDER	REF TEK 130
	LAT	49.8059	GPS	REF TEK GARMIN
	LON	12.1192	POWER	ELECTRICITY
			BATTERY	85Ah

This station is located few hundreds meters from the KTB drilling site, in the seismic vault constructed in the '90s to monitor the seismic activity around the KTB site. Although recently used only as storage room, it was possible to exploit it again for our experiment. The site lies directly on the ZEV metamorfic body drilled till 9 km depth.

Figure A1. Station KW01 technical information.

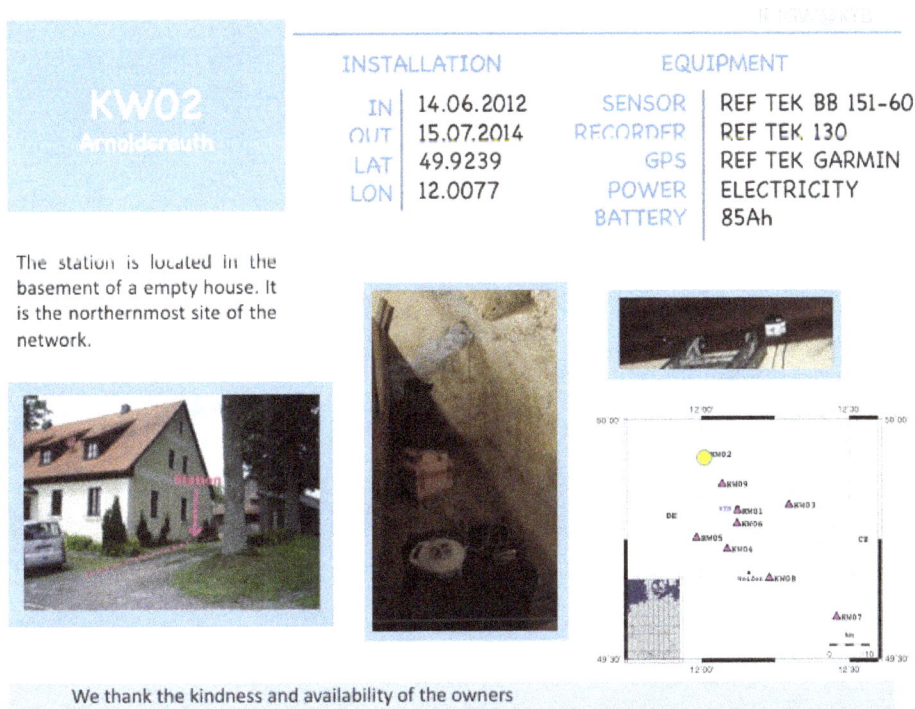

KW02 Arnoldsreuth	INSTALLATION		EQUIPMENT	
	IN	14.06.2012	SENSOR	REF TEK BB 151-60
	OUT	15.07.2014	RECORDER	REF TEK 130
	LAT	49.9239	GPS	REF TEK GARMIN
	LON	12.0077	POWER	ELECTRICITY
			BATTERY	85Ah

The station is located in the basement of a empty house. It is the northernmost site of the network.

We thank the kindness and availability of the owners

Figure A2. Station KW02 technical information.

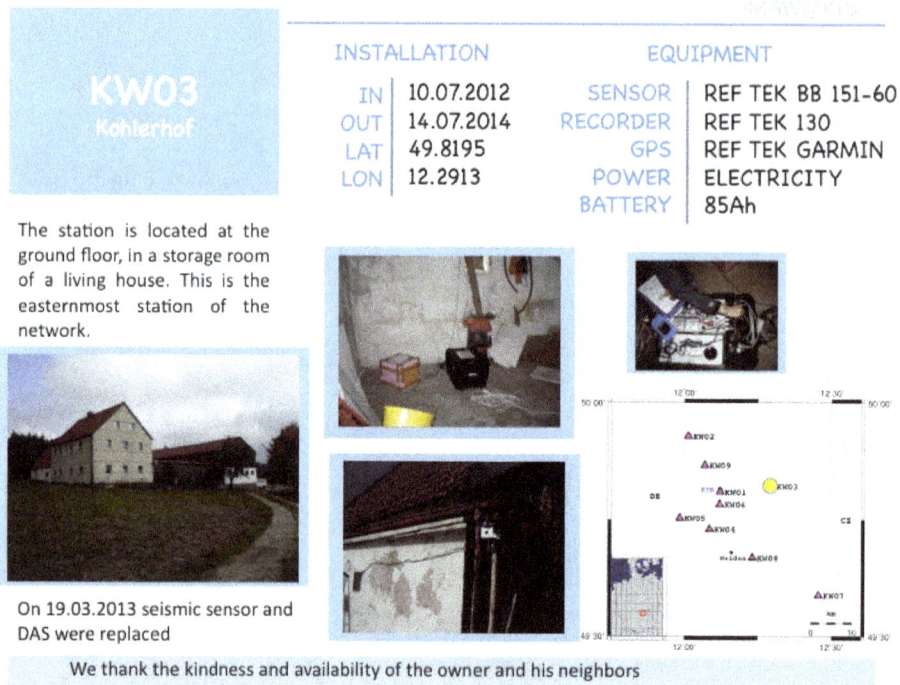

KW03
Köhlerhof

INSTALLATION		EQUIPMENT	
IN	10.07.2012	SENSOR	REF TEK BB 151-60
OUT	14.07.2014	RECORDER	REF TEK 130
LAT	49.8195	GPS	REF TEK GARMIN
LON	12.2913	POWER	ELECTRICITY
		BATTERY	85Ah

The station is located at the ground floor, in a storage room of a living house. This is the easternmost station of the network.

On 19.03.2013 seismic sensor and DAS were replaced

We thank the kindness and availability of the owner and his neighbors

Figure A3. Station KW03 technical information.

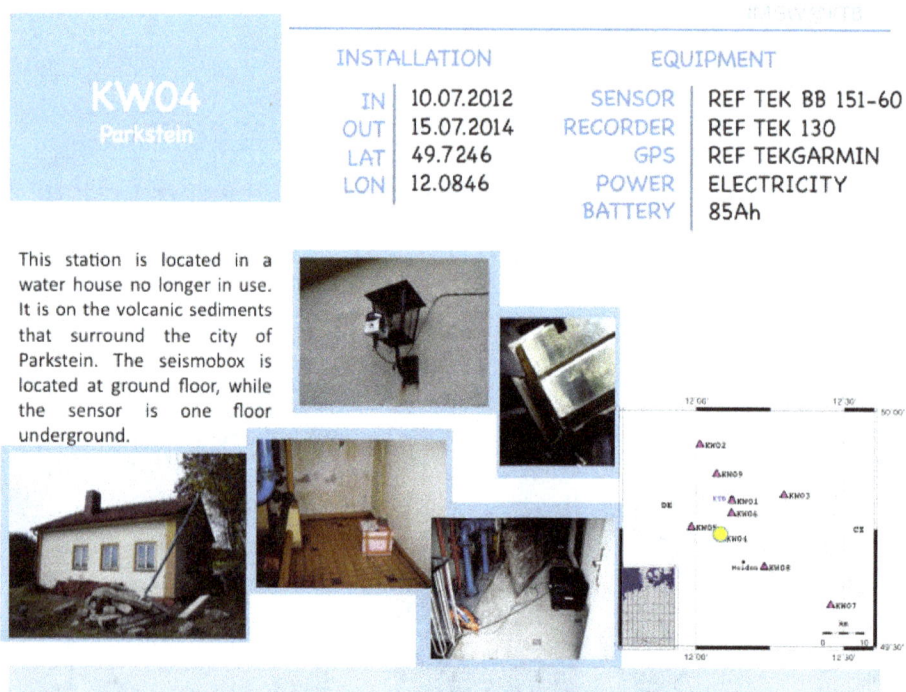

KW04
Parkstein

INSTALLATION		EQUIPMENT	
IN	10.07.2012	SENSOR	REF TEK BB 151-60
OUT	15.07.2014	RECORDER	REF TEK 130
LAT	49.7246	GPS	REF TEKGARMIN
LON	12.0846	POWER	ELECTRICITY
		BATTERY	85Ah

This station is located in a water house no longer in use. It is on the volcanic sediments that surround the city of Parkstein. The seismobox is located at ground floor, while the sensor is one floor underground.

Figure A4. Station KW04 technical information.

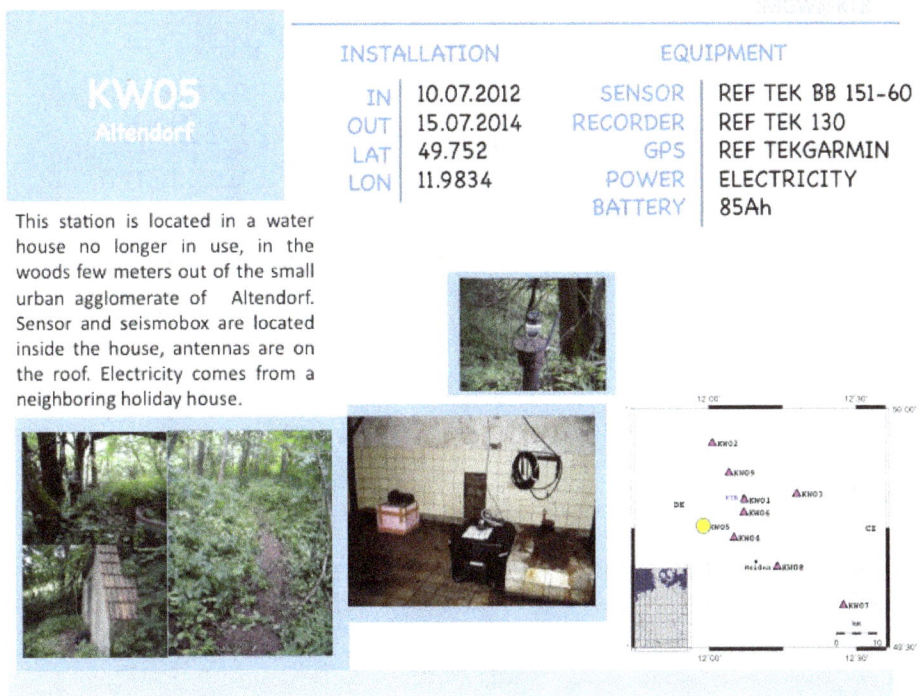

Figure A5. Station KW05 technical information.

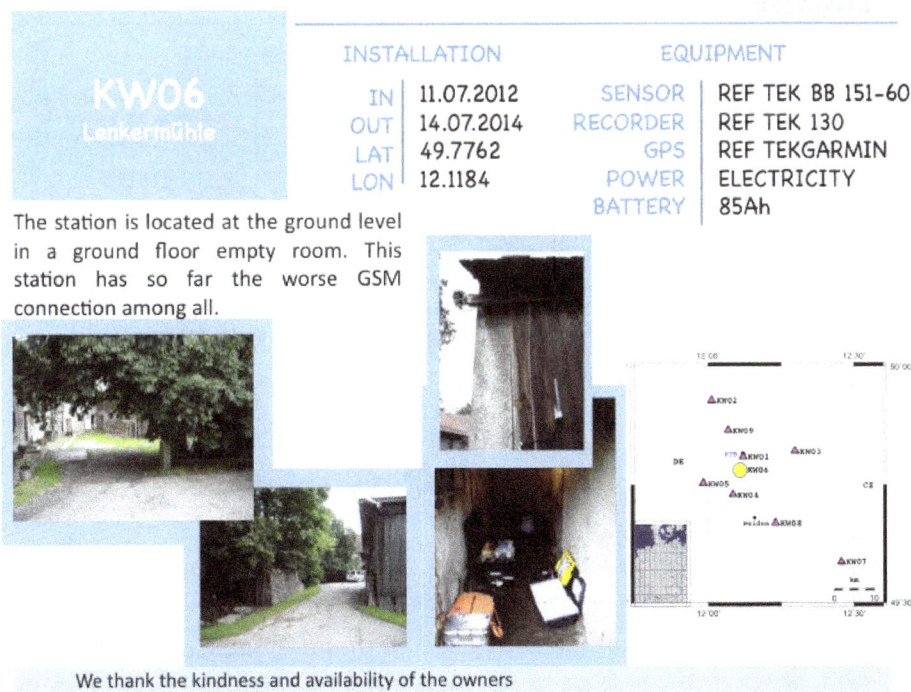

Figure A6. Station KW06 technical information.

KW07
Moosbach

INSTALLATION		EQUIPMENT	
IN	11.07.2012	SENSOR	REF TEK BB 151-60
OUT	14.07.2014	RECORDER	REF TEK 130
LAT	49.5859	GPS	REF TEKGARMIN
LON	12.4533	POWER	ELECTRICITY
		BATTERY	85Ah

This station is located in a functioning water house. Seismobox and sensor are located one floor underground. The sensor was oriented parallel to the wall of the house due to the impossibility to detect the Earth magnetic field there. The recorded N component is thus ~35 deg E rotated respect to the Magnetic North.

GPS problems in the first months of acquisition.
GPS stopped working on jd 293 in 201, restarted normally on jd 078 in 2013 (19.03.2013)

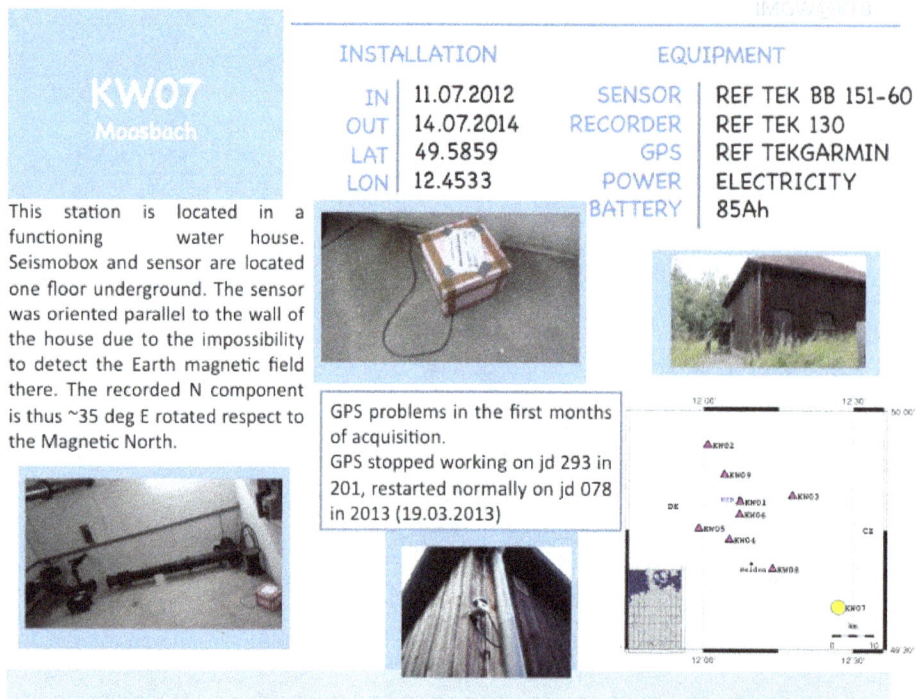

Figure A7. Station KW07 technical information.

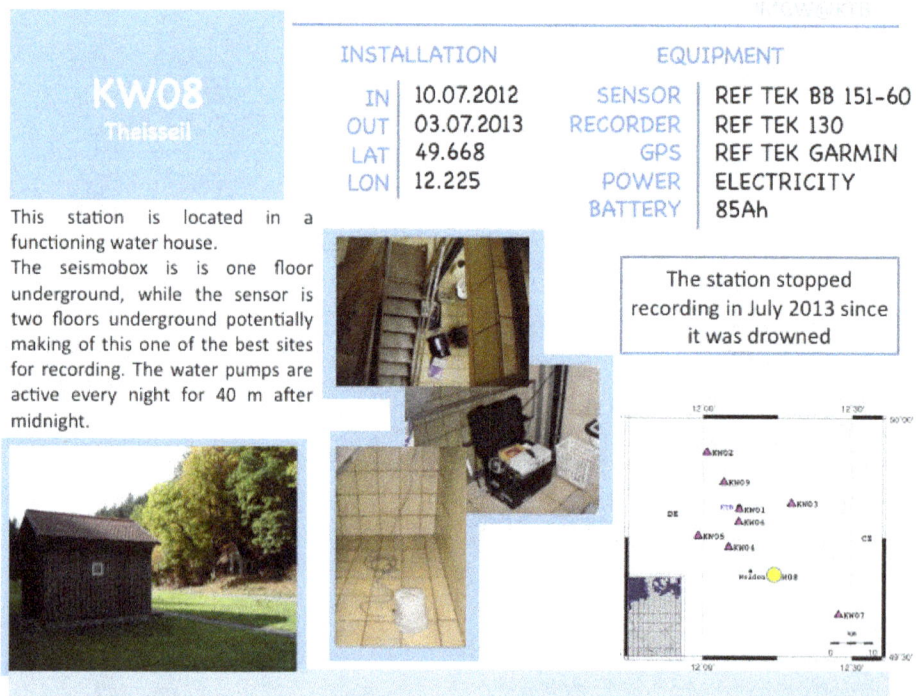

KW08
Theissel

INSTALLATION		EQUIPMENT	
IN	10.07.2012	SENSOR	REF TEK BB 151-60
OUT	03.07.2013	RECORDER	REF TEK 130
LAT	49.668	GPS	REF TEK GARMIN
LON	12.225	POWER	ELECTRICITY
		BATTERY	85Ah

This station is located in a functioning water house.
The seismobox is is one floor underground, while the sensor is two floors underground potentially making of this one of the best sites for recording. The water pumps are active every night for 40 m after midnight.

The station stopped recording in July 2013 since it was drowned

Figure A8. Station KW08 technical information.

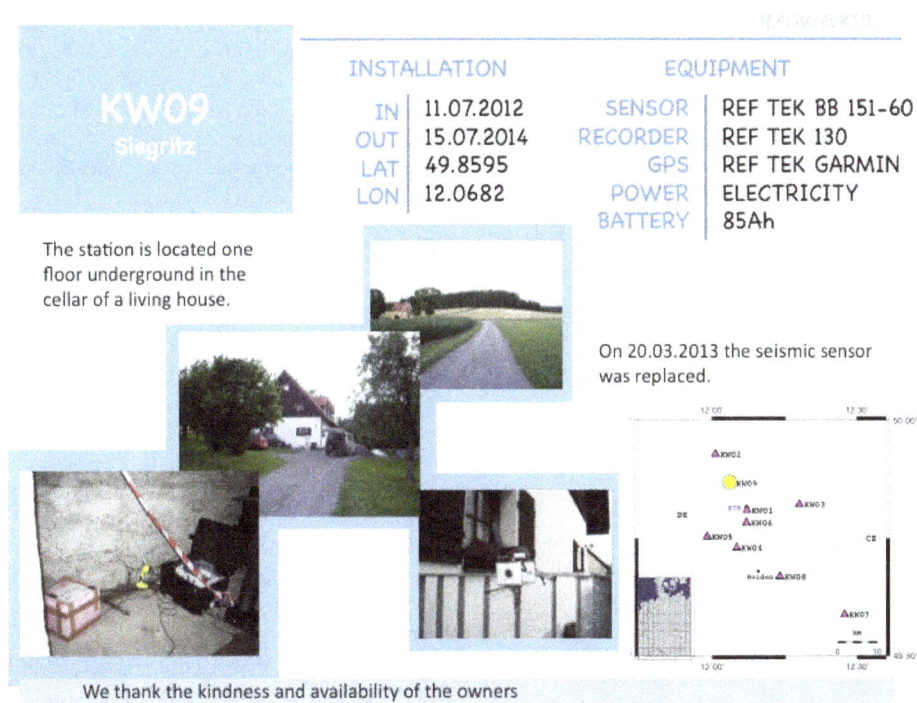

Figure A9. Station KW09 technical information.

Acknowledgements. We thank Hans-Albert Dahlheim for his precious advice, and help during the field campaign. We are also grateful to Franz Holzförster, scientific director of the KTB Geozentrum, for giving us the opportunity to use the spaces of the geo-center. We thank P. Jordakiev, P. Arneitz and A. Gerner for their help during the installation campaign. We acknowledge funding by the Austrian Science Fund (FWF): 24218. Our acknowledgements go to J. Plomerova and T. Plenefisch for their comments that contribute improving our manuscript.

References

Babuska V. and Plomerova J.: Boundaries of mantle-lithosphere domains in the Bohemian Massif as extinct exhumation channels for high-pressure rocks, Gondwana Res., 23, 973–987, doi:10.1016/j.gr.2012.07.005, 2013.

Dahlheim, H. A., Gebrande, H., Schmedes, E., and Soffel H.: Seismicity and stress field in the vicinity of the KTB location, J. Geophys. Res., 102, 493–506, 1997.

DEKORP Research Group: Near-vertical and wide-angle seismic surveys in the Black Forest, SW Germany, J. Geophys., 62, 1–30, 1987.

DEKORP Research Group: Results of the DEKORP 4/KTB Oberpfalz deep seismic reflection investigations, J. Geophys., 62, 69–101, 1988.

DEKORP and Orogenic processes Working Groups: Exhumation of subducted crust – the Saxonian granulites from reflection seismic experiment GRANU' 95. Tectonics, 18, 756–773, 1999.

Eisbacher, G.-H., Lüschen, E., and Wickert, F.: Crustal-scale thrusting and extension in the Hercynian Schwarzwald and Vosges, Central Europe, Tectonics, 8, 1–21, 1989.

Emmermann, R. and Lauterjung, J.: The German Continental Deep Drilling Program KTB: Overview and major results, J. Geophys. Res., 102, 179–201, 1997.

Emmermann, R. and Wohlenberg, J.: The German continental deep drilling program (KTB): Site-selection studies in the Oberpfalz and Schwarzwald, Springer, Berlin-Heidelberg, 553 pp., 1989.

Gebrande, H., Bopp, M., MeichelböCk, M., and Neurieder, P.: 3-D Wide-Angle Investigations in the Ktb Surroundings as Part of The "Integrated Seismics Oberpfalz 1989 (Iso89)", First Results, in: Continental Lithosphere: Deep Seismic Reflections, edited by: Meissner, R., Brown, L., Dürbaum, H.-J., Franke, W., Fuchs, K., and Seifert, F., American Geophysical Union, Washington, DC, doi:10.1029/GD022p0147, 1991.

Haak, V. and Jones, A. G.: Introduction to special section: the KTB deep drill hole, J. Geophys. Res., 102, 175–177, 1997.

Harjes, H. P., Bram, K., Dürbaum, H.-J., Gebrande, H., Hirschmann, G., Janik, M., Klöckner, M., Lüschen, E., Rabbel, W., Simon, M., Thomas, R., Tormann, J., and Wenzel, F.: Origin and nature of crustal reflections: Results from integrated seismic measurements at the KTB superdeep drilling site, J. Geophys. Res.-Sol. Ea., 102, 18267–18288, 1997.

Lüschen, E., Bram, K., Söllner, W., and Sobolev, S.: Nature of seismic reflections and velocities from VSP-experiments and borehole measurements at the KTB deep drilling site in southeast Germany, Tectonophysics, 264, 309–326, 1996.

Marzorati, S. and Bindi, D.: Ambient noise levels in north central Italy. Geochem. Geophys. Geosyst., 7, Q09010, doi:10.1029/2006GC001256, 2006.

Muller, J., Janik, M., and Harjes, H.-P.: The use of wave-field directivity for velocity estimation: Moving source profiling (MSP) experiments at KTB, Pure Appl. Geophys., 156, 303–318, 1999.

Nakamura, Y.: A method for dynamic characteristics estimation of subsurface using microtremor on the ground surface, QR Railway Tech. Res. Inst., 30, 1, 1989.

O'Brien, P. J., Duyster J., Grauert B., Schreyer W., Stoeckhert W., and Weber, K.: Crustal evolution of the KTB drill site: From oldest relics to the late Hercynian granites, J. Geophys. Res.-Sol. Ea., 102, 18203–18220, 1997.

Okaya, D., Rabbel W., Beilecke T., and Hasenclever, J.: P wave material anisotropy of a tectono-metamorphic terrane: An active source seismic experiment at the KTB super-deep drill hole, southeast Germany, Geophys. Res. Lett., 31, L24620, doi:10.1029/2004GL020855, 2004.

Peterson, J.: Observation and modeling of seismic background noise, U.S. Geol. Surv. Tech. Rept., 93–322, 1–95, 1993.

Plenefisch, T., Klinge, K., and Kind, R.: Upper mantle anisotropy at the transition zone of the Saxothuringicum and Moldanubicum in southeast Germany revealed by shear wave splitting. Geophys. J. Int., 144 309–319, 2001.

Plomerová, J., Achauer, U., Babuška, V., Vecsey, L., and BOHEMA working group: Upper mantle beneath the Eger Rift (Central Europe): Plume or asthenosphere upwelling?, Geophys. J. Int., 169, 675–682 doi:10.1111/j.1365-246X.2007.03361.x, 2007.

21

Performance report of the RHUM-RUM ocean bottom seismometer network around La Réunion, western Indian Ocean

S. C. Stähler[1,6], K. Sigloch[2,1], K. Hosseini[1], W. C. Crawford[3], G. Barruol[4], M. C. Schmidt-Aursch[5], M. Tsekhmistrenko[2,5], J.-R. Scholz[4,5], A. Mazzullo[3], and M. Deen[3]

[1]Dept. of Earth Sciences, Ludwig-Maximilians-Universität München, Theresienstrasse 41, 80333 Munich, Germany
[2]Dept. of Earth Sciences, University of Oxford, South Parks Road, Oxford, OX1 3AN, UK
[3]Institut de Physique du Globe de Paris, Sorbonne Paris Cité, UMR7154 – CNRS, Paris, France
[4]Laboratoire GéoSciences Réunion, Université de La Réunion, Institut de Physique du Globe de Paris, Sorbonne Paris Cité, UMR7154 – CNRS, Université Paris Diderot, Saint Denis CEDEX 9, France
[5]Alfred Wegener Institute, Helmholtz Centre for Polar and Marine Research, Am Alten Hafen 26, 27568 Bremerhaven, Germany
[6]Leibniz-Institute for Baltic Sea Research, Seestraße 15, 18119 Rostock, Germany

Correspondence to: S. Stähler (staehler@geophysik.uni-muenchen.de)

Abstract. RHUM-RUM is a German-French seismological experiment based on the sea floor surrounding the island of La Réunion, western Indian Ocean (Barruol and Sigloch, 2013). Its primary objective is to clarify the presence or absence of a mantle plume beneath the Reunion volcanic hotspot. RHUM-RUM's central component is a 13-month deployment (October 2012 to November 2013) of 57 broadband ocean bottom seismometers (OBS) and hydrophones over an area of $2000 \times 2000 \, \text{km}^2$ surrounding the hotspot. The array contained 48 wideband OBS from the German DE-PAS pool and 9 broadband OBS from the French INSU pool. It is the largest deployment of DEPAS and INSU OBS so far, and the first joint experiment.

This article reviews network performance and data quality: of the 57 stations, 46 and 53 yielded good seismometer and hydrophone recordings, respectively. The 19 751 total deployment days yielded 18 735 days of hydrophone recordings and 15 941 days of seismometer recordings, which are 94 and 80 % of the theoretically possible yields.

The INSU seismic sensors stand away from their OBS frames, whereas the DEPAS sensors are integrated into their frames. At long periods (> 10 s), the DEPAS seismometers are affected by significantly stronger noise than the INSU seismometers. On the horizontal components, this can be explained by tilting of the frame and buoy assemblage, e.g. through the action of ocean-bottom currents, but in ad-

dition the DEPAS intruments are affected by significant self-noise at long periods, including on the vertical channels. By comparison, the INSU instruments are much quieter at periods > 30 s and hence better suited for long-period signals studies.

The trade-off of the instrument design is that the integrated DEPAS setup is easier to deploy and recover, especially when large numbers of stations are involved. Additionally, the wideband sensor has only half the power consumption of the broadband INSU seismometers. For the first time, this article publishes response information of the DEPAS instruments, which is necessary for any project where true ground displacement is of interest. The data will become publicly available at the end of 2017.

1 Introduction

RHUM-RUM, short for "Reunion Hotspot and Upper Mantle – Réunions Unterer Mantel", is a German-French experiment that investigates the mantle beneath the Reunion ocean island hotspot from crust to core, using a multitude of seismological and marine geophysical methods (Barruol and Sigloch, 2013). The project also studies the hypothesized interaction between the hotspot and its surrounding mid-ocean ridges (Morgan, 1978; Dyment et al., 2007). The core of the experiment is a deployment of 48 German wideband and 9 French broadband ocean-bottom seismometers (OBS), from the DE-PAS (Deutscher Geräte-Pool für Amphibische Seismologie, managed by AWI Bremerhaven) and INSU (Institut national des sciences de l'Univers) pools respectively (see Table 1 for the data return).

There have been multiple experiments in tectonic settings similar to RHUM-RUM: 35 wideband and broadband OBS from the US OBS Instrument Pool (OBSIP) were deployed by the PLUME Hawaii experiment (Laske et al., 2009; Wolfe et al., 2009) twice for 1 year. Japanese large-scale imaging efforts around an oceanic hotspot were the PLUME Tahiti experiment with 9 Japanese broadband OBS (BBOBS) (Barruol, 2002; Suetsugu et al., 2005) and the TIARES array with again 9 BBOBS around the Society hotspot (Suetsugu et al., 2012). In 2011–2012, 24 German DEPAS OBS were deployed around the Tristan da Cunha hotspot (ISOLDE experiment, Geissler and Schmidt, 2013). Other larger, long-term DEPAS deployments in non-hotspot settings were in the Aegean Sea (EGELADOS, Meier et al., 2007) and in the Gulf of Cadiz (NEAREST, Geissler et al., 2010).

RHUM-RUM has been the largest DEPAS deployment so far in terms of the number of stations deployed ($44 + 4$) and in terms of aperture. This allows to resolve the deep-mantle signature of a plume using seismic tomography, especially when combined with concurrent land deployments. It is the first OBS experiment that specifically tries to use data for waveform tomography. This requires full response information on all instruments and also a high signal-to-noise ratio in the whole frequency range between 0.01 and 1 Hz.

The central component of the experiment was a deployment of 44 wideband OBS from DEPAS, of the so-called "LOBSTER" (Longterm OBS for Tsunami and Earthquake Research) type; 4 from Geomar Kiel, essentially identical to the DEPAS LOBSTERs; and 9 LCPO2000 broadband OBS from INSU, which are based on the "L-CHEAPO" instrument (Low-Cost Hardware for Earth Applications and Physical Oceanography) developed at the Scripps Institution of Oceanography (SIO).

We report on, and compare, the performance of seismometers and hydrophones from the two involved instrument pools, the German DEPAS and the French Parc Sismomètre Fond de Mer of INSU. This is the first side-by-side comparison of instruments from the German and French community OBS pools.

Data from the RHUM-RUM ocean bottom stations (and island stations) will be made freely available at the end of 2017 (Barruol et al., 2011).

This paper reviews the functioning of the OBS network and documents issues encountered in data collection, quality control, and processing. We review the experiment layout in Sect. 2.1, and the two types of OBS employed in Sect. 2.2. The performance of the stations is described in Sect. 3, with a focus on noise levels in Sect. 3.3. Possible reasons for the surprisingly different noise levels are discussed in Sect. 4.

Table 1. Data return in RHUM-RUM experiment.

Data return:	# of stations
Data return on all four channels throughout the entire deployment:	27
Data return on all four channels for only part of the deployment:	18
Only hydrophone data throughout the entire deployment:	1
Only hydrophone data for only part of the deployment:	7
No data returned:	4
Total number of stations deployed:	57
Data days recorded:	
Data days (hydrophones):	18 735
Data days (seismometers):	15 941
Deployment days:	19 751
Percent data recovery (hydrophones):	94 %
Percent data recovery (seismometers):	80 %

Appendix A contains a detailed description of the seismometer instrument responses, Appendix B describes an experiment to estimate clock drift rates and Appendix C contains a station-by-station list of noise levels in three period bands.

2 Experiment setup and instrumentation

2.1 The OBS network

For an overview of the whole network see Fig. 1. The oceanic component of the RHUM-RUM experiment consisted of 57 broadband ocean bottom seismometers deployed over an area of 2000 km $\times$ 2000 km from September 2012 to November 2013. The OBS clustered relatively densely around the island of La Réunion, out to distances of 400–500 km, including the vicinity of Mauritius (Fig. 1). This relative dense coverage was extended eastward to the Central Indian Ridge, in order to investigate hypothesized asthenospheric flow from hotspot to ridge (Morgan, 1978; Dyment et al., 2007). The seismicity in the reliably active South Sandwich subduction zone generates body-wave paths which sample the mantle beneath La Réunion at greater depths. Sampling with opposite azimuth is provided by earthquakes in the subduction zones of the south west Pacific, especially since the OBS network is augmented by RHUM-RUM land stations on Madagascar, and on the Îles Éparses in the Mozambique Channel. A linear, less dense arrangement of OBS followed the strike of the Central Indian and Southwest Indian ridges to the east and south, at 800–1200 km distance from the hotspot. Waves originating from earthquakes in the Alpine-Himalayan orogens and recorded at these stations again sample deeper levels of the mantle beneath La Réunion, but are also used to study the mid-ocean ridges themselves. A dense sub-array

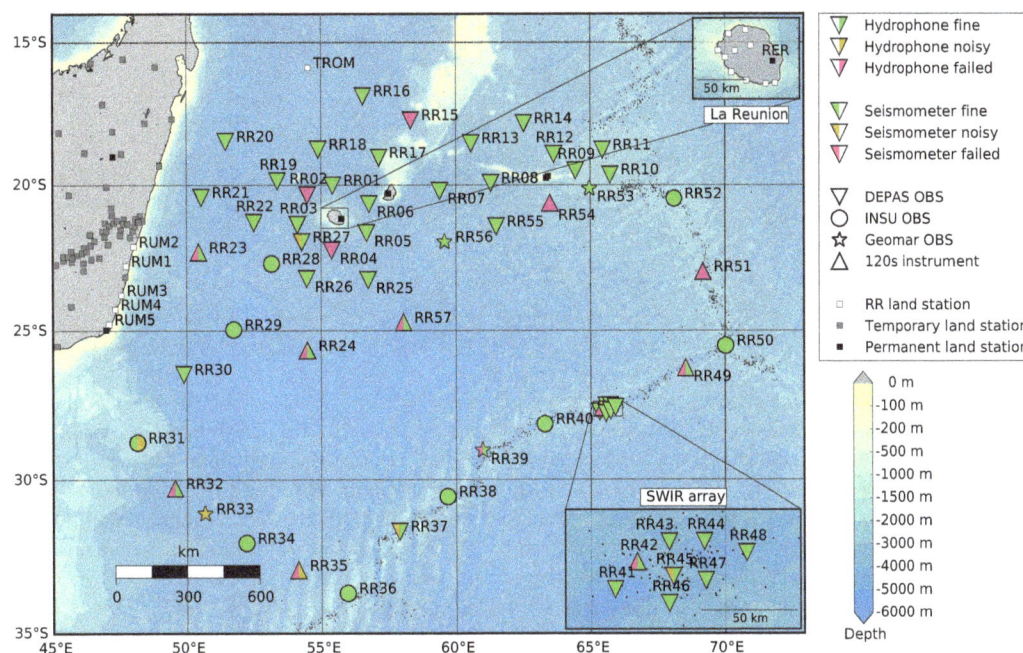

Figure 1. Overview map of the RHUM-RUM ocean bottom seismometer network. OBS are marked by large coloured symbols. Symbol shape marks the station type: DEPAS LOBSTER (inverted triangle), INSU LCPO2000 (circle), Geomar OBS (star). DEPAS instruments with malfunctioning 120 s instruments are marked as regular triangles. Two halves of the inner symbol indicate the functioning of the seismic sensors and hydrophones, respectively. Green indicates good performance; orange, high noise levels; red means the instrument failed to record. White squares indicate temporary land stations as part of the RHUM-RUM network YV, grey square indicate temporary land stations as part of the MACOMO (Wysession et al., 2012) and SELASOMA (Tilmann et al., 2012) projects, which were both installed between 2012 and 2014. Black squares indicate permanent GEOSCOPE stations. Small black dots mark earthquake hypocentres above magnitude 4 between 1981 and 2015, as published by the Preliminary Determination of Epicentres (PDE) bulletin of the US National Earthquake Information Center (NEIC). The seismicity is mainly concentrated on the oceanic ridges. Colour-shaded bathymetry is based on the global 30 arcsec merged bathymetry dataset by Becker et al. (2009), available at: http://topex.ucsd.edu/WWW_html/srtm30_plus.html.

of 8 OBS, referred to as the "SWIR Array", was deployed around an active seamount on the Southwest Indian Ridge in order to investigate the structure and seismicity of this ultra-slow spreading ridge. The sub-array had a footprint of about 70 km × 50 km and was located in segment 8 of the ridge, following the nomenclature of Cannat et al. (1999).

The OBS were deployed in October 2012 by the French research vessel *Marion Dufresne* and were recovered in October/November 2013 by the German research vessel *Meteor*. The instruments spent the intervening 13 months recording on the seafloor.

At each deployment site, the seafloor was surveyed with R/V *Marion Dufresne*'s multi-beam bathymeter and sediment echo sounder before dropping the OBS over board in a location deemed most suitable. The ship left immediately after deployment so that only deployment (and recovery) coordinates are known; no attempt was made to acoustically triangulate the landing positions of the OBS, with the notable exception of the 8 OBS in the densified SWIR Array. In general, OBS recovery positions were found to differ from their deployment positions by no more than a few hundred meters.

2.2 OBS models deployed

Here we give a brief overview of the hardware deployed (see Table 2) and the recording settings used, especially as they relate to the performance assessment of Sect. 3 (see Table 2 for an overview).

2.2.1 LOBSTER

The broadband OBS pool DEPAS (Deutscher Geräte-Pool für Amphibische Seismologie) of the German geophysical community consists of 80 instruments of the LOBSTER type ("Long-term OBS for Tsunami and Earthquake Research"). The OBS were developed in 2005, merging previous design experience mainly by Geomar Kiel (Flueh and Biolas, 1996), the University of Hamburg (Dahm et al., 2002), and the marine engineering firm K.U.M. (Umwelt- und Meerestechnik Kiel). K.U.M. was charged with building 80 LOBSTER units, which were funded by the German Research Foundation (DFG), the Federal Ministry of Education and Research (BMBF) and the Helmholtz Association of National Research Centres (HGF). The Alfred Wegener Institute Bremerhaven houses and maintains the instruments. For

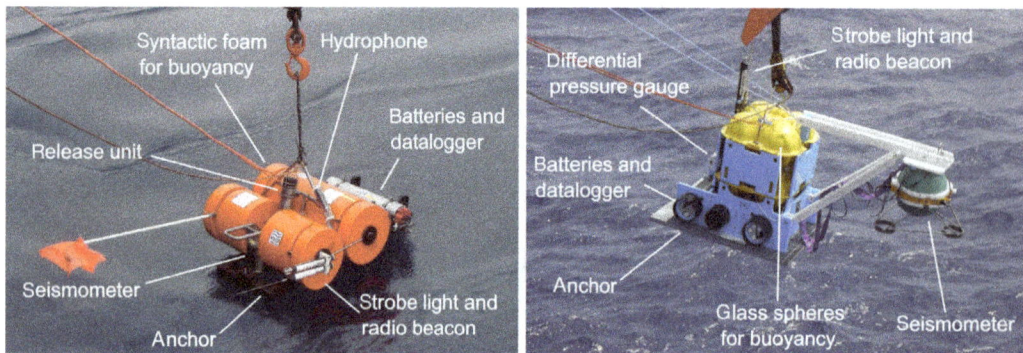

Figure 2. Broadband ocean-bottom seismometers, photographed seconds before deployment. Left panel: one of 48 LOBSTER-type instruments from the German DEPAS pool. The Güralp CMG-OBS40T sensor (corner period 60 s) is fitted in a vertical titanium pressure cylinder between two syntactic foam buoys and wedged against the steel anchor beneath it. Two horizontal titanium cylinders in the background contain the data recorder and the lithium batteries. The broadband hydrophone (corner period 100 s) is strapped to the A-shaped titanium frame that protrudes from the centre of the buoy assemblage. Right panel: one of 9 LCPO2000-BBOBS (Scripps-based) instruments from the French Parc de Sismomètre Fond de Mer pool at INSU. The Nanometrics Trillium sensor (corner period 240 s) is contained in the green sphere, which is dropped (i.e. mechanically separated) from the main frame one hour after arrival on the seabed. The differential pressure gauge is located in the white cylinder behind the frame. Both instruments are equipped with flags, strobe lights and radio beacons to facilitate recovery.

detailed information see http://www.awi.de/depas. The four OBS loaned to RHUM-RUM by Geomar Kiel are essentially identical to the DEPAS OBS.

The modular LOBSTER design (Fig. 2, left panel) is based on an open titanium frame that holds three titanium cylinders (containing the seismic sensor, data acquisition unit, and lithium batteries) and syntactic foam buoys that provide buoyancy for the ascent during recovery. A fourth titanium cylinder contains a mechanical release unit that locks the frame assemblage to a steel anchor until an acoustic release signal is received that initiates detachment from the anchor. The hydrophone is strapped to the frame, as are various recovery aides (a radio beacon, a flash, a flag, and a head buoy).

The titanium tube holding the seismic sensor is seated vertically between two syntactic foam units, and is wedged against the steel anchor by a steel plate, which acts as a lever that is pre-loaded by the mechanical release unit, thus ensuring good seismic coupling to the anchor. The integration of the seismometer into the frame makes the design very sturdy and reduces the number of failure points, but it also means that the seismometer is likely to record any tilt noise created by currents or pressure fluctuations acting on the frame. The orientation of the seismometer channels is fixed with respect to the frame, as it is shown in Fig. 3.

The seismic sensor in most DEPAS units is a three-component wideband Güralp CMG-OBS40T with a corner period of 60 s. The CMG-OBS40T is a lesser-known version of the CMG-40T with reduced power consumption, which is mounted in a gimbal system for usage in OBS. The gimbal system is activated three days after arrival on the seafloor to ensure proper levelling, since the instrument may land in a

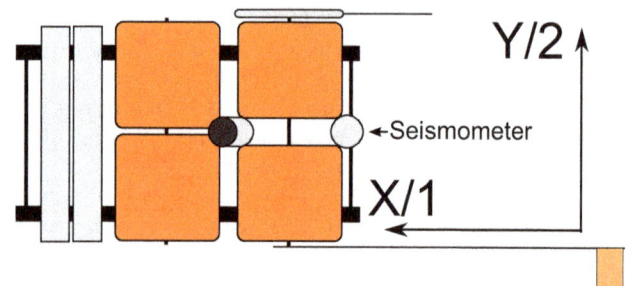

Figure 3. Sketch of a LOBSTER frame with the orientation of the horizontal seismometer channels. The X channel is oriented along the long axis of the LOBSTER, the Y channel 90° clock-wise of it. Positive values in the seismogram correspond to movement in the direction of the arrow. For the vertical (Z) channel, positive values correspond to upward movement. In the RESIF data archive, the X channel is stored as BH1, the Y channel as BH2 and the Z channel as BHZ.

tilted position) and then once every 21 days since the seafloor may settle over time.

The seismometer is sold in versions with different upper corner periods (10, 30, 60 s). All are mechanically identical, but use different feedback mechanisms to control the flat part of the response curve. The 60 s version is used by DEPAS and other OBS pools in Europe (e.g. IDL, Lisbon). Nine out of 48 instruments used in RHUM-RUM featured a prototype, broadband sensor design (corner period of 120 s). All of these nine units failed to level under deep-sea conditions, and repeated, unsuccessful levelling attempts drained the batteries prematurely (see Sect. 3.1).

The DEPAS units were additionally equipped with broadband hydrophones of type HTI-01 and HTI-04-PCA/ULF

Table 2. Comparison of German (DEPAS) and French (INSU-IPGP) OBS types.

Pool	DEPAS	INSU-IPGP
Manufacturer	K.U.M., Kiel	Scripps/INSU-IPGP
OBS type	LOBSTER	LCPO2000-BBOBS
Weight (water/air)	30/400 kg	25/350 kg
Assembly time	30 min (2 persons)	2 h (2 persons)
Transport options	12 in a 20′ container	8 in a 20′ container
Buoyancy	Syntactic foam	Glass spheres
Instrument casing	Titanium	Aluminium
Seismometer	CMG-OBS40T (60/120 s)	Trillium 240OBS (240 s)
Placement	integrated into frame	in external probe
Power consumption	100 mW (seism.)	700 mW (seism.)
	520 mW (recorder)	600 mW (recorder)

manufactured by HighTechInc (corner period 100 s), which usually worked very reliably as long as power was available.

The deepest RHUM-RUM OBS was deployed at 5400 m depth (Table 3), and the standard DEPAS OBS is certified to 6000 m water depth. Two battery tubes can be fitted with up to 180 lithium cells, sufficient for up to 15 months of recording using the settings described below. RHUM-RUM instruments were equipped to record for 13 months at sampling rates of 50 Hz. Eight of the 48 available DEPAS units were of a deep-diving variant certified to 7300 m depth, which has only one battery tube and therefore holds fewer batteries. Most of these instruments were deployed in the SWIR sub-array and typically recorded for 8–9 months at a sampling rate of 100 Hz (higher rate in order to investigate local seismicity). The clocks are supposed to continue running even after the voltage has dropped below the level required for data recording, in order to enable estimates of clock drift even if OBS retrieval is delayed.

2.2.2 The Scripps OBS instrument, INSU instrument pool

The INSU instruments (Fig. 2, right panel) are of the LCPO2000-BBOBS type, which is based on the Scripps Institution of Oceanography (SIO) "L-CHEAPO" design. Three of the instruments were manufactured at SIO and the other six at the INSU-IPGP OBS facility. The data recorder, batteries and release unit are protected in aluminium cylinders. The seismic sensor sits in an aluminium sphere. Buoyancy for recovery is created by hollow glass spheres.

All instruments were equipped with Nanometrics Trillium-240 seismometers with a corner period of 240 s and a differential pressure gauge with a passband between 0.002 to 30 Hz.

The INSU instruments check their level every hour. This caused an electronic spike of approximately 600 counts on the seismometer channels (see Sect. 3). This same spike exists in the 2006–2007 PLUME data set using SIO BBOBS (Laske et al., 2009), although we found no published mention of it. The problem has not been explicitly solved, but the SIO BBOBSs were reprogrammed to only check level once a week after the initial levelling cycle and the INSU BBOBSs are currently being reprogrammed to do the same. Work has been done to remove the hourly spike in the PLUME data (G. Laske, personal communication, 2014) and is being repeated for the RHUM-RUM data: it would be good to publish the correction algorithms, because these instruments probably still have this spike once per week.

The INSU instruments use a differential pressure gauge (DPGs, Cox et al., 1984) rather than a hydrophone. The DPG sits on the lower instrument frame close to the battery cylinder (Fig. 2).

2.3 Instrument responses

Instrument responses specify the transfer functions of seismometers and hydrophones (three seismogram channels and one hydrophone channel per station). The RESIF (RÉseau SIsmologique & géodésique Français) data centre serves this information in the format of StationXML or dataless SEED files.

To our knowledge, detailed meta-data information for DE-PAS OBS has not been published elsewhere. Therefore, we added a detailed discussion of the instrument responses as an appendix to this paper (Sect. A). Figures 4 and 5 show the total responses of instruments and data loggers for hydrophones and seismometers. Figure 6 shows instrument-corrected waveforms. For all seismometer types, instrument correction results in the same P-waveform.

3 Network performance

All 57 OBS were recovered successfully and undamaged. Table 3 summarizes the state of health of all seismometers and hydrophones over the deployment period. For a graphical summary of network performance (see Fig. 7).

Deployments were staggered over four weeks, along the 15 000 km-long cruise track. Recovery took five weeks and proceeded in roughly the same order as deployment, so that all stations spent approximately 13 months on the sea floor. An early end of recording was anticipated for stations RR35, RR41, RR43–RR48, and RR51 because their single battery tube only accommodated batteries for 8–9 months. For other stations, premature end of recording reflects technical issues, as discussed below.

Following the definition of the Cascadia initiative (Sumy et al., 2015), the data recovery was 15 941 data days out of 19 751 deployment days or 80 % for the seismometers, and 18 735 data days or 94 % for the hydrophones (Table 1).

Table 3. Performance summary of the 57 RHUM-RUM OBS and hydrophones. The abbrevation "gz" in the status column refers to the "glitch" on the Z component of the INSU seismograms (see Sect. 3.1). Skew is the measured clock drift in s, i.e. the instrument time at recovery minus the GPS time at recovery ("NA" if unknown because clock stopped early). For DEPAS stations, the number of recording days can exceed the number of deployment days because recording was started on deck prior to deployment. In the comments column, "120 s inst." refers to the new DEPAS sensor type that failed to level, yielding no useful seismometer data; "Geomar" refers to an OBS from Geomar, similar to the DEPAS LOBSTER. Figure 7 summarizes the network's state of health over the deployment period of October 2012 to November 2013.

Station name	Latitude	Longitude	Depth [m]	Deployment date [UTC]	Recovery date [UTC]	End of record [UTC]	Install. time [days]	Record length [days]	s.r. [Hz]	Seismo status	Hydro status	Skew value	Notes
RR01	−20.0069	55.4230	4298	5 Oct 2012	6 Nov 2013	6 Nov 2013	397	397	50	good	good	0.67 s	
RR02	−20.3392	54.4984	4436	5 Oct 2012	6 Nov 2013	5 Oct 2012	396	0	50	failed	failed	NA	
RR03	−21.3732	54.1294	4340	5 Oct 2012	5 Nov 2013	5 Nov 2013	396	396	50	good	good	0.81 s	
RR04	−22.2553	55.3846	4168	5 Oct 2012	5 Nov 2013	7 Oct 2012	396	2	50	failed	failed	NA	
RR05	−21.6626	56.6676	4092	3 Oct 2012	5 Nov 2013	2 Nov 2013	398	395	50	good	good	0.93 s	
RR06	−20.6550	56.7639	4216	3 Oct 2012	7 Nov 2013	31 Oct 2013	399	393	50	good	good	NA	
RR07	−20.1945	59.4058	4370	29 Sep 2012	24 Oct 2013	24 Oct 2013	389	389	50	good	good	0.53 s	
RR08	−19.9259	61.2907	4190	29 Sep 2012	24 Oct 2013	24 Oct 2013	389	389	50	good	good	1.40 s	
RR09	−19.4924	64.4485	2976	30 Sep 2012	25 Oct 2013	25 Oct 2013	389	390	50	good	good	2.18 s	
RR10	−19.6437	65.7558	2310	30 Sep 2012	25 Oct 2013	25 Oct 2013	390	390	50	good	good	0.39 s	
RR11	−18.7784	65.4629	3941	1 Oct 2012	26 Oct 2013	26 Oct 2013	390	390	50	good	good	0.61 s	
RR12	−18.9255	63.6474	3185	1 Oct 2012	26 Oct 2013	26 Oct 2013	390	390	50	good	good	−0.11 s	
RR13	−18.5427	60.5635	4130	2 Oct 2012	27 Oct 2013	9 Oct 2013	390	372	50	good	good	NA	
RR14	−17.8448	62.5299	3420	1 Oct 2012	27 Oct 2013	27 Oct 2013	390	390	50	good	good	2.36 s	
RR15	−17.7402	58.3330	3959	2 Oct 2012	28 Oct 2013	4 Oct 2012	390	1	50	failed	failed	NA	
RR16	−16.8976	56.5335	4426	2 Oct 2012	28 Oct 2013	28 Oct 2013	391	391	50	good	good	1.61 s	
RR17	−19.0427	57.1322	2205	3 Oct 2012	23 Oct 2013	23 Oct 2013	385	385	50	good	good	1.82 s	
RR18	−18.7504	54.8878	4743	6 Oct 2012	29 Oct 2013	29 Oct 2013	388	388	50	good	good	0.36 s	
RR19	−19.8500	53.3805	4901	9 Oct 2012	30 Oct 2013	30 Oct 2013	385	386	50	good	good	1.67 s	
RR20	−18.4774	51.4600	4820	6 Oct 2012	30 Oct 2013	30 Oct 2013	389	389	50	good	good	0.41 s	
RR21	−20.4217	50.5599	4782	7 Oct 2012	31 Oct 2013	31 Oct 2013	389	389	50	good	good	0.27 s	
RR22	−21.3007	52.4994	4920	9 Oct 2012	1 Nov 2013	1 Nov 2013	387	387	50	good	good	0.89 s	
RR23	−22.3290	50.4487	4893	10 Oct 2012	31 Oct 2013	26 Aug 2013	386	320	50	failed	good	NA	120 s inst
RR24	−25.6805	54.4881	5074	22 Oct 2012	3 Nov 2013	8 Oct 2013	376	291	50	failed	good	NA	120 s inst
RR25	−23.2662	56.7249	4759	4 Oct 2012	4 Nov 2013	4 Nov 2013	396	396	50	good	good	0.43 s	
RR26	−23.2293	54.4698	4259	4 Oct 2012	2 Nov 2013	2 Nov 2013	393	393	50	good	good	0.63 s	
RR27	−21.9657	54.2889	4277	5 Oct 2012	5 Nov 2013	19 Jul 2013	396	286	50	noisy	good	NA	
RR28	−22.7152	53.1595	4540	10 Oct 2012	12 Nov 2013	12 Nov 2013	398	397	62.5	good (gZ)	good	3.10 s	INSU
RR29	−24.9657	51.7488	4825	11 Oct 2012	13 Nov 2013	13 Nov 2013	398	397	62.5	good (gZ)	good	3.37 s	INSU
RR30	−26.4861	49.8917	5140	11 Oct 2012	14 Nov 2013	8 Oct 2013	398	361	50	good	good	NA	
RR31	−28.7648	48.1394	2710	12 Oct 2012	15 Nov 2013	15 Nov 2013	398	398	62.5	good (gZ)	noisy	−0.83 s	INSU
RR32	−30.2903	49.5555	4670	12 Oct 2012	15 Nov 2013	6 Nov 2013	398	358	50	failed	good	NA	120 s inst
RR33	−31.1170	50.6835	4904	13 Oct 2012	16 Nov 2013	19 Sep 2013	399	341	50	noisy	noisy	NA	Geomar
RR34	−32.0783	52.2113	4260	13 Oct 2012	16 Nov 2013	16 Nov 2013	399	398	62.5	good (gZ)	good	−1.29 s	INSU
RR35	−32.9694	54.1473	4214	13 Oct 2012	17 Nov 2013	27 May 2013	399	225	50	failed	noisy	NA	120 s inst
RR36	−33.7018	55.9578	3560	14 Oct 2012	17 Nov 2013	17 Nov 2013	399	398	62.5	good (gZ)	good	3.06 s	INSU
RR37	−31.7010	57.8876	4036	14 Oct 2012	18 Nov 2013	19 Oct 2013	399	369	50	noisy	good	NA	
RR38	−30.5650	59.6858	4540	15 Oct 2012	19 Nov 2013	19 Nov 2013	399	399	62.5	good (gZ)	good	−0.06 s	INSU
RR39	−29.0165	60.9755	4700	15 Oct 2012	19 Nov 2013	19 Nov 2013	400	400	50	failed	noisy	NA	Geomar
RR40	−28.1461	63.3020	4750	16 Oct 2012	20 Nov 2013	20 Nov 2013	400	399	62.5	good (gZ)	good	0.19 s	INSU
RR41	−27.7330	65.3344	5430	16 Oct 2012	20 Nov 2013	17 Jun 2013	400	244	100	good	good	NA	
RR42	−27.6192	65.4376	4776	16 Oct 2012	21 Nov 2013	10 Aug 2013	400	298	50	failed	good	NA	120 s inst
RR43	−27.5338	65.5826	4264	16 Oct 2012	21 Nov 2013	15 Jun 2013	401	241	100	good	good	NA	
RR44	−27.5324	65.7480	4548	16 Oct 2012	22 Nov 2013	3 Jun 2013	401	229	100	good	good	NA	
RR45	−27.6581	65.6019	2822	16 Oct 2012	21 Nov 2013	4 Jun 2013	400	138	100	noisy	good	NA	

Table 3. Continued.

Station name	Latitude	Longitude	Depth [m]	Deployment date [UTC]	Recovery date [UTC]	End of record [UTC]	Install. time [days]	Record length [days]	s.r. [Hz]	Seismo status	Hydro status	Skew value	Notes
RR46	−27.7909	65.5835	3640	16 Oct 2012	21 Nov 2013	26 May 2013	400	221	100	good	good	NA	
RR47	−27.6958	65.7553	4582	16 Oct 2012	21 Nov 2013	22 Jun 2013	400	248	100	good	good	NA	
RR48	−27.5792	65.9430	4830	16 Oct 2012	22 Nov 2013	10 Jun 2013	401	237	100	good	good	NA	
RR49	−26.2742	68.5354	4444	17 Oct 2012	23 Nov 2013	6 Nov 2013	401	384	50	failed	good	NA	120 s inst
RR50	−25.5181	70.0222	4100	18 Oct 2012	23 Nov 2013	23 Nov 2013	401	400	62.5	good (gZ)	good	1.74 s	INSU
RR51	−22.9989	69.1911	3463	18 Oct 2012	24 Nov 2013	3 Jan 2013	401	76	50	failed	failed	NA	120 s inst
RR52	−20.4722	68.1094	2880	19 Oct 2012	25 Nov 2013	25 Nov 2013	401	401	62.5	good (gZ)	good	0.97 s	INSU
RR53	−20.1213	64.9664	2940	20 Oct 2012	28 Nov 2013	30 Oct 2013	403	375	50	good	good	NA	Geomar
RR54	−20.6424	63.5082	2499	20 Oct 2012	28 Nov 2013	21 Oct 2013	404	365	50	failed	good	NA	120 s inst
RR55	−21.4417	61.4959	4462	20 Oct 2012	28 Nov 2013	8 Nov 2013	404	383	50	good	good	NA	
RR56	−21.9694	59.5853	4230	21 Oct 2012	29 Nov 2013	29 Jun 2013	404	251	50	good	good	NA	Geomar
RR57	−24.7264	58.0496	5200	21 Oct 2012	3 Nov 2013	31 Oct 2013	378	374	50	failed	good	1.28 s	120 s inst

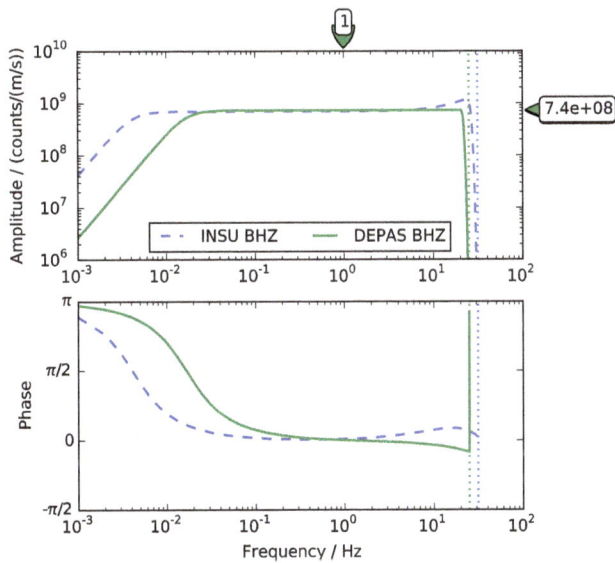

Figure 4. Bode plot of the total instrument responses $G(f)$ as defined in Eq. (A2) of vertical seismometer components, for a DEPAS Güralp CMG-OBS40T seismometer (solid green, station RR26), and for an INSU Trillium-240 (dashed blue, RR28). The corner period is 60 s for DEPAS instruments and 240 s for INSU instruments, which is evident from the amplitude responses. Horizontal channel responses of DEPAS instruments are identical to vertical responses, apart from the channel-specific gain, which varies by a few percent. The horizontal gain of INSU sensors is 1.6×10^8 counts(m s^1)$^{-1}$ compared to of 7.0×10^8 counts(m s^1)$^{-1}$ for the vertical channel. The upper frequency limits (dotted lines) are given by the Nyquist frequencies ($1/2 \times 50$ Hz for RR26 and $1/2 \times 62.5$ Hz for RR28).

Figure 5. Bode plot of the total instrument responses $G(f)$ as defined in Eq. (A2) of a DEPAS HighTechInc HTI-PCA04/ULF hydrophone (solid green, station RR26), and of an INSU differential pressure gauge (dashed blue, RR28). The nominal corner period is 100 s for DEPAS instruments and 500 s for INSU instruments. Dotted lines mark the Nyquist frequencies (see above).

3.1 Instrument failures

Three out of 48 DEPAS stations (RR02, RR04, RR15) delivered neither seismometer nor hydrophone data because their data loggers failed (reason unclear). The seismometers in nine DEPAS stations (RR23, RR24, RR32, RR35, RR42, RR49, RR51, RR54, RR57) featured a redesigned sen-

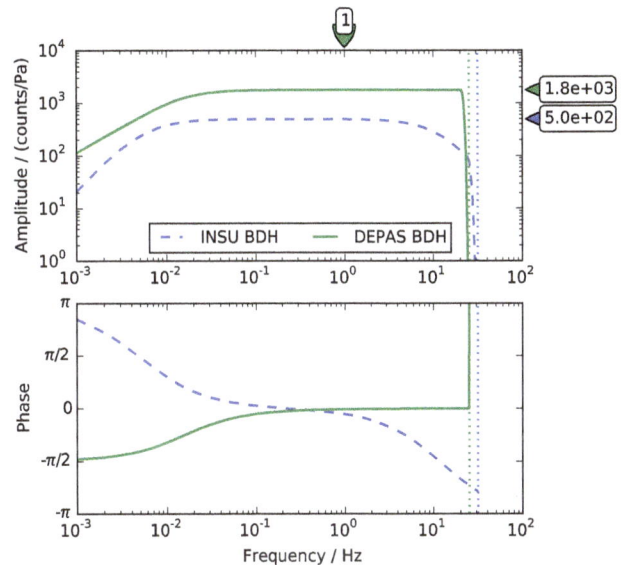

sor/casing package with broader band CMG-OBS40T sensors (120 s), which had previously not been deployed in the deep sea. The levelling mechanisms failed (remained stuck) in all nine stations, for reasons that are still under investigation. Automatic, prolonged attempts to level the sensors drained their batteries prematurely so that the functioning hydrophones also ran out of power 8–9 months into the experiment. DEPAS seismometers RR27 and RR45 recorded, but at high noise levels (reason under investigation). The hydrophones of these stations worked normally. The seismometer in one of the four Geomar stations failed (RR39), and noise levels at Geomar station RR33 are unusually high, al-

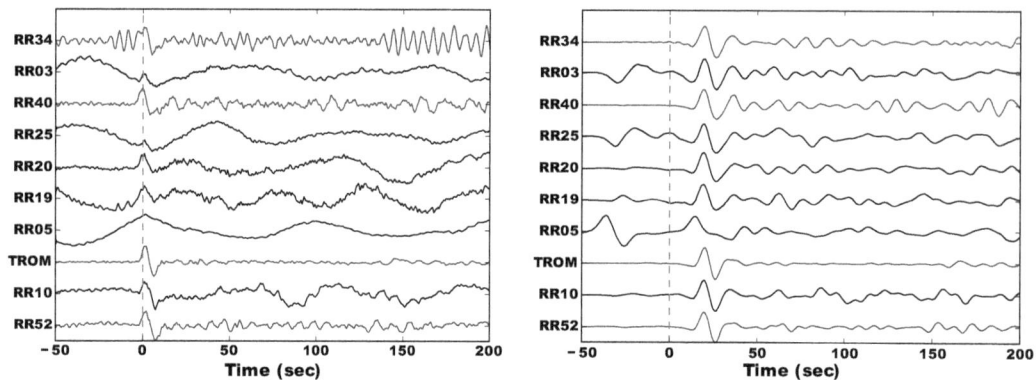

Figure 6. Comparison of broadband (left panel) and bandpass-filtered seismograms for six DEPAS OBS (black), three INSU OBS (RR34, RR40, RR52, blue) and an island station (TROM on Île Tromelin, red) in the northern part of the OBS network (see Fig. 1). All seismograms have been instrument-corrected to displacement, filtered between 1/60 and 3 Hz (the nominal corner frequencies of the least broadband sensor type, the DEPAS OBS) in order to facilitate visual comparison. The waveforms on the right have been bandpass-filtered using a Gabor filter as described in Sigloch (2008, p. 100) with a centre frequency of 1/15 Hz. Waveforms are amplitude-normalized and plotted relative to the theoretical arrival time of a P-wave from a magnitude 6.6 earthquake on 20 April 2013 in Sichuan, China (71° distance, see GEOFON, 2013). This shows that the instrument response has been determined correctly and that even the relatively noisy DEPAS recordings can be used for purposes like waveform tomography. The band-pass filter strongly enhances the P-wave, compared to the wideband traces, where it is lost in the long period noise for most DEPAS stations.

though this might not be due to the sensor. The hydrophones in RR33 and RR39 measured, but at a high noise level.

The 9 INSU stations (RR28, RR29, RR31, RR34, RR36, RR38, RR40, RR50, RR52) were affected by a bug in the data logger software that activated the level-sensing circuitry every 3620 s (roughly every hour). Each such event caused a "glitch" in the seismograms of roughly 1200 s duration, i.e. a characteristic, complex pulse shape, that is very similar but not identical across events. Pulse amplitudes are between 500–800 counts, corresponding to 1.5 μm ground displacement after instrument correction and filtering between 20 and 500 s period. This artefact is rarely visible on horizontal components where noise levels are much higher in this period band, but it exceeds noise amplitudes on the vertical channels by 15 dB. Figure 8 shows that the glitch amplitude is comparable to body wave arrivals of intermediate-size, teleseismic earthquakes, here a M6.6 earthquake at 71° distance. Efforts are under way to suppress this artefact by matched filtering.

The differential pressure gauge in INSU station RR31 had high artefacts roughly every 9000 s. Seismic signals are visible in between, but may be difficult to use. For station RR38, gaps in the data had to be fixed. Although this was carefully done, it is possible that artefacts were introduced.

3.2 Estimation of clock error

The internal clocks of the data recorders are affected by drifts on the order of one second per year. Over 13 months of autonomous recording, drift of this magnitude is nonnegligible for certain applications, such as body-wave tomography. Prior to deployment, each recorder clock was synchronized to GPS, and upon recovery it was compared to

GPS time again, yielding the clock drift or "skew". Assuming that the skew accumulated linearly over the deployment period, the clock error can be corrected for any moment in time. Previous studies (Hannemann et al., 2014; Scholz, 2014) show that linearity is a good first order approximation for the clocks used in the DEPAS instruments. For the LCPO2000 instruments used in the INSU pool, Gouedard et al. (2014) found that drift rates can vary over the course of days. We assume that this effect is cancelling out for longer deployments, therefore RHUM-RUM data at the RESIF data centre are linearly corrected for skew, where available.

Unfortunately a significant number of DEPAS clocks stopped before recovery, so that the skew could not be measured (entries "NA" in Table 3). Clock shutdown was not anticipated even if batteries became weak. At a critical voltage level of 6.0 V (down from 13.0 V), the recorder was programmed to switch off seismometer and hydrophone, allowing its low-consuming clock to continue for several months. The Lithium batteries for long-term deployments have a faster current drop than the alkali batteries for normal deployments, which caused a problem for multiple stations. Superimposed on a gradual voltage decline, the log files show brief, steep voltage drops associated with levelling events every 21 days. Towards the end of the recording period, this led to uncontrolled shutdown of some recorders and clocks, presumably when a drop below critical voltage occurred too suddenly.

Using cross-correlation of ambient noise, Sens-Schönfelder (2008) presented a method to determine the relative clock error between two seismometers a posteriori, which Hannemann et al. (2014) successfully applied to OBS data. Likewise, Scholz (2014) succeeded in estimating

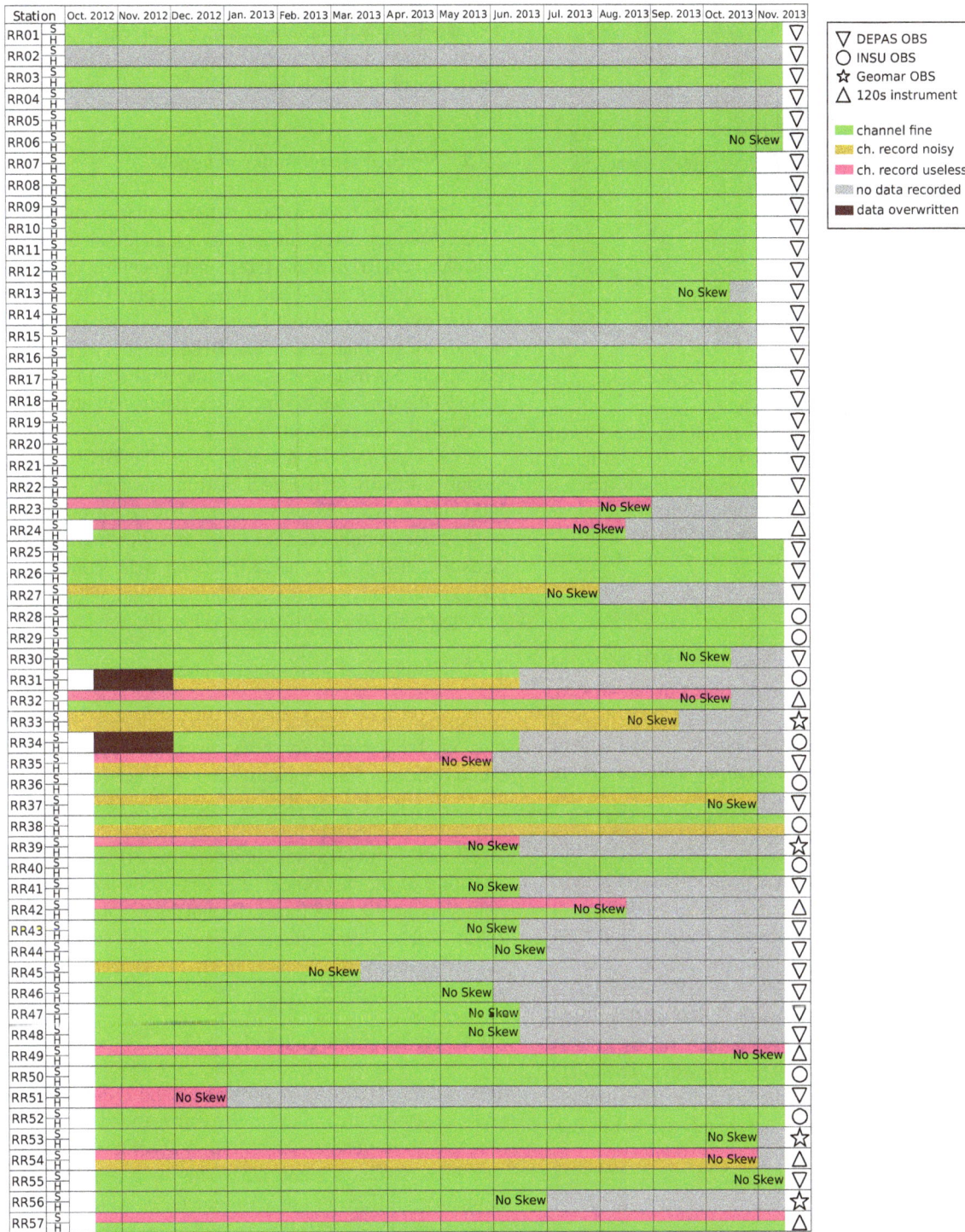

Figure 7. Data availability and quality for all RHUM-RHUM ocean-bottom stations. Green indicates availability of good data. Yellow indicates availability of abnormally noisy data, where earthquakes are visible, but artefacts are so strong, that noise correlation or other advanced analyses will probably fail. Red indicates that the seismometer ("S" in first column) or hydrophone/differential pressure gauge ("H") recorded data that is completely useless for seismological purposes. These time traces will still be archived at RESIF and may be useful for analysis of error sources. Grey shading indicates time intervals when battery power had run out prior to recovery, or where the data logger failed (RR02, RR04, RR15) and no data was recorded at all. Dark red shading indicates time intervals of missing data for INSU stations RR31 and RR34 (overwritten due to erroneous reset of data logger). Station symbols in the last column follow Fig. 1. Inverted Triangles: regular DEPAS LOBSTER (seismometer 60 s corner period, 50 Hz sampling rate); stars: Geomar LOBSTER (60 s, 50 Hz); regular triangles: newer DEPAS LOBSTER (120 s corner period, all seismometers failed); circles: INSU/Scripps instrument (240 s, 62.5 Hz).

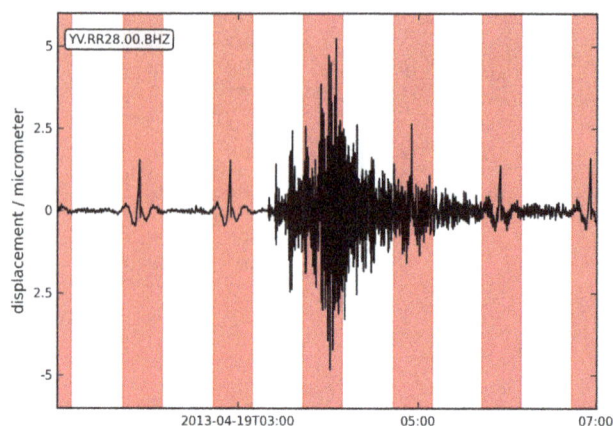

Figure 8. A M6.6 earthquake at 71° distance recorded on the vertical component of INSU OBS RR28. The seismogram has been instrument-corrected to ground displacement and passband-filtered at 20 to 500 s. One red plus one white stripe span 3620 s, slightly more than one hour. The seismogram shows one "glitch" per red shaded interval, i.e. nearly hourly, pulse-like artefacts caused by unintended activation of the sensor levelling mechanism in INSU stations. One glitch is hidden by the surface wave train. The earthquake is the same as in Fig. 6 (66° distance, see GEOFON, 2013).

clock drift for the SWIR sub-array of the RHUM-RUM network (RR42-RR48, inter-station distances of 30–40 km). His results suggest that indeed clock errors accumulated linearly over the installation period. For the remainder of the RHUM-RUM network, inter-station distances were unfortunately found to be too large (> 150 km) to apply this ambient noise method, especially given the high self-noise level of the DEPAS OBS packages.

In an attempt to estimate the clock drift of these 11, otherwise well-functioning OBS a posteriori, we did a dry run of several recorders in the DEPAS lab with batteries and seismometers attached for over a month. Afterwards, we compared the value of the internal clock with GPS time. These experiments reproduced the sign of the clock error (clocks generally ran too slow) but probably not their values, at least not to an accuracy that would be useful in practice. The likely reason is that we did not simulate the low water temperatures on the seafloor. The experiment is described in detail in Appendix B.

3.3 Noise levels

Noise levels can be characterized by Probabilistic Power Spectral Density distributions (PPSDs, McNamara and Buland, 2004) for each of the four sensor components. We obtain PPSDs by computing power spectra on hour-long broadband time series, and by stacking the hourly results over the recording period. Figure 9 shows PPSDs for DEPAS station RR26 (depth 4259 m) and for INSU station RR28 (depth 4540 m), which were deployed at 150 km distance between each other.

We created a poster of PPSDs for all 57 stations and all 4 channels, which is published as a Supplement to this article and shows that the relative noise differences of Fig. 9 are characteristic for INSU versus DEPAS stations more generally.

3.3.1 Vertical seismometer channels

The seismometer spectra are rather similar at short periods but increasingly divergent at periods longer than 5 s. The vertical channel (BHZ) of the INSU instrument has its low-noise notch at 10–30 s period and stays well below the bounds of the (terrestrial) New High Noise Model (Peterson, 1993), to periods longer than 200 s. The BHZ channel of the DEPAS instrument has its low-noise notch around 10–15 s; at longer periods, the noise rapidly increases, rising well above the Peterson High Noise Model.

At 40 s period, the noise level on the BHZ channel is around −125 dB for DEPAS instruments and −155 dB for INSU instruments. These values are before correction for tilt or sea floor compliance (Crawford and Webb, 2000). At periods longer than 20 s, noise levels on BHZ show little amplitude variation over the deployment period, with a variance of roughly 10 dB at most stations (Fig. 9).

3.3.2 Horizontal seismometer channel

Noise on the horizontal seismometer channels is much higher than on the vertical for both instrument types. Horizontal components show mean noise levels between −100 and −115 dB for DEPAS OBS, and around −135 dB for INSU instruments (at 40 s period). The variance is on the order of 20 dB and shows clear seasonal variations (Fig. 10).

Tilting of the instrument, e.g. caused by underwater currents shaking the OBS frame (Duennebier et al., 1981; Trehu, 1985; Webb, 1998) affects the horizontal channels much more than the vertical component, so the higher horizontal noise level is expected.

3.3.3 Hydrophone channel

The spectra of DEPAS hydrophones and INSU differential pressure gauges are rather similar across the entire frequency range, both in general shape and in absolute decibel levels (see Figs. 9, C1, C2 and C3). This is in marked contrast to the large differences in seismometer noise levels between DEPAS and INSU instruments, and again points to a tilt origin or self-noise for the DEPAS seismometer noise, since tilt would hardly affect hydrophone records.

The pressure noise at DEPAS hydrophone RR26 is even slightly lower than at the near-by INSU RR28 (Fig. 9). In general, hydrophone noise levels are approximately 5 dB lower on DEPAS stations than on INSU stations in the period range of 12–40 s (see Fig. C2 in the appendix). This is true for the DEPAS hydrophones in general, with the exception of only a few noisy outliers that had individual problems. The

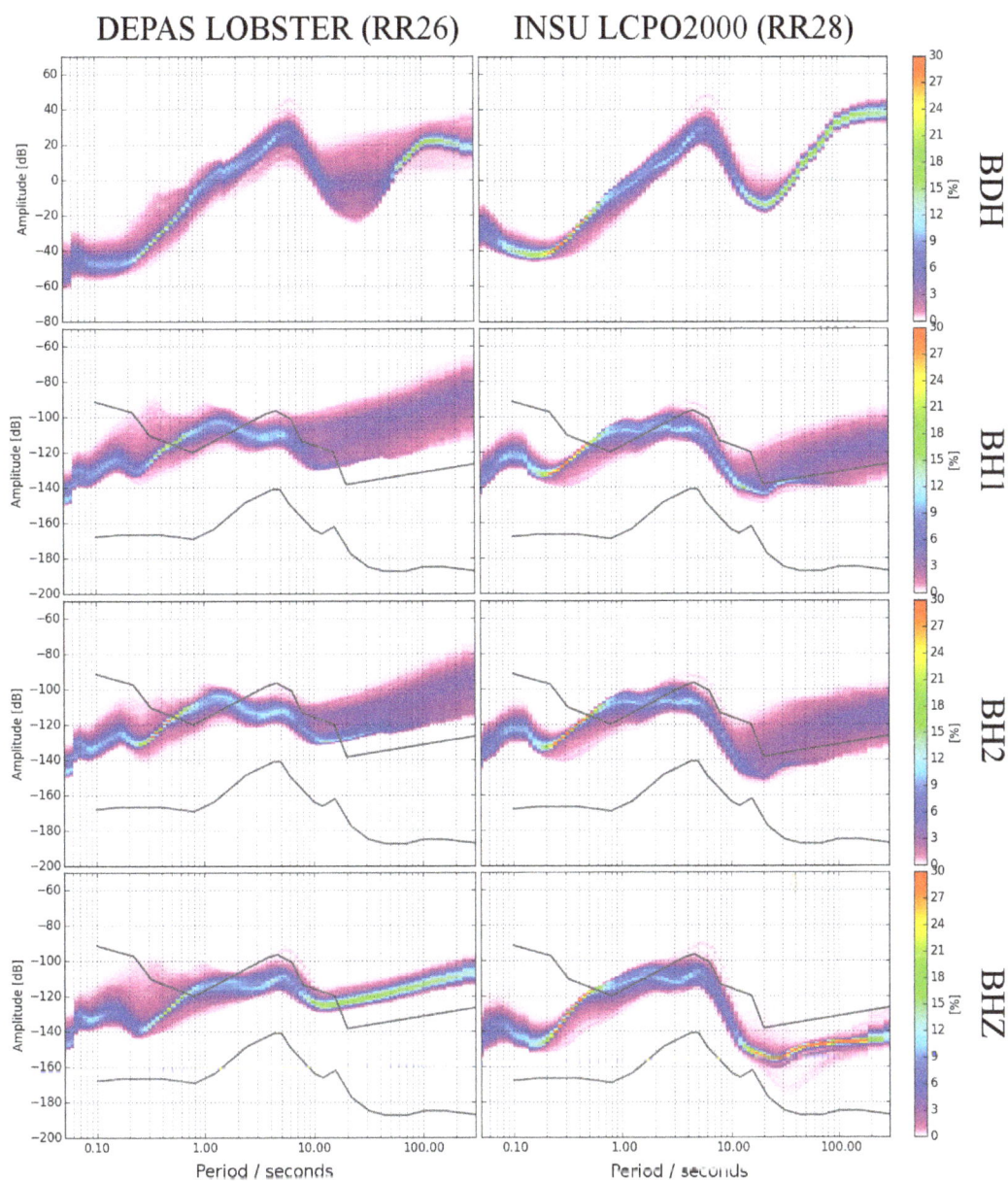

Figure 9. Probabilistic power spectral densities (PPSDs) for a DEPAS station (RR26, left column panels) and an INSU station (RR28, right column panels). PPSDs are composed of hour-long power spectra stacked over the entire deployment interval. Colour marks the frequency of occurrence of different noise levels, where purple indicates relatively rare, and red relatively frequent (McNamara and Buland, 2004). Black curves mark the upper and lower bounds of the New High and Low Noise Model of Peterson (1993). The two instruments were installed within 150 km of each other, in an abyssal plain 300 km south-west of La Réunion island (cf. Fig. 1). At periods longer than 5 s, the INSU seismometers are much quieter than the DEPAS instrument (see Sect. 4). By contrast, the pressure channel BDH of the two models (hydrophone for DEPAS, differential pressure gauge for INSU) shows very similar noise levels. A poster with PPSDs for all stations is available on ResearchGate (Stähler et al., 2015) and as an Supplement to this paper.

overall lower noise level can probably be explained by completely different instrument types (hydrophones on DEPAS versus differential pressure gauges on INSU stations).

3.4 Temporal noise variations

We expect two sources for temporal noise variations: (1) varying wave heights due to storm activity, which affects mostly the microseismic noise band. (2) Water current-induced tilt, which creates long period noise.

Figure 10 shows the temporal evolution of noise levels between October 2012 and October 2013 at DEPAS station RR01 near La Réunion (depth 4298 m), between 2 and 60 s). In the secondary microseismic noise band (2–10 s period), peak noise intervals coincide with cyclone passages during southern summer (blue frames). Cyclones are tropical storms, the Indian Ocean equivalent of hurricanes and typhoons. Their correlation to microseismic noise is most pronounced on the BHZ component. In fact, Davy et al. (2014) were able to track the path of a cyclone across the RHUM-RUM network using recordings of secondary microseismic noise only.

By contrast, peak noise episodes in the 20–60 s band show no clear correlation with cyclone passages. Rather, the highest levels occur during southern winter (March to September), out of cyclone season. Seasonal variations in deep-sea currents might explain tilt noise at these lower frequencies. The HYCOM-based global ocean circulation model (GLBa0.08/expt_90.9) (Cummings, 2005) does predict more episodes of strong currents at RR01 during southern autumn, (Fig. 10 bottom), but its absolute velocity values would appear low for effectively shaking an OBS. However, global ocean circulation models for this region have very poor resolution in the bottom layer, so that true bottom currents may be different. A recent measurement of current profiles at 23° S, 48° E (Ponsoni et al., 2015) suggests that bottom velocities generally do not exceed a few cm per second in the region (L. Ponsoni, personal communication, 2015). Unfortunately, the nearest RHUM-RUM station, (RR23) failed to deliver seismograms for comparison.

4 Discussion of the different noise levels

The relative stronger overall noise on the DEPAS instrument affects the usability of the OBS for waveform tomography and analysis of long-period waveforms. Hence its causes are of interest to future users of the pool and for instrument developers. We discuss four potential differences between the two instrument types:

The gimbal system: if the gimbal system were not stable enough, it could cause additional noise on all components. This hypothesis cannot be proven or falsified, since the CMG-OBS40T cannot be tested outside its gimbal. Experience shows that this would rather cause high-frequency noise.

The data logger: the data loggers of the DEPAS and the INSU OBS could have different self-noise levels. Again, this cannot be tested, since we have no data from other loggers available. But similar to the gimbal system, this would rather affect the high-frequency end of the spectrum, which is similar for both types.

OBS tilt: the integration of the seismometer into the OBS frame makes the DEPAS instruments more susceptible

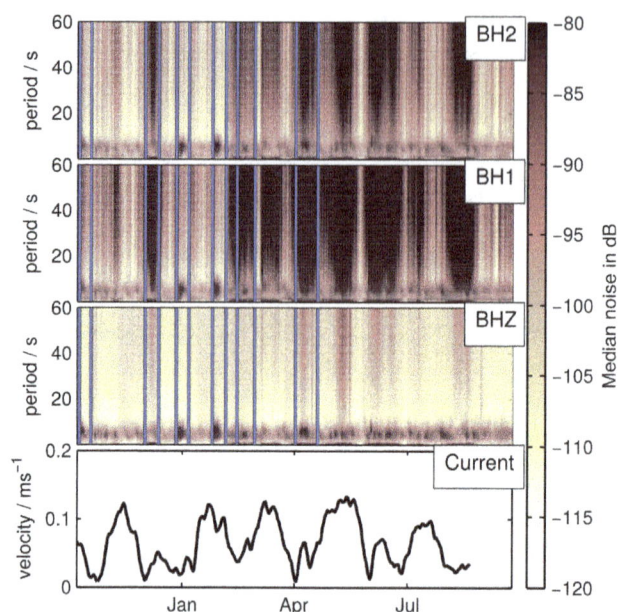

Figure 10. Seasonal changes in the noise levels on OBS RR01 near La Réunion. Spectrograms of noise on the three seismometer components, where noise is plotted as the median of daily probabilistic power spectral densities. Blue boxes mark episodes of cyclone activity, which correlates well with peak noise episodes in the microseismic band (periods around 10 s), especially on the BHZ component. At periods longer than 20 s, seismic noise peaks occur preferentially in southern autumn (February–June), most evident on the horizontal components. The global ocean circulation model HYCOM GLBa0.08/expt_90.9, running from 3 January 2011 to 20 August 2013 predicts more intervals of strong ocean-bottom currents for southern autumn (bottom panel) – qualitatively consistent with the hypothesis that ocean bottom currents cause long-period OBS noise by tilting the seismic sensors.

to current-induced tilt. Seasonal variations on the noise level of the horizontal channels can be seen in Fig. 10 and in the cloudy look of the PPSDs beyond 10 s in Fig. 9. However, tilt noise should affect horizontal channels much more strongly than vertical ones, which is indeed the case for the INSU instruments. For the DEPAS instruments, the vertical noise is too high to be explained by tilt alone.

Seismometer self noise: the CMG-OBS40T is a 60 s wideband instrument, based on the 10 s CMG-40T. While the self noise of the latter is below the New Low Noise Model (NLNM) for periods shorter than 10 s, onshore experiments with one of the CMG-OBS40Ts showed self noise of −140 dB at 10 s period, which is far above the NLNM. This strongly suggests that the reduced power consumption of the OBS40T comes at the price of a significantly increased self-noise level. High self-noise probably explains the larger part of the excessive noise on the vertical channel in our experiment.

To summarize, we expect the high noise level of the DEPAS instruments to be caused by a combination of tilt and instrument self noise, where the former dominates the noise on the horizontal channels and the latter the noise on the vertical channel. The fact that the variability of noise on the horizontal channels is comparable between the two instrument types suggests that the susceptibility to currents is similar, albeit slightly higher on the DEPAS instrument package. The usage of a compact wideband sensor in the LOBSTER instruments has the advantage of a much lower power consumption, at the price of a strongly increased noise level beyond 10 s.

More detailed analysis of the effect of sensor integration would require usage of a more broadband sensor in the DEPAS instrument package.

5 Conclusions

From October 2012 to November 2013, the RHUM-RUM experiment deployed and successfully recovered 48 German DEPAS and 9 French INSU broadband ocean-bottom seismometers around La Réunion, western Indian Ocean, making this the largest deployment of either instrument type, and the only joint experiment. Overall network performance was very satisfactory, but a number of technical issues have been described here, including blocked levelling mechanisms, data logger malfunctioning, and loss of clock synchronization.

For the first time, we publish instrument response information on the DEPAS OBS, which allows to calculate the true ground displacement in a wide frequency range.

This shows that at periods longer than 10 s, the INSU OBS are much quieter than the DEPAS instruments, on all three seismometer components. No such difference in data quality exists for the hydrophones and differential pressure gauges, which both worked extremely reliably. The increased long-period noise on the DEPAS seismometers can be explained by the surprisingly high instrument self-noise on the all channels of the Güralp CMG-OBS40T sensors and partially by a higher susceptibility to current-induced tilt of the whole OBS.

In the microseismic noise band, peak noise intervals can be attributed to tropical storm activity (cyclones), whereas no clear correlation with cyclones was found at lower frequencies, where tilt and self-noise dominates (20–60 s period band). A possible cause for instrument tilt is the action of ocean-bottom currents, which are predicted to peak in southern winter just like the tilt noise, but global ocean circulation models are not sufficiently constrained to test this hypothesis in more detail.

The RHUM-RUM data set has been assigned FDSN network code YV and will be freely available by the end of 2017. Data and detailed StationXML meta-data files are hosted and served by the RESIF data centre in Grenoble (http://portal.resif.fr/?RHUM-RUM-experiment&lang=en).

Appendix A: Instrument responses

While conceptually straightforward, instrument corrections can be non-trivial in practice because filter description can be complex, and their specifications must exactly match the format expected by the software used to apply the corrections.

A1   Seismometers

Assuming that the seismometer is a causal linear time-invariant system, its response can be described by a series of poles p_m and zeros r_n:

$$G_{\text{inst}}(f) = S_{\text{d,inst}} \cdot A_0 \cdot \frac{\prod_{n=1}^{N}(2\pi i f - r_n)}{\prod_{m=1}^{M}(2\pi i f - p_m)}. \quad \text{(A1)}$$

In Eq. (A1), $S_{\text{d,inst}}$ is the sensitivity at reference frequency f_r with dimension counts $(\text{ms}^{-1})^{-1}$. A_0 is a dimensionless normalization constant, which normalizes $G(f)$ to 1 at reference frequency f_r. Following convention, we defined $f_r = 1\,\text{Hz} = (2\pi)^{-1}\,(\text{rad s}^{-1})$. The M poles p_m and N zeros r_n describe the frequency-dependency of the response.

Values for each instrument can be queried sending its serial number email to caldoc@guralp.com. Note that these data sheets contain the frequencies of the poles and zeros in Hz, while the StationXML format prefers them in rad s^{-1}. All DEPAS seismometers that functioned had the same $M = 4$ poles and $N = 2$ zeros as described in Table A1a, with the exception of RR13 that had $M = 5$ (Table A1b) and RR22 with $M = 6$ poles (Table A1c)[1].

Poles and zeros characterize the first, analogue stage of an instrument; subsequent digital filter stages characterize the ADC (Analogue to Digital Converter) and digital processing units of the data recorder. For the seismometers, the analogue filter stages were obtained from the manufacturers Güralp and Nanometrics, and are compared in Fig. 4.

We follow the SEED reference manual's Appendix C (Ahern et al., 2012) to describe the response $G(f)$ in frequency-domain. The total transfer function is the product of complex response functions for the instrument, ADC and FIR decimation stages:

$$G(f) = G_{\text{inst}}(f) \cdot G_{\text{ADC}}(f) \cdot G_{\text{FIR}}(f). \quad \text{(A2)}$$

The gain or sensitivity $S_{\text{d,inst}}$ is channel specific and is determined by Güralp before delivering the instrument. For our instruments, a typical value is $1980\,\text{V}(\text{ms}^{-1})^{-1}$ with an instrument-specific variance of $15\,\text{V}(\text{ms}^{-1})^{-1}$.

The analogue seismometer signal was converted to digital counts by a SEND GEOLON-MCS data logger. This conversion is assumed to have a flat response curve:

---

[1]For the 120 s instruments, the manufacturer lists the same 6 poles and 2 zeros as RR22, which is probably not correct, since they describe a corner period of 60 s. But since none of those recorded data, this should not be a problem to users of the data.

Table A1. (a) 4 poles and 2 zeros of the 60 s Güralp CMG-OBS40T used in the German LOBSTER OBS. Can be applied to all 60 s stations but RR13 and RR22. (b) 5 poles and 2 zeros of the 60 s Güralp CMG-OBS40T used in station RR13. (c) 6 poles and 2 zeros of the 60 s Güralp CMG-OBS40T used in station RR22. (d) 11 poles and 6 zeros of the Trillium 240OBS used in the French OBS at RR38, RR50 and RR52. (e) 11 poles and 6 zeros of Trillium 240OBS with a serial number below 400. Those were used in stations RR28, RR29, RR31, RR34, RR36 and RR40.

	Pole p_m in rad s^{-1}	Zero r_n in rad s^{-1}
(a)		
1/2	$-0.074016 \pm 0.07347\,i$	0
3/4	$-502.65 \pm 596.9\,i$	–
(b)		
1/2	$-0.074016 \pm 0.074016\,i$	0
3	-502.66	–
4	-1005.3	–
5	-1130.98	–
(c)		
1/2	$-0.074016 \pm 0.074016\,i$	0
3	-471.24	–
4/5	$-395.1 \pm 850.69\,i$	–
6	-2199.1	–
(d)		
1/2	$-0.018134 \pm 0.018034\,i$	0
3	-84.4	-72.5
4	$-180.2 + 224.4\,i$	-163.3
5	$-180.2 - 224.4\,i$	-251
6	-725	-3270
7	-1060	–
8	-4300	–
9	-5800	–
10/11	$-4200 \pm 4600\,i$	–
(e)		
1/2	$-0.017699 \pm 0.017604\,i$	0
3	-85.3	-72.5
4	$-155.4 + 210.8\,i$	-159.3
5	$-155.4 - 210.8\,i$	-251
6	-713	-3270
7	-1140	–
8	-4300	–
9	-5800	–
10/11	$-4300 \pm 4400\,i$	–

$$G_{\text{ADC}}(f) = S_{\text{d,ADC}}. \quad \text{(A3)}$$

The sensitivity of this stage is $S_{\text{d,ADC}} = 3.62 \times 10^5$ counts V^{-1}, resulting in an overall sensitivity for the LOBSTER seismometers of roughly 7.4×10^5 counts $(\text{m s}^{-1})^{-1}$ at reference frequency $f_r = 1\,\text{Hz}$ (see Fig. 4).

The decimation of the digital signal to the recording frequency is described by a series of N_{FIR} FIR decimation filters. The kth digital filter stage has L_k coefficients $b_{l,k}$, decimating an input signal of sampling rate Δt_i. The total FIR response is the product of the individual FIR stages:

$$G_{FIR}(f) = \prod_{k=1}^{N_{FIR}} S_{d,FIR,k} \sum_l b_{l,k} e^{2\pi i \Delta t_k}. \tag{A4}$$

For the DEPAS instruments, the decimation from $512\,\text{kHz}$ to 50 or $100\,\text{Hz}$ is described by 8 ($100\,\text{Hz}$) or 9 ($50\,\text{Hz}$) FIR stages of uniform sensitivity $S_{d,FIR,k} = 1$, such that the sensitivity is only affected by the instrument and ADC stages. The coefficients $b_{l,k}$ have been defined by DEPAS and are included in the StationXML and dataless files. They create the sharp cut-off at 90 % of the Nyquist frequency in Figs. 4 and 5.

The INSU Trillium-240OBS seismometers features $M = 12$ poles p_m and $N = 5$ zeros r_n in its analogue stage (see Tables A1d and e). The p_m and r_n were taken from the Trillium-240 user guide, which applies to the 240OBS as well. The sensitivity is $S_{d,inst} = 598.45\,\text{V}(\text{ms}^{-1})^{-1}$. This is half the value specified in the user guide, since the OBS were connected single-ended. The analogue gain is 0.225 for the horizontal channels and 1.0 for the vertical channel, to maximize the vertical sensitivity while avoiding clipping on the horizontal channel. The sensitivity of the CS5321-2 A/D converter is $1\,165\,080\,\text{counts}\,\text{V}^{-1}$, resulting in an overall sensitivity of $6.97 \times 10^7\,\text{counts}(\text{m s}^{-1})^{-1}$ on the horizontal and $1.57 \times 10^8\,\text{counts}/(\text{m s}^{-1})^{-1}$ on the vertical channels, both at reference frequency $f_r = 1\,\text{Hz}$. The decimation from 8000 to $62.5\,\text{Hz}$ is implemented by 7 FIR stages of uniform sensitivity.

A2 DEPAS hydrophones

The responses of the hydrophones and differential pressure gauges are also given by Eq. (A2), though with a different instrument response $G_{inst,h}(f)$, that has to be calculated separately for each instrument, as briefly explained here: a hydrophone measures pressure variations via a piezo element, which has a sensitivity of $S_{d,hyd}$ in $\text{V}\,\text{Pa}^{-1}$. Below its corner frequency (typically in the kHz range), its equivalent circuit is a capacitor C_{hyd}. Together with the input capacity of the amplifier C_{amp}, the system has the total capacitance $C_{total} = \frac{C_{amp} C_{hyd}}{C_{amp} + C_{hyd}}$. With the input impedance R of the sensor, the system forms a high-pass filter with a transfer function

$$G_{inst,h}(f) = S_{d,hyd} \frac{R C_{total} 2\pi i f}{1 + R C_{total} 2\pi i f}, \tag{A5}$$

equivalent to Eq. (A1) with a single pole

$$p_1 = -\frac{1}{R C_{total}} = -\frac{C_{amp} + C_{hyd}}{R C_{amp} C_{hyd}}\,\text{rad}\,\text{s}^{-1} \tag{A6}$$

and one zero $r_1 = 0\,\text{rad}\,\text{s}^{-1}$.

The capacitance C_{hyd} is instrument-specific. The reference value from the manufacturer HighTechInc is $C_{hyd} = 45\,\text{nF}$. Before sale, every hydrophone is calibrated, which showed a mean value $C_{hyd} = 56.3\,\text{nF}$ with a sample standard deviation of $3.5\,\text{nF}$ amongst the 60 instruments in the DEPAS pool. The input resistance R of the data logger was either 210 or $500\,\text{M}\Omega$, depending on the instrument version.

The sensitivity S_h is different for each hydrophone, around $185\,\mu\text{V}\,\text{Pa}^{-1}$ with a sample standard deviation of $8\,\mu\text{V}\,\text{Pa}^{-1}$ amongst the DEPAS instruments. DEPAS supplied us with values for S_d, R and C_{hyd} for each instrument. From those, we calculated poles, zeros and sensitivities, which are listed in the dataless SEED and StationXML files available from the RESIF data centre. Geomar instruments were equipped with a similar hydrophone model, HTI-01-PCA from the same manufacturer. Its nominal values is $C_{hyd} = 50\,\text{nF}$ and since no individually calibrated responses were available, we used the average value of the other HTI-01-PCA in the DEPAS pool, resulting in $S_d = 199.5\,\mu\text{V}\,\text{Pa}^{-1}$ and $p_1 = 0.10774\,\text{rad}\,\text{s}^{-1}$. This applies to the Geomar OBS (RR33, RR39, RR53 and RR56) as well as to RR45 and RR55, where Geomar hydrophones were attached to LOBSTER OBS.

A3 INSU differential pressure gauges

Differential pressure gauges (DPGs, Cox et al., 1984) are hand-manufactured in research laboratories and their sensitivity and low-pass frequency are challenging to calibrate. The DPGs in stations RR28 and RR29 were manually calibrated on land by comparing their impulse response to that of an absolute pressure gauge in a vacuum jar. Since the low-pass frequency is highly dependent on the viscosity of the oil in the gauge and this viscosity may change with temperature and pressure, it is not sure that these values accurately reflect the instrument response at the seafloor, although visual comparison with the DEPAS hydrophone PPSDs does not suggest significant error. The DPGs on the other sensors were not calibrated and the instrument responses given are therefore the same as those for station RR28. This practice is the same as that used by other OBS facilities (e.g. Godin et al., 2013), but it leaves a significant uncertainty in the converted signal amplitudes.

Appendix B: Description of laboratory experiments on the DEPAS clocks

Since the internal clocks of several DEPAS OBS stopped before retrieval, and ambient noise estimation of the clock error proved impossible, we tried to estimate the clock error from laboratory experiments. Hence we re-ran several data recorders after their return to the DEPAS lab at AWI Bremerhaven, in an attempt to measure their clock drifts. Only seven data loggers were available (RR06, RR11, RR41, RR43, RR44, RR45, RR55); the remainder had been redeployed in new experiments. Attached to their original lithium batteries and a seismometer, the recorders were run for 7 days, and then for another 33 days. Table B1 shows the skews measured after the two runs, linearly extrapolated to a hypothetical run time of 365 days.

For 6 out of 7 stations, skew values from the two runs agree to within less than 0.1 s. The exception is RR44, where the skews disagree by more than one second (-0.50 s from the 7-day run, versus $+0.55$ s from the 33-day run). For RR11, a skew of $+0.61$ s had been obtained upon OBS recovery (see Table 3), as compared to -0.15 and -0.21 s in the two lab runs (Table B1), which means mutual consistence to within 0.8 s, an uncertainty as large as the skew estimates themselves. No skew upon recovery was available for the remaining six recorders.

Most lab skew values in Table B1 are rather small in magnitude, compared to skews obtained during the field campaign in Table 3. This pattern is consistent with the direct comparison available for RR11, and hints at a systematic difference between seafloor runs and lab runs. In either setting, the clocks tend to run too fast, as indicated by mostly positive skew values (upon recovery, the elapsed recorder time is larger than the elapsed GPS time). But clocks on the seafloor ran even faster than clocks in the lab. (Note that only DEPAS stations in Table 3 should enter this comparison, since INSU recorders are of a different make.)

The likely shortcoming of our lab experiments is that we did not simulate temperature conditions of the real experiment: a sudden drop from 22 to 4 °C) upon deployment, a constant 4 °C during recording, and sudden warming to 22 °C upon recovery. Solid-state oscillators are known to be temperature dependent, which may explain why our lab experiments could match the field observations qualitatively (correct sign of skew), but probably did not yield the correct skew magnitudes. Hence we assign low confidence to the skew measurements in Table B1 and do *not* apply any skew corrections to RHUM-RUM time series based on these values.

Table B1. Lab measurements of clock skews for seven DEPAS recorders. Two separate runs of 7 and 33 days durations yielded skew measurements that are linearly extrapolated to a hypothetical run of 365 days duration (for convenient comparison to skews measured in the field campaign, Table 3). We assign low confidence to these lab measurements (see text for discussion) and do *not* correct RHUM-RUM time series using these values.

Station	Serial number	Skew prediction for 365 days	
	(data logger)	from 7 day exp.	from 33 day exp.
RR06	060744	0.15 s	0.13 s
RR11	060753	-0.15 s	-0.21 s
RR41	050922	0.3 s	0.23 s
RR43	060702	0.00 s	0.033 s
RR44	060751	-0.5 s	0.55 s
RR45	080104	0.045 s	-0.05 s
RR55	060748	0.0015 s	-0.03 s

Appendix C: Summary charts of noise levels across the RHUM-RUM OBS network

Figures C1 to C3 are graphical summaries of noise statistics for all stations and components, in three different frequency bands:

Fig. C1: microseismic noise band (period range 5–15 s). DEPAS and INSU seismometers record comparable noise levels.

Fig. C2: low-noise notch (period band 15–40 s). The noise level of the INSU seismometers is on average 15 dB lower than the values for the DEPAS instruments.

Fig. C3: long-period band (40–100 s). Both INSU seismometers (corner period 240 s) and the DEPAS seismometers (corner period 60 s) still have nominal instrument sensitivity in this band, but the self-noise of the Güralp instruments used in the DEPAS OBS is pronounced, especially on the BHZ channel.

Probabilistic Power Spectral Densities (cf. Fig. 9) were calculated for all stations and broadband components (BH1, BH2, BHZ, BDH) by stacking hour-long time series. For each of the three frequency bands, we averaged the hourly spectra over the frequencies contained the band of interest, and calculated the median, quartiles, 2.5 % percentile, and 97.5 % percentile power levels of the hourly band averages. These statistics are plotted for all stations, components and frequency bands in Figs. C1 to C3.

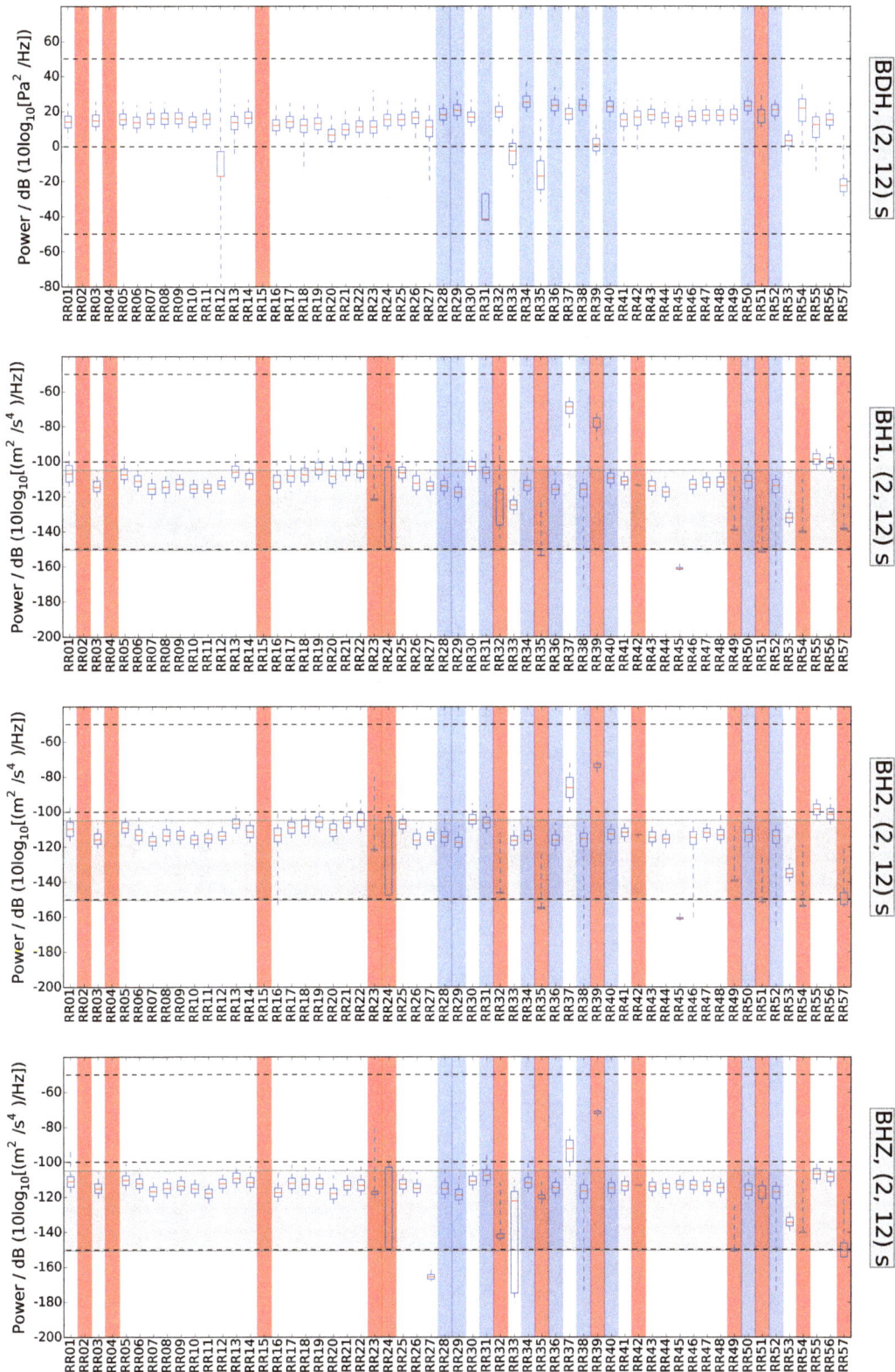

Figure C1. Noise power levels in the microseismic noise band (5–15 s period), on the BH1, BH2, BHZ, and BDH components (4 sub-plots). 57 box plots per panel characterize the 57 RHUM-RUM stations. In each box plot, the red line marks the median power level during the interval of successful recording. Top and bottom edges of the blue box mark the ranges of the two quartiles, and dashed line the range that contains 95 % of all hourly observations in this frequency band (from 2.5 to 97.5 % percentile). Light blue shading indicates INSU stations, all others are DEPAS or Geomar. Red shading indicates failed components. Grey horizontal band marks the power range bracketed by the (terrestrial) New Low Noise and New High Noise Models (Peterson, 1993), in the frequency passband considered here.

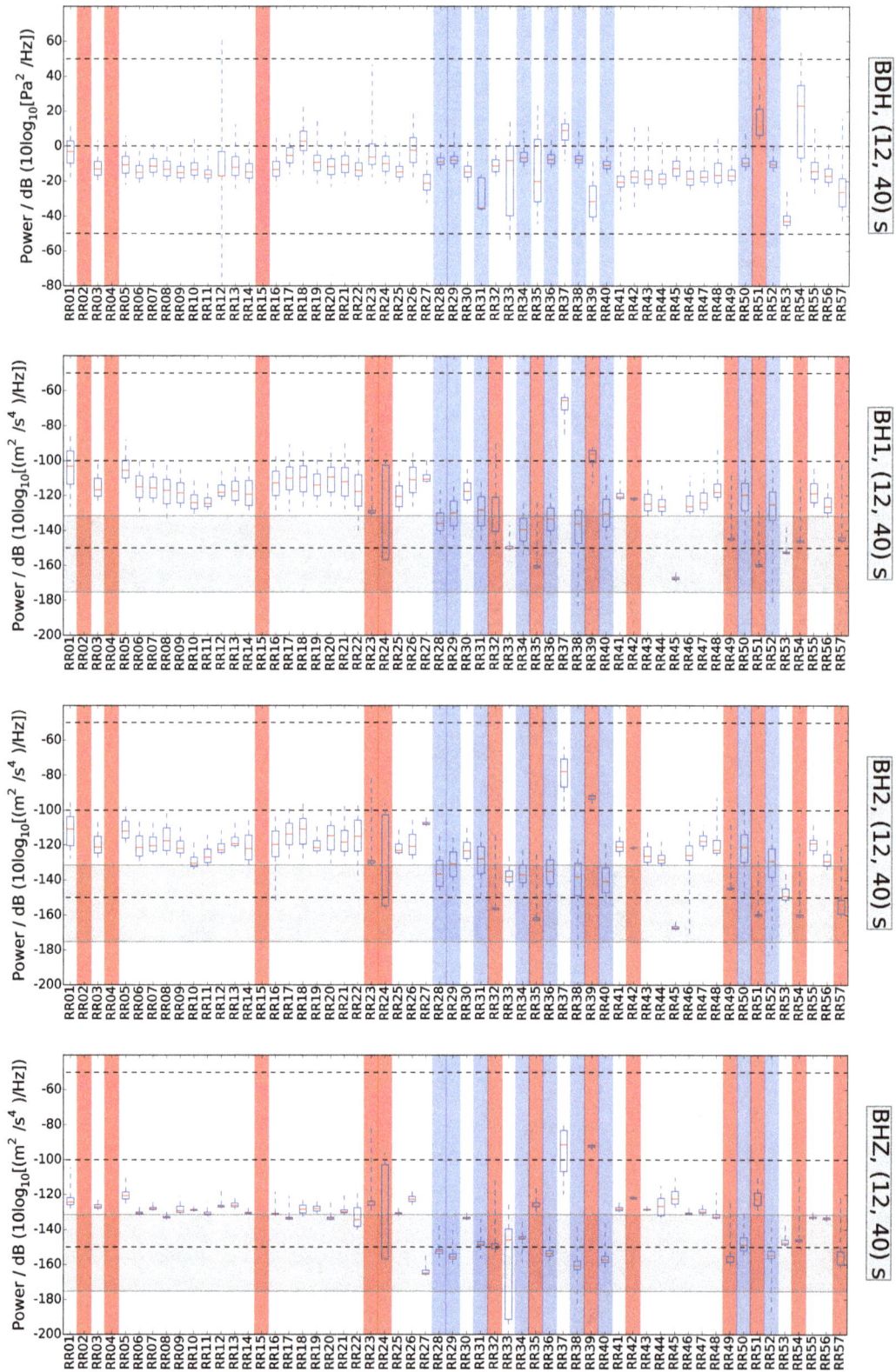

Figure C2. Noise power levels in the band of the low-noise notch (15–40 s period). Refer to the caption of Fig. C1 for explanation.

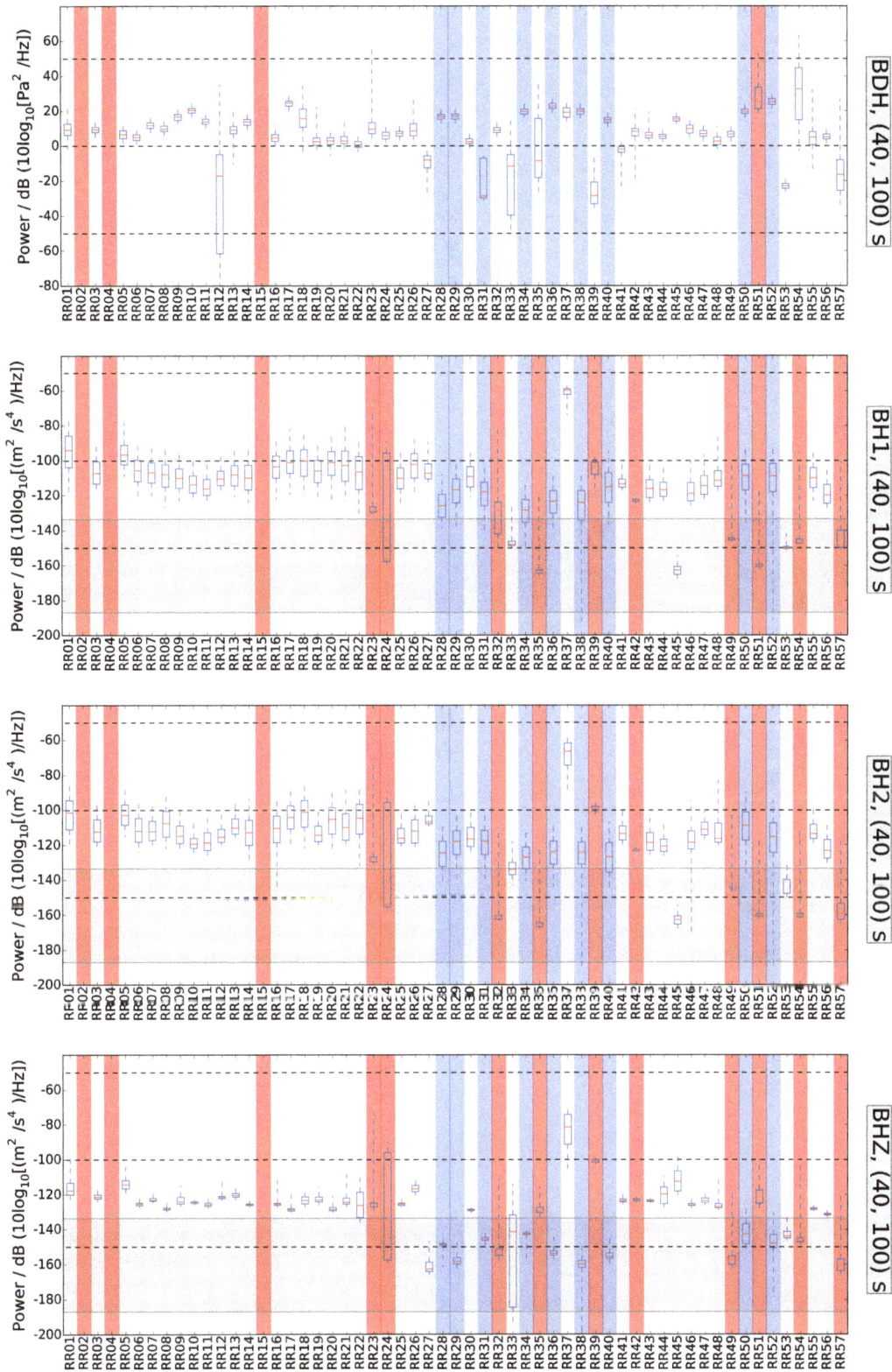

Figure C3. Noise power levels in the long-period band (40–100 s period). Refer to the caption of Fig. C1 for explanation.

Author contributions. K. Hosseini, M. Tsekhmistrenko, K. Sigloch, S. C. Stähler, W. C. Crawford, J.-R. Scholz and A. Mazzullo processed raw data and assessed station performance during and after the OBS recovery cruise. Station meta-data were assembled and verified for the DEPAS instruments by S. C. Stähler and M. C. Schmidt-Aursch, and for the INSU instruments by W. C. Crawford. A. Mazzullo, M. Deen and W. C. Crawford investigated the "glitch" on the INSU instruments. G. Barruol and K. Sigloch designed the RHUM-RUM project, obtained funding for the OBS experiment, and led the cruises. S. C. Stähler and K. Sigloch prepared the manuscript with contributions from all co-authors.

Acknowledgements. RHUM-RUM is funded by Deutsche Forschungsgemeinschaft (grants SI1538/2-1 and SI1538/4-1) and Agence National de la Recherche (project ANR-11-BS56-0013). Additional support is provided by Centre National de la Recherche Scientifique-Institut National des Sciences de l'Univers, Terres Australes et Antarctiques Françaises, Institut Polaire Paul Emile Victor, Alfred Wegener Institute Bremerhaven, and a Marie Curie Career Integration Grant to K. Sigloch. Instruments were provided by "Deutscher Geräte-Pool für Amphibische Seismologie" at Alfred-Wegener-Institut, Bremerhaven, "Parc Sismomètre fond du mer" at INSU/IPGP, and Geomar, Kiel. We thank Erik Labahn, Henning Kirk, and the crews of research vessels *Marion Dufresne* and *Meteor* for excellent support during deployment and recovery. We thank Carlos Corela for preparing the initial compilation of the DEPAS metadata. Ulf Gräwe assisted with downloading HYCOM ocean model data (http://hycom.org). All figures were produced with the ObsPy software, version 0.10.2 (The ObsPy Development Team, 2015). We thank the RESIF data centre in Grenoble, especially Catherine Pequegnat and Pierre Volcke, for hosting the RHUM-RUM data.

RESIF is supported by the French Ministry of Education and Research, by 18 Research Institutions and Universities in France, by the French National Research Agency (ANR) as part of the "Investissements d'Avenir" program (reference: ANR-11-EQPX-0040) and by the French Ministry of Ecology, Sustainable Development and Energy.

References

Ahern, T. K., Casey, R., Barnes, D., Benson, R., and Knight, T.: Seed Reference Manual, Tech. rep., Incorporated Research Institutions for Seismology, http://www.fdsn.org/media/_s/publications/SEEDManual_V2.4.pdf (last access: 19 October 2015), 2012.

Barruol, G.: PLUME investigates South Pacific Superswell, Eos, Trans. Am. Geophys. Union, 83, 511, doi:10.1029/2002EO000354, 2002.

Barruol, G. and Sigloch, K.: Investigating La Réunion hot spot from crust to core, Eos, Trans. Am. Geophys. Union, 94, 205–207, doi:10.1002/2013EO230002, 2013.

Barruol, G., Sigloch, K., and the RHUM-RUM group: RHUM-RUM experiment, 2011–2015, code YV (Réunion Hotspot and Upper Mantle – Réunion's Unterer Mantel) funded by ANR, DFG, CNRS-INSU, IPEV, TAAF, instrumented by DEPAS, INSU-OBS, AWI and the Universities of Muenster, Bonn, La Réunion, doi:10.15778/RESIF.YV2011, 2011.

Becker, J. J., Sandwell, D. T., Smith, W. H. F., Braud, J., Binder, B., Depner, J., Fabre, D., Factor, J., Ingalls, S., Kim, S.-H., Ladner, R., Marks, K., Nelson, S., Pharaoh, A., Trimmer, R., Von Rosenberg, J., Wallace, G., and Weatherall, P.: Global Bathymetry and Elevation Data at 30 Arc Seconds Resolution: SRTM30_PLUS, Mar. Geodyn., 32, 355–371, doi:10.1080/01490410903297766, 2009.

Cannat, M., Rommevaux-Jestin, C., Sauter, D., Deplus, C., and Mendel, V.: Formation of the axial relief at the very slow spreading Southwest Indian Ridge (49° to 69° E), J. Geophys. Res., 104, 22825, doi:10.1029/1999JB900195, 1999.

Cox, C., Deaton, T., and Webb, S.: A Deep-Sea Differential Pressure Gauge, J. Atmos. Ocean. Tech., 1, 237–246, doi:10.1175/1520-0426(1984)001<0237:ADSDPG>2.0.CO;2, 1984.

Crawford, W. C. and Webb, S. C.: Identifying and Removing Tilt Noise from Low-Frequency (< 0.1 Hz) Seafloor Vertical Seismic Data, Bull. Seismol. Soc. Am., 90, 952–963, doi:10.1785/0119990121, 2000.

Cummings, J. A.: Operational multivariate ocean data assimilation, Q. J. Roy. Meteorol. Soc., 131, 3583–3604, doi:10.1256/qj.05.105, 2005.

Dahm, T., Thorwart, M., Flueh, E. R., Braun, T., Herber, R., Favali, P., Beranzoli, L., D'Anna, G., Frugoni, F., and Smriglio, G.: Ocean bottom seismometers deployed in Tyrrhenian Sea, Eos, Trans. Am. Geophys. Union, 83, 309, doi:10.1029/2002EO000221, 2002.

Davy, C., Barruol, G., Fontaine, F. R., Sigloch, K., and Stutzmann, E.: Tracking major storms from microseismic and hydroacoustic observations on the seafloor, Geophys. Res. Lett., 41, 8825–8831, doi:10.1002/2014GL062319, 2014.

Duennebier, F. K., Blackinton, G., and Sutton, G. H.: Current-generated noise recorded on ocean bottom seismometers, Mar. Geophys. Res., 5, 109–115, doi:10.1007/BF00310316, 1981.

Dyment, J., Lin, J., and Baker, E.: Ridge-Hotspot Interactions: What Mid-Ocean Ridges Tell Us About Deep Earth Processes, Oceanography, 20, 102–115, doi:10.5670/oceanog.2007.84, 2007.

Flueh, E. R. and Biolas, J.: A digital, high data capacity ocean bottom recorder for seismic investigations, Int. Underw. Syst. Des., 18, 18–20, 1996.

Geissler, W. H. and Schmidt, R.: Short Cruise Report Maria S. Merian; MSM 24 Walvis Bay – Cape Town, Tech. rep., Leitstelle Deutsche Forschungsschiffe, Hamburg, https://www.ldf.uni-hamburg.de/merian/wochenberichte/wochenberichte-merian/msm22-msm25/msm24-scr.pdf (last access: 19 October 2015), 2013.

Geissler, W. H., Matias, L., Stich, D., Carrilho, F., Jokat, W., Monna, S., Ibenbrahim, A., Mancilla, F., Gutscher, M. A., Sallars, V., and Zitellini, N.: Focal mechanisms for sub-crustal earthquakes in the Gulf of Cadiz from a dense OBS deployment, Geophys. Res. Lett., 37, 7–12, doi:10.1029/2010GL044289, 2010.

GEOFON: Mw 6.6 earthquake, Sichuan China, 2013-04-20 (Moment Tensor Solution), doi:10.5880/GEOFON.gfz2013hrdy, 2013.

Godin, O. A., Zabotin, N. A., Sheehan, A. F., Yang, Z., and Collins, J. A.: Power spectra of infragravity waves in a deep ocean, Geophys. Res. Lett., 40, 2159–2165, doi:10.1002/grl.50418, 2013.

Gouedard, P., Seher, T., McGuire, J. J., Collins, J. A., and van der Hilst, R.: Correction of Ocean-Bottom Seismometer Instrumental Clock Errors Using Ambient Seismic Noise, Bull. Seismol. Soc. Am., 104, doi:10.1785/0120130157, 2014.

Hannemann, K., Krüger, F., and Dahm, T.: Measuring of clock drift rates and static time offsets of ocean bottom stations by means of ambient noise, Geophys. J. Int., 196, 1034–1042, doi:10.1093/gji/ggt434, 2014.

Laske, G., Collins, J. A., Wolfe, C. J., Solomon, S. C., Detrick, R. S., Orcutt, J. A., Bercovici, D., and Hauri, E. H.: Probing the Hawaiian Hot Spot With New Broadband Ocean Bottom Instruments, Eos, Trans. Am. Geophys. Union, 90, 362–363, doi:10.1029/2009EO410002, 2009.

McNamara, D. E. and Buland, R.: Ambient Noise Levels in the Continental United States, Bull. Seismol. Soc. Am., 94, 1517–1527, doi:10.1785/012003001, 2004.

Meier, T., Friederich, W., Papazachos, C., Taymaz, T., and Kind, R.: EGELADOS: a temporary amphibian broadband seismic network in the southern Aegean, in: Geophys. Res. Abstr., 9, 9020, 2007.

Morgan, W. J.: Rodriguez, Darwin, Amsterdam, ..., A second type of Hotspot Island, J. Geophys. Res., 83, 5355, doi:10.1029/JB083iB11p05355, 1978.

Peterson, J.: Observations and Modeling of Seismic Background Noise, Tech. rep., USGS, Albuquerque, New Mexico, 1993.

Ponsoni, L., Aguiar-González, B., Maas, L., van Aken, H., and Ridderinkhof, H.: Long-term observations of the east madagascar undercurrent, Deep-Sea Res. Pt. I, 100, 64–78, doi:10.1016/j.dsr.2015.02.004, 2015.

Scholz, J.-R.: Local seismicity of the segment-8-volcano at the ultraslow spreading Southwest Indian Ridge, Diploma thesis, Technische Universität Dresden, Dresden, 2014.

Sens-Schönfelder, C.: Synchronizing seismic networks with ambient noise, Geophys. J. Int., 174, 966–970, doi:10.1111/j.1365-246X.2008.03842.x, 2008.

Sigloch, K.: Multiple-frequency body-wave tomography, PhD thesis, Princeton, 2008.

Stähler, S. C., Sigloch, K., Barruol, G., and Crawford, W. C.: Noise levels at all stations of the RHUM-RUM OBS network, doi:10.13140/RG.2.1.1374.0886, 2015.

Suetsugu, D., Sugioka, H., Isse, T., Fukao, Y., Shiobara, H., Kanazawa, T., Barruol, G., Schindelé, F., Reymond, D., Bonneville, A., and Debayle, E.: Probing South Pacific mantle plumes with ocean bottom seismographs, Eos, Trans. Am. Geophys. Union, 86, 429, doi:10.1029/2005EO440001, 2005.

Suetsugu, D., Shiobara, H., Sugioka, H., Ito, A., Isse, T., Kasaya, T., Tada, N., Baba, K., Abe, N., Hamano, Y., Tarits, P., Barriot, J.-P., and Reymond, D.: TIARES Project – Tomographic investigation by seafloor array experiment for the Society hotspot, Earth Planets Space, 64, i–iv, doi:10.5047/eps.2011.11.002, 2012.

Sumy, D. F., Lodewyk, J. A., Woodward, R. L., and Evers, B.: Ocean-Bottom Seismograph Performance during the Cascadia Initiative, Seismol. Res. Lett., 86, 1238–1246, doi:10.1785/0220150110, 2015.

The ObsPy Development Team: ObsPy 0.10.2, doi:10.5281/zenodo.17641, 2015.

Tilmann, F., Yuan, X., Rümpker, G., and Rindraharisaona, E.: SELASOMA Project, Madagascar 2012–2014, doi:10.14470/MR7567431421, 2012.

Trehu, A. M.: A note on the effect of bottom currents on an ocean bottom seismometer, Bull. Seismol. Soc. Am., 75, 1195–1204, 1985.

Webb, S. C.: Broadband seismology and noise under the ocean, Rev. Geophys., 36, 105, doi:10.1029/97RG02287, 1998.

Wolfe, C. J., Solomon, S. C., Laske, G., Collins, J. A., Detrick, R. S., Orcutt, J. A., Bercovici, D., and Hauri, E. H.: Mantle Shear-Wave Velocity Structure Beneath the Hawaiian Hot Spot, Science, 326, 1388–1390, doi:10.1126/science.1180165, 2009.

Wysession, M., Wiens, D., Nyblade, A., and Rambolamanana, G.: Investigating Mantle Structure with Broadband Seismic Arrays in Madagascar and Mozambique, AGU Fall Meet. Abstr., p. B2591, 2012.

"Improving seismic networks performances: from site selection to data integration" (EGU2015 SM1.2/GI1.5 session)

D. Pesaresi[1], H. Pedersen[2], and Y. Starovoit[3]

[1]OGS (Istituto Nazionale di Oceanografia e di Geofisica Sperimentale), Trieste, Italy
[2]RESIF (Réseau sismologique & géodésique français), Grenoble, France
[3]CTBTO (Comprehensive Nuclear-Test-Ban Treaty Organization), Vienna, Austria

Correspondence to: D. Pesaresi (dpesaresi@inogs.it)

Abstract. The number and quality of seismic stations and networks in Europe continually improves, nevertheless there is always scope to optimize their performance. In this EGU2015 SM1.2/GI1.5 session we welcomed contributions from all aspects of seismic network installation, operation and management. This includes site selection; equipment testing and installation; planning and implementing communication paths; policies for redundancy in data acquisition, processing and archiving; and integration of different data sets including GPS and OBS.

1 Introduction

The history of seismic networks sessions at European Geosciences Union (EGU) general assemblies started in 2010 with the SM1.3 "Seismic Centers Data Acquisition" session (Pesaresi, 2011), where the convener Damiano Pesaresi supported by the Orfeus Data Center (ODC) Director co-convener Reinoud Sleeman chaired a session of 7 oral and 16 poster presentations. Later in the same year a similar session was held at the XXXII European Seismological Commission (ESC) General Assembly: "SD1, 3 Seismic centers data acquisition", conveners D. Pesaresi and R. Sleeman, with 15 oral presentations.

The history of these sessions continued in 2011 with the EGU2011 SM1.3/G3.8/GD3.7/GI-19/TS8.7 "Improving seismic networks performances: from site selection to data integration" session (EGU2011 SM1.3/G3.8/GD3.7/GI-19/TS8.7 Improving seismic networks performances: from site selection to data integration, 2011), where the convener Damiano Pesaresi supported by the co-conveners John Clinton and Robert Busby chaired a session of 9 oral and 20 poster presentations; in 2012 with the EGU2012

SM1.3/GI1.7 "Improving seismic networks performances: from site selection to data integration" session (Pesaresi and Vernon, 2013), where the convener Damiano Pesaresi supported by the co-convener Frank Vernon chaired a session of 6 oral and 22 poster presentations; in 2013 with the SM1.4/GI1.6 "Improving seismic networks performances: from site selection to data integration" session (Pesaresi and Busby, 2013), where the convener Damiano Pesaresi supported by the co-convener Robert Busby chaired a session of 6 oral and 13 poster presentations; and in 2014 with the EGU2014 SM1.2/GI3.7 "Improving seismic networks performances: from site selection to data integration" session (Pesaresi et al., 2015), where the convener Damiano Pesaresi supported by the co-conveners John Clinton and Helle Pedersen chaired a session of 12 oral and 27 poster presentations.

2 The EGU2015 SM1.2/GI1.5 session

In the EGU2015 SM1.2/GI1.5 "Improving seismic networks performances: from site selection to data integration" session (EGU2015 SM1.2/GI1.5 Improving seismic networks performances: from site selection to data integration, 2015) the convener Damiano Pesaresi supported by the co-conveners Helle Pedersen and Yuri Starovoit chaired a session (Fig. 1) of 20 posters (Table 1).

The 20 presentations came from 13 countries (Austria, Kazakhstan, Malta, Greece, Chile, USA, France, Finland, Bulgaria, Germany, UK, Italy, Slovenia), in four continents (Europe, Asia, South America, North America), which fits well to the goals of the European Geosciences Union.

The solicited presentations in this session were the following:

Table 1. Poster programme for the EGU2015 SM1.2/GI1.5 session.

EGU Abstract ref.	Title	Authors
EGU2015-1611	Ground Truth and Application for the Anisotropic Receiver Functions Technique – Test site KTB: the installation campaign	Irene Bianchi, Mario Anselmi, Maria-Theresia Apoloner, Ehsan Qorbani, Katalin Gribovszki, and Götz Bokelmann
EGU2015-3702	Ebreichsdorf 2013 Earthquake Series: Relative Location	Maria-Theresia Apoloner and Götz Bokelmann
EGU2015-3708	Modeling and Detection of Regional Depth Phases at the GERESS Array	Maria-Theresia Apoloner and Götz Bokelmann
EGU2015-4516	Seismic monitoring of Central Asia territory in KNDC res	Aidyn Mukambayev and Natalia Mikhailova
EGU2015-4523	Recent developments in the setting up of the Malta Seismic Network	Matthew Agius, Pauline Galea, and Sebastiano D'Amico
EGU2015-6232	The Hellenic Seismological Network Of Crete (HSNC): Validation and results of the 2013 aftershock sequences	Georgios Chatzopoulos, Ilias Papadopoulos, and Filippos Vallianatos
EGU2015-7561	Field Installation and Real-Time Data Processing of the New Integrated SeismoGeodetic System with Real-Time Acceleration and Displacement Measurements for Earthquake Characterization Based on High-Rate Seismic and GPS Data	Leonid Zimakov, Michael Jackson, Paul Passmore, Jared Raczka, Marcos Alvarez, and Sergio Barrientos
EGU2015-7985	Comparative Noise Performance of Portable Broadband Sensor Emplacements	Justin Sweet, Eliana Arias-Dotson, Bruce Beaudoin, and Kent Anderson
EGU2015-9164	Sources of high frequency seismic noise: insights from a dense network of ~ 250 stations in northern Alsace (France)	Jerome Vergne, Antoine Blachet, Maximilien Lehujeur and the EstOF Team
EGU2015-9965	AlpArray Austria – Illuminating the subsurface of Austria and understanding of Alpine geodynamics	Florian Fuchs, Götz Bokelmann, Irene Bianchi, Maria-Theresia Apoloner, and AlpArray Working Group
EGU2015-11264	Automatic data processing and analysis system for monitoring region around a planned nuclear power plant	Timo Tiira, Outi Kaisko, Jari Kortström, Tommi Vuorinen, Marja Uski, and Annakaisa Korja
EGU2015-11506	Automatic classification of seismic events within a regional seismograph network	Timo Tiira, Jari Kortström, and Marja Uski
EGU2015-11525	Introduction of digital object identifiers (DOI) for seismic networks	Peter Evans, Angelo Strollo, Adam Clark, Tim Ahern, Rob Newman, John Clinton, Catherine Pequegnat, and Helle Pedersen
EGU2015-11614	Local network deployed around the Kozloduy NPP – a useful tool for seismological monitoring	Dimcho Solakov, Stela Simeonova, Liliya Dimitrova, Krasimira Slavcheva, Plamena Raykova, Maria Popova, and Ivan Georgiev
EGU2015-12129	Preliminary performance report of the RHUM-RUM OBS network	Simon C. Stähler, Wayne Crawford, Guilhem Barruol, Karin Sigloch, and Schmidt-Aursch Mechita
EGU2015-12279	The Austrian National Network 2014	Nikolaus Horn, Helmut Hausmann, and Yan Jia
EGU2015-12334	"SeismoSAT" project results in connecting seismic data centres via satellite	Damiano Pesaresi, Wolfgang Lenhardt, Markus Rauch, Mladen Živčić, Rudolf Steiner, Michele Bertoni, and Heimo Delazer
EGU2015-13196	Seismic noise recorded by seafloor observatories at Mediterranean sites	Mariagrazia De Caro, Stephen Monna, Francesco Frugoni, Laura Beranzoli, and Paolo Favali
EGU2015-14387	Impact of sensor installation techniques on seismic network performance	Geoffrey Bainbridge, Michael Laporte, Dario Baturan, and Wesley Greig
EGU2015-14813	UMTS rapid response real-time seismic networks: implementation and strategies at INGV	Aladino Govoni, Lucia Margheriti, Milena Moretti, Valentino Lauciani, Gianpaolo Sensale, Augusto Bucci, and Fabio Criscuoli

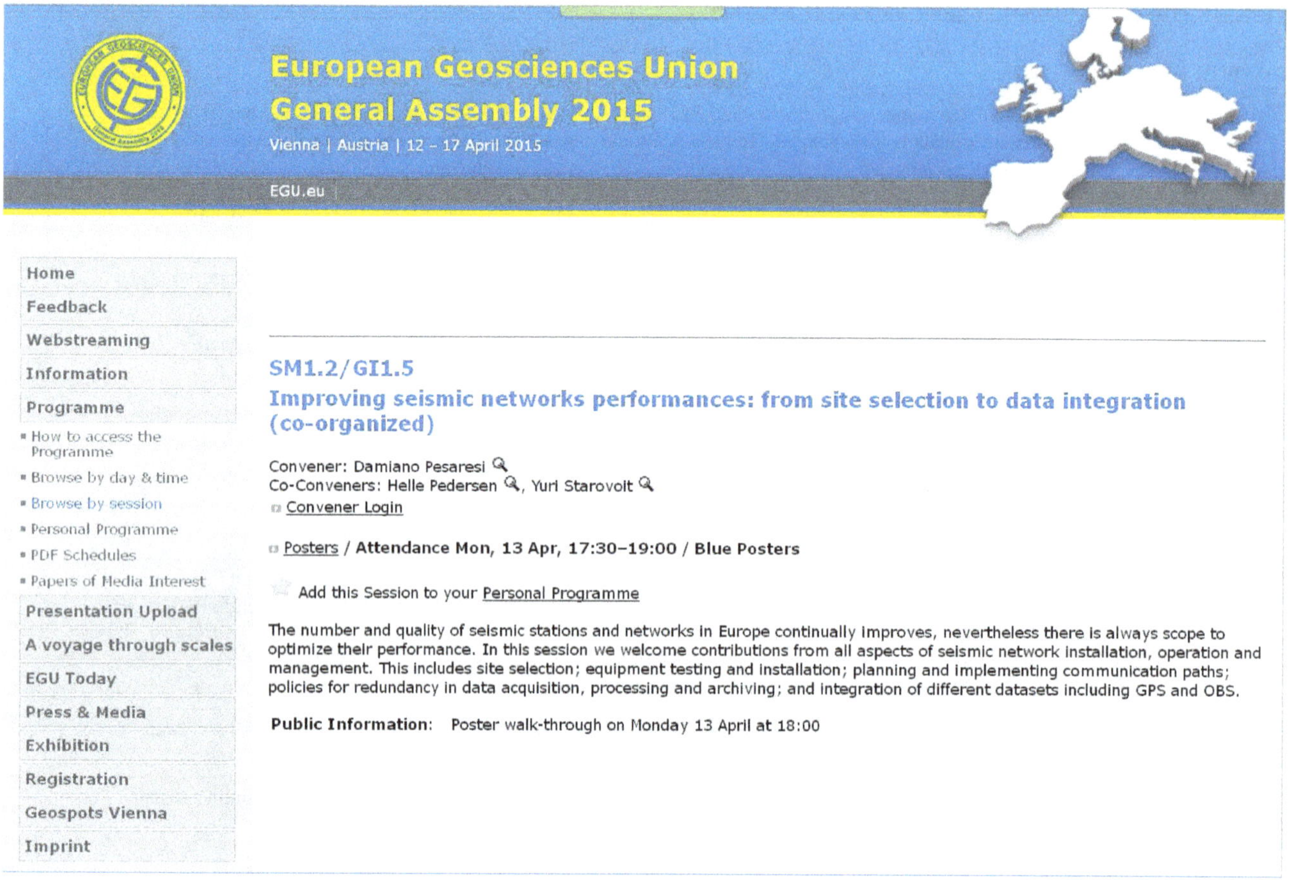

Figure 1. EGU2015 SM1.2/GI1.5 session (from the EGU2015 home page).

i. "Comparative Noise Performance of Portable Broadband Sensor Emplacements", by Justin Sweet, Eliana Arias-Dotson, Bruce Beaudoin, and Kent Anderson (Sweet et al., 2015);

ii. "Sources of high frequency seismic noise: insights from a dense network of ~250 stations in northern Alsace (France)", by Jerome Vergne, Antoine Blachet, Maximilien Lehujeur and the EstOF Team (Vergne et al., 2015);

iii. "AlpArray Austria – Illuminating the subsurface of Austria and understanding of Alpine geodynamics", by Florian Fuchs, Götz Bokelmann, Irene Bianchi, Maria-Theresia Apoloner, and AlpArray Working Group (Fuchs et al., 2015);

iv. "Introduction of digital object identifiers (DOI) for seismic networks", by Peter Evans, Angelo Strollo, Adam Clark, Tim Ahern, Rob Newman, John Clinton, Catherine Pequegnat, and Helle Pedersen (Evans et al., 2015);

v. "Impact of sensor installation techniques on seismic network performance", by Geoffrey Bainbridge,

Michael Laporte, Dario Baturan, and Wesley Greig (Bainbridge et al., 2015).

3 Conclusions

The quality and quantity of presentations made at the EGU2015 SM1.2/GI1.5 session satisfied the expectations of the convener and co-conveners and fit the goals of the European Geosciences Union.

This year, for the first time, the number of presentations at the seismic networks session decreased; however, the same is true for the overall number of presentations of the entire EGU2015 General Assembly. Therefore, the conveners are still confident that the path they followed in organizing such sessions at the yearly EGU General Assembly is a valid one, since there is need in the seismological community worldwide to present and discuss different solutions to common problems in running seismic networks.

Acknowledgements. The authors wish to thank the authors of the EGU2015 SM1.2/GI1.5 session presentations, especially those who made the effort to publish their presentations in these proceedings in *Advances in Geosciences.* The authors also especially thank the former EGU Seismology Division President (and future EGU Seismology Division Vice-President) Charlotte Krawczyk, for her continuous strong support of the seismic networks sessions at the EGU, and welcome the new EGU Seismology Division President Paul Martin Mai.

References

EGU2011 SM1.3/G3.8/GD3.7/GI-19/TS8.7 Improving seismic networks performances: from site selection to data integration, http://meetingorganizer.copernicus.org/EGU2011/session/7340 (last access: 6 May 2015), 2011.

EGU2015 SM1.2/GI1.5 Improving seismic networks performances: from site selection to data integration, http://meetingorganizer.copernicus.org/EGU2015/session/17368 (last access: 5 May 2015), 2015.

Bainbridge, G., Laporte, M., Baturan, D., and Greig, W.: Impact of sensor installation techniques on seismic network performance, EGU General Assembly, Vienna, Austria, 12–17 April 2015, EGU2015-14387, 2015.

Evans, P., Strollo, A., Clark, A., Ahern, T., Newman, R., Clinton, J., Pequegnat, C., and Pedersen, H.: Introduction of digital object identifiers (DOI) for seismic networks, EGU General Assembly, Vienna, Austria, 12–17 April 2015, EGU2015-11525, 2015.

Fuchs, F., Bokelmann, G., Bianchi, I., Apoloner, M.-T., and AlpArray Working Group: AlpArray Austria – Illuminating the subsurface of Austria and understanding of Alpine geodynamics, EGU General Assembly, Vienna, Austria, 12–17 April 2015, EGU2015-9965, 2015.

Pesaresi, D.: The EGU2010 SM1.3 Seismic Centers Data Acquisition session: an introduction to Antelope, EarthWorm and SeisComP, and their use around the World, Ann. Geophys.-Italy, 54, 1–7, doi:10.4401/ag-4972, 2011.

Pesaresi, D. and Busby, R.: EGU2013 SM1.4/GI1.6 session: "Improving seismic networks performances: from site selection to data integration", Adv. Geosci., 36, 1–5, doi:10.5194/adgeo-36-1-2013, 2013.

Pesaresi, D. and Vernon, F.: EGU2012 SM1.3/GI1.7 session: "Improving seismic networks performances: from site selection to data integration", Adv. Geosci., 34, 1–4, doi:10.5194/adgeo-34-1-2013, 2013.

Pesaresi, D., Clinton, J., and Pedersen, H.: Preface: Improving seismic networks performances: from site selection to data integration (EGU2014 SM1.2/GI3.7 session), Adv. Geosci., 40, 19–25, doi:10.5194/adgeo-1-19-2015, 2015.

Sweet, J., Arias-Dotson, E., Beaudoin, B., and Anderson, K.: Comparative Noise Performance of Portable Broadband Sensor Emplacements, EGU General Assembly, Vienna, Austria, 12–17 April 2015, EGU2015-7985, 2015.

Vergne, J., Blachet, A., Lehujeur, M., and the EstOF Team: Sources of high frequency seismic noise: insights from a dense network of ~ 250 stations in northern Alsace (France), EGU General Assembly, Vienna, Austria, 12–17 April 2015, EGU2015-9164, 2015.

23

Automatic data processing and analysis system for monitoring region around a planned nuclear power plant

Jari Kortström, Timo Tiira, and Outi Kaisko

Institute of Seismology, Department of Geosciences and Geography, University of Helsinki, Finland

Correspondence to: Jari Kortström (jari.kortstrom@helsinki.fi)

Abstract. The Institute of Seismology of University of Helsinki is building a new local seismic network, called OBF network, around planned nuclear power plant in Northern Ostrobothnia, Finland. The network will consist of nine new stations and one existing station. The network should be dense enough to provide azimuthal coverage better than $180°$ and automatic detection capability down to M_L -0.1 within a radius of 25 km from the site.

The network construction work began in 2012 and the first four stations started operation at the end of May 2013. We applied an automatic seismic signal detection and event location system to a network of 13 stations consisting of the four new stations and the nearest stations of Finnish and Swedish national seismic networks. Between the end of May and December 2013 the network detected 214 events inside the predefined area of 50 km radius surrounding the planned nuclear power plant site. Of those detections, 120 were identified as spurious events. A total of 74 events were associated with known quarries and mining areas. The average location error, calculated as a difference between the announced location from environment authorities and companies and the automatic location, was 2.9 km. During the same time period eight earthquakes between magnitude range 0.1–1.0 occurred within the area. Of these seven could be automatically detected. The results from the phase 1 stations of the OBF network indicates that the planned network can achieve its goals.

1 Introduction

Pyhäjoki at the eastern coast of the Bay of Bothnia is a potential area for a new nuclear power plant. The area is characterized by low-active intraplate seismicity, with earthquake magnitudes rarely exceeding 4.0. Specific safety guide of International Atomic Energy Agency (IAEA, 2010) states that when a nuclear power plant site is evaluated a network of sensitive seismographs, having a recording capability for microearthquakes, should be installed to acquire more detailed information on potential seismic sources. The operation period of the network should be sufficiently long to obtain a comprehensive earthquake catalogue for seismotectonic interpretation (IAEA, 2010), and the monitoring of natural hazards shall be commenced no later than the start of construction and shall be continued up until decommissioning (IAEA, 2003). The data processing, reporting and network operation are advised to be linked to the national or regional networks. Earthquakes recorded within and near such a network should be carefully analyzed in connection with seismotectonic studies of the near region (IAEA, 2010).

Tiira et al. (2015) outlined a plan for a local seismograph network OBF to be installed around the Pyhäjoki Nuclear Power Plant (PNPP). An optimal configuration of ten seismograph stations was proposed. The ten broad-band 3-C stations will be within 50 km from the planned power plant. The authors state that the proposed network should be dense enough to fulfill the IAEA (2010) requirements of azimuthal coverage better than $180°$ and automatic event detection capability down to $\sim M_L$ -0.1 within a radius of 25 km from the site. The earthquake location accuracy was anticipated to be 1–2 km for horizontal coordinates within 25 km distance from the PNPP and the annual number of earthquakes detected was estimated to be 2 ($M_L \geq \sim -0.1$) within 25 km radius and 5 ($M_L \geq \sim -0.1-\sim 0.1$) within 50 km radius from the PNPP (Tiira et al., 2015).

Institute of Seismology of University of Helsinki (ISUH) started to build the proposed network in 2012. Building of the OBF network was divided into two phases. Phase 1 included four new stations and they started operation at the end of May 2013. The rest of the network started operation in fall 2015. Valtonen et al. (2012) reported the details on site selection, construction and instrumentation of the stations. All the stations were installed on the ground surface. The main obstacles in the site selection were lack of bedrock outcrops in the vicinity of simulated station locations and willing landowners to lease the land. These facts slightly changed the configuration of the realized network compared to the simulated optimal configuration. However, the changes did not impair the azimuthal coverage or event detection capability of the network (Valtonen et al., 2012). Furthermore in a report by Kortström et al. (2012) ISUH defined a seismic monitoring strategy for PNPP site area. Following the IAEA (2010) guidelines, real-time monitoring of the site area is integrated with the automatic detection and location process operated by ISUH. The monitoring strategy is introduced in Sect. 2.

The objective of this study is to evaluate the detection and location capability of the planned network. We will evaluate whether the realized network meets the requirements set to the simulated network. Moreover, we will investigate whether the four new stations already improve detection capability and location accuracy of PNPP site area compared to national network. In this study we focus on automatic processing of the data collected from the new stations. Automatic analysis results are presented from the network consisting of the four new stations and the existing nearby stations of the national seismic networks of Finland and Sweden.

2 Seismic monitoring strategy for OBF network area

Continuous waveform data from OBF network will be transmitted in real-time via internet to ISUH data server, where they will be transferred to the routine data analysis and archival systems. Thereafter, the recordings are processed with automatic software suitable for detecting micro-earthquakes in an intraplate seismotectonic environment. The main objective of the automated event processing is to distinguish seismic events from the continuous seismograph recordings for further analyses. Basically, the routine breaks up into three sub-tasks: (1) detection of seismic signals from the background noise; (2) identification and association of detected seismic phases; (3) location and identification of seismic events, determination of source depths and magnitudes.

Detector programs are typically based on STA / LTA ratio methods of which many dates back to works of Allen (1978 and 1982) and Withers et al. (1998). Parameters steering the detector program must be fine-tuned as per the events of interest and noise conditions at the stations. For example, prior to STA / LTA detection process the recordings are usually filtered with band-pass filters that are optimized for the frequency content of the target events (teleseismic, regional, or local). Several methods are also available for identification of the detected phases and for their association to single events

(e.g. Tong, 1995; Withers et al., 1999; Satriano et al., 2008). The optimal method depends on both the geographical extent of the network and the type and density of stations. The current ISUH system is described in Sect. 3.

The automatic processing of OBF network data starts with the existing ISUH software. Due to differences in monitoring target between ISUH and OBF networks, the processing is separated into two parallel processes. For regional monitoring purposes, data from two sub-stations of OBF network will be included in the automatic event detection and location system of ISUH. This system monitors events above the magnitude threshold of the national network, currently $\sim M_L 0.9$ (Tiira et al., 2015) in the PNPP site area. The inclusion of all the data from the dense OBF network into the sparse regional network data is not meaningful, because it would increase the number of noise detections in the system that is optimized for sparse regional network.

In the second process earthquakes smaller than threshold magnitude of the national network within PNPP area are searched with separate event processing system customized for OBF network. This system will utilize OBF stations and the nearest online stations of the Finnish and Swedish national networks.

3 Event detection and location system

As summarized in Sect. 2, automatic processing of the OBF network data is separated into two parallel systems, aimed for regional and local monitoring. Both systems utilize ISUH automatic event processing software but with different processing parameters and network configurations. The system is based on network processing of three-component (3-C) stations. At a single 3-C site the detection is done with basic STA / LTA-detector. The code used is an implementation of Ruud and Husebye (1992), which, in turn, uses "predicted coherence" measure of Roberts et al. (1989). The original Ruud and Husebye (1992) code produced fully automatic single station bulletin. Location of regional events was based on back azimuth and travel time difference of automatically identified P and S phase detections.

In the ISUH automatic processing of single 3-C stations the association and identification of local and regional P and S phases is handled differently. The aim of single station processing is not to produce reliable event bulletins but to build event seeds for network processing. The event seeds are formed as follows: (1) every detection in a detection log is considered as a possible P phase signal. (2) Every detection following the possible P phase within a certain time window in the same detection log is considered a possible S phase belonging to the same event as the preceding P. (3) Event seeds are formed by computing location for all possible P and S pairs using P and S signal onset times and back azimuth estimate.

As a result single station event detection log contains a large amount of false events intermingled with real event detections. The next step is to associate phases from the rest of the network with all event seeds. This association is done by calculating theoretical P and S arrival times for every station of the network and for every event seed of single station event logs. A pick in single station's detection log is associated with an event seed as P or S phase if its time matches with theoretical arrival time of P or S phase in a given time window. At this point the theoretical arrival times are calculated for all local or regional phases, namely $Pg/Pb/Pn$ and $Sg/Sb/Sn$ phases. Pg and Sg are direct waves in the upper crust, Pb and Sb are waves in the lower crust or along the Conrad discontinuity, and Pn and Sn are waves refracted below the Mohorovicic discontinuity (Willmore, 1979). An event seed is selected for further processing if the number of associated phases exceeds a threshold value.

At this stage multiple solutions may occur for the same event. Two or more solutions are considered originating from the same event if their location and time matches within certain limits. The most stable solution is selected for another phase association round, which is done with stricter association rules. After the second association round, the events are ready for final location, identification and magnitude determination. The 2-step association method allows the application of low detection thresholds at single stations, a necessary condition for detecting very weak seismic signals. A simplified flow chart of the network processing is shown in Fig. 1.

In both single 3-C station and network processing the location of the events are done with HYPOSAT location program (Schweitzer, 2001). HYPOSAT can use versatile data to obtain source location. ISUH system utilizes single arrival times of P or S phases, travel time difference of P and S phases at the same station, and back azimuth observations. The 1-D velocity model used by ISUH to locate of seismic events in the Fennoscandian shield is shown in Table 1. Automatic solutions include origin time, latitude and longitude of the epicenter, focal depth, magnitude, travel-times of signals at each station and error statistics for each of the estimated parameters. The depth of the event is always constrained to zero during the iterations, because accuracy of the automatic arrival times is not good enough for depth estimation. In addition, more than 95 % of the local and regional events in this area are near surface explosions. Fixed depth often helps achieving more stable solutions for the other source parameters as well. For events identified as earthquakes, the source depth is estimated during interactive analysis. That is, the location program is allowed to iterate the depth with carefully picked phase onsets and/or estimate the depth using depth phases.

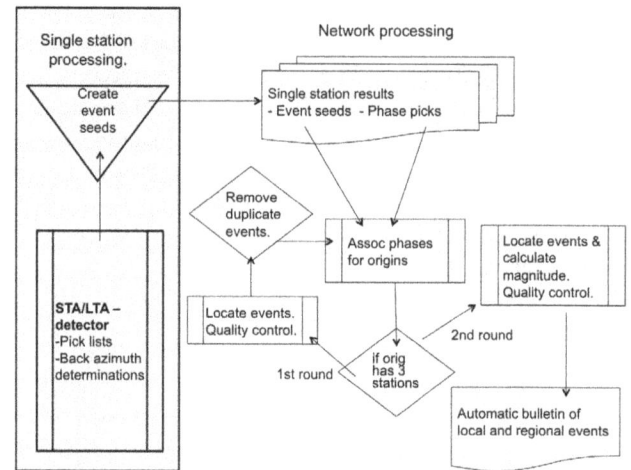

Figure 1. A simplified flow chart of ISUH automatic event processing system.

Table 1. The 1-D velocity model used in the location procedure.

Layer thickness (km)	Vp (km s^{-1})	Vs (km s^{-1})
15.0	6.20	3.62
25.0	6.70	3.84
40.0	8.03	4.64
	8.50	4.75

4 Preliminary data processing system for OBF network

At the end of May 2013 the first four new stations of OBF network started operation. We started the automatic data processing already with the first four stations in order to gain experience of stations' performance as soon as possible. The network used for automatic processing comprised 13 stations: five stations in PNPP area (OBF0, OBF4, OBF6, OBF7 and OUF) and the closest eight stations from the national seismic networks of Finland (OUL, KEF, SUF, TOF, VAF) and Sweden (BURU, KALU, UMAU). The station locations are presented in Fig. 2 together with the predicted minimum magnitude thresholds calculated for this setup. The threshold magnitude map was calculated with same method as in the initial network simulations of Tiira et al. (2015). They derived a relation between event magnitude and maximum distance at which both P and S phase can be automatically detected for that event. Data for the modeling were automatic locations of earthquakes from ISUH database (Tiira et al., 2015). The maps are calculated by forming a $0.1° \times 0.1°$ grid over the network area. Every grid point is a possible earthquake epicenter from which the distances to the stations are calculated. The distance to the third closest station is then converted to the minimum detectable magnitude at every grid point. This ensures that there should be phase readings from

Table 2. Detection parameters for Ruud and Husebye (1992) signal detector. First two filter channels are intended for S type signals and latter two for P type signals. STA / LTA detection threshold 1 (Th1) is used together with coherence threshold (Coh) and threshold 2 (Th2) alone. Duration (Dur) is minimum signal duration to accept detection.

Low cut [Hz]	High cut [Hz]	Step length [s]	STA length [s]	LTA length [s]	Th1	Coh	Th2	Dur [s]
2.0	10.0	0.08	0.16	0.64	5.1	0.9	3.0	0.8
6.0	20.0	0.05	0.10	0.40	5.1	0.9	3.0	0.8
5.0	20.0	0.08	0.16	0.64	3.0	0.3	3.35	0.8
15.0	40.0	0.05	0.10	0.40	3.0	0.3	3.55	0.8

at least three stations, as required by ISUH automatic location process.

All OBF stations are equipped with similar instruments, that is, Nanometrics Trillium Compact seismometers and EarthData PS-6 digitizers. The sampling rate of the continuous data acquisition is set to 250 sps. For data acquisition we use normal miniPC computers with Linux operating system and SeisComP3 software. We applied the processing system described in Sect. 3 with some modifications: The event seeds were searched only from the five stations within the target area and higher frequency bands for signal detection were applied to the new OBF stations. The main detection parameters for initial signal detection are presented in Table 2.

During the test period of the system, from the end of May to December 2013, we focused mainly to the location accuracy obtained with the system. For that purpose explosions from 50 km radius from PNPP site were identified and associated to known mines and quarries. Information about blasting sites and times were obtained from environmental authorities and companies carrying out such works. The location error was calculated as a difference of announced location and computed automatic location. Accuracy of automatic event solutions of OBF network was also compared to automatic event solutions obtained by regional network. We utilized explosions from the most active open pit quarry in the area, the Laivakangas gold mine.

5 Results

During end of May–December 2013 the automatic analysis system of the OBF network produced 214 event solutions inside the predefined area of 50 km radius around the PNPP site. Of those solutions 94 were real seismic events and 120 were identified as spurious events, that is, events generated by erroneous phase and/or noise associations. The number of real automatic event detections almost doubled in the area compared to the regional monitoring system, which produced 51 automatic event detections for the same time period.

Location accuracy tests were done to a total of 74 events, which could be identified with known place and time. The events and blasting sites are shown in Fig. 3 and the results in Table 3. The average location error of automatic locations was 2.9 km. An experienced seismic analyst also located the

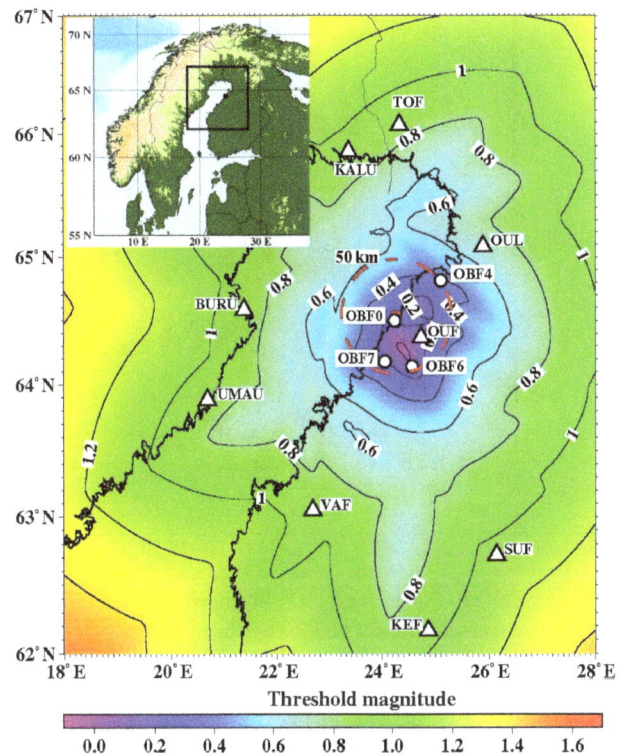

Figure 2. Online seismic stations used by ISUH in the automatic event processing iof PNPP area. A circle denotes a 3-C station of the OBF-network. A triangle denotes a 3-C station of the permanent national networks of Finland and Sweden. Background map is the predicted threshold magnitude calculated for the network shown on the map. The red dashed circumference of 50 km radius confines the target area of the network. The inset map shows the location of the study area and the PNPP site is marked with a black square.

events interactively. Average location error of these locations for the same events was 1.2 km. In the area between stations OUF, OBF6 and OBF7, which is a near complete part of the network, the average location error was 2.4 km for automatic solutions and 0.8 km for manual solutions. Altogether 19 explosions from the Laivakangas gold mine were automatically located with both local and regional systems. The average location errors of OBF network and regional network were 3.3 and 5.4 km, respectively.

Table 3. Results of location accuracy tests.

Location error [km]	Automatic locations	Manual locations
50 km radius from PNPP	2.9	1.2
Area between OBF6, OBF7 & OUF	2.4	0.8
	OBF network	Regional network
Laivakangas gold mine, automatic locations	3.3	5.4

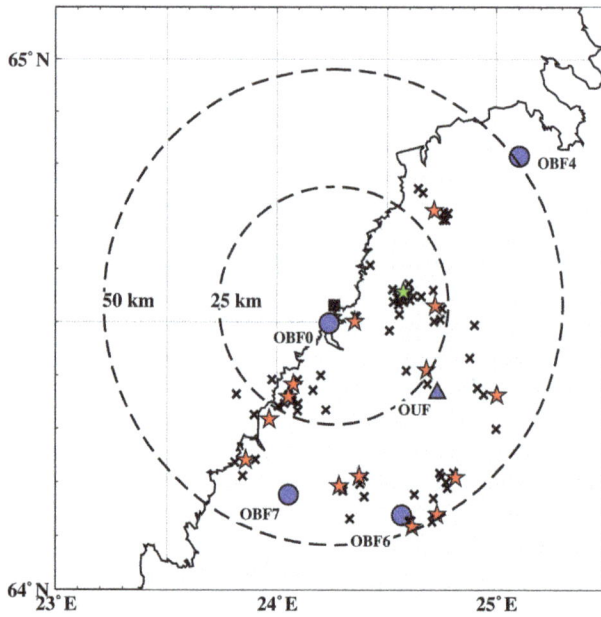

Figure 3. Automatic locations of events from known blasting sites. Blasting sites are marked with red stars except the Laivakangas gold mine, which is marked with green star. Automatic event locations are marked with black crosses. Blue circles are the new seismic stations in the OBF area and blue triangle the pre-existing station OUF. The dashed circles define areas of 25 and 50 km radii from the PNPP (black square).

After the visual inspection of the automatic analysis results, three of the detected events were identified as earthquakes. The local magnitudes of these events were 0.5, 0.8 and 1.0 calculated using a local magnitude scale by Uski and Tuppurainen (1996). The regional network had also detected the two largest earthquakes. Finally, the seismograms of OBF network were scanned visually to pick weak events missed by the system. As a result five additional earthquakes were found within 50 km radius of PNPP site. Magnitudes of these earthquakes ranged from 0.1 to 0.4. Locations of these undetected earthquakes were compared to the predicted detection threshold map of the first stage OBF network, which suggests that the magnitude 0.4 earthquake should have been automatically detected. Figure 5 is shows the recordings of the magnitude 0.4 earthquake with manual phase picks. The automatic processing related to these undetected earthquakes

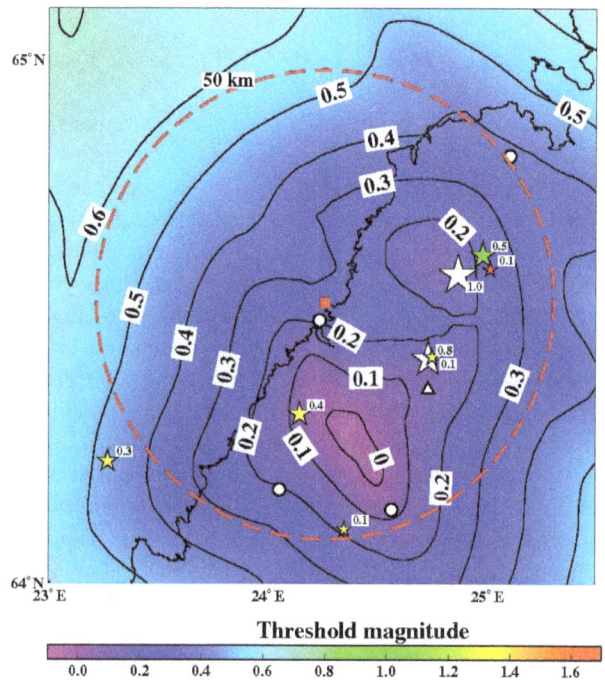

Figure 4. Locations of earthquakes (stars) that occurred at the PNPP area between the end of May and December 2013. Earthquakes marked with white stars were detected automatically by both regional and local OBF networks. Green star denotes earthquake that was detected automatically only by the OBF network. Earthquakes marked with yellow stars were detected by the OBF network after readjustment of detection parameters. Earthquake marked with red star remained undetected. White circles denote the new OBF stations and white triangle is the pre-existing station OUF at the area. Dashed red circle defines the area of 50 km radius from the PNPP site (red square). Background shows the predicted magnitude threshold map for the network used in local automatic system (cf. Fig. 2).

was examined closely. As a result adjustments were made to STA and LTA window lengths of single station processing and phase association rules of network processing. Recordings of earthquakes were reprocessed with new settings and now four more earthquakes were automatically detected. As a result the automatic system detected smaller earthquakes than magnitude threshold map predicts (Fig. 4). Details of the

Table 4. Source parameters for earthquakes recorded by OBF in 2013. Origin time, latitude and longitude in WGS-84 coordinate system, depth (F = fixed during the location process) and local magnitude M_L.

	Date	Time, UTC	Lat (°N)	Lon (°E)	Depth (km)	M_L
1	26.05.2013	00:23:02.5	64.585	24.867	24.5	1.0
2	16.07.2013	02:07:49.4	64.428	24.748	24.0	0.1
3	27.07.2013	12:52:45.8	64.230	23.269	10.7	0.3
4	08.11.2013	03:24:29.7	64.102	24.350	6.8	0.1
5	07.12.2013	20:44:13.1	64.623	24.979	24.7	0.5
6	07.12.2013	20:44:19.5	64.596	25.012	20F	0.1
7	16.12.2013	17:31:49.8	64.424	24.731	8.9	0.8
8	17.12.2013	09:43:03.0	64.319	24.146	25.7	0.4

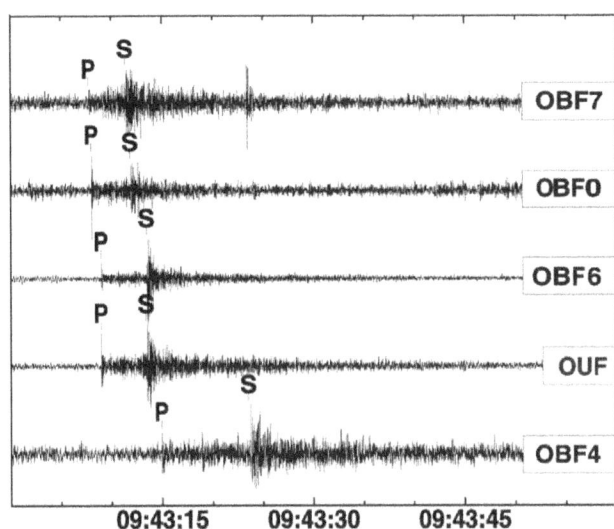

Figure 5. Vertical recordings of the M_L 0.4 earthquake in Kalajoki 17 December 2013 at the OBF-stations. The recordings are presented with increasing station to epicenter distance. The arrival times of first P and S arrivals are marked. Recordings are filtered with a 2–30 Hz band pass filter.

all earthquakes recorded by the OBF network in the PNPP area are given in Table 4.

Background noise of the new sites was also studied. Short test measurements had been already made during site surveys, which suggested that the noise levels are comparable to other locations in Finland (Valtonen et al., 2012; Korja et al., 2011). The analysis of the new sites confirmed that the overall background noise is at the same good level as in the most of the Finnish stations. Analyses were made using pqlx-software (McNamara and Boaz, 2005). Figure 6a–e shows the probability density functions of the four new stations and pre-existing station OUF at the area. Differences between OUF and new OBF stations at low frequency (long period) part of the spectra are due to the different seismometers used in these stations. OUF is equipped with Streckeisen STS-2 seismometer while the Nanometrics Trillium Compact seismometers of the OBF stations are less sensitive at low fre-

quencies. There are also variations in high frequency part of the noise spectra between the stations. The new OBF stations have higher sampling rate (250 sps) compared to station the OUF (100 sps). Thus, noise spectra of new OBF stations contain more site-specific variations from short distances.

6 Discussion

IAEA 3.30 (2010) guidelines state that when a nuclear power plant site is evaluated a network of sensitive seismographs having a recording capability for micro-earthquakes should be installed to acquire more detailed information on potential seismic sources. In areas with high seismicity the active faults are seen clearly when large number of earthquakes are located evenly inside of error margins of potential fault lines. In such areas low detection threshold is not a necessity. In areas of low seismicity both low detection threshold and good location accuracy are substantial when trying to map active faults using earthquake locations. When location accuracy is good, even single events can be associated with certain faults, whereas location errors measured in many kilometers make the single observations useless in seismotectonic interpretations (Korja et al., 2011).

Location accuracy gained with phase 1 stations of the OBF network is comparable to earlier studies (e.g. Bondár et al., 2004; Korja et al., 2011; Uski et al., 2011) with similar conditions. Our results show that the overall location accuracy of manual locations is already at the level predicted by the network simulations (Tiira et al., 2015). Furthermore, the location accuracy results indicate that the accuracy of both automatic and manual locations is best at the near complete part of the network, that is, in between of stations OUF, OBF6 and OBF7. It is a known fact that the location accuracy can be improved with accurate local velocity model and sufficient number of evenly spaced seismic stations near the source (e.g. Bondár et al., 2004; Uski et al., 2011). If the source is outside the network, in other words the maximum azimuthal gap of event location is greater than 180°, the location accuracy decreases rapidly (Tiira et. al., 2015). In addition location accuracy can also be improved by applying relative location methods (e.g. Waldhauser and Ellsworth, 2000; Fehler et

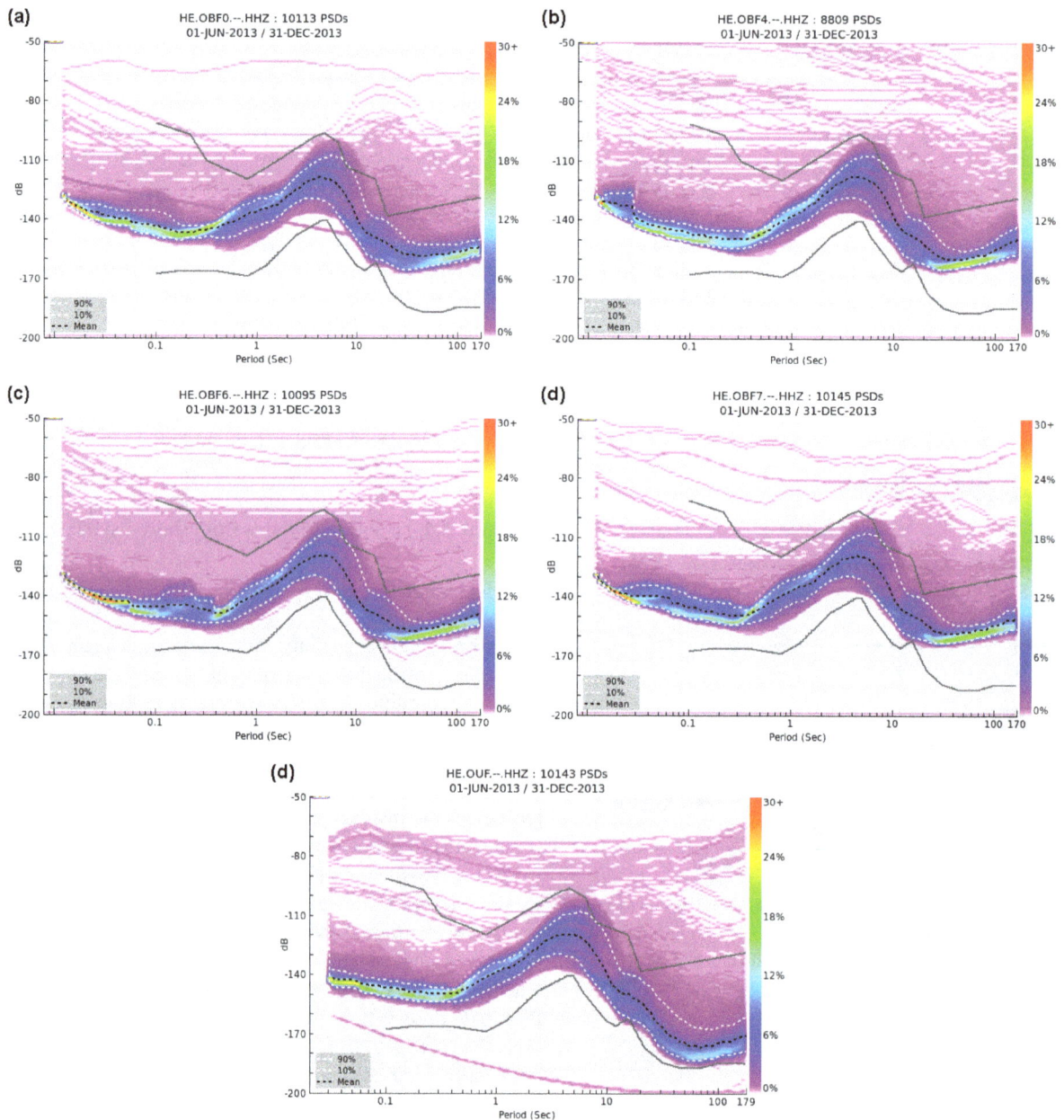

Figure 6. A spectral density function at station OBF0 (**a**), OBF4 (**b**), OBF6 (**c**), OBF7 (**d**) and OUF (**e**). The background noise at the station is seen as regularly observed energy levels (blue, green, yellow and red). The more scarce energy levels (pink and violet) are associated with sudden disturbances such as signals of seismic events. The grey lines denote the global average of low and high noise level models (Peterson, 1993).

al., 2000; Castellanos and Van der Baan, 2013). In summary, our results with phase 1 stations of the OBF network implies that the complete OBF network can achieve a good location accuracy within 50 km radius of PNPP site, at least at the in-land part of the network.

The new stations helped to find six small ($M_L < 0.8$) earthquakes, which would probably have remained undetected using national network only. The number of $M_L > 0$ earthquakes found in 2013 is in fact greater than what was pre-

dicted by Tiira et al. (2015). Most of the small earthquakes would still have remained undetected without visual scanning of the waveforms. The automatic analysis system applied to the phase 1 stations of OBF network performed quite poorly. The false alarm rate was relatively high and the system missed one earthquake it should have detected according to predicted detection capability. This was somewhat expected as the applied system was not properly tuned for local distances. Nevertheless, preliminary results look promising for the ultimate goal of the network. Our experience from the regional processing of Finnish networks and contributing stations from neighboring networks has proven that event association rejects spurious events effectively when number of stations increases. Thus, we expect the false alarm rate to decrease when the network is complete and number of stations doubled from five to ten. At the same time we expect that number of automatically detected earthquakes will increase significantly.

The automatic data processing system used in Sweden and Iceland, called SIL (Bödvarsson and Lund, 2003), has similarities with our automatic processing software. The software uses STA / LTA detector to initial phase picking and the processing is divided in single station and multi station parts. The SIL system was originally developed to monitor automatically microearthquake activity around Icelandic volcanoes where it has shown the capability of automatic evaluation of more than 1500 earthquakes per day (Bödvarsson and Lund, 2003). In Sweden the SIL system is applied to the data of the Swedish National Seismic Network (SNSN) that consists of more than 60 stations with station spacing of about 100 km. The network is sufficiently sensitive to record all earthquakes down to a magnitude 0.5 within the network (Bödvarsson, 2011).

Lindblom (2011) studied microearthquakes along Pärvie fault of Northern Sweden. She set up a temporary network to improve detection capability and location accuracy of the SNSN. The network comprised seven temporary stations and a subset of suitable SNSN stations in the study area. She used the SIL system to process the data and as a result, like in our study, the number of spurious events was large. Also the explanation of false alarms is similar to our study: "the network's standard automatic event detection and location procedures were not optimal for the temporary station data set" (Lindblom, 2011). Nevertheless, Lindblom (2011) could increase the number of automatically detected earthquakes with temporary stations compared to the SNSN. Moreover, by using time-domain cross-correlation detector the number of initial automatic detections was doubled (Lindblom, 2011). Waveform correlation is a well-established technique to identify close-lying events, which produce very similar waveforms at same stations if the source mechanism is similar. Waveform correlation is especially powerful in detecting signals that fall below the detection capability of energy based detectors as shown in previous studies like Gibbons et al. (2007) in Norway and Withers et al. (1999) in New Mexico.

The INGV National Seismic Network in Italy has undergone large improvements since the beginning of the new millennium, as described by Amato and Mele (2008). The improvements included tripling of on-line stations from less than 100 to more than 200 and having three component broad-band sensors for the majority of the stations. Compared to the year 2000, the INGV network doubled the number of earthquake detections in year 2006. Also the magnitude completeness of the whole network region improved from 1.7 in year 2006 to 2.4 in year 2000. Quality of automatic locations was found to be good, meaning less than 10 km difference to manual location, inside the network with the improved network (Amato and Mele, 2008). These findings support our findings that with sufficient number of well-situated stations the OBF network can achieve its goals.

The development of automatic processing system of the OBF network will continue in coming years. The current system needs refining of parameters in every stage of processing and new methods should be introduced. In the future usage of cross-correlation detector could increase the number of detected small earthquakes and the location accuracy can be further improved with relative location methods. These methods will be applied after sufficient amount of ground truth data from earthquakes and chemical explosions has been recorded with all stations of the completed network.

Acknowledgements. Seismic stations operated by the Institute of Seismology, University of Helsinki, the Sodankylä Geophysical Observatory, University of Oulu, and the Uppsala University have provided online waveform data for this study. The maps in Figs. 2–4 were prepared with the Generic Mapping Tools (GMT; Wessel and Smith, 1991, 1998).

References

Allen, R.: Automatic earthquake recognition and timing from single traces, B. Seismol. Soc. Am., 68, 1521–1532, 1978.

Allen, R.: Automatic phase pickers: their present use and future prospects, B. Seismol. Soc. Am., 72, S225–S242, 1982.

Amato, A. and Mele, F. M.: Performance of the INGV National Seismic Network from 1997 to 2007, Ann. Geophys.-Italy, 51, 2–3, 2008.

Bödvarsson, R.: Swedish National Seismic Network (SNSN) – A short report on recorded earthquakes during the third quarter of the year 2011. Report P-10-33, Swedish Nuclear Fuel and Waste Management Co, Stockholm, Sweden, 15 pp., 2011.

Bödvarsson, R. and Lund, B.: The SIL seismological data acquisition system – as operated in Iceland and in Sweden, in: Methods and Applications of Signal Processing in Seismic Network Operations, edited by: Takanami, T. and Kitagawa, G., Springer Verlag, Berlin, Germany, number 98 in Lecture Notes in Earth Sciences, 2003.

Bondár, I., Myers, S. C., Engdahl, E .R., and Bergman, E. A.: Epicentre accuracy based on seismic network criteria, Geophys. J. Int., 156, 483–496, 2004.

Castellanos, F. and Van der Baan, M.: Microseismic event locations using the Double-Difference algorithm, CSEG Recorder, 38, 26–36, 2013.

Fehler, M., Phillips, W. S., House, L., Jones, R. H., Aster, R., and Rowe, C.: Improved Relative Locations of Clustered Earthquakes Using Constrained Multiple Event Location, B. Seismol. Soc. Am., 90, 775–780, 2000.

Gibbons, S., Sorensen, M., Harris, D., and Ringdal, F.: The detection and location of low magnitude earthquakes in northern Norway using multichannel waveform correlation at regional distances, Phys. Earth Planet. In., 160, 285–309, doi:10.1016/j.pepi.2006.11.008, 2007.

International Atomic Energy Agency (IAEA): Site Evaluation for Nuclear Installations. Safety Requirements Series No. NS-R-3, Vienna, Austria, 28 pp., 2003.

International Atomic Energy Agency (IAEA): Seismic hazards in site evaluation for nuclear installations. Specific safety guide SSG-9, Vienna, Austria, 60 pp., 2010.

Korja, A., Kortström, J., Lindblom, P., Mäntyniemi, P., Uski, M., and Valtonen, O.: Seismisten aineistojen kerääminen ja tulkinta Pyhäjoen alueelta, Report M210E2010, Geological Survey of Finland, 35 pp., 2011.

Kortström, J., Uski, M., Tiira, T., and Korja, A.: Data processing and analysis system for the Pyhäjoki seismic network, Report T-86, Institute of Seismology, University of Helsinki, Helsinki, Finland, 15 pp., 2012.

Lindblom, E.: Microearthquake Study of End-glacial Faults in Northern Sweden, Licentiate thesis, Uppsala Universitet, Uppsala, Sweden, available at: http://uu.diva-portal.org/smash/get/diva2:456600/FULLTEXT01.pdf (last access: 9 March 2016), 2011.

McNamara, D. E. and Boaz, R. I.: Seismic Noise Analysis System, Power Spectral Density Probability Density Function: Stand-Alone Software Package, United States Geological Survey Open File Report, NO. 2005-1438, 30 pp., 2005.

Peterson, J.: Observations and modelling of background seismic noise, Open-file report 93-322, US Geological Survey, Albuquerque, New Mexico, USA, 1993.

Roberts, R. G., Christoffersson, A., and Cassidy, F.: Real-time event detection, phase identification and source location estimation using single station three-component seismic data, Geophys. J. Int., 97, 471–480, 1989.

Ruud, B. O. and Husebye, E. S.: A new three component detector and automatic single-station bulletin production, B. Seismol. Soc. Am., 82, 221–237, 1992.

Satriano, C., Lomax, A., and Zollo, A.: Real-time evolutionary earthquake location for seismic early warning, B. Seismol. Soc. Am., 98, 1482–1494, 2008.

Schweitzer, J.: HYPOSAT – An Enhanced Routine to Locate Seismic Events, Pure Appl. Geophys., 158, 277–289, 2001.

Tiira, T., Uski, M., Kortström, J., Kaisko, O., and Korja, A.: Local seismic network for monitoring of a potential nuclear power plant area, J. Seismol., 20, 397–417, 2015.

Tong, C.: Characterization of seismic phases – an automatic analyser for seismograms, Geophys. J. Int., 123, 937–947, 1995.

Uski, M. and Tuppurainen, A.: A new local magnitude scale for Finnish seismic network, Tectonophysics, 261, 23–37, 1996.

Uski, M., Tiira, T., Grad, M., and Yliniemi, J.: Crustal seismic structure and depth distribution of earthquakes in the Archean Kuusamo region, Fennoscandian Shield, J. Geodyn., 53, 61–80, 2011.

Valtonen, O., Lindblom, P., Komminaho, K., Kortström, J., Keskinen, J. Smedberg I., and Korja, A.: Suunnitelma Pyhäjoen seismisestä paikallisverkosta. Report T-85, Institute of Seismology, University of Helsinki, Helsinki, Finland, 19 pp., 2012.

Waldhauser, F. and Ellsworth, W.: A double-difference earthquake location algorithm: Method and application to the northern Hayward fault, B. Seismol. Soc. Am. 90, 1353–1368, 2000. Wessel, P. and Smith, W. H. F.: Free software helps map and display data, EOS Trans. AGU, 72, 441–448, 1991.

Wessel, P. and Smith, W. H. F.: New, improved version of Generic Mapping Tools released, Eos Trans. AGU, 79, p. 579, 1998.

Willmore, P. L.: Manual of Seismological Observatory Practice. World Data Center A for Solid Earth Geophysics, Report SE-20, September 1979, Boulder, CO, USA, 165 pp., 1979.

Withers, M., Aster, R., Young, C., Beiriger, J., Harris, M., Moore, S. and Trujillo, J.: A comparison of select trigger algorithms for automated global seismic phase and event detection, B. Seismol. Soc. Am., 88, 95–106, 1998.

Withers, M., Aster, R., and Young, C.: An automated local and regional seismic event detection and location system using waveform correlation. B. Seismol. Soc. Am., 89, 657–669, 1999.

24

Site selection for a countrywide temporary network in Austria: noise analysis and preliminary performance

F. Fuchs, P. Kolínský, G. Gröschl, M.-T. Apoloner, E. Qorbani, F. Schneider, and G. Bokelmann

Department of Meteorology and Geophysics, University of Vienna, Althanstraße 14, UZA 2, 1090 Vienna, Austria

Correspondence to: F. Fuchs (florian.fuchs@univie.ac.at)

Abstract. Site selection is a crucial part of the work flow for installing seismic stations. Here, we report the preparations for a countrywide temporary seismic network in Austria. We describe the specific requirements for a multi-purpose seismic array with 40 km station spacing that will be operative approximately three years. Reftek 151 60 s sensors and Reftek 130/130S digitizers form the core instrumentation of our seismic stations which are mostly installed inside abandoned or occasionally used basements or cellars. We present probabilistic power spectral density analysis to assess noise conditions at selected sites and show exemplary seismic events that were recorded by the preliminary network by the end of July 2015.

1 Introduction

Site selection is a crucial part of the work flow for installing seismic stations. Detailed instructions on site scouting and preparations are formulated e.g. by Trnkoczy et al. (2012), yet the specific requirements for each seismic station depend on the scientific aim. Obviously, permanent broadband seismic stations require more thorough and cost-intensive site preparations than temporary or short period stations. Temporary station networks are often designed for a balance between data quality and project budget and thus can be realized in many ways. Installations for detection of local seismicity will e.g. aim to minimize high frequency anthropogenic noise and thus try to avoid deployments near populated and especially industrial areas. In turn, experiments that focus on long period seismic data suffer less from anthropogenic noise, but the broadband sensors require proper thermal insulation and shielding from atmospheric pressure fluctuations (Bormann and Wielandt, 2012; Forbinger, 2012).

Consequently, the preferred type of seismic site often depends on experiment duration, scientific target and available budget. In this manuscript, we describe our preparations for a multi-purpose countrywide network of broadband stations in Austria, for a temporary deployment of approximately three years.

2 Scientific goals, network layout and station design

2.1 Scientific goals

The sites we describe in this manuscript are testing sites for the Austrian part of the international AlpArray seismic network (Kissling et al., 2014). AlpArray is a unique European transnational research initiative in which 43 research institutes from 15 countries join their expertise to advance our knowledge about the structure and evolution of the lithosphere beneath the entire Alpine area. AlpArray will shed light on the detailed geological structure and geodynamical evolution of the Alps to answer outstanding questions e.g. on slab geometry and subduction polarity under the Eastern Alps. While the primary scope of AlpArray is fundamental research the unique dataset will also improve our knowledge about near-surface geologic structures and help to assess the seismic hazard in the Alpine area. The scientific goals of the AlpArray project are manifold and among others include e.g. Alpine geodynamics, crustal and mantle imaging, seismic anisotropy, as well as regional and local seismic activity. Hence, temporary seismic stations installed in the framework of AlpArray should be multi-purpose stations that perform reasonably well both for frequencies above and below the microseism peaks.

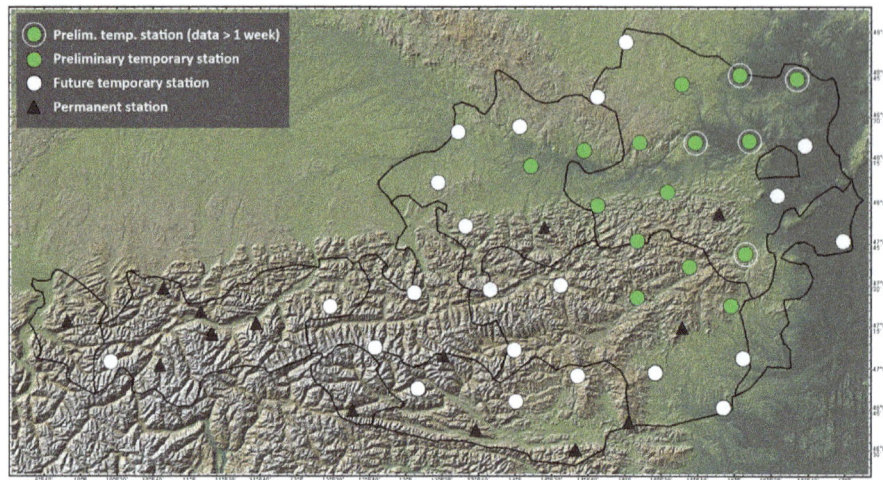

Figure 1. Prospective layout for the countrywide temporary network in Austria. Black Triangles mark existing permanent stations. White dots mark future temporary broadband stations and green dots mark existing preliminary testing stations. The five stations discussed in detail in Section 3 are marked with an additional circle.

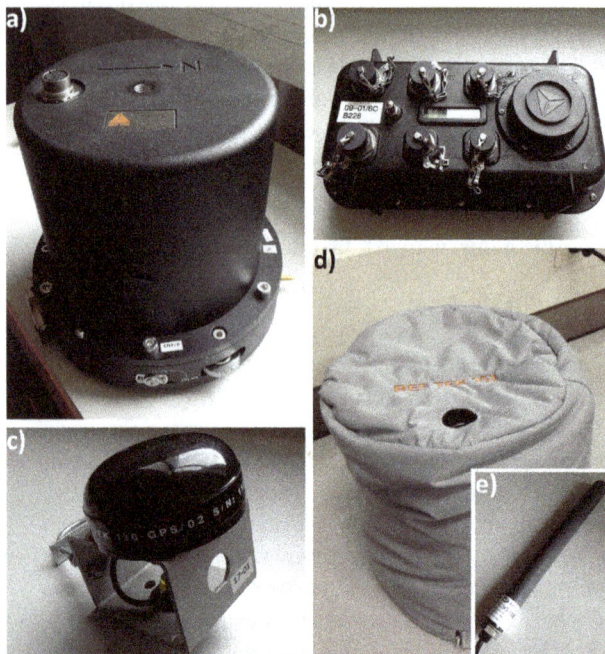

Figure 2. Equipment used for the installations described in this manuscript. (**a**) Reftek 151 60 s sensor, (**b**) Reftek 130/130S digitizer, (**c**) Reftek 130 GPS antenna, (**d**) textile thermal insulation cover for the sensor, (**e**) mobile network antenna.

2.2 Network layout

The AlpArray temporary seismic network is designed to complement existing permanent seismic stations. In Austria the Austrian Central Institute for Meteorology and Geodynamics operates 15 permanent stations (Fig. 1). The additional temporary seismic stations densify this network to achieve a uniform coverage with approximately 40 km interstation spacing. Central coordinates for all temporary AlpArray stations were computed to obtain homogeneous coverage throughout the entire array and all stations must be installed within a 3 km radius around the central coordinates. This constraint usually limits the choice of potential installation sites.

2.3 Station design

One seismic station comprises the following components (Fig. 2): a 60 s broadband sensor Reftek 151 "Observer" together with a Reftek 130 or Reftek 130S 24 bit digitizer with > 136 dB dynamic range (at 100 Hz sampling rate), a continuous mode Reftek 130 GPS, a Digi WAN 3G mobile router for telemetry and a 100 Ah battery. For stations not connected to the power grid 100 W solar panels charge the supply battery. Especially in Alpine regions snow coverage during winter may, however, prevent power supply through solar panels. In this case fuel cells can act as backup power source when the batteries are drained below a given threshold. For a temporary deployment of approximately three years, our specific requirements for seismic sites inside the 3 km radius were the following:

- *Seismic noise*: Average noise levels should be 20 dB lower than the New High Noise Model (NHNM) (Peterson, 1993) on all components within the 1–20 Hz frequency range. For long periods (30–200 s range) average noise levels on the vertical component should be 20 dB lower than the NHNM while on horizontal components noise levels should only be 10 dB less than the NHNM. This accounts for the strong sensitivity of horizontal components to e.g. long-period surface tilt from atmospheric pressure fluctuations. Consequently,

Figure 3. Station 01: **(a)** aerial view of the surroundings with potential noise sources highlighted, **(b)** outside view of installation, **(c)** inside view of installation.

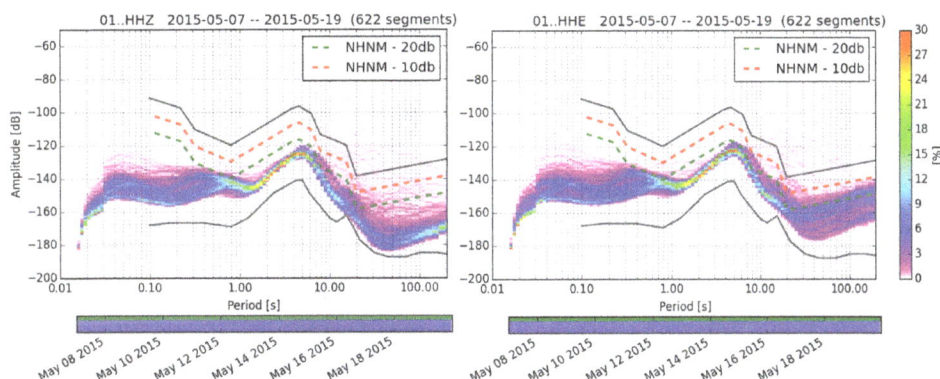

Figure 4. Probabilistic power spectral density of 12 days in spring 2015 of data for Station 01. Left panel: vertical component (HHZ), right panel: horizontal E–W component (HHE).

for near-surface stations, noise on horizontal components is usually stronger than on the vertical. Avoiding long-period noise on horizontal components requires advanced site preparation (Forbinger, 2012) which is usually out of scope for temporary deployments.

– *Accessibility and safety*: All sites should be accessible by car and safe in terms of theft or flood risk and all parts of the station shall not be exposed to any risk of potential damage. Additionally, the terms and conditions of the instrument insurance require the seismic stations to be indoors in spaces that can be locked. The surroundings of the site should not significantly change over the course of three years.

– *Power supply*: Most parts of Austria experience snow fall during winter and thus for many sites power supply through solar panels cannot be guaranteed. Hence,

we prefer sites were power supply from the regular 50 Hz/230 V power grid is possible.

– *Connectivity*: For monitoring purposes all seismic stations should send live data using the mobile network. Minimum requirement is sufficient signal strength and stability to transmit state-of-health data, while preferably continuous 100 Hz waveform should be transmitted. For our instrumentation and 100 Hz waveform data in STEIM1 compression format, the amount of data to transmit is approximately 30 Megabytes day^{-1} for seismically quiet sites and 50 Megabytes day^{-1} for noisy sites. Thus, for 100 Hz real-time waveform streams a mobile bandwidth of 5–10 kbits s^{-1} should be sufficient, which can even be achieved in GSM networks. In fact, stability of the mobile connection is more important than bandwidth.

Figure 5. Station 05: **(a)** aerial view of the surroundings with potential noise sources highlighted, **(b)** outside view of installation, **(c)** inside view of installation.

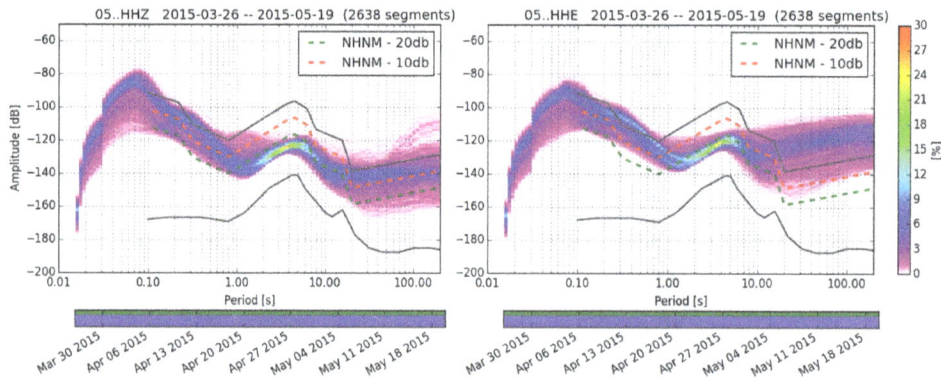

Figure 6. Probabilistic power spectral density of 54 days in spring 2015 of data for Station 05. Left panel: vertical component (HHZ), right panel: horizontal E–W component (HHE).

Following the site requirements listed above, typical installation sites for our broadband instruments are basements in abandoned or occasionally used houses and huts (see Figs. 3–11). In various regions throughout Austria wine cellars and occasionally castles or bunkers may be used for seismic installations. The sensor is placed on solid ground - preferably flat bedrock, but more commonly concrete floors. If no such ground is available, we build a concrete base approximately 15–20 cm thick and of 60×60 cm size. The sensors are covered with textile bags fabricated from microfleece material with primaloft insulation (Fig. 2) and styrofoam boxes (Figs. 3–11) for thermal insulation. To minimize air circulation the bottom of the styrofoam boxes is sealed with silicon.

By end of July 2015, 15 temporary stations are running in test operation (Fig. 1). Stations were installed in two phases – five stations were deployed in spring and another ten stations in summer. In the following we compare five of the 15 currently operating sites in detail and discuss noise levels in the light of site surroundings and highlight possible noise sources. We restrict our report to the five sites of the first installation phase since continuous data over more than one week is not available for the stations installed in the second phase. Probabilistic power spectral density graphs were created with the ObsPy toolbox (Krischer et al., 2015) following the procedure of McNamara and Buland (2004).

Figure 7. Station 06: (**a**) aerial view of the surroundings with potential noise sources highlighted, (**b**) outside view of installation, (**c**) inside view of installation.

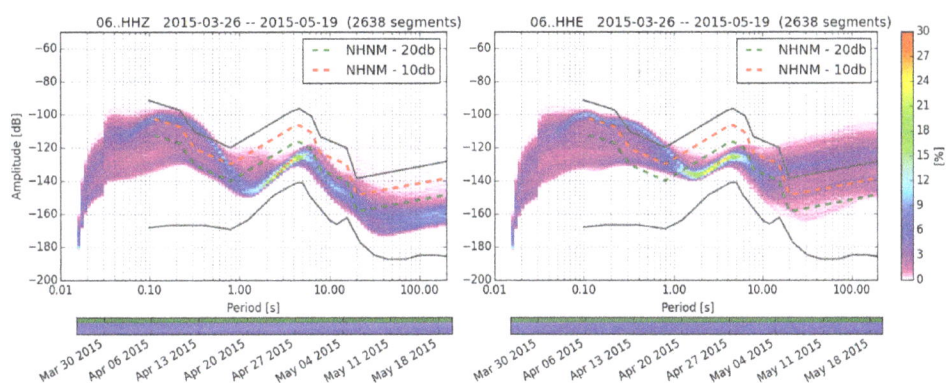

Figure 8. Probabilistic power spectral density of 54 days in spring 2015 of data for Station 06. Left panel: vertical component (HHZ), right panel: horizontal E–W component (HHE).

3   Site analysis

3.1   Site 01

Station 01 is located in an old bunker near the village of Falkenstein, Lower Austria and power supplied by a solar panel only. It is approximately 600 m from inhabited houses, 300 m from a minor road with little traffic and surrounded by vineyards (Fig. 3). The bunker that hosts the seismic station is built upon and likely connected to a 200 m long outcrop of bedrock which used to be a quarry. Cables for the GPS and mobile antennas are lead to the surface through ventilation pipes, which were sealed with expanding foam. The location inside the bunker provides good ground contact and minimizes daily temperature changes. Consequently, site 01 is by far the most seismically quiet station in the network to date (Fig. 4). Long-period (30–200 s) noise levels on the vertical component are close to the New Low Noise Model (NLNM)

(Peterson, 1993). On the E–W component, long-period noise is substantially higher than on the vertical and approximately 20 dB less than the NHNM. We note that in the same period range, noise on the N–S component is 10 dB stronger than on the E–W, which may be a sensor leveling effect. The elevated long-period noise on horizontal components compared to the vertical is likely due to ground tilt. In the higher frequencies (> 1 Hz), noise levels are comparable on all components and approximately 10 dB higher than the NLNM. This shows that anthropogenic noise from the nearby roads and village is small.

3.2   Site 05

Station 05 is located in a small stone shelter inside the village Schmida adjacent to the floodplain of the Danube river in Lower Austria (Fig. 5). It is supplied from the power grid. The site is few meters away from a road, 600 m from a high-

Figure 9. Station 08: **(a)** aerial view of the surroundings with potential noise sources highlighted, **(b)** outside view of installation, **(c)** inside view of installation.

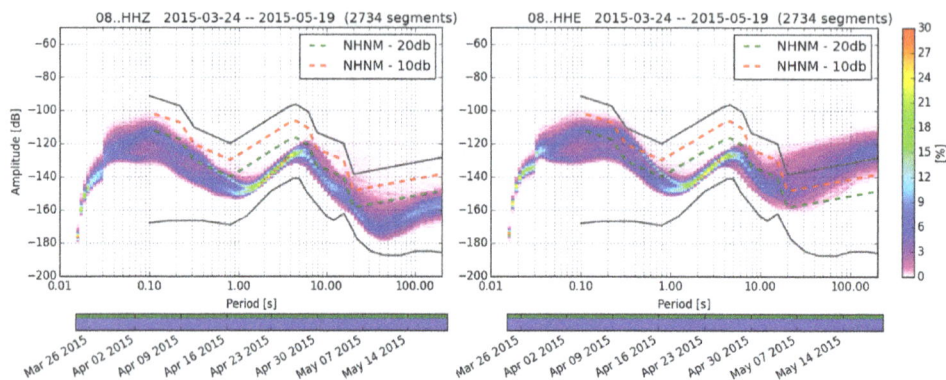

Figure 10. Probabilistic power spectral density of 54 days in spring 2015 of data for Station 08. Left panel: vertical component (HHZ), right panel: horizontal E–W component (HHE).

way and 5 km from the Danube. The concrete base of the shelter provides good ground contact, yet ground in the entire area is made of loose sediments dominated by the nearby river floodplain. Consequently, noise levels on Station 05 are poor (Fig. 6). Long-period vertical noise levels are in the order of the NHNM while horizontal components suffer from substantially stronger noise. We suspect that the site experiences strong long-period tilt from passing cars. Traffic is most likely also responsible for the elevated high-frequency noise in the order of the NHNM on all three components. In fact, during night times (8:00 p.m. to 6:00 a.m.) noise levels are generally 10 dB lower on all components than at daytime (6:00 a.m. to 8:00 p.m.) which confirms anthropogenic noise as the main noise source. At night, noise levels on the vertical component are close to NHNM −20 dB for both long-period and high-frequency bands. Despite the generally high noise levels at this site (especially during daytime), record-

ings could still be used for identification of both teleseismic and local seismic events (Figs. 13 and 14).

3.3 Site 06

Station 06 is placed inside an abandoned wine cellar approximately 5 m below the surface (Fig. 7) outside the village Kleinriedenthal in Lower Austria. Power supply is from the grid. Since the ground consists of loose soil, we built a concrete base to put the sensor on. The site lies 200 m outside a small village, 300 m from a road and is surrounded by vineyard agriculture. Despite the underground location noise levels are high (Fig. 8). Long-period vertical noise is steadily well below NHNM −20 dB while horizontal noise levels strongly vary in amplitude and on average fall around the NHNM. This behavior is unexpected since the underground installation should minimize both temperature fluctuations and effects of surface tilt. One reason for the strong

Figure 11. Station 09: **(a)** aerial view of the surroundings with potential noise sources highlighted, **(b)** outside view of installation, **(c)** inside view of installation.

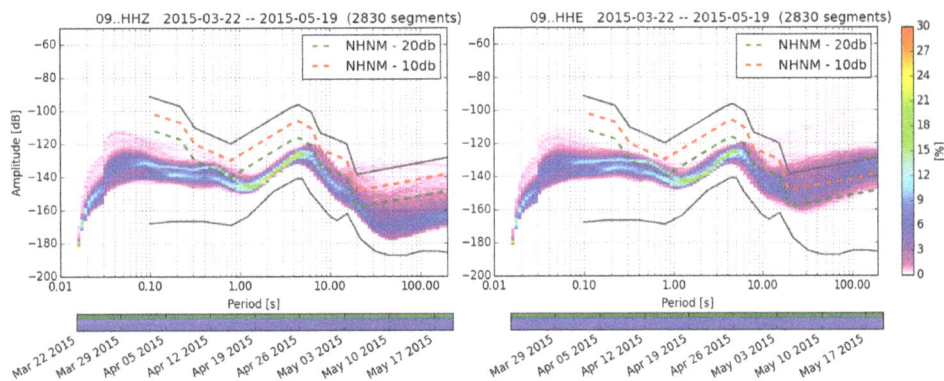

Figure 12. Probabilistic power spectral density of 54 days in spring 2015 of data for Station 09. Left panel: vertical component (HHZ), right panel: horizontal E–W component (HHE).

susceptibility of the horizontal components to long-period noise might be sensor mis-leveling. During its installation the sensor was put on the concrete base which might not have been entirely solid at the time (two days after construction) and thus the sensor or the base may have tilted. Tilted sensors will show much stronger horizontal noise. In fact, upon inspection about 2 months after installation, the sensor leveling did considerably change since the time of installation. The high frequency noise also distributes around the NHNM but also down to −20 dB less. The high frequency noise appears to split into a major (close to NHNM) and a minor (close to NHNM −20 dB) branch, which indicates that the elevated noise levels may be due to anthropogenic noise. While we did not expect huge impact of traffic for this site, a comparison of day to night data in fact confirms very clearly that high frequency noise is primarily anthropogenic and as such probably due to traffic. Both on vertical and horizontal

components, nighttime noise levels are more than 20 dB less than during daytime for higher-frequencies and 10 dB less for longer periods. In addition to traffic from the closest road (300 m) a huge commercial and recreation facility about 3 km NE of the site may contribute to the observed noise levels.

3.4 Site 08

Station 08 is placed near a flood protection facility inside a several meter deep artificial trough next to the village Tiefenfucha in Lower Austria (Fig. 9). It is supplied from the power grid. The station is 200 m from the village, 1.5 km from a highway and the Danube river and surrounded mostly by vineyards. Trees (approximately 5–10 m high) surround the site. The sensor is placed atop a base made of concrete which fills the foundation of a small wooden hut, that serves as shelter for the station. An artificial water stream may flow several m from the station after periods of strong precipita-

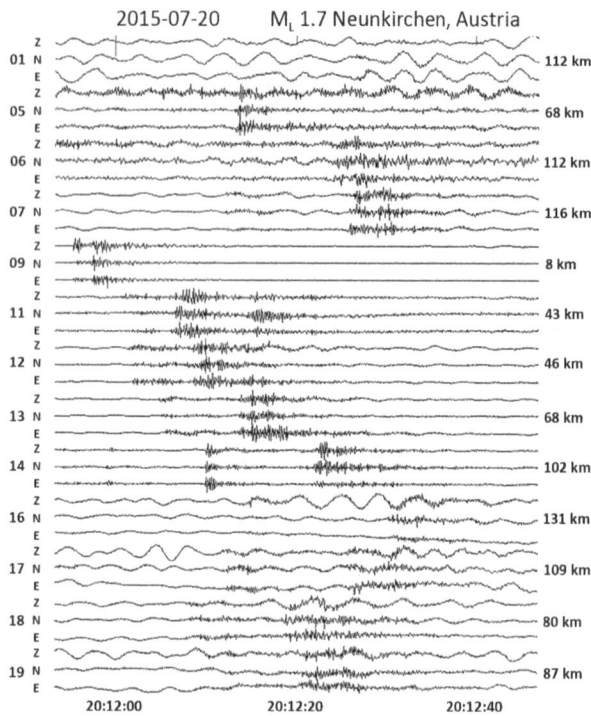

Figure 13. Waveform recordings (unfiltered) of all test sites deployed at the time of a local M_l 1.7 earthquake near Neunkirchen, Lower Austria. Station 09 is closest to the epicenter. Note, that even on very noise sites such as 05 (68 km distance) and 06 (112 km distance) S wave arrivals can be recognized. Without filtering, the signal on the 112 km distant bunker site 01 is dominated by microseism noise. The earthquake signal can be recovered on all 13 stations operative at the time. For most stations both P and S arrivals could be picked.

Figure 14. Waveform recordings (unfiltered) of all test sites deployed at the time of a teleseismic M_w 6.5 earthquake in the Caribbean Sea approximately 7700 km from the network. Multiple phase arrivals and clear surface waves can be identified on all stations operative at the time.

tion but is almost non-existing during dry conditions. Long-period noise levels (Fig. 10) are well below NHNM −20 dB on the vertical component but significantly higher on horizontal components. High-frequency noise splits into two branches on all three components which likely reflects the effect of day and night anthropogenic noise or workday to weekend variations. Despite the comparably small distance between the station and the closest houses, high-frequency noise levels are reasonably well below NHNM −10 dB or even NHNM −20 dB.

3.5 Site 09

Station 09 is located inside a rarely used storage cellar that is built into the slope of a hill (Fig. 11) and supplied from the power grid. The site is close to a secluded family house and 800 m from any larger settlement (Hafning, Lower Austria), 1.5 km from a highway and mainly surrounded by forest. The installation inside the cellar reduces surface effects which results in low long-period noise levels (Fig. 12). While the vertical component long-period noise level is approximately only 10 dB higher than the NLNM, horizontal noise-

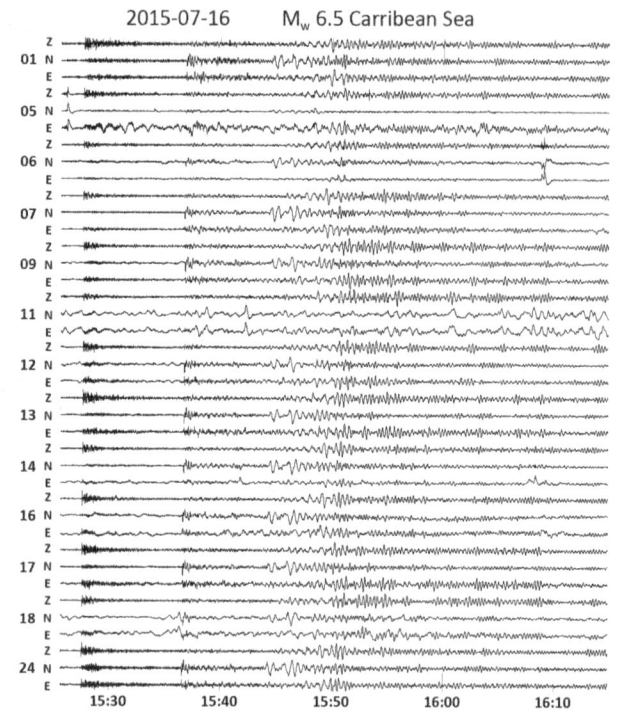

levels fall around NHNM −10 dB but are still within the desired range. High-frequency noise on all components separates into two branches which, however, fall below NHNM −20 dB. This likely reflects the little anthropogenic noise originating from the nearby house. Among the sites where continuous data is available for more than one week, Site 09 performs second best to the installation inside the bunker (Site 01).

4 Network performance and first selected seismic events

Several of the sites discussed above experience substantially higher long-period noise levels on the horizontal components than on the vertical components. Yet, most of the sites show comparably low long-period noise levels on the vertical component, in the range of 20 dB lower than the NHNM. With one exception (wine cellar) all of our sites are surface or near-surface installations and we did not attempt for sophisticated protection against pressure variations and surface tilt such as pressure sealed sensor covers or very stiff gabbro baseplates (Forbinger, 2012). Consequently, elevated long-period noise on the horizontal components due to local surface tilt induced by pressure gradients (Bormann and Wielandt, 2012) is expected and unavoidable for the type of installation described here. Still, because of the additionally strong noise in higher-frequency bands, stations 05 and 06 may be relocated for the final deployment within the AlpArray framework.

Except for the sites discussed above, most of the preliminary stations by end of July 2015 were only running for less than seven days with telemetry data. Hence, we did not attempt to analyze noise levels for these stations. However, we can still get a first impression of the site qualities by checking the waveform data of earthquakes that have been recorded by the network. As examples we show here the unfiltered recordings (not corrected for instrument response, all sensors are of similar specifications) of a local M_l 1.7 earthquake near Neunkirchen, Lower Austria (Fig. 13) and a teleseismic event (Fig. 14) with M_w 6.5 at approximately 7700 km distance in the Caribbean Sea. Note, that the local event occurred during potentially more quiet evening times, while the teleseismic event was recorded during more noisy mid-day time. However, long-period teleseismic signals are not strongly affected by anthropogenic high-frequency noise. Both events can be well recognized on all stations even on unfiltered waveform data. Several seismic phases may be identified on closer inspection. Thus, almost all preliminary stations perform reasonably well and meet the multi-purpose requirements of capturing both local and distant seismic events that will allow for various geodynamical studies.

5 Conclusions

In the framework of the upcoming AlpArray project we described our preparations for temporary seismic broadband installations in Austria. Following the specific project requirement (stations should perform reasonably well in the two frequency bands 1–20 Hz and 30–200 s) a typical installation comprises microfleece and styrofoam covered sensors in the basements of unutilized houses or huts. One station is located inside an abandoned bunker. By the end of July 2015, 15 stations are in testing operation, five of which are operating since spring 2015. For these five we presented probabilistic power spectral densities that allow for a first noise characterization of the sites. While generally most stations perform reasonably well in a range of NHNM −10 dB to NHNM −20 dB on the vertical component, the bunker station is exceptionally quiet and two of the three stations near or inside villages suffer from elevated anthopogenic noise and are thus considered for replacement. Since most of our sites are surface or near surface installations, horizontal noise levels are generally higher by approximately 10–20 dB than the vertical. Still, first events recorded with the complete set of 15 test sites are well resolved on all stations which indicates that following the site selection and preparation that we describe in this manuscript can result in seismic stations that perform reasonably well for both local and teleseismic events.

Data availability

Seismic data used for this manuscript is not publicly accessible by decision of the AlpArray working group. Waveform data from the preliminary station tests may be available upon request directed to the corresponding author of this manuscript (florian.fuchs@univie.ac.at).

Acknowledgements. AlpArray Austria is funded by the FWF Austrian Science Fund project number P 26391. We acknowledge planning and organization of the AlpArray coordinators Edi Kissling, György Hetenyi, Irene Molinari and John Clinton at ETH Zürich, Switzerland, who created the AlpArray seismic network layout. We thank Johann Huber for technical assistance and Dimitri Zigone for help in the field. We thank all involved Austrian communities, forest administrations and individuals for their help during site scouting. We thank Aladino Govoni and an anonymous reviewer for their suggestions to improve the manuscript.

References

Bormann, P. and Wielandt, E.: Seismic Signals and Noise, in: New Manual of Seismological Observatory Practice 2 (NMSOP2), edited by: Bormann, P., Deutsches GeoForschungsZentrum GFZ, Potsdam, Germany, 1–62, doi:10.2312/GFZ.NMSOP-2_ch4, 2012.

Forbinger, T.: Recommendations for seismometer deployment and shielding, in: New Manual of Seismological Observatory Practice 2 (NMSOP-2), edited by: Bormann, P., Deutsches GeoForschungsZentrum GFZ, Potsdam, Germany, 1–10, doi:10.2312/GFZ.NMSOP-2_IS_5.4, 2012.

Kissling, E., Hetenyi, G., and AlpArray Working Group: AlpArray – Probing Alpine geodynamics with the next generation of geophysical experiments and techniques, Geophysical Research Abstracts, EGU General Assembly 2014, Vienna, Austria, 16, EGU2014–7065, 2014.

Krischer, L., Megies, T., Barsch, R., Beyreuther, M., Lecocq, T., Caudron, C., and Wassermann, J.: ObsPy: a bridge for seismology into the scientific Python ecosystem, Computational Science & Discovery, 8, 014003, doi:10.1088/1749-4699/8/1/014003, 2015.

McNamara, D. E. and Buland, R. P.: Ambient noise levels in the continental United States, B. Seismol. Soc. Am., 94, 1517–1527, doi:10.1785/012003001, 2004.

Peterson, J.: Observations and modeling of seismic background noise, USGS Open-File report, 93–322, 1993.

Trnkoczy, A., Bormann, P., Hanka, W., Holcomb, L. G., Nigbor, R. L., Shinohara, M., Shiobara, H., and Suyehiro, K.: Site Selection, Preparation and Installation of Seismic Stations, in: New Manual of Seismological Observatory Practice 2 (NMSOP-2), edited by: Bormann, P., Deutsches GeoForschungsZentrum GFZ, Potsdam, Germany, 1–139, doi:10.2312/GFZ.NMSOP-2_ch7, 2012.

25

Precipitation response to El Niño/La Niña events in Southern South America – emphasis in regional drought occurrences

Olga Clorinda Penalba[1,3] and Juan Antonio Rivera[2,3]

[1]Departamento de Ciencias de la Atmósfera y los Océanos (DCAO/FCEN), Universidad de Buenos Aires, Buenos Aires, C1428EGA, Argentina
[2]Instituto Argentino de Nivología, Glaciología y Ciencias Ambientales (IANIGLA/CONICET), Mendoza, 5500, Argentina
[3]Instituto Franco-Argentino para el Estudio del Clima y sus Impactos (UMI IFAECI/CNRS), Buenos Aires, C1428EGA, Argentina

Correspondence to: Juan Antonio Rivera (jrivera@mendoza-conicet.gob.ar)

Abstract. The ENSO phenomenon is one of the key factors that influence the interannual variability of precipitation over Southern South America. The aim of this study is to identify the regional response of precipitation to El Niño/La Niña events, with emphasis in drought conditions. The standardized precipitation index (SPI) was used to characterize precipitation variabilities through the 1961–2008 period for time scales of 3 (SPI3) and 12 (SPI12) months. A regionalization based on rotated principal component analysis allowed to identify seven coherent regions for each of the time scales considered. In order to identify the regional influence of El Niño and La Niña events on the SPI time series, we calculated the mean SPI values for the El Niño and La Niña years and assessed its significance through bootstrap analysis. We found coherent and significant SPI responses to ENSO phases in most of the seven regions considered, mainly for the SPI12 time series. The precipitation response to La Niña events is characterized with regional deficits, identified with negative values of the SPI during the end of La Niña year and the year after. During El Niño events the precipitation response is reversed and more intense/consistent than in the case of La Niña events. This signal has some regional differences regarding its magnitude and timing, and the quantification of these features, together with the assessment of the SST composites during drought conditions provided critical baseline information for the agricultural and water resources sectors.

1 Introduction

The El Niño-Southern Oscillation (ENSO) phenomenon is the dominant mode of coupled atmosphere-ocean variability on interannual time scales in several regions of the world (Trenberth and Stepaniak, 2001). One of the regions with larger impacts associated with extreme precipitation and ENSO events is Southern South America (SSA). A pioneer study performed by Ropelewski and Halpert (1987) on global scale, based on their previous research for North America (Ropelewski and Halpert, 1986), identified a clear ENSO signal in precipitation patterns over SSA. This signal was characterized with an increase in precipitation over central-east Argentina, Uruguay and Southern Brazil during the summer following the development of El Niño conditions. With focus in South America and surrounding oceans, Aceituno (1988) also showed that the SO-related changes in the large-scale circulation lead to a vast diversity of anomalous regional precipitation regimes. This result was further verified by Grimm et al. (2000), who conducted an analysis of precipitation variability associated with El Niño and La Niña phases through a regional approach. The most important signal usually occurs during the austral summer and autumn after the year of occurrence of El Niño events and during the spring after La Niña years. Penalba et al. (2005) analyzed the probability of exceeding the median during both phases of ENSO, showing coherence between El Niño (La Niña) occurrence and precipitation excess (deficit) over central-eastern Argentina. Nevertheless, this study showed a high degree of regional variability in ENSO-related precipitation, which has to be taken into account in precipitation forecasts over large areas.

Impacts of El Niño and La Niña events were evident in the hydrological (Boulanger et al., 2005) and agricultural (Podestá et al., 1999) sectors over SSA and particularly over northeastern Argentina, the southernmost part of Brazil and Uruguay. The precipitation excess associated to El Niño events contribute to the observed excess conditions in soil water content (Spescha et al., 2004) and an increase in about 10 % in soil water storage (Penalba et al., 2014a). River overflows and floods were related with above-normal precipitation linked to El Niño years were extensively documented, particularly over La Plata Basin and its main water courses (Camilloni and Barros, 2003; Chamorro, 2006; Antico et al., 2015). Regarding the trends in precipitation, Barros et al. (2008) demonstrated that half of the annual trends over northeastern Argentina, southern Brazil and Paraguay were associated to the El Niño phase. Even when there is an evidence of the existence of a core region regarding the signal of ENSO in precipitation patterns, other areas of SSA showed a modulation in hydroclimatic variables associated with El Niño and La Niña events. For instance, streamflow excess over the central Andes of Argentina was mainly related to El Niño conditions, while streamflow deficiencies correspond mostly to La Niña occurrences (Compagnucci and Vargas, 1998; Prieto et al., 1999; Rivera et al., 2015). Minetti et al. (2003) showed that La Niña events of 1988–89 and 1995–96 changed the increasing trend slope in annual precipitation over northwest Argentina. In central Argentina, Pasquini et al. (2006) showed that only 3 out of 12 rivers exhibit an ENSO signal, indicating that is not possible to connect the discharge anomalies in the Pampas ranges with the events in the tropical Pacific. The signal of ENSO was observed also in the precipitation records over the Patagonia region (Paruelo et al., 1998; Russián et al., 2010), in combination with a signal from the Southern Annular Mode (SAM) (Silvestri and Vera, 2003; González and Vera, 2010).

This large number of regional studies used different precipitation databases and periods of record, and compared observations from different climatic regions. Moreover, the significance of the ENSO signal in the precipitation records was not often tested, and its length and intensity is also regionally dependent. Based on these considerations, this paper aims to advance in the understanding of the temporal variability of precipitation over SSA at a regional level, with emphasis on the response of precipitation to El Niño/La Niña occurrences. The standardized precipitation index (SPI) is used to characterize precipitation fluctuations over the study area, given that it allows to compare precipitation records from locations with different climate conditions (Moreira et al., 2008). Moreover, the SPI can be calculated for a variety of time scales (Hayes et al., 1999), allowing the quantification of short- and long-term precipitation shortages and excess. Emphasis will be given to the assessment of drought conditions, which are especially important in regions where economic

activities are highly dependent on water resources, and particularly affect nations heavily reliant on agriculture, including both subsistence and highly intensive and high technology agricultural practices, as the case of SSA. In this sense, Vicente-Serrano et al. (2011) showed that during La Niña events there is an increase in the probability of drought occurrences in several regions of SSA. Droughts by nature are a regional phenomenon and commonly cover large areas during long periods of time; therefore, it is important to study these events in a regional context (Hisdal and Tallaksen, 2003). These results together with the analysis of SST composites will provide reference information for different sectors related to agriculture and water resources, which can enhance the understanding of seasonal precipitation forecasts.

2 Data

2.1 Precipitation data

The database for the study consists of 56 high-quality monthly precipitation time series that cover the period 1961–2008, obtained through the CLARIS LPB database (Penalba et al., 2014b). These stations are located in the portion of South America south of 20° S, have less than 2 % of missing data and were subjected to quality control procedures, as described in Penalba et al. (2014b). Additionally, a homogeneity control using the Standard Normal Homogeneity Test (Alexandersson, 1986) allowed to identify inhomogeneities in 4 of the 56 precipitation time series, which correspond to climatic jumps and not to instrumental factors. Gap-filling routines were applied through linear regressions with reference time series.

These high-quality time series were used to calculate the SPI, developed by McKee et al. (1993) for drought definition and monitoring. This index quantifies the number of standard deviations that accumulated precipitation in a given time scale deviates from the average value of a location in a particular period. The SPI is a powerful, flexible index that is simple to calculate, it has been recommended by the Lincoln Declaration on Drought Indices (Hayes et al., 2011) and according to Penalba and Rivera (2015) is the most adequate meteorological drought index for SSA. For the calculation of the SPI, the accumulated precipitation time series were divided in 12 monthly series of 48 years, each of them were fitted to a two-parameter gamma probability density function. This distribution appropriately fits the accumulated precipitation in the study region for time scales between 1 and 12 months, which was verified through the Anderson-Darling goodness-of-fit test (Anderson and Darling, 1952) for a confidence level of 95 % (Penalba and Rivera 2012, 2016; Penalba et al., 2016). The 12 probability density functions for each time scale and period were translated to 12 cumulative density functions. Given that the gamma distribution is undefined for $x = 0$, the relative frequency of precipitation

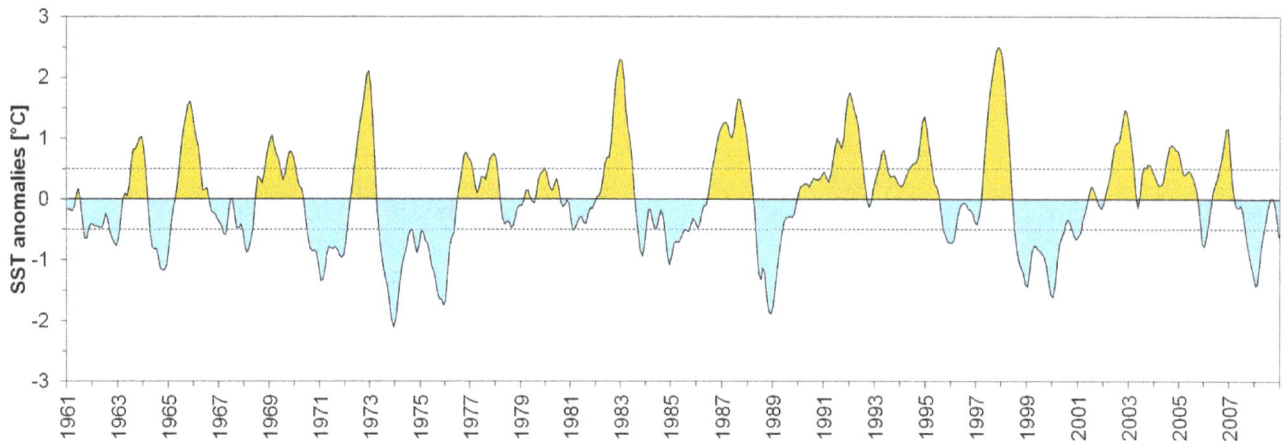

Figure 1. Time series of the Oceanic Niño Index for the period 1961–2008. Dotted lines indicate the ±0.5 °C thresholds.

containing zero values (q) was considered for the cumulative density function:

$$H(x) = q + (1-q)G(x), \qquad (1)$$

where $G(x)$ is the gamma cumulative density function. The value of q decreases as the time scale for the accumulation of precipitation increases, being zero for time scales longer than 7-months. This value also shows a seasonal variation; for example, in the case of 3-month accumulation, the higher values of q are observed during the dry season – winter of the southern hemisphere – in semi-arid regions, with a maximum value of 0.1. Finally, an equi-probability transformation from the cumulative density functions to the standard normal distribution with the mean of 0 and the variance of 1 were performed to obtain the SPI.

A detailed description of the calculation of the SPI can be found in Lloyd-Hughes and Saunders (2002). Given that droughts will impact different sectors and activities depending on the time scale over which precipitation deficits accumulate (Edwards and McKee, 1997), in this work, the SPI was computed on time scales of 3 (SPI3) and 12 months (SPI12), representing short-term and long-term droughts, respectively. Short-term droughts used to affect the agricultural sector, while long-term droughts have impacts on the water resources (Guttman, 1999). Both sectors are extremely important in SSA.

2.2 ENSO index

Several indicators have been developed for the definition and monitoring of ENSO conditions, such as the Southern Oscillation Index (SOI, Trenberth, 1984), the Multivariate ENSO Index (MEI, Wolter and Timlin, 1993) and the Trans-Niño Index (Trenberth and Stepaniak, 2001). In this work, the Oceanic Niño Index (ONI), developed by the Climate Prediction Center (CPC) from the National Oceanic and Atmospheric Administration (NOAA), was used to define El

Niño/La Niña conditions. The ONI was obtained from the Extended Reconstructed SST v3b (Smith et al., 2008) as a 3-month moving average applied to the SST anomalies over El Niño 3.4 region, located in 5° N–5° S, 120–170° W. A detail of the areas chosen to monitor el SST anomalies related to ENSO can be found in Penland et al. (2010). SST anomalies were calculated based on the 1971–2000 baseline period. El Niño and La Niña events can be defined as 5 consecutive overlapping 3-month period at or above the +0.5 °C anomaly for warm (El Niño) events and at or below the −0.5 °C anomaly for cold (La Niña) events. Finally, El Niño years between 1961 and 2008 are 1963, 1965, 1968, 1972, 1976, 1982, 1986, 1991, 1994, 1997, 2002, 2004, 2006; and La Niña years are 1962, 1964, 1967, 1970, 1973, 1984, 1988, 1995, 1998, 2000 and 2007. It should be noted that the El Niño and La Niña years remain the same if we consider other indices like El Niño 3.4 or the SOI, and minor differences can be observed on a monthly basis, due to the different definition of each index. Nevertheless, changing the baseline period for the calculation of the SST anomalies can introduce some differences in the definition of ENSO years (not shown). Figure 1 shows the temporal evolution of the ONI during the period 1961–2008.

3 Methodology

Using a regional average provides a time series that is a better representation of large-scale processes, and it is easier to deal with one index series (Schonher and Nicholson, 1989). Moreover, it gives a more accurate estimation of the probabilities of observed precipitation (Guttman, 1999). In order to obtain the regional features of precipitation time series, we used the Rotated Principal Components Analysis (RPCA) (Richman, 1986). This methodology was applied in S-mode, which allows to obtain a spatial regionalization and temporal patterns of precipitation in the studied domain. Varimax rotation (Kaiser, 1958) was applied to obtain consistent spatial patterns, retaining the orthogonality and enhancing the interpretability of the results. The variables to be substituted by the principal components derived from the RPCA are the 56 meteorological stations, represented by their SPIs. The scree test of Cattell (1966) was adopted to decide how many PCs to retain in order to separate signal to noise. The maximum loading approach, in which each station is assigned to the component upon which it loads most highly (Comrie and Glenn, 1998), was used to delineate the regions. For both time scales of 3 and 12 months, seven homogeneous regions were obtained, which are climatically and geographically consistent.

Once the SPI homogeneous regions were obtained, the seven regional SPI time series for each of the time scales considered were used to calculate the composites during El Niño and La Niña years (year 0) and the years before (−) and after (+) the events, in line with the methodology proposed by Ropelewski and Halpert (1987). Thus, the analysis of the aggregate composites covers a period of 36 months. SPI average values for each region for the 13 El Niño and 11 La Niña events during the 1961–2008 were calculated, for each of the 36 months centered in the year of beginning of the event, for both SPI3 and SPI12 time series. Since the SPI is standardized, the anomalies can simply be obtained from an average of the values in the years considered. The use of regional composites will allow to identify the El Niño/La Niña signal in a better way than using individual locations, which may only detect anomalies related to a local phenomenon (Chiew et al., 1998). As an example, Fig. 2 shows the composite of SST anomalies over the El Niño 3.4 region during El Niño and La Niña events for the 36 months considered for the analysis of the signal in precipitation. An approximately symmetric behavior between the composites of El Niño and La Niña years is observed, as previously shown by Vicente-Serrano et al. (2005). The largest SST anomalies occur during the summer following the start of the event, typically during the months of November (0), December (0) and January (+) (Fig. 2). For the southern hemisphere seasons, the onset of the El Niño/La Niña events occurs in late autumn or during the beginning of winter and lasts until autumn of the following year. It is expected a precipitation response during these months, although when precipitation is accumulated over several months the response can be delayed. The inter-quartile interval shows that there is a larger dispersion in the composite of the SST anomalies once El Niño event has finished and before the beginning of La Niña conditions. It is expected that this uncertainty also affects the precipitation responses at a regional level.

In order to identify significant changes in the 36 monthly anomalies, we used a bootstrap resampling procedure (Efron and Tibshirani, 1993). The bootstrap is a computing-intensive statistical method that provides a confidence band around the SPI composites. Its advantage is that it is less restricted by parametric assumptions than more traditional approaches (Mudelsee, 2011). The confidence intervals were

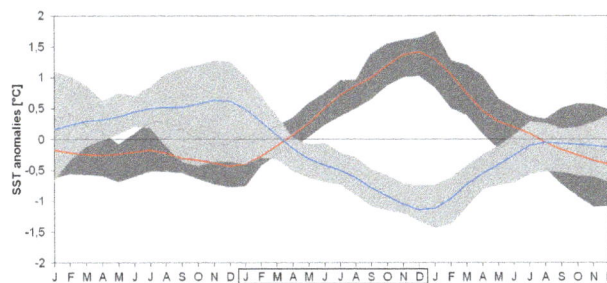

Figure 2. Composite of monthly SST anomalies for El Niño (red line) and La Niña (blue line) during the 1961–2008 period. The 36 month composite starts at January (−) and continues until December (+). Dark (light) grey bands around the composite anomalies represent the inter-quartile interval for El Niño (La Niña).

based on 1000 resamples for the 95 % significance level; therefore, they are given by the interval between the 50th and the 95th largest values. By comparing the values of the monthly SPI composites with the percentile-based confidence interval constructed with the resampling of the SPI values, we can assess if the composites show a significant El Niño/La Niña signal.

4   Results

4.1   El Niño and La Niña signal in precipitation over SSA: SPI3

The regionalization of the 56 SPI3 time series resulted in a total of seven homogeneous regions: CES (Central-East South); P + PN (Pampas and Northern Patagonia); CW (Central-West); CEN (Central-East North); NE (Northeast); SP (Southern Patagonia) and NW (North-West) (Fig. 3). The regional SPI3 patterns account for different temporal variations in precipitation and were ordered taking into account the explained variance of each mode. Positive (negative) values account for above (below) average precipitation periods, and its intensity is proportional to its value, either positive for the excess or negative for deficits. The existence of relatively dry periods which are common in several of the regions is evident, although high frequency variations present in the time series are relevant. Drought events during 1960s and the beginning of 1970s stand out, as the events of late 1980s and mid-1990s (Fig. 3). The spatial extension of these drought events was assessed by Rivera and Penalba (2014), typically covering more than 60 % of the study area.

Figure 4 shows the SPI3 composites during a 36-month period centered in El Niño (left panel) and La Niña (right panel) years for the seven homogeneous regions. It is observed that the months where the influence of El Niño events in the precipitation is significant (marked with red) has considerable regional variations, nevertheless, in all the regions the SPI significant composites are positive (Fig. 4, left panel). For

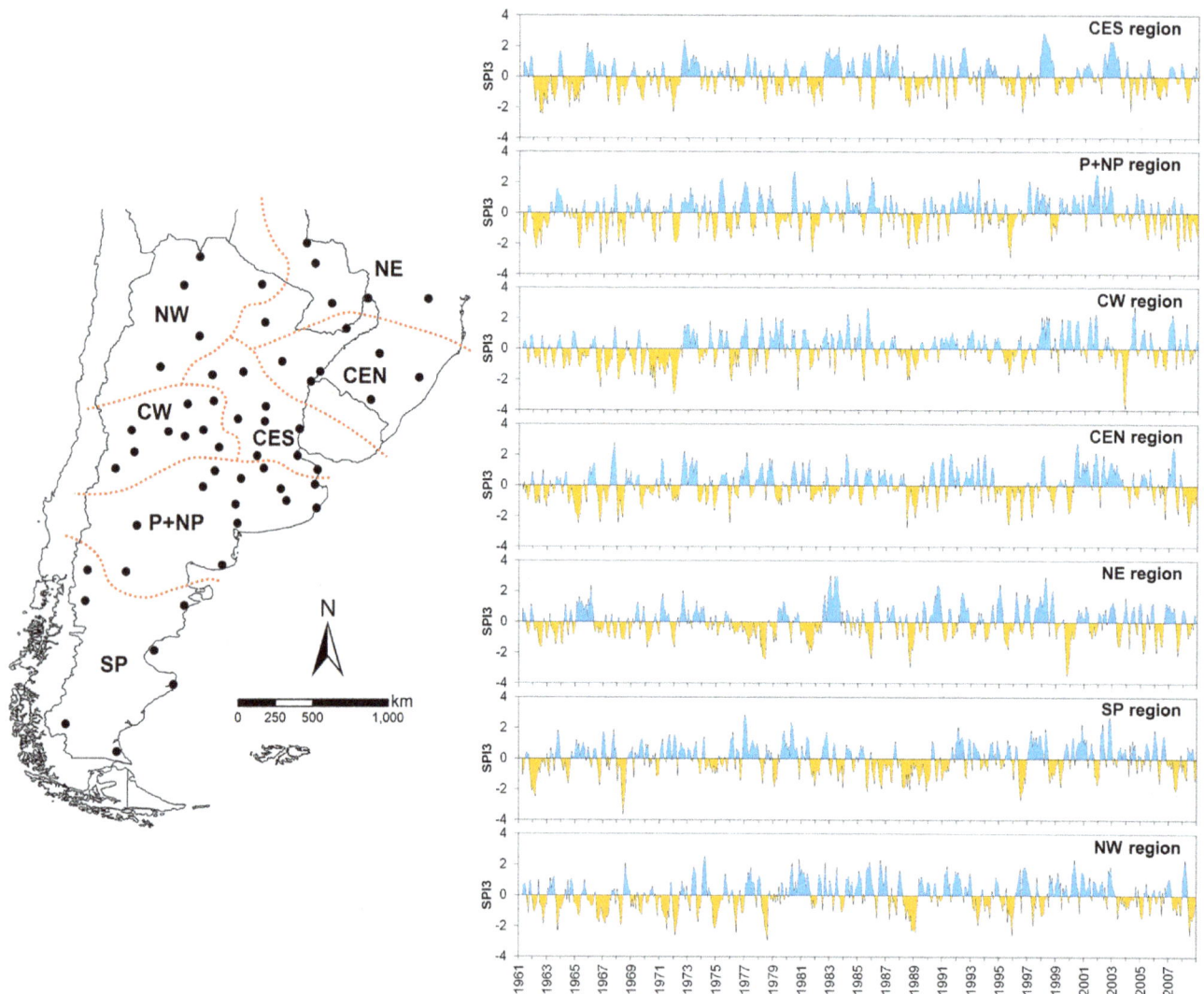

Figure 3. Regions obtained through RPCA applied to the SPI3 time series (left) and its corresponding regional temporal patterns (right).

the CES region, the El Niño influence is significant between December (0) to May (+), although there is a significant positive anomaly during August (0). In the P + NP region the El Niño signal is observed from the month of August (0) to April (+), in line with the average temporal behavior of SST anomalies shown in Fig. 2. The SPI3 in the CW region shows significant anomalies during the period August (0)–February (+). The most consistent response is observed in the CEN region, with significant SPI3 anomalies between October (0) to August (+). The amplitude of the anomalies is close to one standard deviation during the summer months of the year following El Niño development, i.e., when the magnitude of the SST anomalies during El Niño years have the higher values (see Fig. 2).

In the NE region, the El Niño events have a significant influence in precipitation during the period between November (0) and June (+), in line with the signal observed in the adja-

cent CEN region but with less intensity. In the case of the SP, the response is observed during the months of September (0) to January (+), while for the NW region the El Niño signal is observed between August (0) and December (0). In general, a greater influence of El Niño in the precipitation anomalies were observed in the SPI3 composites of the CEN, NE and CES regions, both in magnitude and duration. Moreover, the temporal evolution of the significant SPI3 composites in these regions is similar to the pattern of the SST anomalies over the tropical Pacific (Fig. 2). It is noteworthy that in the regions of the central portion of SSA during the year preceding El Niño development there is a tendency to have precipitation deficits, something previously identified by Grimm et al. (2000).

The right panel of Fig. 4 shows the regional SPI3 composites during the 36-month period centered in La Niña years. In most of the regions the influence of La Niña events in pre-

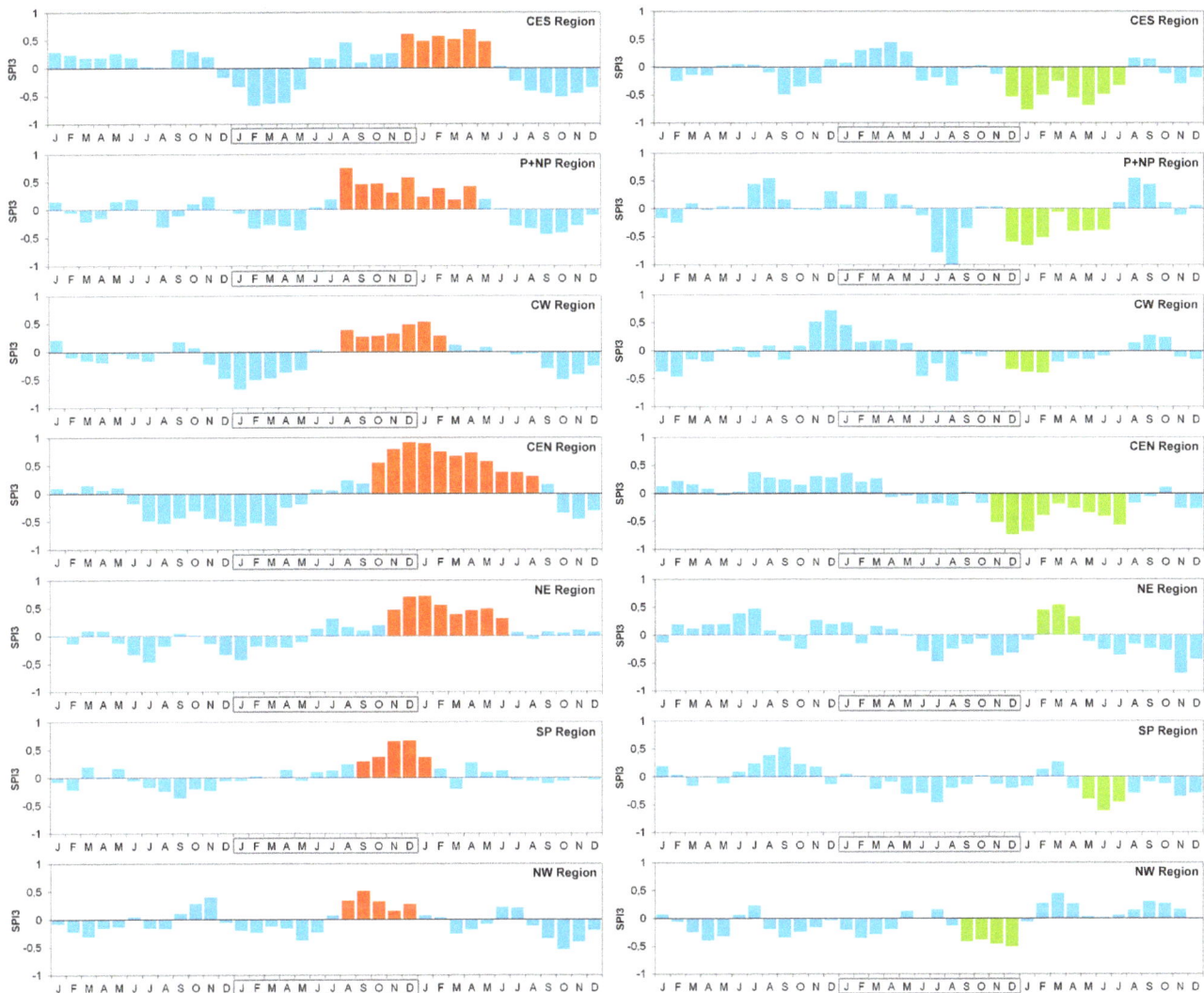

Figure 4. Regional aggregate SPI3 composites during El Niño (left) and La Niña (right) events. The months in the box refer to El Niño or La Niña years (0). Red (green) bars highlight the months of possible El Niño (La Niña) influence on the SPI3 time series.

cipitation is opposite to the one observed in the case of El Niño, with negative anomalies in precipitation. However, the response is not often clear in time and is detailed below. The CES region shows a signal associated with La Niña events between the months of December (0) to July (+), which is two months larger than in the case of El Niño signal. In the case of the P + NP region, the SPI3 significant anomalies are observed between December (0) and June (+), with negative SPI3 anomalies. Even when there is another period of significant negative anomalies, between the months of July (0) and September (0), it is considered that the signal is not consistent over time since the anomalies during October (0) and November (0) are positive. The months of August (+) and September (+) have significant positive SPI3 anomalies, which indicates that there is a strong seasonality associated to La Niña occurrences over the P + NP region. The influence of La Niña events in the precipitation of the CW region

is observed in the austral summer after the La Niña development, although there are significant negative anomalies in the winter of the year 0. In the CEN region, the negative anomalies between November (0) to July (+) can be identified as La Niña influence. The NE region is characterized by both positive and negative significant anomalies associated to La Niña, indicating a strong seasonality in the SPI3 composites. This pattern is also a result of different impacts associated to La Niña years, with some years with above average precipitation and others with deficit conditions. The most consistent response is observed during the months of February–March–April (+), with positive SPI3 anomalies. The months of July, November and December of both the years 0 and + shows negative significant anomalies but do not form a consistent response over time. Moreover, most of the months between autumn (0) to the year following La Niña onset shows negative SPI3 composites, in line with the general regional be-

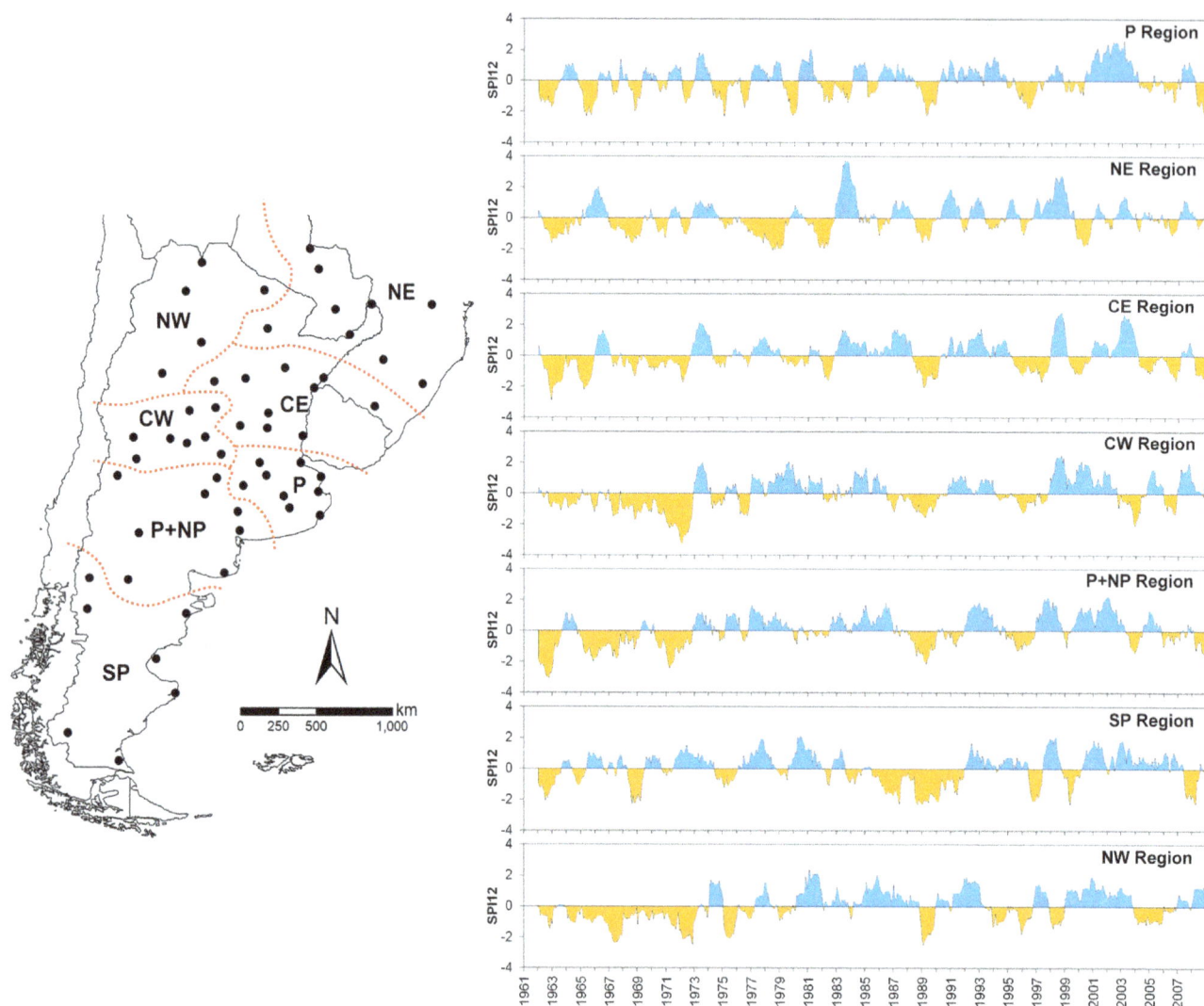

Figure 5. Same as Fig. 3 for the SPI12 time series.

havior and highlighting the heterogeneous precipitation response during La Niña events. The SP is characterized by a signal in precipitation during the months of May–June–July (+), while in the NW region this is observed in the months of September (0) to December (0). In the NW region significant positive and negative anomalies are observed in March (+) and September (+). The most relevant influence, both in time and intensity, of La Niña events in the regional precipitation for the time scale of 3 months is observed in the CES and CEN regions.

4.2 El Niño and La Niña signal in precipitation over SSA: SPI12

In the case of the regionalization of the SPI12 time series, seven homogeneous precipitation patterns were obtained (Fig. 5). The regions NW, NE, CW, P + NP and SP possess a similar spatial extent regarding the regionalization

obtained for the SPI3. The Central-East (CE) and Pampas (P) appear as a combination of the regions CEN and CES, and CES and P + NP, respectively. Figure 5 shows the regional SPI12 time series, ordered by the explained variance from the RPCA. Given the reduction of the high frequency variability in comparison with the SPI3 temporal variations, regional dry and wet periods can be identified in a better way. Dry periods stand out during the 1960s and 1970s in several of the regions, which affected with drought conditions up to 60 % of the locations over the study area (Rivera and Penalba, 2014). Comparing these regional patterns with the ones obtained for the SPI3, it is observed that SPI responds more slowly to changes in precipitation, with less frequent but longer dry and wet periods.

Figure 6 shows the 36-month composites of the SPI12 during El Niño (left panel) and La Niña (right panel) years and its significant signal. It is noted that, even when the region-

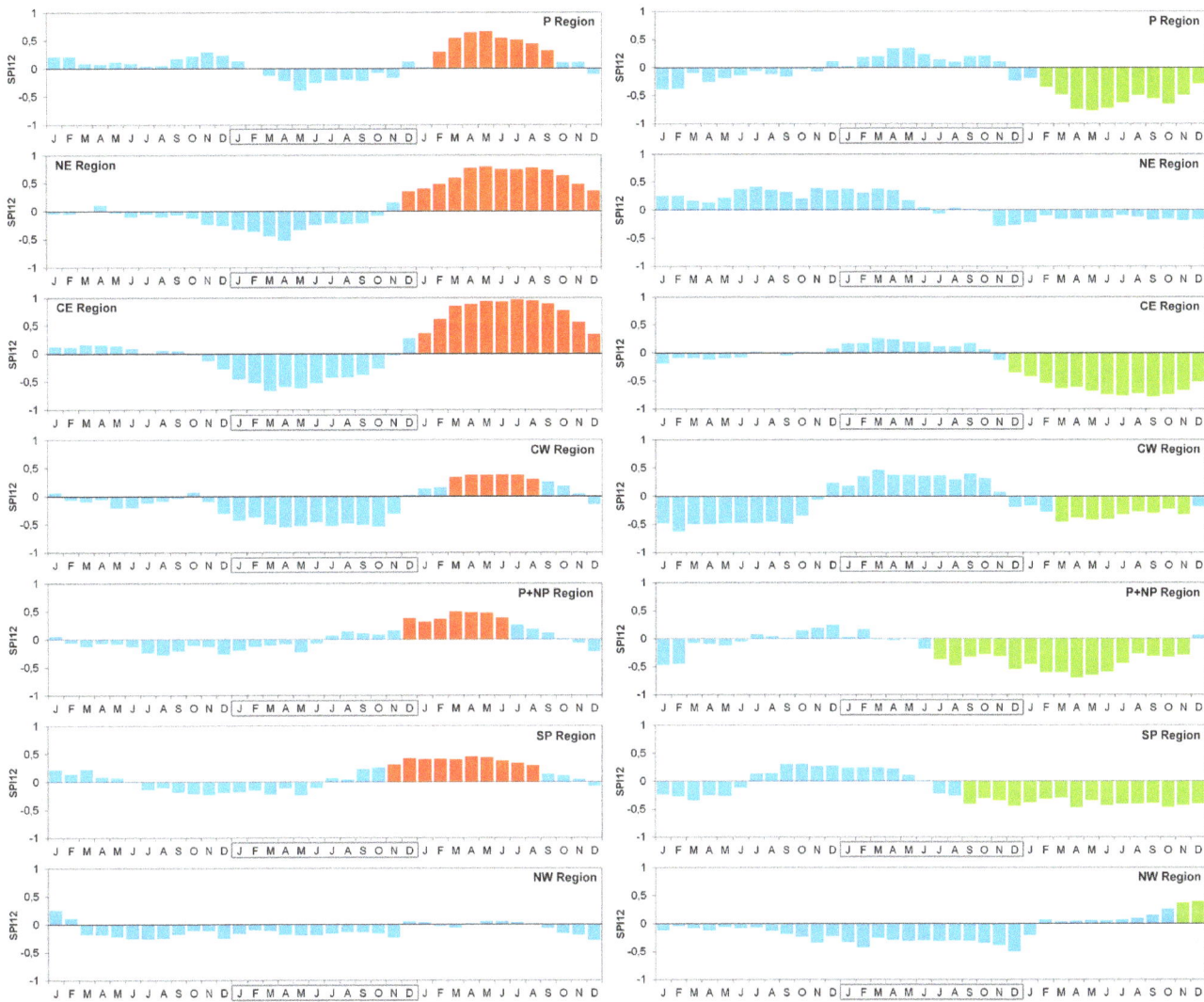

Figure 6. Same as Fig. 4 for SPI12.

alization is different from the obtained for the SPI3 time series and a direct comparison is not feasible, the response is more consistent in time and space, with a well-defined signal in most of the regions, associated to the filtering of the high-frequency variations of precipitation. Also, due to the increase in the time scale in which the accumulated precipitation is considered, the response is delayed in time with respect to the results observed for SPI3. In coherence with the findings for the SPI3 regional assessment, the sign of the anomalies of precipitation in the case of El Niño events is positive, while is negative in the case of La Niña events. The composite of SPI12 values during El Niño events over the P region presents a significant response between February (+) to September (+). The NE region has the longest temporal El Niño signal, from December (0) to December (+). Significant anomalies in the CE region occur during all the year following the development of El Niño and the intensity

is higher compared to other regions, with SPI12 composite values close to 1. The wet period from March (+) to August (+) in the CW region is associated with El Niño signal, although it has small amplitude. In the P + NP region the signal is observed between December (0) to June (+), while the SP shows significant precipitation anomalies during November (0) to August (+). There is no evidence of El Niño influence on SPI12 time series in the NW region, which indicates that the accumulation of precipitation on time scales longer than 3 months creates a mixture of processes that influence regional precipitation patterns generating a poor signal associated with El Niño. It is remarkable the occurrence of consistent negative SPI12 composite values during the development of El Niño events, mainly in NE, CE and CW regions.

In the case of precipitation response to La Niña events (Fig. 6, right panel), the signal in P region is significant between February (+) and December (+), being 3 months longer than El Niño signal. The difference between the precipitation response to El Niño and La Niña events is noteworthy in the NE region, where during La Niña conditions only November (0) shows a significant signal. Taking into account that the region has a spatial extension similar to the results from the SPI3 regionalization (Fig. 3), this poor signal can be attributed to different precipitation patterns recorded during La Niña years, which led to an inconsistent response over the region. In the CE region, the response is similar to the one observed during El Niño events, but opposite in sign, extending through all the year +. Precipitation over the CW region shows a coherent signal from March (+) to November (+), although the anomalies in August (+) and October (+) are not significant. The precipitation response to La Niña in P + NP region is noticeable extended over time with respect to the El Niño years, from July (0) to November (+). SP region shows a La Niña signal in precipitation patterns during the whole year + and the spring of the year 0, but with moderate values of the SPI12 composites. Finally, it is observed that the response in the NW region is restricted to the months of November (+) and December (+), with positive SPI12 composites during most of the year +. When the signal of El Niño and La Niña are compared, it is evident that there is a signal in precipitation that is more extended in time during La Niña years. This could be linked to the occurrences of La Niña events extending beyond one year, which can extend the precipitation response.

4.3   SST composites during regional drought conditions

Based on the results of the previous sections, it was observed that the influence of La Niña events resulted in the occurrence of precipitation deficits with different regional responses, both in duration and magnitude for time. This is associated to a teleconnection pattern that leads to negative moisture anomalies and subsidence over much of SSA (Grimm and Ambrizzi, 2009). These conditions favored the development of drought conditions, as reported by Grimm et al. (2000); Minetti et al. (2007); Chen et al. (2010) and Rivera and Penalba (2014), among others. In order to identify the influence of La Niña in the development of regional drought conditions over SSA, we analyzed the SST composites over the surrounding oceans during the months with SPI values below −1.0, a threshold usually chosen to define droughts (Lloyd-Hughes and Saunders, 2002). Similar assessments were carried out by Phillips and Denning (2007) for the analysis of extreme daily precipitation and its relationship with synoptic weather conditions and by Sienz et al. (2007), who analyzed the relationship between extreme SPI values and its relationship with mid-level circulation patterns, among others.

Figure 7 shows the composites of the SST anomalies for the months with drought conditions during the period 1961–2008 based on the regional SPI3 time series. Analogous results can be obtained for the SST composites based on the

months with SPI12 values below −1.0, with an increase in the magnitude of the anomalies, considering a lag of approximately 7 months (not shown). This is in line with the lag observed between the La Niña peak (Fig. 2) and the maximum in the SPI12 signal during the following year (Fig. 6). The configuration of SST anomalies associated with drought events shows a clear La Niña pattern for most of the regions, with negative SST anomalies along the tropical Pacific. This stands out for the CES, P + NP, CW, CEN and NE regions. The NE region shows a clear link between drought occurrences and cold SST anomalies, a result that was not obvious considering the SPI3 composites from Fig. 4.

The spatial pattern of SST anomalies during drought conditions for the CES region has its greatest negative anomalies in the Niño 3 and Niño 1 + 2 region, while the pattern of the CW region has its major anomalies just in the Niño 3 region (Fig. 7). SST anomalies related to droughts over CEN region are located mainly over Niño 3.4 region, while in the case of the NE region the higher SST anomalies in the Niño 3 and Niño 1 + 2 regions. Droughts over SP also show a similar spatial pattern, although the magnitude of the anomalies is lower, located mainly in the Niño 1 + 2 region (Fig. 7). The spatial pattern of SST anomalies during droughts events over the NW region shows small areas with significant values, indicating a weak relationship to La Niña conditions. This is consistent with the results from the previous sections regarding the regional precipitation response for both SPI3 and SPI12 regional time series. Another interesting feature is the presence of negative SST anomalies along the coast of Brazil related to regional drought occurrences over CES and CEN regions. Due to the proximity of the SST anomalies to the above mentioned regions, localized cooling conditions leading to subsidence and lack of precipitation can influence the regional drought occurrences. This region of cold anomalies is not evident in the composites based on SPI12 time series (not shown), indicating that some areas of the South Atlantic Ocean modulate the seasonal precipitation patterns over SSA, as previously shown by Berri and Bertossa (2004) and Taschetto and Wainer (2008).

If we consider the composites for the months in which more than 30 % of the meteorological stations over SSA show drought conditions – defined as critical dry months (Krepper and Zucarelli, 2012; Penalba and Rivera, 2016), we obtain for the SPI3 a SST anomalies pattern similar to the one related to the CEN region and for the SPI12 a spatial pattern that resembles the results for the P + NP region (result not shown). This verifies that La Niña events can trigger drought conditions simultaneously over several regions, as is shown in Vicente-Serrano et al. (2011) for different time scales. The analysis can be extended to different drought categories, which are defined for different SPI thresholds, for instance, moderate ($-1.5 < \mathrm{SPI} \leq -1.0$), severe ($-2.0 < \mathrm{SPI} \leq -1.5$) and extreme ($\mathrm{SPI} \leq -2.0$) drought conditions. Nevertheless, it should be noted

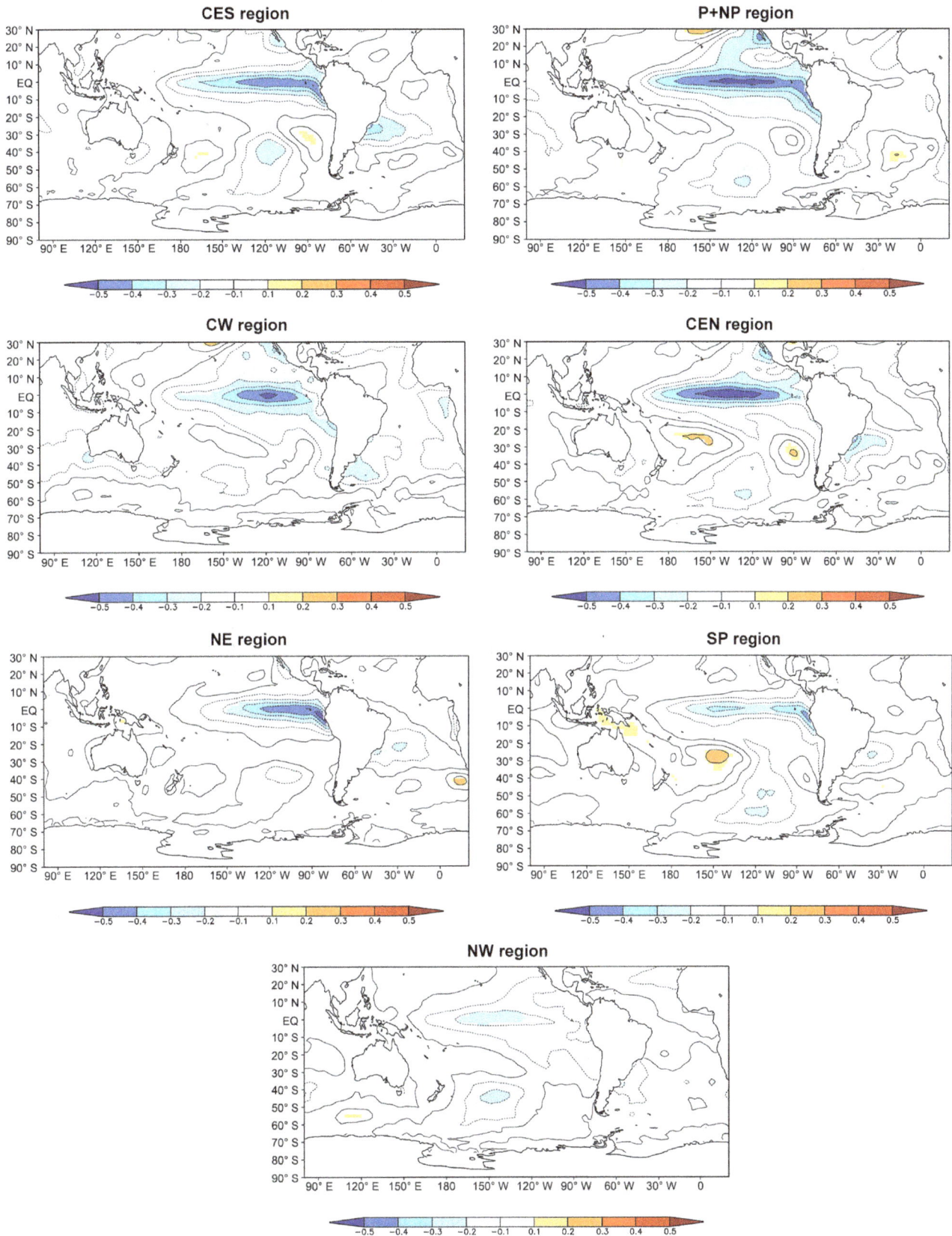

Figure 7. Composites of SST anomalies during the months with drought conditions for each of the seven regional SPI3 patterns. Shaded areas are significant at the 95 % level.

that the sample size can be reduced considerably, leading to limitations in the results.

5 Discussion and conclusions

The ENSO phenomenon has a strong impact on precipitation in SSA, both at seasonal to interannual time scales, associated to its two phases: El Niño and La Niña. Links between precipitation patterns and ENSO signal were assessed extensively both at regional and global scales, in order to improve the monitoring and forecasting of its impacts. However, some gaps remain regarding the ENSO prediction and its expected remote impacts, as stated in Kirtman et al. (2013). In order to advance in the knowledge between El Niño and La Niña occurrences and the regional patterns of precipitation, this research analyzed the precipitation response to the ENSO phases taking into account the regional features of precipitation over SSA. In order to quantify precipitation departures, we selected the SPI, given that it allows the comparison of time series from regions with different climate conditions. Other studies using for example the PDSI faced problems to compare individual time series (Piechota and Dracup, 1996), one of the many limitations of the index (Hayes et al., 1999). Percentiles based on an appropriate probability distribution were commonly used to identify El Niño/La Niña signal on hydroclimatological records (Dracup and Kahya, 1994; Karabork and Kahya, 2003). Nevertheless, the SPI is widely considered the most robust and effective index (Vicente-Serrano et al., 2005), proved to be a good indicator of both wet (Seiler et al., 2002) and dry (Penalba and Rivera, 2015) events, and it can be calculated at different time scales, allowing the monitoring of different drought types. The magnitude of departure from zero represents a probability of occurrence so that decisions can be made based on this SPI value (Hayes et al., 1999). When the time scale is short, for instance 3-month, the SPI reflect the seasonality of the data and is more appropriate to identify drought impacts on agriculture (Moreira et al., 2008). For time scale of 12 months, the SPI can reflect changes that are relevant for hydrological drought analyses and applications. Therefore, it is relevant to analyze the signal of El Niño/La Niña events in precipitation at different time scales, in order to discuss possible differences and the applicability of the findings.

In order to identify the regional features of precipitation, RPCA were applied to the 56 SPI time series at time scales of 3 and 12 months. This approach differs to the typical assessment based on harmonical fit to the precipitation composites and vectorial coherence, following the works of Ropelewski and Halpert (1986, 1987). The regionalization allowed to obtain seven homogeneous regions for both time scales, which are climatically and geographically consistent, that can be useful to obtain not only the El Niño and La Niña signals, but also other forcings related to precipitation over SSA. Through a bootstrap resampling procedure, sig-

nificant SPI composites were identified for both El Niño and La Niña, with a reversal in sign of the SPI composites during the El Niño phase (wet conditions) compared to those during La Niña phase (dry conditions). The consistency of the signal shows differences between the two phases, being more consistent during El Niño events and less consistent during La Niña, particularly for SPI3. The length and intensity of the signal is regionally dependent. In the case of the SPI3, the most relevant signal associated to El Niño occurrences was obtained for the precipitation time series of the CEN, CES, P + NP and NE regions (Fig. 4), ranging from 6 to 11 months. The SPI3 response to La Niña events shows that the most relevant influence, both in time and intensity, was observed in the CES and CEN regions. However, there is a clear seasonality in the SPI3 composites, mainly over the CES, CEN and P + NP regions, which can lead to agricultural impacts if different flavors of La Niña are not assessed. This is also evident in the response associated to the SPI12 over the NE region (Fig. 6), and in other regions and variables, like the snowpack variations and the streamflow records over the Central Andes (Masiokas et al., 2006). The SPI12 response to La Niña events has an average duration of approximately one year, being longer than for El Niño events. This could be linked to the occurrences of La Niña events that last longer than one year, which can extend the precipitation response. Mechanisms related to this kind of sustained La Niña events still not fully understood (Kirtman et al., 2013); therefore, advances in this topic can be helpful to understand the different impacts on regional precipitation, especially for drought conditions.

It is difficult to make a contrast between the results obtained in this research and those reported in the literature for the study region, given the different methodologies applied, the different periods considered and the time scales used for the accumulation of precipitation. However, the consistency in the sign of the SPI composites during El Niño/La Niña events is in agreement with the results obtained by previous research.

The analysis of SST composite anomalies during regional drought conditions shows a clear La Niña pattern for most of the regions, with a cold tongue extending over the Tropical Pacific (Fig. 7). Even when the most used El Niño definition is based on the El Niño 3.4 region (Trenberth, 1997), the area of strong SST anomalies associated to drought conditions has some regional variations. For example, for some of the precipitation regions a better response can be obtained analyzing the SST anomalies in the El Niño 1 + 2 area, like for the NE region. This indicate that the precipitation response can be dependent upon the box in which the El Niño index is constructed. Moreover, a separation between typical El Niño pattern and El Niño Modoki (Ashok et al., 2007) can also be helpful in understanding the differences in the precipitation responses. In the case of the SST composites associated with months with drought over the CEN and CES regions, it is evident a contribution of the South Atlantic Ocean in the modulation of seasonal SPI3, with cooling conditions that can lead to subsidence and below average precipitation. Further research is needed to determine if this local cooling responds to remote Tropical Pacific conditions or if it is a different source of seasonal variability. In this sense, to consider only the ENSO as the solely responsible for the precipitation temporal variability over the SSA prevents the development of a successful prediction tool, hence, other modes of variability should be considered on several time scales.

Acknowledgements. We thank S. M. Vicente-Serrano and one anonymous referee for their valuable comments and critical reading of the manuscript. The research leading to these results received funding from the European Community's Seventh Framework Programme (FP7/2007–2013: CLARIS LPB. A Europe–South America Network for Climate Change Assessment and Impact Studies in La Plata Basin), by the University of Buenos Aires (Grant UBA-20020130200142BA) and the Argentinean Council of Scientific and Technical Research (Grant PIP 0227).

References

Aceituno, P.: On the functioning of the Southern Oscillation in the South American sector – Part I: Surface climate, Mon. Weather Rev., 116, 505–524, 1988.

Alexandersson, H.: A homogeneity test applied to precipitation data, J. Climatol., 6, 661–675, 1986.

Anderson, T. W. and Darling, D. A.: Asymptotic theory of certain "Goodness of Fit" criteria based on stochastic processes, Ann. Math. Stat., 23, 193–212, 1952.

Antico, A., Torres, M. E., and Diaz, H. F.: Contributions of different time scales to extreme Paraná floods, Clim. Dynam., doi:10.1007/s00382-015-2804-x, online first, 2015.

Ashok, K., Behera, S. K., Rao, S. A., Weng, H., and Yamagata, T.: El Niño Modoki and its possible teleconnection, J. Geophys. Res., 112, C11007, doi:10.1029/2006JC003798, 2007.

Barros, V. R., Doyle, M. E., and Camilloni, I. A.: Precipitation trends in southeastern South America: relationship with ENSO phases and with low-level circulation, Theor. Appl. Climatol., 93, 19–33, 2008.

Berri, G. J. and Bertossa, G. I.: The influence of the Tropical and Subtropical Atlantic and Pacific oceans on precipitation variability over Southern Central South America on seasonal time scales, Int. J. Climatol., 24, 415–435, 2004.

Boulanger, J.-P., Leloup, J., Penalba, O., Rusticucci, M., Lafon, F., and Vargas, W.: Low-frequency modes of observed precipitation variability over the La Plata Basin, Clim. Dynam., 24, 393–413, 2005.

Camilloni, I. and Barros, V.: Extreme discharge events in the Paraná River and their climate forcing, J. Hydrology, 278, 94–106, 2003.

Cattell, R. B.: The screen test for the number of factors, J. Multiv. Behav. Res., 1, 245–276, 1966.

Chamorro, L.: Los principales usos y problemas de los recursos hídricos, in: El Cambio climático en la cuenca del Plata, Buenos Aires, edited by: Barros, V., Clarke, R., and Silva Dias, P., CIMA/CONICET, 111–123, 2006.

Chen, J. L., Wilson, C. R., Tapley, B. D., Longuevergne, L., Yang, Z. L., and Scanlon, B. R.: Recent La Plata basin drought conditions observed by satellite gravimetry, J. Geophys. Res., 115, D22108, doi:10.1029/2010JD014689, 2010.

Chiew, F. H. S., Piechota, T. C., Dracup, J. A., and McMahon, T. A.: El Niño/Southern Oscillation and Australian rainfall, streamflow and drought: Links and potential for forecasting, J. Hydrology, 204, 138–149, 1998.

Compagnucci, R. H. and Vargas, W. M.: Interannual variability of Cuyo rivers streamflow in Argentinean Andean Mountains and ENSO events, Int. J. Climatol., 18, 1593–1609, 1998.

Comrie, A. C. and Glenn, E. C.: Principal components-based regionalization of precipitation regimes across the southwest United States and northern Mexico, with an application to monsoon precipitation variability, Clim. Res., 10, 201–215, 1998.

Dracup, J. A. and Kahya, E.: The relationships between U.S. streamflow and La Niña events, Water Resour. Res., 30, 2133–2141, 1994.

Edwards, D. C. and McKee, T. B.: Characteristics of 20th century drought in the United States at multiple time scales, Atmospheric Science Paper No. 634, Climatology Report No. 97-2, Department of Atmospheric Sciences, Colorado State University, Fort Collins, CO, 1997.

Efron, B. and Tibshirani, R. J.: An Introduction to the Bootstrap, Chapman & Hall, International Thomson Publication, New York, USA, 1993.

González, M. H. and Vera, C. S.: On the interannual winter rainfall variability in Southern Andes, Int. J. Climatol., 30, 643–657, 2010.

Grimm, A. M. and Ambrizzi, T.: Teleconnections into South America from the Tropics and Extratropics on Interannual to Intraseasonal Timescales, in: Past Climate Variability in South America and Surrounding Regions, edited by: Vimeux, F., Dev. Paleoenviron. Res., 14, 159–191, doi:10.1007/978-90-2672-9_7, Springer, 2009.

Grimm, A. M., Barros, V. R., and Doyle, M. E.: Climate variability in Southern South America associated with El Niño and La Niña events, J. Climate, 13, 35–58, 2000.

Guttman, N. B.: Accepting the standardized precipitation index: A calculation algorithm, J. Am. Water Resour. As., 35, 311–322, 1999.

Hayes, M. J., Svoboda, M. D., Wilhite, D. A., and Vanyarkho, O. V.: Monitoring the 1996 drought using the standardized precipitation index, B. Am. Meteorol. Soc., 80, 429–438, 1999.

Hayes, M., Svoboda, M., Wall, N., and Widhalm, M.: The Lincoln Declaration on Drought Indices: Universal meteorological drought index recommended, B. Am. Meteorol. Soc., 92, 485–488, 2011.

Hisdal, H. and Tallaksen, L. M.: Estimation of regional meteorological and hydrological drought characteristics: a case study for Denmark, J. Hydrol., 281, 230–247, 2003.

Kaiser, H. F.: The Varimax criterion for analytic rotation in factor analysis, Psychometrika, 23, 187–200, 1958.

Karabork, M. C. and Kahya, E.: The teleconnections between the extreme phases of the Southern Oscillation and precipitation patterns over Turkey, Int. J. Climatol., 23, 1607–1625, 2003.

Kirtman, B., Anderson, D., Brunet, G., Kang, I.-S., Scaife, A. A., and Smith, D.: Prediction from Weeks to Decades, in: Climate Science for Serving Society – Research, Modeling and Prediction Priorities, edited by: Asrar, G. R. and Hurrell, J. W., 205–235, Springer, New York, USA, 2013.

Krepper, C. M. and Zucarelli, V.: Climatology of Water Excess and Shortages in the La Plata Basin, Theor. Appl. Climatol., 102, 13–27, 2012.

Lloyd-Hughes, B. and Saunders, M. A.: A drought climatology for Europe, Int. J. Climatol., 22, 1571–1592, 2002.

Masiokas, M., Villalba, R., Luckman, B., Le Quesne, C., and Aravena, J. C.: Snowpack variations in the Central Andes of Argentina and Chile, 1951–2005: Large-scale atmospheric influences and implications for water resources in the region, J. Climate, 19, 6334–6352, 2006.

McKee, T. B., Doesken, N. J., and Kleist, J.: The relationship of drought frequency and duration to time scales, in: Proceedings of the Eight Conference on Applied Climatology, Anaheim, CA, American Meteorological Society, 179–184, 1993.

Minetti, J. L., Vargas, W. M., Poblete, A. G., Acuña, L. R., and Casagrande, G.: Non-linear trends and low frequency oscillations in annual precipitation over Argentina and Chile, 1931–1999, Atmósfera, 16, 119–135, 2003.

Moreira, E. E., Coelho, C. A., Paulo, A. A., Pereira, L. S., and Mexia, J. T.: SPI-based drought category prediction using log-linear models, J. Hydrol., 354, 116–130, 2008.

Mudelsee, M.: The Bootstrap in Climate Risk Analysis, in: Extremis – Disruptive Events and Trends in Climate and Hydrology, edited by: Kropp, J. and Schellnhuber, H.-J., 44–58, ISBN: 978-3-642-14863-7, 2011.

Paruelo, J. M., Beltrán, A., Jobbágy, E., Sala, O. E., and Golluscio, R. A.: The climate of Patagonia: general patterns and controls on biotic processes, Ecología Austral., 8, 85–101, 1998.

Pasquini, A. I., Lecomte, K. L., Piovano, E. L., and Depetris, P. J.: Recent rainfall and runoff variability in central Argentina, Quat. Int., 158, 127–139, 2006.

Penalba, O. C. and Rivera, J. A.: Uso de la distribución de probabilidades gamma para la representación de la precipitación mensual en el Sudeste de Sudamérica, Cambios espacio-temporales en sus parámetros, in: Proceedings of the XI CONGREMET, Mendoza, Argentina, 28 May–1 June, 2012.

Penalba, O. C. and Rivera, J. A.: Comparación de seis índices para el monitoreo de sequías meteorológicas en el sur de Sudamérica, Meteorológica, 40, 33–57, 2015.

Penalba, O. C. and Rivera, J. A.: Regional aspects of future precipitation and meteorological drought characteristics over Southern South America projected by a CMIP5 multi-model ensemble, Int. J. Climatol., 36, 974–986, 2016.

Penalba, O. C., Beltran, A., and Messina, C.: Monthly rainfall in central-eastern Argentina and ENSO: a comparative study of rainfall forecast methodologies, Rev. Bras. Agrometeorologia, 13, 49–61, 2005.

Penalba, O. C., Pántano, V. C., Spescha, L. B., and Murphy, G. M.: ENSO impact on dry sequences during different phenological periods in the east-northeast of Argentina, III International Conference on ENSO: Bridging the gaps between global ENSO science and regional processes, extremes and impacts, Guayaquil, Ecuador, 12–14 November, 2014a.

Penalba, O. C., Rivera, J. A., and Pántano, V. C.: The CLARIS LPB database: constructing a long-term daily hydro-meteorological dataset for La Plata Basin, Southern South America, Geosci. Data J., 1, 20–29, 2014b.

Penalba O. C., Rivera J. A., Pántano, V. C., and Bettolli, M. L.: Extreme rainfall and hydric condition in Southern La Plata Basin and the associated atmospheric circulation, Clim. Res., doi:10.3354/cr01353, online first, 2016.

Penland, C., Sun, D.-Z., Capotondi, A., and Vimont, D. J.: A brief introduction of El Niño and La Niña, in: Climate dynamics: Why does climate vary?, edited by: Sun, D.-Z. and Bryan, F., Geophys. Monogr. Ser., 189, 53–64, American Geophysical Union, Washington, USA, 2010.

Piechota, T. C. and Dracup, J. A.: Drought and regional hydrologic variation in the United States: Associations with the El Niño-Southern Oscillation, Water Resour. Res., 32, 1359–1373, 1996.

Phillips, I. D. and Denning, H.: Winter daily precipitation variability over the South West Peninsula of England, Theor. Appl. Climatol., 87, 103–122, 2007.

Podestá, G. P., Messina, C. D., Grondona, M. O., and Magrin, G. O.: Associations between grain crop yields in central-eastern Argentina and El Niño-Southern Oscillation, J. Appl. Meteorol., 38, 1488–1498, 1999.

Prieto, M. R., Herrera, R., and Dussel, P.: Historical Evidences of the Mendoza River Streamflow Fluctuations and their Relationship with ENSO, Holocene, 9, 473–471, 1999.

Richman, M.: Rotation of Principal Components, J. Climatol., 6, 293–335, 1986.

Rivera, J. A. and Penalba, O. C.: Trends and spatial patterns of drought affected area in Southern South America, Climate, 2, 264–278, 2014.

Rivera, J. A., Araneo, D. C., and Penalba, O. C.: Evaluación del período de crisis hídrica 2010–2014 en la región de Cuyo, XII Congreso Argentino de Meteorología, Mar del Plata, Argentina, 26–29 May, 2015.

Ropelewski, C. F. and Halpert, M. S.: North American precipitation and temperature patterns associated with the El Niño/Southern Oscillation (ENSO), Mon. Weather Rev., 114, 2352–2362, 1986.

Ropelewski, C. F. and Halpert, M. S.: Global and regional scale precipitation patterns associated with the El Niño/Southern Oscillation, Mon. Weather Rev., 115, 1606–1626, 1987.

Russián, G. F., Agosta, E. A., and Compagnucci, R. H.: Variabilidad interanual a interdecádica de la precipitación en Patagonia Norte, Geoacta, 35, 27–43, 2010.

Schonher, T. and Nicholson, S. E.: The relationship between California rainfall and ENSO events, J. Climate, 2, 1258–1269, 1989.

Seiler, R. A., Hayes, M., and Bressan, L.: Using the standardized precipitation index for flood risk monitoring, Int. J. Climatol., 22, 1365–1376, 2002.

Sienz, F., Bordi, I., Fraedrich, K., and Schneidereit, A.: Extreme dry and wet events in Iceland: observations, simulations and scenarios, Meteorol. Z., 16, 9–16, 2007.

Silvestri, G. E. and Vera, C. S.: Antarctic Oscillation signal on precipitation anomalies over southeastern South America, Geophys. Res. Lett., 30, 2115, doi:10.1029/2003GL018277, 2003.

Smith, T. M., Reynolds, R. W., Peterson, T. C., and Lawrimore, J.: Improvements to NOAA's Historical Merged Land-Ocean Surface Temperature Analysis (1880–2006), J. Climate, 21, 2283–2296, 2008.

Spescha, L., Forte Lay, J., Scarpati, O., and Hurtado, R.: Los excesos de agua edáfica y su relación con el ENSO en la región Pampeana, Rev. Facultad de Agronomía, 24, 161–167, 2004.

Taschetto, A. S. and Wainer, I.: The impact of the subtropical South Atlantic SST on South American precipitation, Ann. Geophys., 26, 3457–3476, doi:10.5194/angeo-26-3457-2008, 2008.

Trenberth, K. E.: Signal versus Noise in the Southern Oscillation, Mon. Weather Rev., 112, 326–332, 1984.

Trenberth, K. E.: The definition of El Niño, B. Am. Meteorol. Soc., 78, 2771–2777, 1997.

Trenberth, K. E. and Stepaniak, D. P.: Indices of El Niño evolution, J. Climate, 14, 1697–1701, 2001.

Vicente-Serrano, S. M.: El Niño and La Niña influence on droughts at different timescales in the Iberian Peninsula, Water Resour. Res., 41, W12415, doi:10.1029/2004WR003908, 2005.

Vicente-Serrano, S. M., López-Moreno, J. I., Gimeno, L., Nieto, R., Morán-Tejada, E., Lorenzo-Lacruz, J., Beguería, S., and Azorin-Molina, C.: A multiscalar global evaluation of the impact of ENSO on droughts, J. Geophys. Res., 116, D20109, doi:10.1029/2011JD016039, 2011.

Wolter, K. and Timlin, M. S.: Monitoring ENSO in COADS with a seasonally adjusted principal component index, Proceedings of the 17th Climate Diagnostics Workshop, Norman, Oklahoma, NOAA/NMC/CAC, NSSL, Oklahoma Clim. Survey, CIMMS and the School of Meteor., Univ. of Oklahoma, 52–57, 1993.

26

Interannual variability of the midsummer drought in Central America and the connection with sea surface temperatures

Tito Maldonado[1,2,4]**, Anna Rutgersson**[2]**, Eric Alfaro**[3,4,5]**, Jorge Amador**[3,4]**, and Björn Claremar**[2]

[1]Centre for Natural Disaster Science, Uppsala University, Villav. 16, 752 36, Uppsala, Sweden
[2]Department of Earth Sciences, Uppsala University, Villav. 16 752 36, Uppsala, Sweden
[3]School of Physics, University of Costa Rica, San Pedro de Montes de Oca, 11501-2060 San Jose, Costa Rica
[4]Center for Geophysical Research, University of Costa Rica, San Pedro de Montes de Oca, 11501-2060 San Jose, Costa Rica
[5]Centre for Research in Marine Sciences and Limnology, University of Costa Rica, San Pedro de Montes de Oca, 11501-2060 San Jose, Costa Rica

Correspondence to: Tito Maldonado (tito.maldonado@geo.uu.se)

Abstract. The midsummer drought (MSD) in Central America is characterised in order to create annual indexes representing the timing of its phases (start, minimum and end), and other features relevant for MSD forecasting such as the intensity and the magnitude. The MSD intensity is defined as the minimum rainfall detected during the MSD, meanwhile the magnitude is the total precipitation divided by the total days between the start and end of the MSD. It is shown that the MSD extends along the Pacific coast, however, a similar MSD structure was detected also in two stations in the Caribbean side of Central America, located in Nicaragua. The MSD intensity and magnitude show a negative relationship with Niño 3.4 and a positive relationship with the Caribbean low-level jet (CLLJ) index, however for the Caribbean stations the results were not statistically significant, which is indicating that other processes might be modulating the precipitation during the MSD over the Caribbean coast. On the other hand, the temporal variables (start, minimum and end) show low and no significant correlations with the same indexes.

The results from canonical correlation analysis (CCA) show good performance to study the MSD intensity and magnitude, however, for the temporal indexes the performance is not satisfactory due to the low skill to predict the MSD phases. Moreover, we find that CCA shows potential predictability of the MSD intensity and magnitude using sea surface temperatures (SST) with leading times of up to 3 months. Using CCA as diagnostic tool it is found that during June, an SST dipole pattern upon the neighbouring waters to

Central America is the main variability mode controlling the inter-annual variability of the MSD features. However, there is also evidence that the regional waters are playing an important role in the annual modulation of the MSD features. The waters in the PDO vicinity might be also controlling the rainfall during the MSD, however, exerting an opposite effect at the north and south regions of Central America.

1 Introduction

The geographical features of Central America imprint the characteristics of the regional climate and weather at the isthmus. The region is conformed by a large and high mountain system surrounded by the Pacific and Atlantic oceans, which induces the maritime climate conditions governing in the region (Taylor and Alfaro, 2005). The interaction between the easterly winds (trades), which in turns are the dominant wind regime, and topography divide the region into two different climate areas: the Pacific located leeward of the main wind and the Caribbean presenting windward conditions.

The annual rainfall cycle for the entire region has already been well documented for Central America (Alfaro, 2002), and for the Eastern Tropical Pacific (ETP, Magaña et al., 1999; Amador et al., 2006). In the Pacific region the annual precipitation cycle exhibits a bimodal behaviour (Magaña et al., 1999; Taylor and Alfaro, 2005; Amador et al., 2006). The first precipitation maximum occurs during May–June when the nearby ocean waters have warmed to around

29 °C over the eastern Pacific warm pool, and the easterly winds have decreased, enhancing convection in the region. During July-August a relative precipitation minimum is observed, along with a decrease in the sea surface temperatures (SSTs) over the eastern Pacific warm pool, and an increase of the easterly winds. This reduction in the rainfall is known as the midsummer drought (MSD, Magaña et al., 1999; Amador et al., 2006). The second precipitation maximum occurs during August–September–October (ASO) and is accompanied by a reduction of the trades and a relative increase of the SSTs up to 28.5 °C over the eastern Pacific warm pool. It is also during this quarter of the year that the hurricane season is more active in the Tropical North Atlantic (Amador et al., 2006).

On the other side, upon the Caribbean coast of Central America the annual precipitation cycle contrasts the one in the Pacific mainly during the boreal winter months, when wetter and more humid conditions are found over the eastern coast of Central America (Taylor and Alfaro, 2005; Amador et al., 2006). A similar bimodal cycle, however, has been extensively reported over the Greater Antilles (Chen and Taylor, 2002; Taylor et al., 2002; Spence et al., 2004; Ashby et al., 2005). The reduction in rainfall experienced during the summer months over the western Caribbean is argued to have a similar pattern to that in the Pacific (Martin and Schumacher, 2011). Little is known about the origins of the MSD in both, the Pacific and Caribbean. The processes involved in such rainfall decrease are most probably different to those operating in the western Central America and Mexico, however, both MSDs appear to share the Caribbean low-level jet (CLLJ) as main process increasing the moisture flux in the Caribbean, which suppresses convection and decreases rainfall (Magaña et al., 1999; Wang, 2007; Muñoz et al., 2008; Amador, 2008; Whyte et al., 2008). In this study, we focus on the variability of the MSD in the Pacific, nevertheless, we also examine two stations in the Caribbean side, to contrast the features present in both MSDs.

The development of the MSD has been explained in terms of the interaction of changes in the divergent (convergent) low-level winds over the warm waters nearby Mexico and Central America (Magaña et al., 1999). Recent studies such as Karnauskas et al. (2013) and Herrera et al. (2015) have proposed different hypothesis about the origin of the MSD. Karnauskas et al. (2013) argue that the MSD originates as a response to one single precipitation-enhancing mechanism occurring twice, contrary to a suppressing mechanism. The latitudinal dependence in the two peaks of precipitation surge as response to the biannual crossing of the solar declination, which leads to the two peaks in convective instability and hence rainfall. This hypothesis, nevertheless, does not explain the almost simultaneous occurrence of the MSD in both southern Mexico and Central America (Magaña et al., 1999). On the other hand, Herrera et al. (2015) describe the origins of the MSD studying the air-sea interaction between the CLLJ (Amador, 1998, 2008; Muñoz et al., 2008; Maldon-

ado et al., 2016) and the ocean waters in the neighbourhood of the Pacific coast. Their findings show that the CLLJ peaks in July, provokes a cooling of the SSTs over the Pacific and a westward shift of low-level moisture convergence that combined with subsidence produce the MSD.

The MSD is important for the region since it develops mostly every year (see Fig. 4 in Magaña et al., 1999) in areas with high industrial and agricultural activities, which in turn, are highly populated (Alfaro, 2014). Most of the population lives in the Central American Pacific slope. Although the MSD is present almost every year, it shows high variability in space and time (Amador, 2008; Alfaro, 2014), and the understanding of this variability is still a scientific challenge (Amador et al., 2006; Amador, 2008; Herrera et al., 2015). The beginning, and duration are uncertainty factors for different socio-economic sectors such as the agricultural and potable water supply. On the other hand, one can expect that extreme MSD events impact negatively reducing the water resources in zones with such characteristics (Hernández and Fernández, 2015; Alfaro, 2014; Solano, 2015).

A better understanding of the variability of the MSD would then contribute to the knowledge about the weather and climate dynamics in the region, also to the information given to the economic and social sectors. Water resources, agriculture, and hydro-electrical power production are benefited in terms of obtaining better data for planning, mitigation and prevention to severe dry conditions, e.g. the droughts reported in the North Pacific region of Costa Rica during July 2014 (Chinchilla-Ramírez, 2014).

Previous studies have analysed the influence of the SST anomalies in the precipitation regime in Central America. Enfield and Alfaro (1999) have shown that the magnitude of the precipitation during the rainy season (from May to October) in Central America is highly modulated by the SST anomalies at both oceans. Alfaro (2007) and Fallas-López and Alfaro (2012b) studied the predictability of the rainfall using canonical correlation analysis (CCA) with SST as predictor. Both studies found that the SST modify the response on precipitation in different ways during each maximum, the first rainfall peak being modulated mainly by the SST over the tropical Atlantic, and the secondary maximum being controlled by a dipole in the SST anomalies over the neighbour oceans to the Central American coast. Moreover, Fallas-López and Alfaro (2012a) found that the MSD intensity is dependent on several global climate indexes such as the Atlantic Multi-Decadal Oscillation (AMO), Niño 3 (N3), and Pacific Decadal Oscillation (PDO). Maldonado and Alfaro (2011) and Maldonado et al. (2013) utilised CCA to explore the distribution of precipitation extreme events during the second precipitation maximum finding that during the second rainfall maximum, the total precipitation and temporal distribution of precipitation events are controlled by a dipole in the SSTs between the Pacific and the Caribbean-North Atlantic waters. Given the above evidence, it would be expected that SST is involved in the MSD variability at inter-annual scales.

In fact, Alfaro (2014) found that warm (cool) conditions of Niño 3.4 tend to be associated to drier (wetter) MSD events in some regions of the North Pacific and Central Valley of Costa Rica.

Other climate processes occur simultaneously with the MSD, and could also alter the precipitation during June–August (JJA) e.g the intensification of the North Atlantic Sub-tropical High (NASH, Wang, 2007), the Western Hemisphere Warm Pools (WHWM, Wang and Enfield, 2001, 2003), the North American Monsoon (Vera et al., 2006) and the CLLJ (Amador, 1998, 2008). Recently, Hidalgo et al. (2015) also found that the latitudinal position of the precipitation centre of mass is correlated with the intensity of the CLLJ, i.e a weaker (stronger) jet is associated to the northward (southward) shift of the latitudinal precipitation centre of mass, resulting in rainfall above normal during JJA that might be also associated with variations of the MSD events.

With that said, the targets of this study are first to contribute to the understanding of the climate in the region, and second to provide a systematic method for MSD forecasting. This study consists of an analysis of the connection between the features of the MSD observed in the Pacific and Caribbean coast of Central America and the anomalies in the SST of the regional waters by means of a combination of principal component analysis (PCA) and canonical correlation analysis (CCA). The precipitation observations, however, are first examined in order to determine the data quality and to characterised the MSD per station. The MSD is described in terms of start, timing, end, duration, depth (intensity) and total precipitation (magnitude). This representation is also used to produce annual indexes of those variables. Previous reports have already characterised the existence of the MSD in Costa Rica (Ramírez, 1983; Hernández and Fernández, 2015; Alfaro, 2014; Solano, 2015), nonetheless, in this study the portray is expanded to include the entire the region.

This manuscript is organised as follows: in Sect. 2 is described all the information relevant to the databases used in this study. Section 3 is a explanation of the methodology applied in this analysis. In Sect. 4 the result and Sect. 5 the discussion and concluding remarks are presented.

2 Data

A total of 25 gauge stations with daily observations of precipitation provided by the meteorological services in Central America are used. Their location is shown in Fig. 1, and the information about the coordinates of each station is given in Table 1. The average annual cycle of precipitation for the selected period in every station is shown in Fig. 2. The MSD signal is noticeable in all the stations, around julian day 200 (July–August), except for Bluefields and Puerto Cabezas stations (in the Caribbean), where the rainfall reduction is observed near the julian day 300 (August–September). Note that the majority (23) of the stations are situated along the Pacific coast with only two stations located along the Caribbean

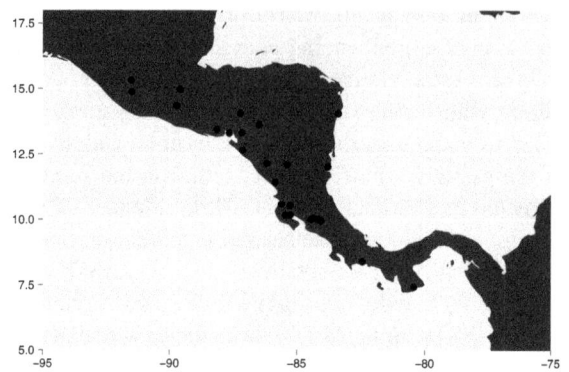

Figure 1. Spatial distribution of the gauge stations in Central America.

coast. The two Caribbean stations possess relevant features to describe the precipitation behaviour in the Caribbean slope, such as their location at the CLLJ exit and their contrasting annual precipitation cycle to that in the Pacific. Thus, in order to achieve a better representation of the complete region, we consider the Caribbean stations for the analysis. Since each meteorological station has a distinct time coverage, a common time series length is determined according to the percentage of missing data, and in this case, we pick a period in which stations do not surpass the 40 % of missing data. Therefore, the selected time series length covers from 1961 to 2012, ensuring that each station posses at least 60 % of the data. The gaps in the time series are filled using the methodology described in Alfaro and Soley (2009), which uses autoregressive models of order 1. This method can filter persistent signals comparable to the length of the filter, and the estimated values are consistent with the statistical properties of the time series without external superposition of the data.

The extended reconstructed sea surface temperatures (ERSSTv3b, Xue et al., 2003; Smith et al., 2007) are used in this study. The SST anomalies are constructed using a combination of observed data along with models and historical sampling grids. This global database has a horizontal resolution of 2.5° by 2.5°. The domain bounded by -22 to $63°$ N and 111 to $15°$ E is considered in order to capture the signal of the most important climate variability modes for the Central American isthmus such as El Niño Southern Oscillation (ENSO), the Pacific Decadal Oscillation (PDO), the Atlantic Multi decadal Oscillation (AMO), the North Atlantic Oscillation (NAO), and the Tropical North Atlantic (TNA), which in turns, have shown to be relevant in terms of variability of rainfall during the season JJA (Fallas-López and Alfaro, 2012a, b). The SST anomalies are used as predictors in the CCA models.

The Niño 3.4 index (Trenberth, 1997) provided by the International Research Institute for Climate and Society (IRI, 2015) is used to estimate the relationship between this index

Table 1. Geographical position of the gauge stations in degrees north for latitude and degrees west for longitude.

Station	LAT	LON	Country	Region	Num. of years without MSD
La Argentina	10.03	84.35	Costa Rica	Pacific	3
Fabio Baudrit	10.00	84.43	Costa Rica	Pacific	2
Juan Santa Maria	10.00	84.17	Costa Rica	Pacific	3
Liberia	10.58	85.58	Costa Rica	Pacific	9
Nicoya	10.15	85.45	Costa Rica	Pacific	5
Santa Cruz	10.17	85.25	Costa Rica	Pacific	8
Bagaces	10.53	85.25	Costa Rica	Pacific	18
CIGEFI	9.94	84.04	Costa Rica	Pacific	21
Bluefields	12.02	83.78	Nicaragua	Caribbean	19
Ocotal	13.62	86.47	Nicaragua	Pacific	6
Chinandega	12.63	87.13	Nicaragua	Pacific	9
Juigalpa	12.10	85.37	Nicaragua	Pacific	7
Managua	12.14	86.16	Nicaragua	Pacific	5
Puerto Cabezas	14.05	83.38	Nicaragua	Caribbean	9
Rivas	11.44	85.83	Panama	Pacific	13
David	8.40	82.43	Panama	Pacific	8
Los Santos	7.42	80.42	Panama	Pacific	5
San Miguel	13.43	88.15	El Salvador	Pacific	3
Asuncion Mita	14.33	89.71	Guatemala	Pacific	11
Huehuetenango	15.32	91.50	Guatemala	Pacific	4
LaborOvalle	14.87	91.48	Guatemala	Pacific	12
La Fragua	14.96	89.58	Guatemala	Pacific	13
Amapala	13.29	87.65	Honduras	Pacific	18
Choluteca	13.32	87.15	Honduras	Pacific	9
Tegucigalpa	14.06	87.22	Honduras	Pacific	9

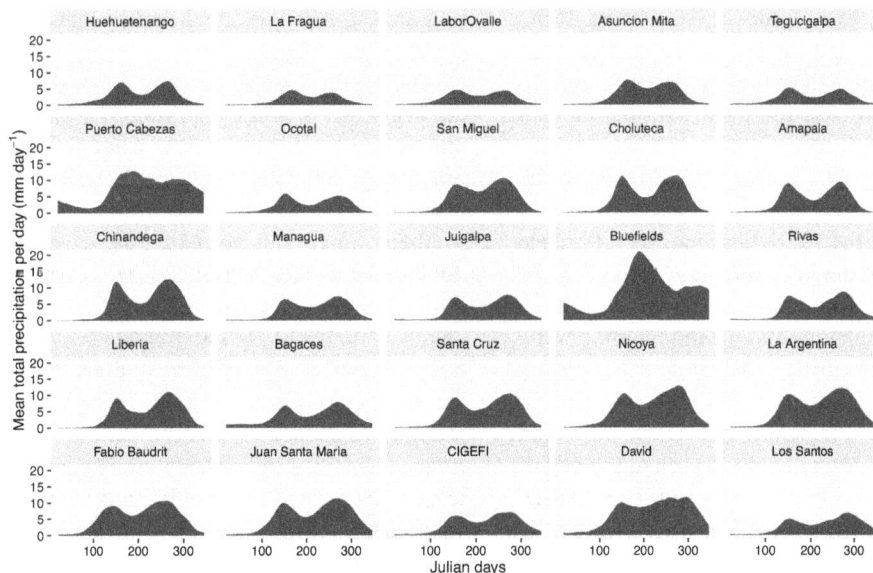

Figure 2. Annual cycle of precipitation in every station for the period 1961–2012.

and relevant features of the MSD by means of empirical contingency tables for conditional data.

Horizontal wind data at 925 hPa is provided by the NCEP/NCAR reanalysis (Kalnay et al., 1996; Kistler et al., 2001) which has a horizontal resolution of 2.5° by 2.5°. The wind data is used to calculate the CLLJ index as in Amador (2008) and Maldonado et al. (2016).

3 Methods

Based on the methodology developed in Alfaro (2014), and also applied by Solano (2015), the daily precipitation times series are filtered using a running triangular weight average with a window of 31 days, to avoid or minimise interruptions of the MSD due to weakening of the trades and/or approaching of the ITCZ, as suggested by Ramírez (1983) and Alfaro (2014). An algorithm to systematically identify the features of the MSD is applied to the filtered daily precipitation time series. Figure 3 shows the schematics of this approach. This algorithm seeks for the timing of each phase of the MSD (start, minimum, and end), besides the intensity and magnitude of the MSD. Notice that this algorithm provides an annual value of each quantity, which are used as indexes to characterise the MSD later in this study. The start of the MSD is considered as the moment when the decrease in precipitation initiates, usually after May–June. The end of the MSD is when the precipitation stops increasing, normally taking place around September–October. The minimum occurs in between the start and end of the MSD. The MSD intensity is defined as the minimum rainfall detected during the MSD (or the depth of the valley in Fig. 3), meanwhile the magnitude is the total precipitation divided by the total number of days between the start and end of the MSD. The description of the whole MSD detecting process is explained as follows: first, the algorithm scans for the precipitation minimum in the filtered time series within a reasonable range of days for the existence of the MSD, determined by the climatologies of each station, i.e. typically from 1 June to 30 August. If the precipitation minimum falls outside this period, it is not considered a MSD event. Second, the nearest inflection points to the minimum are sought to determine the start and the end of the MSD. The shortest distance allowed between the inflection points and the minimum is 5 days. Consequently, the shortest MSD would have a duration of 10 days. Moreover, the inflection points have the restriction that the difference of precipitation between them and the minimum should be at least 20 % per day. It is worth mentioning, that not all the years present the inflection points and the minimum. For the scope of this study, the years missing any of the MSD phases are removed. The causes for those fails in the detection of the MSD phases are several, however, they can be attributed to anomalous years with a extended dry season at the beginning of the year, combined with an earlier and severe MSD, or anomalous drier condition during the secondary precipitation maximum, years with no MSD at all, and in most cases due to missing data in the observations. In Table 1 the number of years without MSD is shown per station. A statistical summary of the features of the MSD is shown in Fig. 4.

The contingency tables method seeks for a significant predictive relationship between two variables, being one independent and the other dependent (Alfaro, 2007; Fallas-López and Alfaro, 2012a). The time series are divided into three categories: below normal, normal and above normal (BN, N,

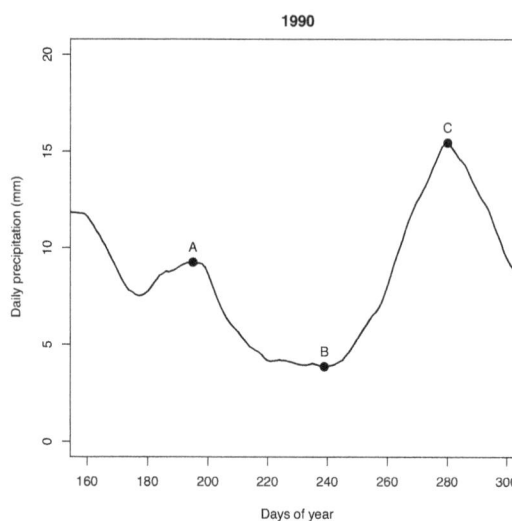

Figure 3. Filtered daily time series for the observed precipitation in station La Argentina during 1990. The dots represent the beginning (A), minimum (B) and end (C) of the MSD. In this year the MSD started by day 195 (14 July), reached the minimum at day 239 (27 August) and ended by day 280 (7 October).

AN, respectively) using the percentiles 33rd and 67th. Fallas-López and Alfaro (2012a) have shown that using these categories provides a good measure of the possible states of El Niño and the MSD indexes, besides, the information obtained is easy to use for the public. Thus, the contingency tables are estimated for conditional events to study the dependence of the MSD features (intensity and magnitude) with ENSO events (independent variable). As mentioned above, the years in which the algorithm does not detect any of the phases or no MSD at all, are dismissed. The statistical significance is estimated by means of bootstrapping with no replace (Gershunov and Barnett, 1998). With bootstrapping, a synthetic time series normally distributed is generated from random noise, then, we estimated contingency tables for the artificial time series. This process is repeated several times to obtain a number of random frequencies normally distributed, which in this case is 1000 repetitions, thus, we obtain 1000 contingency tables per station. If the value of the observed absolute frequency is greater than the 90th of the synthetic frequencies, then the observation is significant at 0.1 significance level. The same can be said for the 0.05 and 0.01 of significance level.

As a tool to investigate the variability of the MSD, we use Canonical correlation analysis (CCA, Wilks, 2011) which is a statistical technique that searches for pairs of patterns in two multivariate data sets (fields), and constructs sets of transformed data variables by projecting the original data onto those patterns. These new variables maximise the interrelationships between the two fields. The new variables can be used, analogously to the regression coefficient in the multiple regression. The new variables or vector weights are

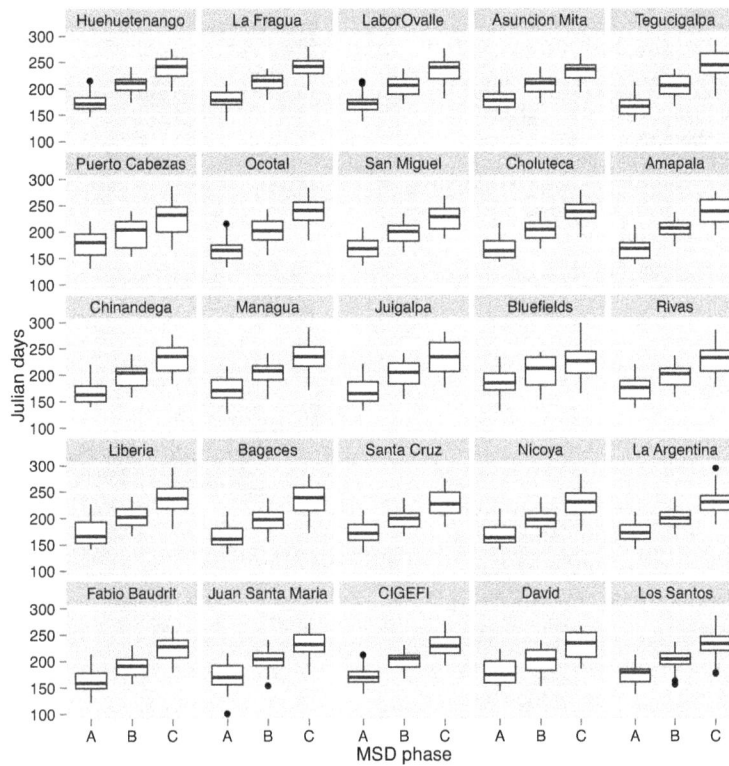

Figure 4. Box-plot of the indexes describing the MSD timing. In the x axis are each of the events defining the MSD, start (A), minimum (B), and end (C). In the y axis are the time-coordinates in Julian days. The upper and lower limits of each box correspond to the first and third quartiles (the 25th and 75th percentiles). The upper (lower) whisker extends from the hinge to the highest (lowest) value that is within 1.5 times inter-quartile range of the hinge. Data beyond the end of the whiskers are outliers and plotted as dots. The median is the bold line in each box. Note that the panels are distributed from north to south, left to right and top to bottom.

also known as canonical vectors, and their projection with their respective fields as canonical variates. CCA can be useful in different applications, such as: (i) to obtain diagnostic aspects of the coupled variability of two fields, in the case when the time series of observations of the two fields are simultaneous, or (ii) to perform statistical forecasts, in the case when the time series of observations of one field precede the other (e.g. Alfaro, 2007; Maldonado and Alfaro, 2011; Fallas-López and Alfaro, 2012b; Maldonado et al., 2013).

In this study CCA is used as a diagnostic tool, however, models for statistical forecasting can be built for future use. Consequently, CCA is employed in order to analyse the relationship between the SSTs monthly anomalies (SSTA, also denoted field X) with each of the indexes (start, minimum, end, intensity and magnitude of MSD) estimated to represent each of the features of the MSD (each index would be the field Y, for the corresponding model), that is, we seek for a statistical relationship between the large and regional scale features of the SSTA and the characteristics of the MSD in each of the stations (local scale). The above CCA methodology is based on Maldonado et al. (2013) and it is implemented as follows: the fields (SSTA, MSD indexes) are first reduced by means of principal component analy-

sis (PCA) to assure stability in the CCA parameters. A maximum of 17 EOFs and CCA modes in the filtering stage are allowed. This threshold was suggested by Gershunov and Cayan (2003) and Alfaro (2007) to avoid overparameterisation. The optimal combination of EOFs and CCA modes are calculated by means of the goodness index (R^2). Notice that any set EOFs will produce unique CCA modes for that specific set, then, once the best combination of EOFs is determined, that set EOF is capturing the maximum variability in each field (X and Y), separately, for each specific CCA model (intensity and magnitude). The maximum possible number of CCA modes, however, is determined first by the minimum number of EOFs between both fields. Then with the goodness skill, the maximum number of CCA modes is found for the best fit to avoid any overparameterisation in the model $\hat{Y} = b^T \cdot X$, where the elements of b are the ordinary least-squares regression coefficients computed with CCA, and $\hat{Y}$ is the predicted value of Y. The R^2 is computed using cross-validation models with 1-month window from 1961–2012 for each station in all the models. This metric also allows identifying the best month to predict any of the MSD features. It is worth mentioning that at the end the models would not necessarily have 17 EOF and CCA modes.

Stations having indexes with more than 30 % of missing data (i.e. fails detecting MSD events) are discarded. In total, 21 stations are left for the CCA models. Precipitation data are transformed to a normal distribution using percentiles, to achieve better performance in the EOF. The 21 stations having less than 30 % of missing data are filled using the long-term means of each index. Note that filling the data in this step is not the same procedure described above.

4　Results

4.1　General features of the MSD

Figure 4 shows a statistical summary of the timing of the MSD phases. Each phase exhibits subtle differences throughout the stations. Those discrepancies suggest to study this phenomenon separately per station. For example, from Fig. 4, one can see that in the southern part of Central America (Panama, Costa Rica, Nicaragua) the MSD tends to start, develop and end earlier than in the northern part (El Salvador, Guatemala, Honduras), in agreement with previous studies such as Alfaro (2002) and Herrera et al. (2015). In the southern stations, typically, the start of the MSD is detected by 20 June, the minimum is reached by 19 July and the end by 20 August, meanwhile at the north the start is observed by 22 June, the minimum by 24 July and the end around 24 August. Note that the difference can be up to 20 days in each phase from south to north if the stations are analysed separately. The results in Fig. 4 also show the high temporal and spatial variability existent throughout the region agreement with Alfaro (2002) and Amador (2008), but that feature is also observed at local scales as noted by Ramírez (1983); Hernández and Fernández (2015); Alfaro (2014), and Solano (2015).

The algorithm captures similar MSD structure in stations located over the eastern Nicaragua (Bluefields and Puerto Cabezas) at the Caribbean region. In addition, the stations located at the Caribbean coast present later development of the MSD, starting in average by 29 June, the minimum in 21 July, and the end at 14 August, therefore, being later and shorter in the Caribbean than in the Pacific side of Central America, showing a marked difference between the eastern and western Central America. This MSD structure over the Caribbean has been previously studied by Martin and Schumacher (2011), which concluded that the MSD in the Caribbean is controlled by different external forcing than the one operating in the Pacific sector.

This observed latitudinal difference in the timing in each MSD phase is examined using a t test (not shown). The results, nevertheless, show no statistical difference among the stations at 90 % of significance level, except for the start index in Bluefields station, indicating that in the Caribbean slope the MSD is out-of-phase at least at the start. Thus, it can be argued that the MSD over the Pacific occur simultaneously as found by Magaña et al. (1999).

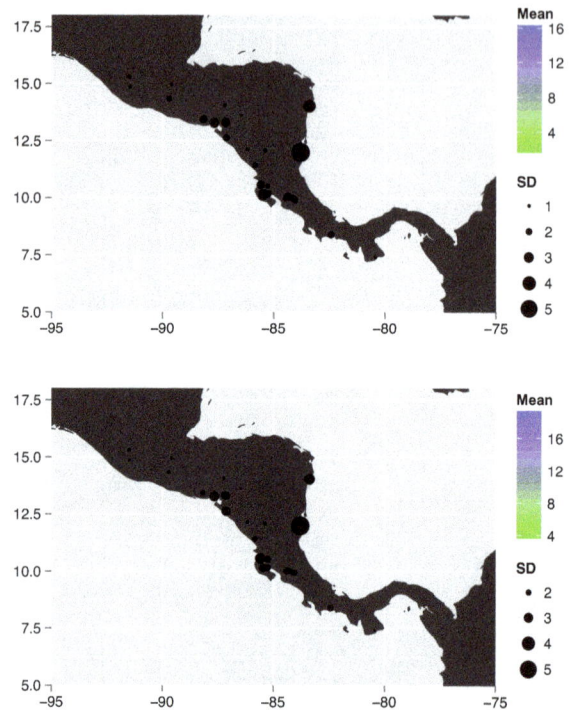

Figure 5. Mean and standard deviation for the MSD intensity (mm, top) and magnitude (mm day^{-1}, bottom) for the period 1961–2012. Note the color scale is not the same in both figures, since the intensity and magnitude are not necessarily of the same order of magnitude. The dots size represents the standard deviation.

The intensity and magnitude of the MSD also show high spatial variability (Fig. 5), with a strong contrast between the Pacific and the Caribbean coasts. The Pacific side presents drier MSDs than in the Caribbean basin. In the western Central America, the mean intensity is 4.06 mm ($\sigma = 1.85$ mm) and mean magnitude of 6.50 mm day^{-1} ($\sigma = 2.00$ mm day^{-1}), with the highest score detected in David, Panama with an intensity of 7.70 mm ($\sigma = 2.01$ mm), and a magnitude of 10.45 mm ($\sigma = 2.20$ mm day^{-1}). On the Caribbean side, Bluefields and Puerto Cabezas show the highest records in intensity and magnitude. The mean intensity recorded in Bluefields is 16.64 mm ($\sigma = 5.50$ mm), and Puerto Cabezas 8.81 mm ($\sigma = 3.68$ mm), whereas the mean magnitude scored is 19.93 mm day^{-1} ($\sigma = 5.39$ mm day^{-1}) and 11.31 mm day^{-1} ($\sigma = 3.30$ mm day^{-1}). As above, using t test is determined that the differences in the MSD intensity and magnitude indexes amongst the stations are statistically significant at 90 % of significance level.

Inter-annual variability of the MSD intensity and magnitude is examined in Fig. 6. Correlations with Niño 3.4 index are shown in Fig. 6a, c. Notice that all the stations show a significant negative correlation, except one (San Miguel, El Salvador), nevertheless in both cases the correlation is relatively low (0.24 for intensity and 0.18 for magnitude), hence, this result can be discarded. Tables 2 and 3 show a summary

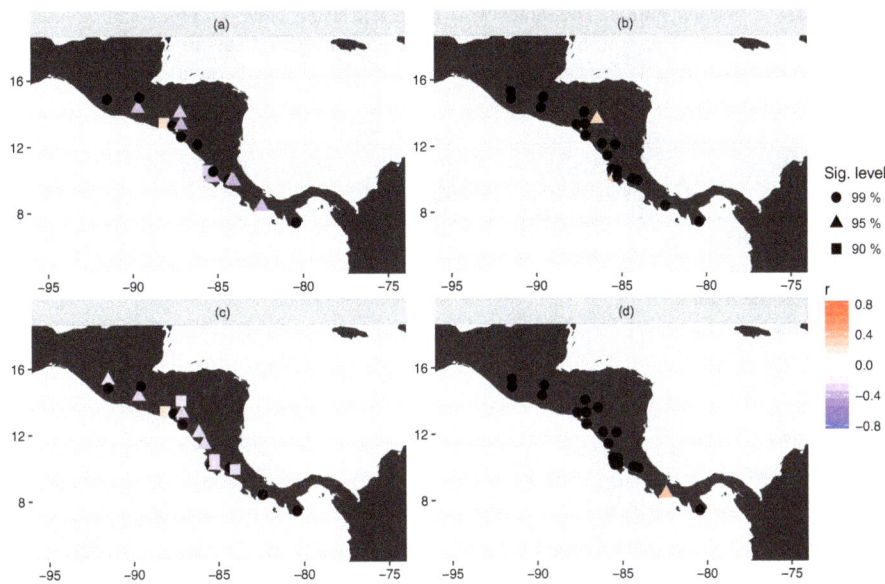

Figure 6. Top Pearson correlation between MSD intensity and (**a**) CLLJ index and (**b**) Niño 3.4 indexes. Bottom the same metric but for the MSD magnitude with (**c**) CLLJ and (**d**) Niño 3.4 indexes. Niño 3.4 index is taken for June. The CLLJ index is estimated from the daily time series, using the definition by Amador (2008) for July.

of the contingency tables calculated. In both cases, the extreme values of the MSD intensity and magnitude present a connection with ENSO events, however, those conditions are not observed in all the stations, despite some of them have a significant correlation with Niño 3.4 (e.g. Liberia and Juigalpa). Notice that the results for the Caribbean stations are not statistically significant. Alfaro (2014) and Solano (2015) have found similar results for stations in Costa Rica; and Fallas-López and Alfaro (2012a) reported the same relationship with other stations in Central America. This association between the SST anomalies in the Niño 3.4 region and the MSD intensity and magnitude over the west cost of Central America can be connected with low-level wind anomalies over the Caribbean Sea. Previous studies have found a relationship between the fluctuation of the CLLJ and precipitation during the summer months (Wang, 2007; Amador, 2008). Figure 6b, d show the correlation of the intensity and magnitude of the MSD with the CLLJ index. Similarly, most of the stations show positive significant correlations, except the stations in the Caribbean. It is known that in the summer months (JJA) during warm (cold) ENSO episodes the CLLJ is stronger (weaker) than normal (Amador, 2008) with a correlation of -0.53 (p value < 0.01). Hidalgo et al. (2015) also noted that during stronger (weaker) CLLJ the ITCZ is located south (north) from the normal position, leading to a reduction (increase) in the total precipitation during boreal summer (JJA), thus, affecting the MSD. Therefore, the bottom line of this result is that the ENSO events modulate the intensity and magnitude of the MSD. It is worth to mention again that from these results, the ENSO events explain only the inter-annual variability of the MSD in the Pacific coast.

The same estimations were done for the other variables (start, minimum, end), however, the results do not shown significant correlations with El Niño and CLLJ index (not shown).

4.2 Canonical correlation analysis

Models based on CCA are tailored for each of the indexes describing the features of the MSD, defined in Sect. 3. Using the SST anomalies as predictor (X field), CCA identifies SST patterns that are related to the perturbations of the MSD features (Y field, predictant). The goodness index (R^2) is shown in Fig. 7. The intensity and magnitude show the best skill score compared to the other indexes. These results show that the models to study the timing of the MSD phases have a poor performance using the CCA technique. Consequently, those models are not considered in the further analysis. Both CCA models for MSD intensity and magnitude show one of the highest values of R^2 in April, which means in operational terms, information concerning the MSD intensity and magnitude can be retrieve up to 3 months in advance of an event. This would be valuable for preparation and planning of the societal, economical and agricultural activities during the MSD period. Notice that both models also shown the highest results in July, concurrently with the existence of the MSD, and the CLLJ.

In this study, we analyse the SSTA patterns identified by the model in June, since for that month, CCA models show a relative high R^2, and there is also a 1-month leading time for forecasting the MSD. Besides, the SST do not change significantly from June to July, when the CCA models have the best performance, thus, the SST in July is expected to be close to

Table 2. Summary of contingency tables for each station. The tables were estimated using Niño 3.4 index and the mean intensity per year. The categories were defined using the 33rd (below normal, BN) and 67th (above normal, AN). Years in which the algorithm failed detecting any of the phases of the MSD, were removed. The star (*) represents cases in which the observed condition in column 2 or 3 dominates, and has statistical significance > 90 %.

Station	SST AN and Intensity BN	SST BN and Intensity AN	r (Kendall)	p value
La Argentina	*	*	−0.38	0.00
Fabio Baudrit			−0.05	0.62
Juan Santa Maria	*	*	−0.34	0.00
Liberia			−0.21	0.05
Nicoya		*	−0.24	0.02
Santa Cruz	*	*	−0.26	0.01
Bagaces		*	−0.30	0.01
CIGEFI		*	−0.30	0.02
Bluefields			−0.04	0.77
Ocotal	*		−0.27	0.01
Chinandega	*		−0.33	0.00
Juigalpa			−0.21	0.04
Managua	*	*	−0.39	0.00
Puerto Cabezas			−0.02	0.87
Rivas			−0.13	0.26
David	*	*	−0.32	0.00
Los Santos		*	−0.22	0.03
San Miguel			0.25	0.01
Asuncion Mita			−0.22	0.04
Huehuetenango	*		−0.14	0.16
Labor Ovalle	*		−0.35	0.00
La Fragua	*	*	−0.42	0.00
Amapala		*	−0.40	0.00
Choluteca	*		−0.28	0.01
Tegucigalpa	*	*	−0.26	0.02

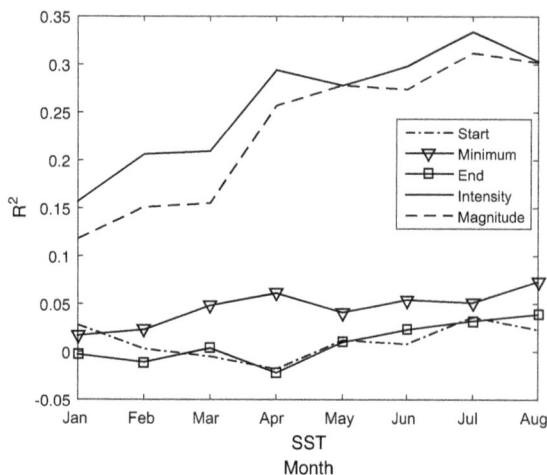

Figure 7. Goodness index (R^2) estimated as the average of the Pearson correlation between synthetic time series generated by cross-validation models and the observed MSD features per station.

the SST in June. The best combination of EOF and CCA modes for the MSD intensity are $X = 14$, $Y = 4$ EOFs and

2 CCA modes, and for the MSD magnitude $X = 11$, $Y = 6$ EOFs and 3 CCA modes, meaning that for each model (for intensity and magnitude respectively) the best fit is achieved when using 2 and 3 modes (canonical variates) respectively, capturing the maximum influence of the SST on the precipitation field, and specifically in the modulation of the MSD. Figure 8 shows the Pearson correlation between the predicted time series generated using cross-validation models and the observations of both intensity and magnitude in each station. Each case, the station exhibits relative high significant correlation, about 0.35 in average.

The X loadings (correlation between the canonical vectors of the SST and the SSTA) of the first mode controlling the MSD intensity shows a bipolar pattern in the correlation with the SSTAs surrounding Central America (Fig. 9a, b). The highest positive correlations are found over the Pacific in the El Niño region, whereas the highest negative correlation are found over the Tropical North Atlantic and near to the Brazil coast. The influence of this variability mode on the precipitation in Central America has been previously studied by Enfield and Alfaro (1999); Maldonado and Alfaro (2011) and Maldonado et al. (2013), but for the secondary

Table 3. Same as Table 2 but for MSD magnitude.

Station	SST AN and Magnitude BN	SST BN and Magnitude AN	r (Kendall)	p value
La Argentina	*		−0.38	0.00
Fabio Baudrit			−0.04	0.71
Juan Santa Maria	*	*	−0.36	0.00
Liberia	*		−0.17	0.10
Nicoya	*		−0.26	0.01
Santa Cruz	*	*	−0.23	0.03
Bagaces		*	−0.23	0.06
CIGEFI		*	−0.23	0.07
Bluefields			−0.06	0.66
Ocotal			−0.06	0.55
Chinandega	*	*	−0.32	0.00
Juigalpa			−0.08	0.43
Managua		*	−0.25	0.01
Puerto Cabezas			0.02	0.87
Rivas			−0.24	0.03
David	*	*	−0.32	0.00
Los Santos	*		−0.26	0.01
San Miguel			0.19	0.06
Asuncion Mita	*		−0.28	0.01
Huehuetenango	*	*	−0.21	0.01
Labor Ovalle	*		−0.35	0.00
La Fragua	*		−0.29	0.01
Amapala		*	−0.33	0.01
Choluteca	*		−0.26	0.01
Tegucigalpa			−0.21	0.05

Table 4. Correlations between X mode times series and Oceanic El Niño Index (ONI, NOAA, 2016), El Niño Modoki Index (EMI, JAM-STEC, 2016), Pacific Decadal Oscillation (PDO, NOAA, 2016), Tropical North Atlantic (TNA, NOAA, 2016), and Atlantic Multidecadal Oscillation (AMO, NOAA, 2016). The differences ONI-TNA and ONI-AMO are calculated with the normalized indexes. Correlations are marked with * and ** for 0.05 and 0.01 of significance level respectively.

Variable	Mode	ONI	EMI	PDO	TNA	AMO	ONI-TNA	ONI-AMO
Intensity	1	0.55**	0.07	−0.12	−0.62**	0.18	0.82**	0.56**
	2	−0.42**	0.00	0.16	−0.55**	0.52**	0.09	−0.41**
Magnitude	1	0.70**	0.11	−0.01	−0.45**	0.21	0.81**	0.71**
	2	−0.02	0.03	0.39*	−0.17	0.26	0.10	−0.02
	3	−0.33*	0.15	−0.38**	−0.66**	0.11	0.23	−0.32*

peak of precipitation during August-October (ASO), thus, this results shows that the rainfall during the MSD is governed by the same variability mode present during the highest precipitation season in Central America. On the other hand, for the MSD magnitude, the first mode exerts more influence of the SSTA over the El Niño region, and the regional waters close the west Mexican coast. Notice that in both cases also high positive correlations are found near the western coast of Mexico revealing that both models are affected by the influence of regional waters, varying with the same phase that the superficial temperatures in the El Niño region. The Y loadings in both models are negatively correlated with most of

the stations (Fig. 9c, d). That means for a state in which the water over the Pacific is warmer (colder) than normal, plus the SST over the Atlantic colder (warmer) than normal, the intensity and magnitude of the MSD drier (wetter). This results also connects the MSD period with the second rainfall peak, that is, SSTs conditions for drier MSDs, would persist leading to less rainfall during the second peak, following the results in Maldonado et al. (2013). The temporal scores (canonical variates) of this mode in both models show that this mode has mainly inter-annual variations (Fig. 9e, f).

The X loadings of the second mode are shown in Fig. 10a and b. The MSD intensity is dominated by nega-

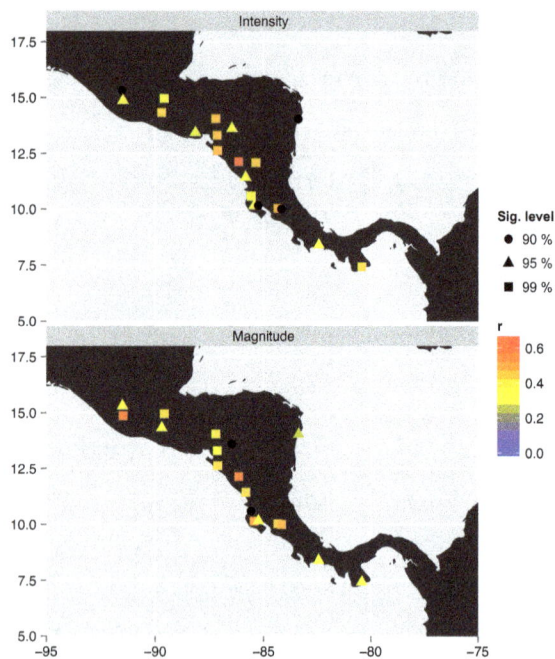

Figure 8. Pearson correlation of the CCA models with the time series of intensity (top) and magnitude (bottom) of MSD.

tive correlations with SST anomalies in the Tropical North Atlantic, the Caribbean and Gulf of Mexico and the eastern Pacific, being more important than the first region. The X loadings for the MSD magnitude on the other hand, reveal a tripolar configuration in the North Pacific. This tripolar setting is located in the region where the PDO develops. The PDO has shown more influence in the southern USA and northern Mexico than in Central America (Muñoz et al., 2010; Maldonado et al., 2016), and that could be the reason that its influence is observed in the second mode. The Y loadings in both cases (Fig. 10c, d) show a non-uniform distribution of the correlation. For the MSD intensity (Fig. 10c), the stations are positively correlated with this mode, meanwhile, the MSD magnitude (Fig. 10d) shows a clear division between northern and southern Central America, being the former region (Guatemala, El Salvador, Honduras and west Nicaragua) negatively correlated and the latter region (Costa Rica, Panama and east Nicaragua) positively correlated. The temporal scores for the intensity model exhibit mainly interannual variability with a negative trend after 1990 (Fig. 10e), while the temporal scores of the magnitude (Fig. 10f) present a decadal variation, that again possibly relates to the influence of the PDO.

The MSD magnitude is the only feature with a third mode, the X loadings (Fig. 11a) reveal that this mode is dominated by regional SSTs in the Pacific and Tropical North Atlantic. The north and south Pacific show a dipole being positive and negative correlations respectively. The Y loadings (Fig. 11b) show a north/south division being positively correlated to the

north and negatively to the south, while the temporal scores (Fig. 11c) reveal an interannual variation of this mode.

Table 4 shows the correlation between the SST modes for each CCA model and the indexes associated to the maximum correlation found using CCA. The results from this table confirm our previous observations, while the first order modes in the intensity and magnitude can be mainly driven by El Niño 3.4 and Tropical North Atlantic, the combination of these indexes, and even with low frequency events such as the AMO turn to be more important. Higher order modes of the magnitude show an important association with PDO which is also a low frequency event.

5 Discussion and conclusions

The MSD is characterised using daily time series of rainfall for 25 stations located mainly in the Pacific coast of Central America during the 1961–2012 period. An algorithm to detect the MSD phases is developed in order to compare the timing of the MSD start, date of minimum, and end in each station. This algorithm clearly captures the development of the MSD over the Pacific, but also a similar MSD-like structure in two stations located in the Caribbean coast. The results confirm previous observations showing the high spatial and temporal variability of the MSD, thus, suggesting the need of a more detailed analysis. We, thus, analyse the features of the MSD individually per station, using a combination of empirical analysis (contingency tables) and a more sophisticated statistical method as the canonical correlation analysis to diagnose the variability of the MSD in terms of changes in the SSTA of the surrounding waters.

The MSD intensity and magnitude are correlated with the Niño 3.4 and the CLLJ index in stations located over the western side of Central America. These results reveal a significant negative correlation in almost all the stations with ENSO phases, i.e. warm (cold) anomalies in the Niño 3.4 region, corresponding to an enhanced (decreased) CLLJ intensity, are associated with drier (wetter) MSD. The results of the correlation with the CLLJ show also a positive association of such process, indicating that with a stronger (weaker) jet the MSD intensifies (decreases). However, other processes such the migration of the ITCZ, the North American Monsoon and perturbations in the SLP fields over the Amazon could also be engaged with anomalous MSD (Hidalgo et al., 2015). The two stations placed in the Caribbean side do not reveal results statistically significant, hence, suggesting that other elements are involved explaining the annual rainfall variability during their MSD period.

The outputs of CCA show that the MSD intensity is mainly modulated by a bipolar configuration in the SST anomalies formed between the Pacific and the Tropical North Atlantic and Caribbean sea, meanwhile, the first mode controlling the MSD magnitude shows more influence from El Niño and the regional waters close to the western coast of Mexico and Central America. This variability mode in both models has

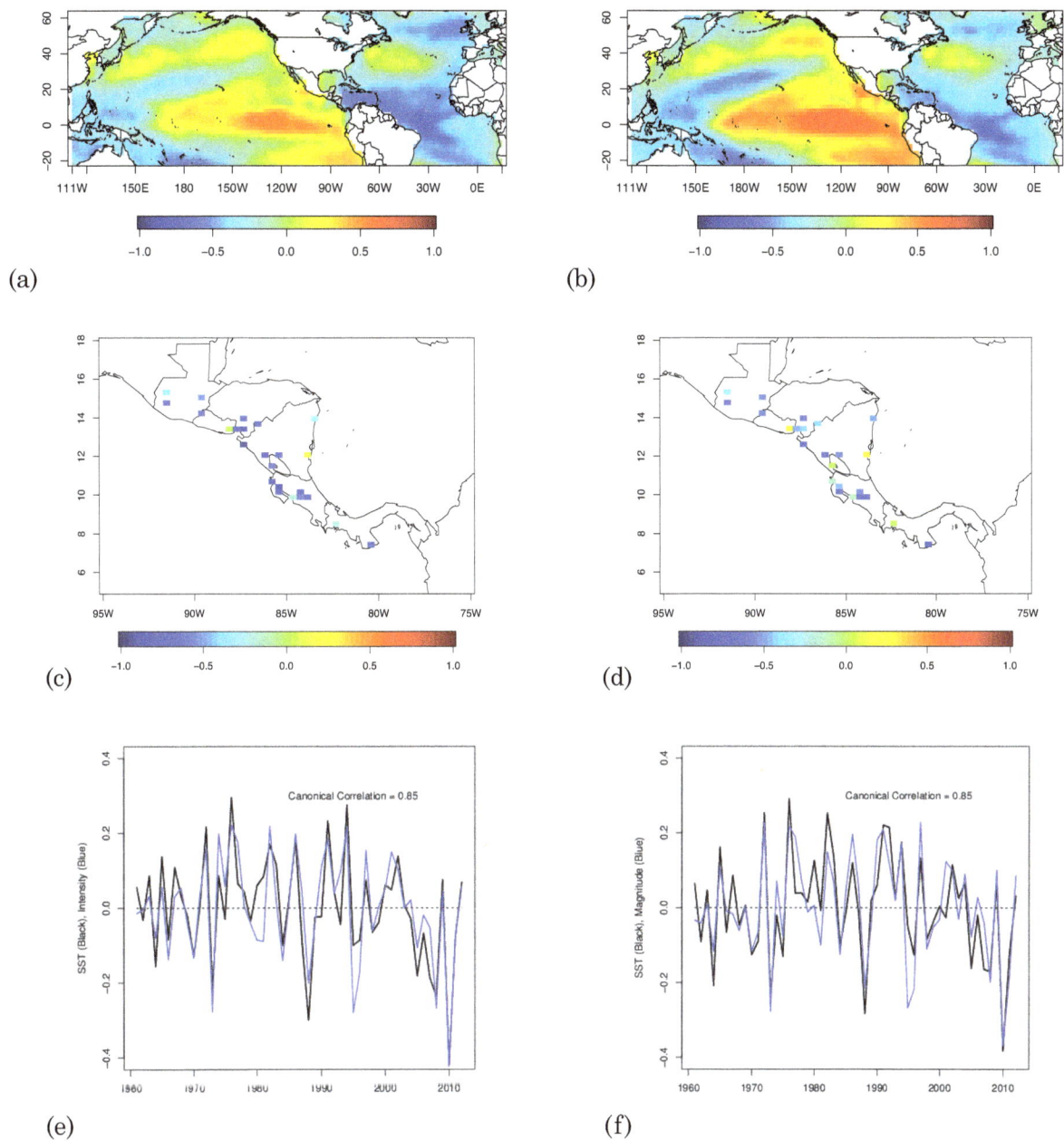

Figure 9. CCA mode 1 for both MSD intensity and magnitude. In the left column are the X (upper) and Y (middle) loadings, and the time scores (bottom) for the intensity. In the right column the same but for the magnitude.

an inter-annual scale, with negative effects in precipitation, i.e. when this mode is positive for both intensity and magnitude, a reduction in rainfall is observed in almost all the stations. Similar SST dipole has been reported to influence the anomalies in rainfall during the months of the highest precipitation events (August–November), and it has been associated with intensification/reduction of the trades, leading to a decrease/increase of precipitation in Central America during the second rainfall period (Alfaro, 2007; Maldonado et al., 2013), however, these results show this mode is present

prior to the quarter ASO and is also modulating the precipitation during boreal summer.

The second mode in both models presents more complex structures, and it is difficult to distinguish a general pattern affecting the intensity as well as magnitude of the MSD. For the intensity model, the SST show a negative correlation pattern over the equatorial Pacific and North Atlantic waters, nevertheless, the latter region dominates; the time series for this mode show a negative trend after 1990. Meanwhile, for the magnitude, a tripolar configuration over the vicinity of

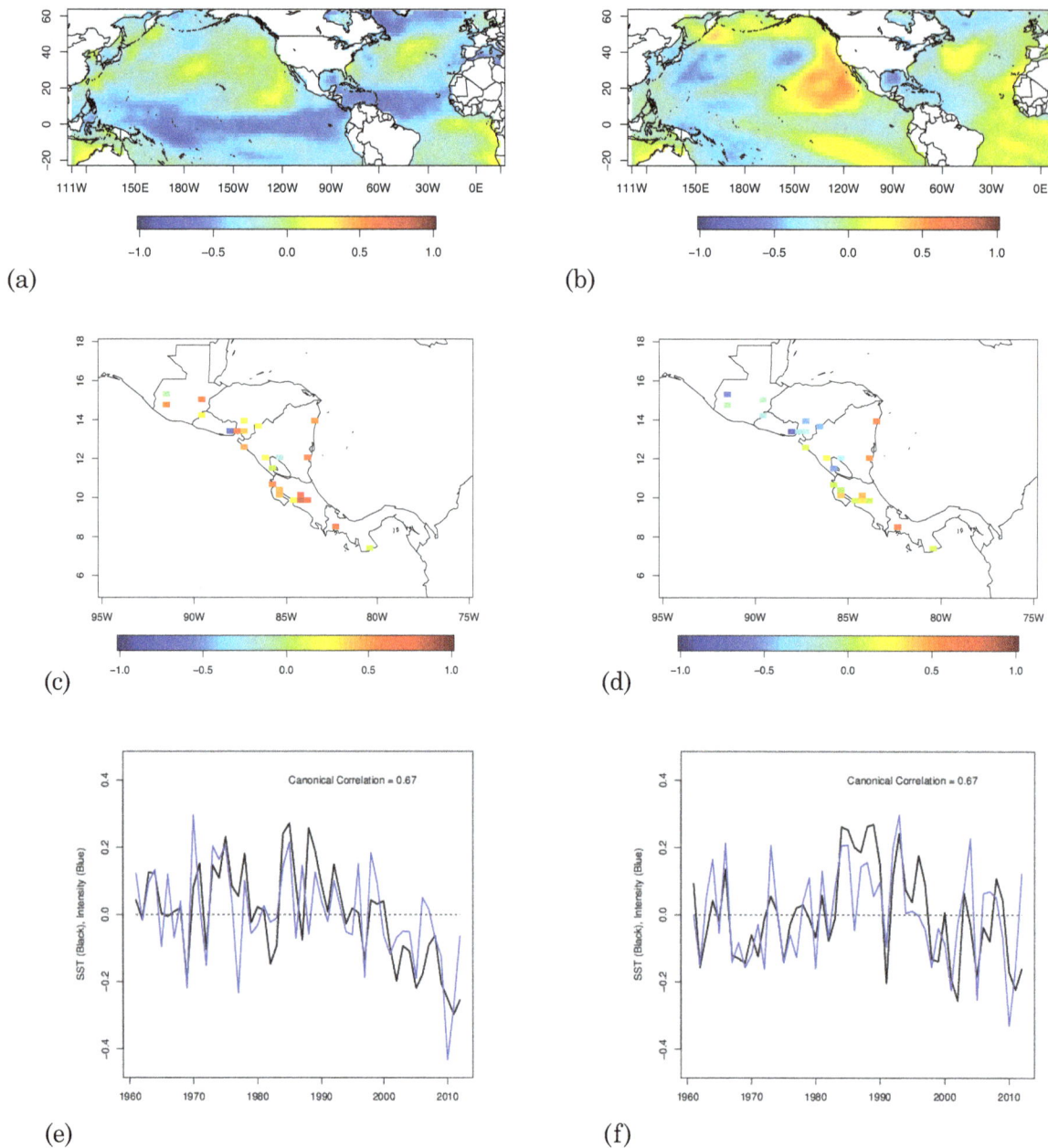

(a)

(b)

(c)

(d)

(e)

(f)

Figure 10. Same as Fig. 9 but for Mode 2.

the PDO development region is found. Also negative correlations are found over the Gulf of Mexico. The influence of the PDO should be taken with caution and needs more analysis since the PDO has shown more influence in the southern USA and northern Mexico than in Central America. The time series exhibited inter-decadal variability. As previously mentioned, the correlations between both variables (intensity and magnitude) and this mode, however, do not depict a clear pattern in the stations; for the MSD intensity the correlations reveal a positive association with the SSTA, meanwhile, for the MSD magnitude shows a division be-

tween the north and south Central America, being the north (Guatemala, Honduras, El Salvador and west Nicaragua) affected negatively and the south and Caribbean (Costa Rica, Panama and east Nicaragua) positively. It is clear that the second mode is not controlling the precipitation in both models in the same way. The North Atlantic waters become more important for the MSD intensity, hence, could be associated with the same controlling mechanism present during the first precipitation maximum (Alfaro, 2007; Fallas-López and Alfaro, 2012b). While for the MSD magnitude the tripolar configuration might suggest the influence of the PDO, also noted

(a)

(b)

(c)

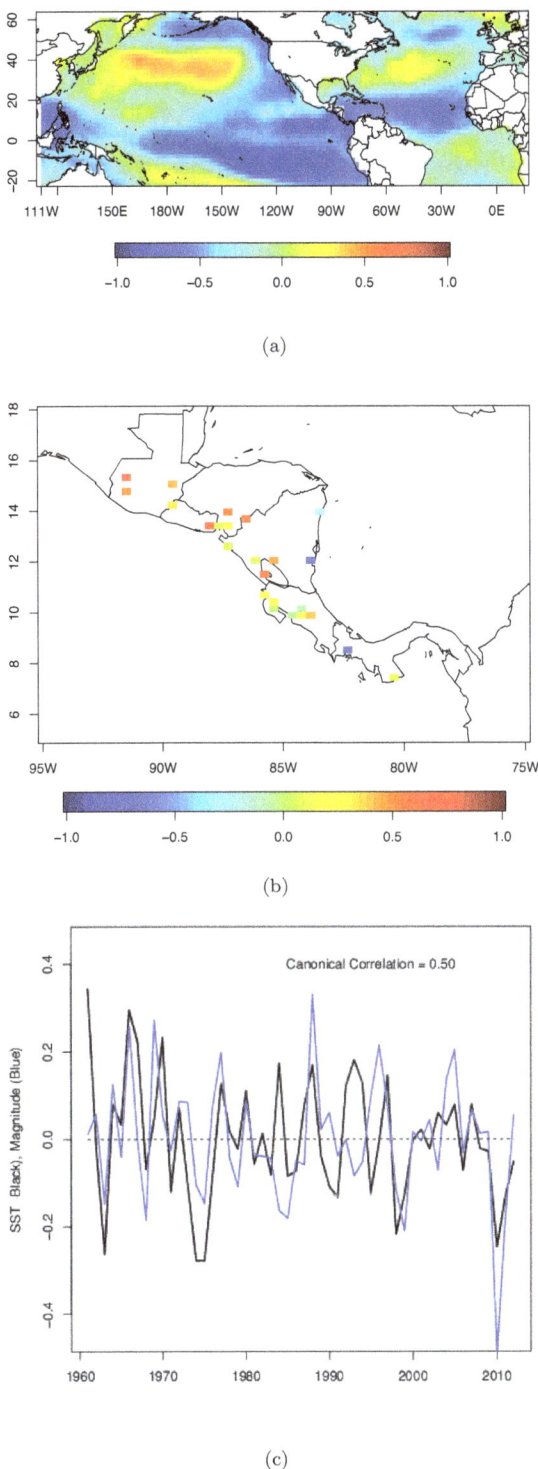

Figure 11. CCA mode 3 for MSD magnitude. X (**a**) and Y (**b**) loadings, and (**c**) the time scores.

in the contrasting effects of this mode between north and south Central America.

The MSD magnitude is the only variable with higher CCA modes. The third mode reveals the influence of regional wa-

ters close to the Central America Pacific and a dipole formed between the north (positive) and south (negative) Pacific, affecting with opposite sign the eastern (negative) and western (positive) part of Central America. This mode presents an inter-annual variability scale. The CCA allows identifying patterns of the SST affecting variables describing the MSD such as the intensity and magnitude, however, it shows a poor performance related to the temporal variables. Another benefit achieved using CCA was the identification of particular months suitable for prediction of the intensity and magnitude of the MSD, being capable of forecasting up to 3 months in advance, which is a reasonable time in terms of practical matters related to prevention and planning for the season. Finally, it is worth pointing out that this analysis also provide a systematic method to study the MSD features, that can be used for statistical forecasts of such phenomenon in an operative context also.

Data availability

Niño 3.4 index is available at http://iridl.ldeo.columbia.edu/SOURCES/.Indices/.nino/.NCEP/.NINO34/.
PDO, TNA, and AMO indexes are available at http://www.esrl.noaa.gov/psd/data/climateindices/list/,
and EMI index is available at http://www.jamstec.go.jp/frcgc/research/d1/iod/enmodoki_home_s.html.en.

Acknowledgements. This research was carried out within the CNDS research school, supported by the Swedish International Development Cooperation Agency (Sida) through their contract with the International Science Programme (ISP) at Uppsala University (contract number: 54100006). The authors would like to thank the Centre for Natural Disaster Science (CNDS) in Uppsala University, the National Centers for Environmental Prediction (NCEP), and the National Center for Atmospheric Research (NCAR) for the reanalysis data. We would also like to acknowledge support via project VI-805-A9-532, CIGEFI-UCR, provided via a SIDA-CSUCA agreement and project 805 B6 143, supported by UCR, CONICIT and MICITT.

References

Alfaro, E.: Some Characteristics of the Annual Precipitation Cycle in Central America and their Relationships with its Surrounding Tropical Oceans, Tópicos Meteorológicos y Oceanográficos, 9, 1–13, https://www.imn.ac.cr/documents/10179/20907/T%C3%B3picos+Meteorol%C3%B3gicos+y+Oceanogr%C3%A1ficos+-+2000-2 , 2002.

Alfaro, E.: Uso del análisis de correlación canónica para la predicción de la precipitación pluvial en Centroamérica, Ingeniería y Competitividad, 9, 33–48, http://bibliotecadigital.univalle.edu.co/xmlui/handle/10893/1622, 2007.

Alfaro, E.: Caracterización del "veranillo" en dos cuencas de la vertiente del Pacífico de Costa Rica, América Central (Characterization of the Mid Summer Drought in two Pacific slope river basins of Costa Rica, Central America), International Journal of Tropical Biology, 62, 1–15, available at: https://www.academia.edu/9493294 (last access: 22 April 2016), 2014.

Alfaro, E. and Soley, J.: Descripción de dos métodos de rellenado de datos ausentes en series de tiempo meteorológicas, Revista de Matemáticas: Teoría y Aplicaciones, 16, 59–74, available at: http://dx.doi.org/10.15517/rmta.v16i1.1419 (last access 22 April 2016), 2009.

Amador, J. A.: A Climatic Feature of the Tropical Americas: The Trade Wind Easterly Jet, Tópicos Meteorológicos y Oceanográficos., 5, 91–102, available at: https://www.imn.ac.cr/documents/10179/20907/T%C3%B3picos+Meteorol%C3%B3gicos+y+Oceanogr%C3%A1ficos+-+1998-2 (last access: 22 April 2016), 1998.

Amador, J. A.: The Intra-Americas Sea Low-level Jet Overview and Future Research, Ann. NY Acad. Sci., 1146, 153–188, doi:10.1196/annals.1446.012, 2008.

Amador, J. A., Alfaro, E. J., Lizano, O. G., and Magaña, V. O.: Atmospheric forcing of the eastern tropical Pacific: A review, Prog. Oceanogr., 69, 101–142, doi:10.1016/j.pocean.2006.03.007, 2006.

Ashby, S. A., Taylor, M. A., and Chen, A. A.: Statistical models for predicting rainfall in the Caribbean, Theor. Appl. Climatol., 82, 65–80, doi:10.1007/s00704-004-0118-8, 2005.

Chen, A. A. and Taylor, M. A.: Investigating the link between early season Caribbean rainfall and the El Niño+ 1 year, Int. J. Climatol., 22, 87–106, doi:10.1002/joc.711, 2002.

Chinchilla-Ramírez, G.: Resumen Meteorológico Julio 2014, Boletín Meteorológico Mensual, https://www.imn.ac.cr/documents/10179/14639/JULIO (last access: 22 April 2016), 2014.

Enfield, D. B. and Alfaro, E. J.: The Dependence of Caribbean Rainfall on the Interaction of the Tropical Atlantic and Pacific Oceans, J. Climate, 12, 2093–2103, doi:10.1175/1520-0442(1999)012<2093:TDOCRO>2.0.CO;2, 1999.

Fallas-López, B. and Alfaro, E. J.: Uso de herramientas estadísticas para la predicción estacional del campo de precipitación en América Central como apoyo a los Foros Climáticos Regionales. 1: Análisis de tablas de contingencia, Revista de Climatología, 12, 61–79, http://webs.ono.com/reclim7/reclim12e.pdf, 2012a.

Fallas-López, B. and Alfaro, E. J.: Uso de herramientas estadísticas para la predicción estacional del campo de precipitación en América Central como apoyo a los Foros Climáticos Regionales, 2: Análisis de Correlación Canónica, Revista de Climatología, 12, 93–105, http://webs.ono.com/reclim8/reclim12g.pdf, 2012b.

Gershunov, A. and Barnett, T.: ENSO influence on intraseasonal extreme rainfall and temperature frequencies in the contiguous United States: Observations and model results, J. Climate, 11, 1575–1586, doi:10.1175/1520-0442(1998)011<1575:EIOIER>2.0.CO;2, 1998.

Gershunov, A. and Cayan, D. R.: Heavy Daily Precipitation Frequency over the Contiguous United States: Sources of Climatic Variability and Seasonal Predictability, J. Climate, 16, 2752–2765, doi:10.1175/1520-0442(2003)016<2752:HDPFOT>2.0.CO;2, 2003.

Hernández, K. and Fernández, W.: Estudio de la evaporación para el cálculo del inicio y la conclusión de la época seca y lluviosa en Costa Rica, Tópicos Meteorológicos y Oceanográficos, 14, 18–26, 2015.

Herrera, E., Magaña, V., and Caetano, E.: Air-sea interactions and dynamical processes associated with the midsummer drought, Int. J. Climatol., 35, 1569–1578, doi:10.1002/joc.4077, 2015.

Hidalgo, H. G., Durán-Quesada, A. M., Amador, J. A., and Alfaro, E. J.: The Caribbean Low-Level Jet, the Inter-Tropical Convergence Zone and Precipitation Patterns in the Intra-Americas Sea: A Proposed Dynamical Mechanism, Geogr. Ann. A, 97, 41–59, doi:10.1111/geoa.12085, 2015.

International Research Institute for Climate and Society IRI, Earth Institute, Columbia University: NCEP NINO34 from Indices nino, New York, USA, Dataset: http://iridl.ldeo.columbia.edu/SOURCES/.Indices/.nino/.NCEP/.NINO34/, 2015.

Japan Agency for Marine-Earth Science and Technology (JAMSTEC): Modoki ENSO, Japan, Dataset: http://www.jamstec.go.jp/frcgc/research/d1/iod/enmodoki_home_s.html.en, 2016.

Kalnay, E., Kanamitsu, M., Kistler, R., Collins, W., Deaven, D., Gandin, L., Iredell, M., Saha, S., White, G., Woollen, J., Zhu, Y., Leetmaa, A., Reynolds, R., Chelliah, M., Ebisuzaki, W., Higgins, W., Janowiak, J., Mo, K. C., Ropelewski, C., Wang, J., Jenne, R., and Joseph, D.: The NCEP/NCAR 40-Year Reanalysis Project, B. Am. Meteorol. Soc., 77, 437–471, doi:10.1175/1520-0477(1996)077<0437:TNYRP>2.0.CO;2, 1996.

Karnauskas, K. B., Seager, R., Giannini, A., and Busalacchi, A. J.: A simple mechanism for the climatological midsummer drought along the Pacific coast of Central America, Atmósfera, 26, 261–281, doi:10.1016/S0187-6236(13)71075-0, 2013.

Kistler, R., Collins, W., Saha, S., White, G., Woollen, J., Kalnay, E., Chelliah, M., Ebisuzaki, W., Kanamitsu, M., Kousky, V., van den Dool, H., Jenne, R., and Fiorino, M.: The NCEP-NCAR 50-Year Reanalysis: Monthly Means CD-ROM and Documentation, B. Am. Meteorol. Soc., 82, 247–267, doi:10.1175/1520-0477(2001)082<0247:TNNYRM>2.3.CO;2, 2001.

Magaña, V., Amador, J. A., and Medina, S.: The Midsummer Drought over Mexico and Central America, J. Climate, 12, 1577–1588, doi:10.1175/1520-0442(1999)012<1577:TMDOMA>2.0.CO;2, 1999.

Maldonado, T. and Alfaro, E.: Predicción estacional para ASO de eventos extremos y días con precipitación sobre las vertientes Pacífico y Caribe de América Central, utilizando análisis de correlación canónica, InterSedes, 12, 78–108, http://www.intersedes.ucr.ac.cr/ojs/index.php/intersedes/article/view/301, 2011.

Maldonado, T., Alfaro, E., Fallas-López, B., and Alvarado, L.: Seasonal prediction of extreme precipitation events and frequency of rainy days over Costa Rica, Central America, using Canonical Correlation Analysis, Adv. Geosci., 33, 41–52, doi:10.5194/adgeo-33-41-2013, 2013.

Maldonado, T., Rutgersson, A., Amador, J., Alfaro, E., and Claremar, B.: Variability of the Caribbean low-level jet during boreal winter: large-scale forcings, International J. Climatol., 36, 1954–1969, doi:10.1002/joc.4472, 2016.

Martin, E. R. and Schumacher, C.: The Caribbean Low-Level Jet and Its Relationship with Precipitation in IPCC AR4 Models, J. Climate, 24, 5935–5950, doi:10.1175/JCLI-D-11-00134.1, 2011.

Muñoz, E., Busalacchi, A. J., Nigam, S., and Ruiz-Barradas, A.: Winter and Summer Structure of the Caribbean Low-Level Jet, J. Climate, 21, 1260–1276, doi:10.1175/2007JCLI1855.1, 2008.

Muñoz, E., Wang, C., and Enfield, D.: The Intra-Americas Sea springtime surface temperature anomaly dipole as fingerprint of remote influence, J. Climate, 23, 43–56, doi:10.1175/2009JCLI3006.1, 2010.

National Oceanic and Atmospheric Administration (NOAA): Earth System Research Laboratory, Physical Science Division, Boulder, Colorado, Datasets: http://www.esrl.noaa.gov/psd/data/climateindices/list/, 2016.

Ramírez, P.: Estudio Meteorológico de los Veranillos en Costa Rica, Nota de investigación 5, Instituto Meteorológico Nacional, Ministerio de Agricultura y Ganadería, San José, Costa Rica, 1983.

Smith, T., Reynolds, R., Peterson, T. C., and Lawrimore, J.: Improvements to NOAA's Historical Merged Land-Ocean Surface Temperature Analysis (1880–2006), J. Climate, 21, 2283–2296, 2007.

Solano, E.: Análisis del comportamiento de los períodos caniculares en Costa Rica en algunas cuencas del Pacífico Norte y del Valle Central entre los años 1981 y 2010, Tesis de Grado, Licenciatura, Escuela de Física, Universidad de Costa Rica, San José, Costa Rica, 2015.

Spence, J. M., Taylor, M. A., and Chen, A. A.: The effect of concurrent sea-surface temperature anomalies in the tropical Pacific and Atlantic on Caribbean rainfall, Int. J. Climatol., 24, 1531–1541, doi:10.1002/joc.1068, 2004.

Taylor, M. A. and Alfaro, E. J.: Climate of Central America and the Caribbean, in: Encyclopedia of World Climatology, edited by: Oliver, J. E., Springer, the Netherlands, 183–186, 2005.

Taylor, M. A., Enfield, D. B., and Chen, A. A.: Influence of the tropical Atlantic versus the tropical Pacific on Caribbean rainfall, J. Geophys. Res.-Oceans, 107, 10-1–10-14, doi:10.1029/2001JC001097, 2002.

Trenberth, K. E.: The Definition of El Niño, B. Am. Meteorol. Soc., 78, 2771–2777, doi:10.1175/1520-0477(1997)078<2771:TDOENO>2.0.CO;2, 1997.

Vera, C., Higgins, W., Amador, J., Ambrizzi, T., Garreaud, R., Gochis, D., Gutzler, D., Lettenmaier, D., Marengo, J., and Mechoso, C. R.: Toward a unified view of the American monsoon systems, J. Climate, 19, 4977–5000, http://journals.ametsoc.org/doi/pdf/10.1175/JCLI3896.1, 2006.

Wang, C.: Variability of the Caribbean Low-Level Jet and its relations to climate, Clim. Dynam., 29, 411–422, doi:10.1007/s00382-007-0243-z, 2007.

Wang, C. and Enfield, D. B.: The Tropical Western Hemisphere Warm Pool, Geophys. Res. Lett., 28, 1635–1638, doi:10.1029/2000GL011763, 2001.

Wang, C. and Enfield, D. B.: A Further Study of the Tropical Western Hemisphere Warm Pool, J. Climate, 16, 1476–1493, doi:10.1175/1520-0442(2003)016<1476:AFSOTT>2.0.CO;2, 2003.

Whyte, F. S., Taylor, M. A., Stephenson, T. S., and Campbell, J. D.: Features of the Caribbean low level jet, Int. J. Climatol., 28, 119–128, doi:10.1002/joc.1510, 2008.

Wilks, D. S.: Statistical Methods in the Atmospheric Sciences, Academic Press, Amsterdam, the Netherlands, Boston, USA, Volume 100, 3rd edn., 2011.

Xue, Y., Smith, T., and Reynolds, R.: Interdecadal changes of 30-yr SST normals during 1871-2000, J. Climate, 16, 1601–1612, doi:10.1175/1520-0442-16.10.1601, 2003.

Trans-national earthquake early warning (EEW) in north-eastern Italy, Slovenia and Austria: first experience with PRESTo at the CE³RN network

M. Picozzi[1]**, L. Elia**[1]**, D. Pesaresi**[2]**, A. Zollo**[1]**, M. Mucciarelli**[2]**, A. Gosar**[3]**, W. Lenhardt**[4]**, and M. Živčić**[3]

[1]RISSC, Università "Federico II" di Napoli – AMRA, Naples, Italy
[2]CRS, OGS (Istituto Nazionale di Oceanografia e di Geofisica Sperimentale), Trieste, Italy
[3]ARSO – Agencija Republike Slovenije za Okolje, Ljubljana, Slovenia
[4]ZAMG – Zentralanstalt für Meteorologie und Geodynamik, Vienna, Austria

Correspondence to: M. Picozzi (matteo.picozzi@unina.it)

Abstract. The region of central and eastern Europe is an area characterised by a relatively high seismic risk. Since 2001, to monitor the seismicity of this area, the OGS (Istituto Nazionale di Oceanografia e di Geofisica Sperimentale) in Italy, the Agencija Republike Slovenije za Okolje (ARSO) in Slovenia, the Zentralanstalt für Meteorologie und Geodynamik (ZAMG) in Austria, and the Università di Trieste (UniTS) have cooperated in real-time seismological data exchange. In 2014 OGS, ARSO, ZAMG and UniTS created a cooperative network named the Central and Eastern European Earthquake Research Network (CE³RN), and teamed up with the University of Naples Federico II, Italy, to implement an earthquake early warning system based on the existing networks. Since May 2014, the earthquake early warning system (EEWS) given by the integration of the PRESTo (PRobability and Evolutionary early warning SysTem) alert management platform and the CE³RN accelerometric stations has been under real-time testing in order to assess the system's performance. This work presents a preliminary analysis of the EEWS performance carried out by playing back real strong motion recordings for the 1976 Friuli earthquake ($M_W = 6.5$). Then, the results of the first 6 months of real-time testing of the EEWS are presented and discussed.

1 Introduction

With the aim of monitoring the seismic activity in the eastern sector of the Alps, since 2001 OGS (Istituto Nazionale di Oceanografia e di Geofisica Sperimentale) in Udine (Italy), the Agencija Republike Slovenije za Okolje (ARSO) in Ljubljana (Slovenia), the Zentralanstalt für Meteorologie und Geodynamik (ZAMG) in Vienna (Austria), and the University of Trieste (UniTS) have been collecting, analysing, archiving and exchanging seismic data in real time. The data exchange has proven to be effective and very useful in the case of seismic events at the borders between Italy, Austria and Slovenia, where the poor coverage of individual national seismic networks precluded a precise earthquake location. The usage of common data from the integrated networks improves significantly the overall capability of real-time event detection and rapid characterisation in this area. Furthermore, in 2014, OGS, ARSO, ZAMG and UniTS signed a memorandum of understanding naming the cooperative network as the Central and Eastern European Earthquake Research Network (CE³RN) (Bragato et al., 2014).

Recently, in order to extend the seismic monitoring in north-eastern Italy, Slovenia and southern Austria towards earthquake early warning applications, OGS, ARSO and ZAMG teamed up with the RISSC-Lab group (http://www.rissclab.unina.it) of the Department of Physics at the University of Naples Federico II in Italy.

An earthquake early warning system (EEWS) is a real-time system integrating seismic networks and software capable of performing real-time data telemetry and analysis in order to issue alert messages within seconds from the origin of an earthquake and before the destructive S-waves generated by the event reach the users. When accompanied by appropriate training and preparedness of the population, an EEWS is an effective and viable tool for reducing the exposure of a population to seismic risk (e.g. Allen et al., 2009; Hoshiba, 2013; Picozzi et al., 2015a). The application of EEWS is nowadays increasing and several countries around the world have already developed EEWS, or are on the verge of doing so. Japan, Taiwan, Mexico, Romania and California, for example, already have operational EEWSs (e.g. Horiuchi et al., 2005; Wu and Zhao, 2006; Espinosa-Aranda et al., 2009; Böse et al., 2007, 2009). EEWSs are also under development and testing in other regions of the world, such as Italy, Turkey, Spain, and China (Satriano et al., 2010; Zollo et al., 2014; Alcik et al., 2009; Peng et al., 2011; Picozzi et al., 2014, 2015b).

The collaboration among OGS, ARSO, ZAMG and RISSC-LAB focuses on testing the EEW platform PRESTo (probabilistic and evolutionary early warning system: http://www.prestoews.org) in north-eastern Italy, Slovenia and Austria at the network CE^{3}RN, and represents, to our knowledge, the first worldwide attempt of implementing a trans-national EEWs. PRESTo is a stand-alone software system that processes live accelerometric streams from a seismic network to promptly provide probabilistic and evolutionary estimates of location and magnitude of detected earthquakes while they are occurring, as well as shaking prediction at the regional scale (Satriano et al., 2011).

Since 2014 PRESTo has run on OGS, ARSO and ZAMG data, by collecting and analysing in real time the data streams from 20 stations (Fig. 1).

In the following, first, we briefly present the CE^{3}RN project, and we summarise the characteristics of EEWS and PRESTo. Then, we present the results of a test carried out by playing back the waveforms of the strong motion data of the $M_W = 6.5$, 1976 Friuli earthquake, and, finally, we report on the performance of the EEW system during this preliminary testing phase.

2 The CE^{3}RN project

The region of central and eastern Europe is an area characterised by a relatively high seismicity. The active seismogenic structures and the related potentially destructive events are located in the proximity of the political boundaries between several countries existing in the area. An example is the seismic region between north-eastern Italy (Friuli-Venezia Giulia, Trentino-Alto Adige and Veneto), Austria (Tyrol, Carinthia) and Slovenia. So, when a destructive earthquake occurs in the area, all three countries are possibly affected. In the year 2001, the institutes OGS, ARSO, ZAMG, and UniTS signed an agreement for real-time seismological data exchange in the south-eastern Alps region. Soon after, the Interreg IIIa Italia-Austria Trans-National Seismological Networks in the South-Eastern Alps and FASTLINK projects started. The main goal of these projects was the creation of a transfrontier network for the common seismic monitoring of the region for scientific and civil defense purposes.

The OGS, ZAMG and ARSO seismic networks present many similarities. While there is a variety of sensor typologies in use (i.e. from strong motion to (very) broadband), all the stations are equipped with Quanterra data loggers (Q_{6180}, Q_{4120}, Q_{730} and Q_{330}), and similar strong motion sensors are used almost at each seismic station of the single networks. As shown by Stein and Reimiller (2014), the stations equipped with data logger Q330 are capable of delivering data with a latency of less than 1 s, and therefore are suitable for early warning applications. The use of similar instrumentation facilitated a very important consequence of Interreg project Trans-National Seismological Networks in the South-Eastern Alps, and the adoption of common software suite Antelope from Boulder Real-Time Technologies (BRTT), for seismic data real-time acquisition, archiving, analysis and exchange. It is in fact straightforward, given that all the involved institutions use the same data acquisition software, to extend the single networks' seismic monitoring capabilities to the entire transfrontier network, thus acting like an extended virtual network. All the involved partners exchange waveforms and parametric data in real time through a network of bi-directional data links, mainly via the Internet, interconnecting all data centres.

During the recent past years, the high-quality data recorded by the trans-national network have been used by the involved institutions for their scientific research, for institutional activities and for civil defence services. Several common international projects have been realised with success.

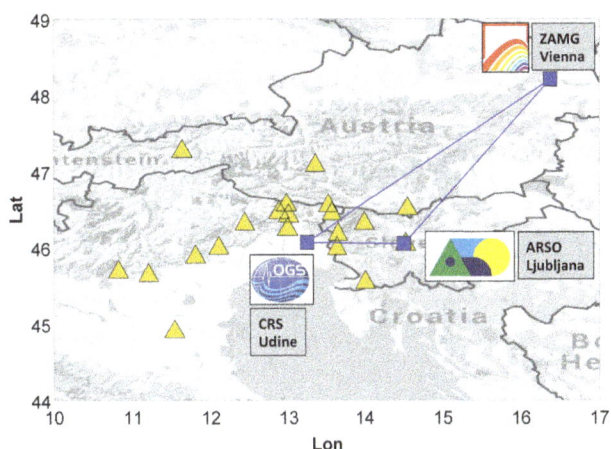

Figure 1. CE^{3}RN institutions involved in the EEW experiment (blue squares), real-time accelerometric stations (yellow triangles).

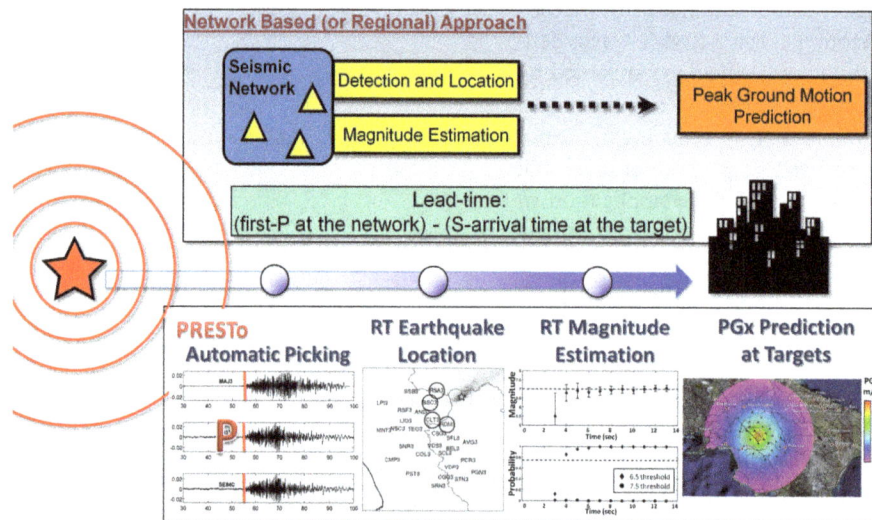

Figure 2. Schematic representation of the regional approach for EEW (modified from Satriano et al., 2011), and overview of the analyses carried out by the PRESTo software system for the real-time event characterisation and ground motion level at target site prediction.

The instrumentation has been continuously upgraded and the installations quality improved, as well as the data transmission efficiency.

In 2014, OGS, ARSO, ZAMG and UniTS signed a memorandum of understanding named the Central and Eastern European Earthquake Research Network (CE^{3}RN) cooperative network (Bragato et al., 2014). CE^{3}RN represents an excellent example of international high-quality research infrastructure and the starting point for the enlargement of the transfrontier network to all countries and their seismological institutions of the central and eastern Europe region. Furthermore, one of the main goals of the CE^{3}RN is to intensify the cooperation between these institutions through common research activities and preparation of common international projects.

On 11 November 2014, the CE^{3}RN partnership was enlarged to also include the Croatian Seismological Survey (CSS) of the University of Zagreb in Croatia.

3   Earthquake early warning systems and PRESTo

EEWS typically follows two basic approaches: "regional" (or network based), and "on-site" (or a single station). Regional EEWS are based on the use of a seismic network located near one or more expected epicentral areas, whose aim is to detect and locate an earthquake, and to determine its magnitude from the analysis of the first few seconds of the arriving P-waves at multiple stations close to the epicentre (Satriano et al., 2011). On the other hand, on-site EEWS are based on seismic sensors deployed directly at the target site and exploit only the information carried by the faster early P-waves to infer the larger shaking related to the incoming S and surface waves.

One key parameter for an EEWS is the lead time, i.e. the time available to perform safety measures at distant targets once an earthquake has been promptly detected and characterised, and an alarm has been issued. The lead time for regional EEWS is defined as the travel-time difference between the arrival of the first S-waves at the target site and the P-waves recorded in the source area, after accounting for the necessary computation and data transmission times. In on-site EEWS, the lead time is equal to the difference in S- and P-wave arrival times at the target itself.

Recently, Zollo et al. (2010) showed that the two approaches can be profitably integrated within a unique system that allows the early estimation of the potential damage zone (PDZ) associated with an event. Clearly, the integration of regional and on-site approaches is particularly useful whenever target sites are threatened by more than one seismic source area, and the latter are located at variable distances from the target sites. An exhaustive review of the concepts, methods, and physical basis of EEWS has been presented by Satriano et al. (2010).

PRESTo is a free and open source, highly configurable and easily portable platform for earthquake early warning (Iannaccone et al., 2010). PRESTo processes the real-time accelerometric data streams from the stations of a seismic network to promptly detect the P-wave arrival, provide the probabilistic and evolutionary estimates of location and magnitude of earthquakes while they are occurring, as well as the shaking prediction on a regional scale (Fig. 2). Alarm messages containing the continuously updated estimates of source and ground motion at target parameters, and their associated uncertainties, are sent over the Internet, and can thus also reach distant vulnerable infrastructures before the arrival of destructive waves, enabling the activation of automatic safety procedures. Following the idea proposed by Zollo et al. (2010), PRESTo implements both a regional and an on-site approach.

In its regional configuration (Fig. 2), PRESTo uses (a) a phase detector and picker algorithm, which is optimised for real-time seismic monitoring and EEW (Lomax et al., 2012); (b) a location algorithm, RTLoc (Satriano et al., 2008), which locates earthquakes using information from both triggered and not-yet-triggered stations, and which provides a fully probabilistic description of the hypocentre coordinates and origin time; (c) the RTMag algorithm (Lancieri and Zollo, 2008), a Bayesian approach that uses the peak displacement (Pd) measured on the first seconds of the high-pass-filtered signal on short time windows of P-waves (i.e. 2 and 4 s) and S-waves (i.e. 1 or 2 s), and empirical correlation laws between this latter parameter and the final earthquake magnitude (M); and (d) finally, ground motion prediction equations (GMPE) that allow one to predict the peak ground motion at target sites and at seismic stations using EEW location and magnitude estimates.

The regional approach to early warning is integrated with an on-site, threshold-based method for the definition of independent local alert levels at each station. To this aim, the dominant period, τ_c, and the peak displacement in a short time window after the first P-arrival time, Pd, are simultaneously measured at each station, independently of the rest of the seismic network. As shown by Zollo et al. (2010), Pd can be correlated with the final PGV and consequently with the modified Mercalli intensity (I_{MM}), which is a measure of the expected damage, while τ_c can be correlated with the earthquake magnitude. These two parameters are compared with threshold values that define a decisional table with four alert levels, declaring the expected earthquake effects nearby the station or at distant sites. The alert level can be used to initiate safety measures at each site independently of the regional processing. At the same time, on the regional scale, the local alert levels, as they become available, can be combined with the estimated source parameters to define the extent of the potential damage zone (PDZ), i.e. the area in which the highest intensity levels are expected (Zollo et al., 2010).

Since 2009, PRESTo has been under real-time experimentation in southern Italy on the data streams of the Irpinia Seismic Network (ISNet). Moreover, in order to analyse its performance in different seismic hazard contexts and seismic networks of varying extensions, PRESTo is also currently operating in other seismological centres (e.g. at the Korean Institute of Geoscience and Mineral Resources, KIGAM, in South Korea, at the Kandilli Observatory and Earthquake Research Institute, KOERI, in Turkey, and at the National Institute of Research and Development for Earth Physics, NIEP, in Romania). In addition, the feasibility study of a nation-wide early warning system in Italy using the National Accelerometric Network (RAN) and PRESTo is in progress.

Figure 3. Snapshot of the PRESTo system during the playback of the $M_W = 6.5$, 1976 Friuli earthquake, at the instant when three stations have triggered and the first alert is issued.

4 EEW analysis of the 1976 Friuli earthquake data

One of the first tests that we carried out was devoted to verifying what could have been the performance of PRESTo in the case of the 1976, $M_W = 6.5$ Friuli earthquake in northern Italy (Carulli and Sleiko, 2005). To this aim, we realised an off-line run of the algorithm (i.e. a playback) of this earthquake using the historical recordings downloaded by ITACA 2.0 (Luzi et al., 2008; Pacor et al., 2011). The playback was run considering the network geometry of 1976, but assuming the existence at the time of the hardware and the management software necessary for the real-time data streaming to the OGS's seismological centre of Udine.

Figure 3 shows a snapshot of the first event detection and characterisation provided by PRESTo at the instant when only three stations have triggered and the first alert is issued. Although based on few initial P-measurements, the early magnitude estimation with only two stations ($M_L = 6.8$) is close to the final value (i.e. 6.5) as inferred from authoritative catalogues.

The blind zone is the region where S-waves arrive before the first alert is issued, and it corresponds to the circular area where no lead time is available and no safety actions can be undertaken. Given the station's available density at that time, we observe that the blind zone has a radius of 36 km. Despite the fact that, under such conditions, the municipalities affected by the most severe damage level could not have been alerted, the comparison with the macroseismic field estimated by Giorgetti (1976) shows that some of the municipalities in the area of intensity VII and most of those in the area of intensity VI could have potentially received an alert (Fig. 3). For instance, at the city of Pordenone (falling within the area of intensity VII and located about 65 km from the epicentre), we measure a lead time of about 9 s. Furthermore, for the area included within isoseismal level VI (i.e. where the perceived ground shaking level is strong), the lead time could have been between about 15 and 20 s (e.g. 14 s for Trieste, and 21 s for Treviso, Fig. 3). Considering the network geometry that exists nowadays, we estimated that, for an event with the same epicentre of the 1976 one, the blind zone radius may shrink to about 22 km. For instance, in the case of Pordenone, the lead time might increase to about 13 s.

Figure 4 shows, still for the playback of the Friuli 1976 earthquake, the PDZ obtained estimated from the Pd measurements at the instant when the first four stations have triggered. Interestingly, despite the PDZ not showing the complex shape of isoseismal level VII, this was somehow expected given the few stations available for the analysis; in first approximation this early estimation of the damage extension matches reasonably well with the size of the observed damage zone by Giorgetti (1976). As shown by Colombelli et al. (2012) on Japanese data examples, whenever a dense network of stations is available, the PDZ maps can reproduce the extension of the damage area well (i.e. the area for which the observed macroseismic intensity is larger than VII).

5　PRESTo performance on CE³RN

Since the beginning of 2014, PRESTo (version 0.2.7) has been under experimentation in the transnational area including north-eastern Italy, Slovenia and Austria. During this preliminary test phase, in order to avoid overloading the Antelope system managing the CE³RN, a dedicated SeisComP server (SeisComP, 2009) has been set up at the OGS CRS data centre in Udine with the aim of collecting and converting in SeedLink (Heinloo, 2000) the data of 20 accelerometric stations from the Antelope system (Fig. 1), and pushing them towards a dedicated PRESTo system at RISSC-Lab in Naples.

After an initial period during which we tested different setups of the system parameters, since the end of March 2014 we have been experimenting with the velocity model used for routine earthquake analysis and bulletin production at OGS (OGS, 1995–2013); a minimum number of five stations required to trigger within 12 s for event declaration; the coefficients of the empirical correlation laws between the peak displacement (Pd) measured on short time windows of P-waves and the earthquake magnitude (M) estimated by Lancieri and Zollo (2008); and the Akkar and Bommer (2007) ground motion prediction equation.

Since the station distribution has a key role in determining the resilience of a system, that is to say the network rapidity in issuing EEW alerts, we estimated for the CE³RN network the time of the first alert and the blind-zone extent when three stations have triggered (Fig. 5). The analysis was carried out considering a grid of virtual seismic sources (i.e. a node each of 0.05° × 0.05° for a total of 9801 nodes) with a fixed depth at 6.4 km.

Following Picozzi et al. (2015b), the time of the first alert is defined as the time when P-waves reached the third station

Figure 4. Same as Fig. 3 but showing the PDZ (pink area) corresponding to real-time estimation of the area with macroseismic intensity equal to or higher than VII.

of the network. Furthermore, the BZ is defined as the sum of three delays: (1) the time of the first alert, (2) a fixed delay for the telemetry and computation equal to 2 s, selected according to the value recorded with PRESTo at the ISNet accelerometric network in southern Italy over a long period of testing (Satriano et al., 2011), and (3) the constraint of having 2 s long P-wave time windows at an N-1 station used by RTLoc, which is the needed information for RTMag to estimate the magnitude. This latter constraint is due to the fact that at the instant when RTLoc locates an event with N stations, RTMag provides the first magnitude estimation using N-1 stations, under the condition that they recorded at least 2 s of P-waves. Finally, the sum of these three times is converted in the radius of BZ by multiplying it by the S-wave velocity, assuming that this latter value is equal to $3\,\mathrm{km\,s^{-1}}$.

Figure 5a shows that the time of the first alert is less than or equal to 10 s for the central area of the network, which includes the Friuli 1976 earthquake's epicentre and the Italian–Slovenian boundary, where the station's density is high. The first alert time, and the smallest as well as the larger values, are in general elongated approximately in the east–west direction, according to the network geometry. Also, the blind-zone map shows a similar trend, having the smallest values (i.e. below 25 km) in the Friuli 1976 earthquake's epicentre area, with larger values towards the network boundaries (Fig. 5b).

Concerning the real-time testing of the EEWS, since the end of May 2014, that is to say when a stable configuration of the EEWS was found, PRESTo (version 0.2.7) detected in real time 23 earthquakes, while one event was missed (i.e. event no. 21, Table 1).

Figure 6 shows that the performance of the system in locating the earthquakes has in general been very good, with 18 events out of 23 located within 10 km of the authoritative value. Only in one case is the discrepancy between 10 and 50 km, and in four cases it is larger than 50 km. Concerning

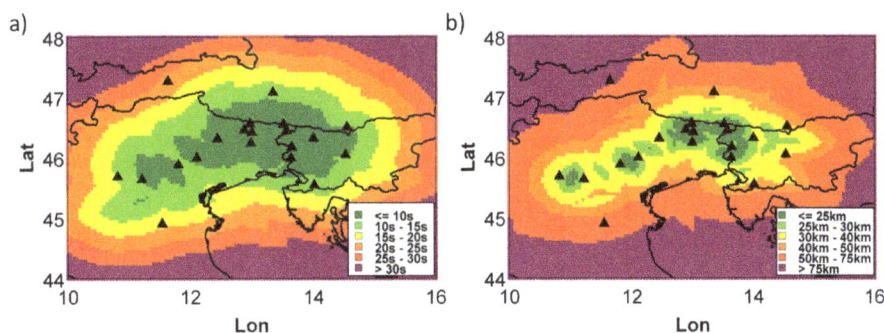

Figure 5. Distribution of time of the first alert (**a**) and dimension of the blind zone (**b**) for the grid of synthetic sources.

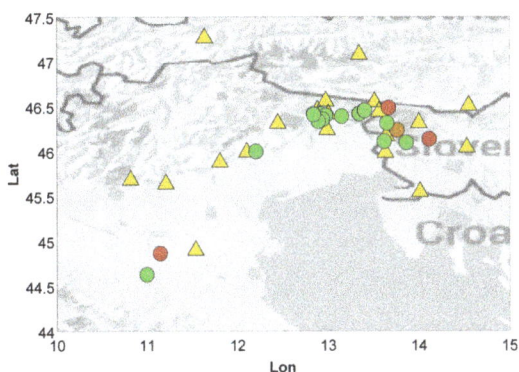

Figure 6. CE^{3}RN stations (yellow). Location error within 10 km (green), between 10 and 50 km (orange), and larger than 50 km (red).

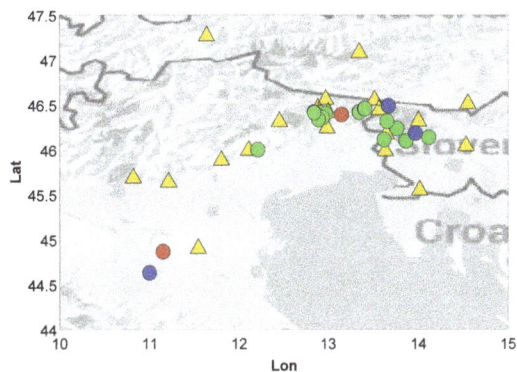

Figure 7. Same as Fig. 1 but showing the correctly detected (green), missed (red), and false (blue) events.

the depth estimation, it must be kept in mind that 90 % of the events in this region are related to a seismogenic layer placed at a depth of about 8 km (Gruppo di lavoro MPS, 2004). The peculiar distribution of events in depth, together with the observation that, given the Pd vs. M relationship adopted, location discrepancies of the order of 15 km determine a magnitude error within 0.5 magnitude units, led the depth estimation to be, for the moment, a parameter of minor importance in our experiment. Similarly to what was already observed in the Irpinia region (Satriano et al., 2011), the hypocentral locations for the events inside the network are generally well constrained starting from the very first estimates. For the events outside the network, the azimuth is well determined, but there is typically a larger uncertainty in the distance.

In order to quantitatively assess the EEWS performance, we compared the EW magnitude (M_{EW}) with the authoritative one (M_{BULL}), and we declared success (S) when M_{EW} falls within a ± 0.5 interval around M_{BULL}, missed (M) when M_{EW} is lower than $M_{BULL} - 0.5$ units, and false (F) when M_{EW} is higher than $M_{BULL} + 0.5$ units. Table 1 shows that the system had 17 successful detections (70.8 %), 3 false detections (12.5 %), and 4 missed events (16.7 %), of which

2 were detected but with underestimated magnitude, 1 was a M_B, 5 occurred in Greece (event no. 20, http://cnt.rm.ingv.it), and 1 was not detected (event no. 2). In general, we observed that both the mis-detection and the wrong location and magnitude estimation occurred when the events were located out of the network, or where the latter has a lower station density (i.e. no. 16, no. 20, and no. 21, Table 1). On the contrary, Fig. 7 shows that when the events occur in the area of higher station density, which also corresponds to higher seismic risk areas, the estimation of EEW magnitude is generally correct. Figure 8 shows, as an example, the good detection of event no. 1 (Table 1) that occurred in Slovenia.

Recently, on the occasion of the $M = 4.1$ event that occurred nearby the town of Udine, Italy (i.e. event no. 24 of 30 January 2015; Table 1), we observed that PRESTo provided a correct location, but estimated an EW magnitude 0.6 units less than the authoritative one (3.5 M_{EW}, 4.1 M_{BULL}). The location being accurate, we guessed that the discrepancy between the early warning and the bulletin magnitude estimates might be related to the parameters of the peak displacement (Pd) vs. M relationships. We decided to investigate this case in more detail by using the recordings of this event to run an off-line PRESTo playback. In particular, the playback was run using new parameters of the

Table 1. Earthquake detected by PRESTo at CE³RN during the period from May to December 2014. The early warning (EW) estimates are compared with those of the OGS-CRS bulletin (BULL; from OGS, 1995–2013). EEW performance: success (S), missed (M), false (F).

ID	Date (yyyy-mm-dd) and time (UTC)	M_{BULL} (±0.3)	Lon_{BULL} (°)	Lat_{BULL} (°)	M_{EW}	Lon_{EW} (°)	Lat_{EW} (°)	$M_{EW} - M_{BULL}$	Time first info loc. & M (s)	EEW perf.
1	29 May 2014 07:24:18.63	3.8	13.862	46.098	3.5	13.8511	46.0967	−0.3	10.0	S
2	2 Jun 2014 02:15:03.02	2.0	12.915	46.414	2.0	12.9662	46.4153	0	13.8	S
3	19 Jun 2014 11:26:21.40	2.6	14.114	46.137	2.8	14.1955	45.4527	0.2	15.5	S
4	24 Jun 2014 22:43:25.39	2.7	13.762	46.237	2.8	13.9952	46.6362	0.1	40.2	S
5	29 Jun 2014 18:39:32.15	2.1	12.916	46.414	1.9	12.8762	46.3952	−0.2	50.2	S
6	5 Jul 2014 15:01:14.57	1.7	13.342	46.418	1.8	13.3719	46.4447	0.1	32.8	S
7	5 Jul 2014 15:47:05.50	1.6	13.344	46.418	1.7	13.3719	46.4547	0.1	51.6	S
8	7 Jul 2014 10:50:38.87	2.8	12.206	46.001	3.1	12.3071	45.9876	0.3	14.1	S
9	20 Jul 2014 14:44:13.58	2.4	13.668	46.486	3.7	10.2293	45.0643	1.3	71.5	F
10	25 Jul 2014 06:32:00.58	1.9	12.972	46.398	1.6	12.9662	46.3952	−0.3	68.1	S
11	8 Aug 2014 12:14:16.38	2.6	12.917	46.361	3.1	12.9325	46.3502	0.5	13.2	S
12	1 Sep 2014 00:50:52.70	–	–	–	2.3	13.9646	46.1859	–	23.9	F
13	12 Sep 2014 15:50:52.85	2.2	13.401	46.455	2.0	13.4283	46.4495	−0.2	21.2	S
14	12 Sep 2014 15:53:45.06	2.0	13.405	46.455	2.0	13.4735	46.4694	0	57.0	S
15	18 Sep 2014 14:24:41.45	2.2	12.937	46.356	1.8	12.9325	46.3502	−0.4	11.5	S
16	5 Oct 2014 07:09:23.00	2.5	10.997	44.631	3.9	11.0858	44.6298	1.4	180.5	F
17	22 Nov 2014 03:22:35.41	1.9	13.650	46.316	2.0	13.6748	46.3382	0.1	103.9	S
18	5 Dec 2014 09:11:36.31	2.8	12.835	46.418	2.2	12.8357	46.4183	−0.6	11.5	M
19	7 Dec 2014 08:00:32.35	1.8	13.620	46.113	2.1	13.6208	46.1138	0.3	92.1	S
20	11 Dec 2014 22:26:02.39	4.9	20.444	38.478	3.7	14.5266	44.6075	−1.2	44.6	M
21	12 Dec 2014 07:01:25.00	3.5	11.146	44.866		–	–	–		M
22	18 Jan 2015 14:42:23.98	2.9	12.890	46.335	2.7	12.8538	46.3351	−0.2	12.0	S
23	22 Jan 2015 15:34:35.27	1.7	12.838	46.408	1.3	12.8311	46.4151	−0.4	10.2	S
24	30 Jan 2015 00:45:48.51	4.1	13.148	46.391	3.5	13.1463	46.3751	−0.6	8.4	M

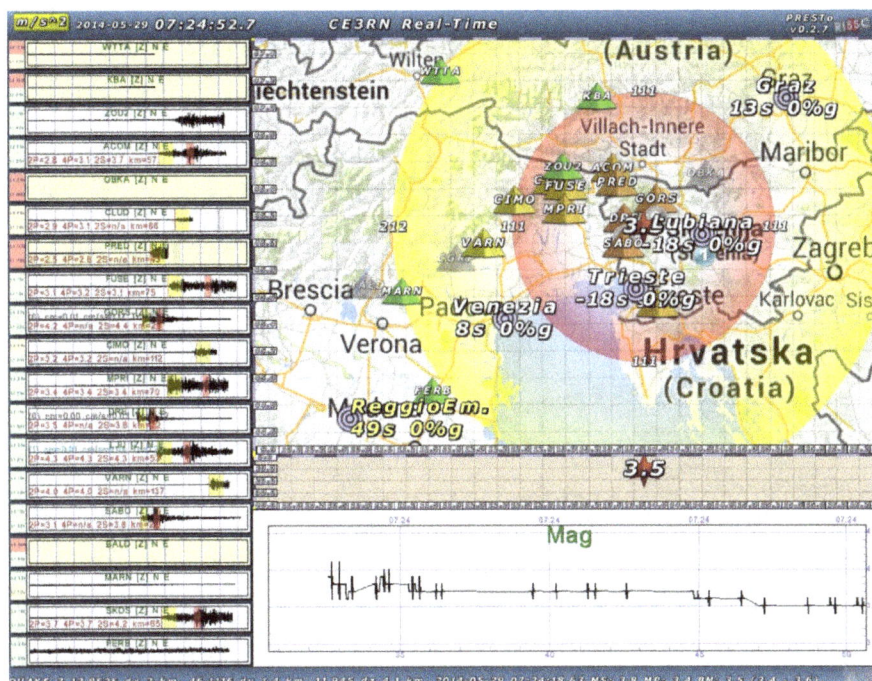

Figure 8. Snapshot of the PRESTo system during the 29 May 2014 $M_L = 3.8$ Slovenian earthquake (event no. 1, Table 1).

Pd vs. M relationship derived from the local magnitude law used by the INGV. Figure 9 shows that the new law provides potential M_{EW} estimates in better agreement with the M_{BULL} (4.2 M_{EW}, 4.1 M_{BULL}). However, it is worth mentioning that the magnitude law used by the INGV is the one computed for the southern California region, which mostly adheres to actual Italian data for station–hypocentre distances greater than 100 km, whereas it overestimates the local magnitude at closer stations (M. Di Bona, personal communication, 2015; http://iside.rm.ingv.it). More tests on this point are needed before drawing a conclusion.

Concerning the few wrong event characterisations, we guess that the low magnitude of the events might have played a major role. Indeed, small magnitudes lead to a low signal-to-noise ratio of the recordings, which in turn makes the real-time analysis more difficult than in the case of moderate size events. This issue can be overcome by considering velocity streams, a feature that we included in the newest version of PRESTo (PRESTo 0.2.8; http://www.prestoews.org) and that in the near future will also be adopted at CE³RN.

The time when the first EEW information on the location and magnitude of the earthquake was available is also reported in Table 1, as the time after the first P arrival detected at a CE³RN station (Fig. 10). We observe that, in 10 cases, the EEW information is available within 15 s (the minimum value of 8.4 s has been observed for event no. 24 of 30 January 2015), while in 13 cases the delay was larger than 15 s (the maximum vale was 180.5 s for event no. 16 of 22 November 2014). The spatial distribution of the delays

(Fig. 10) highlights that, for EEW purposes, the reasons for the larger telemetry delays of stations in the Slovenian sector should be better investigated.

6 Conclusions

This work presents the preliminary results of a feasibility study carried out with EEW platform PRESTo in the high seismic hazard region including north-eastern Italy, Slovenia and Austria, where the 1976 Friuli earthquake occurred.

Results from the offline analysis using the software platform for EW, PRESTo, indicate that, despite the network geometry at that time being rather poor, the EEWS could have been potentially very useful. Indeed, we estimated that the blind-zone radius could have been of the order of 36 km, and that municipalities located within the intensity VI and VII areas could have potentially benefited from an alert. Of course, implementing an EEWS requires, besides these scientific aspects, many further issues to be taken into consideration. For instance: the definition of actions that end users could effectively put into operation within the available lead time for the reduction of their exposure to the seismic risk; cost–benefit analysis of the aforementioned actions; the definition, test, and validation of the procedures which allow the implementation of these mitigation actions; a comprehensive campaign of information on what has to be done; and, finally, a clear attribution of the responsibilities.

Interestingly, we also found that, in the case of a large event with a similar epicentre to the 1976 Friuli earthquake,

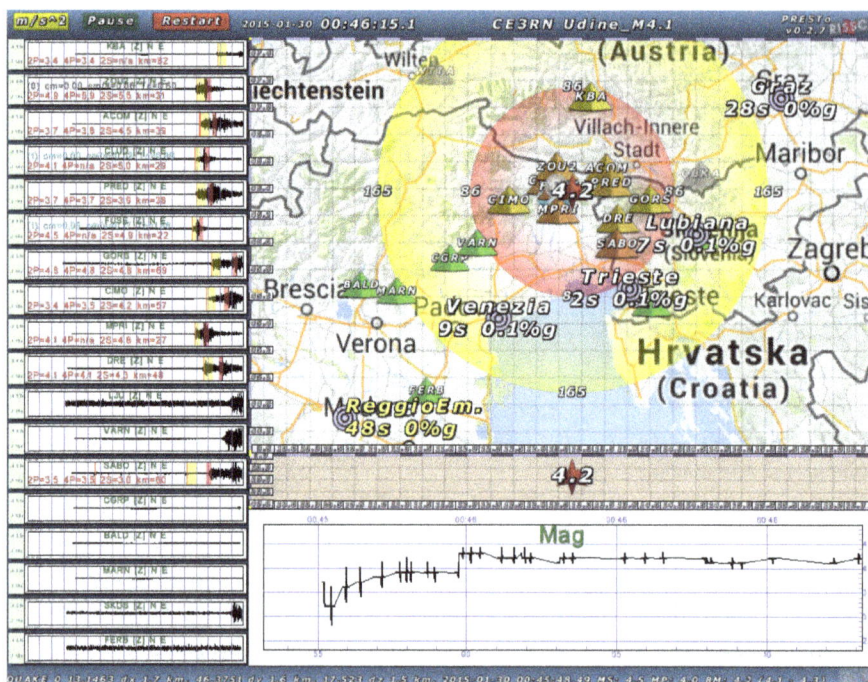

Figure 9. Snapshot of the playback of the PRESTo system during the 30 January 2015 $M_L = 4.1$ earthquake (event no. 24, Table 1) using new parameters of the $\log(Pd)$ vs. M relationship.

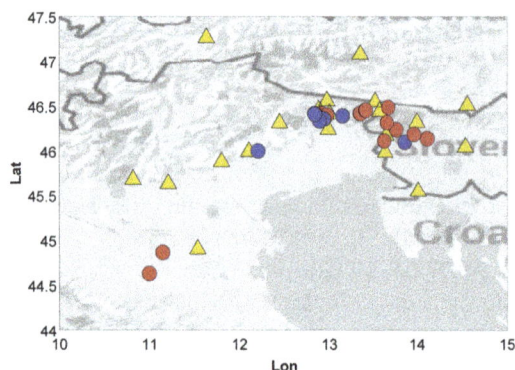

Figure 10. CE^{3}RN stations (yellow). Delays less than 15 s (blue) or larger (red).

the performance of the EEWS would improve, considering the actual CE^{3}RN network configuration. In particular, for such a scenario, we found that three major centres in the region (i.e. Pordenone, Trieste, and Ljubljana) could fall within isoseismal level VI (i.e. experiencing a strong ground shaking) but potentially benefit from a lead time longer than 10 s. As discussed by Goltz (2002), when the population is trained to rapidly respond and take protective measures (e.g. duck and cover, turn off gas burners, move away from windows or equipment, etc.), even fewer than 10 s can help to reduce the risk of injury from an earthquake's secondary effects.

During the period May–December 2014, PRESTo detected in real time 23 earthquakes in the magnitude range 1.7 to 4.1, of which 14 were correctly detected, while 4 and 3 events resulted in missed and false alerts, respectively. Despite the testing period still being too short to come up with definitive conclusions, it seems that the EEWS given by the integration of PRESTo and CE^{3}RN is efficient with respect to earthquakes that occur nearby the area with higher station density. Nevertheless, more testing and an improvement in the system are necessary to cope with events occurring out of the network, and in general where it has a lower station density. With respect to this last issue, we are evaluating to increase the network density, including in the EEWS also stations with velocimetric sensors.

The testing period of the EEW system is carried out primarily with the goal of highlighting the existence of weak points (i.e. in the hardware, network management and analysis software with respect to the seismicity of the area). In fact, besides the specific characteristics of an EEW algorithm, the performance of an EEW system strongly depends also on technological issues, like for example the efficiency of the data telemetry and the seismic noise level at the stations. For this reason, especially these latter two aspects will be studied in the next tests of the EEWS. Of course, the realisation of the EEWS in the area monitored by CE^{3}RN will be accompanied by an extensive activity of communication and training, specifically tailored for both the population and the different stakeholders.

Besides the standard application of EEW, the use of PRESTo in the area surveyed by CE^{3}RN will give a potential benefit to local civil protection agencies. In the case of a very strong shock, the standard monitoring network equipped with modern BB sensors has a saturation zone that may hamper immediate response (e.g. see Fig. 2 from Faenza et al., 2011) in a radius of the order of 100 km. This means that the epicentral location is available when the strongest S-wave phase has already affected the area. On the contrary, an EEW system may broadcast the information to civil protection centres before the strong ground motion can cause potential failure or hampering of the communication system. Hence, civil protection would have the information necessary to act immediately, according to the severity of the situation.

Acknowledgements. We would like to thank the Associate Editor J. Clinton, C. Cauzzi and an anonymous reviewer for their comments and suggestions that allowed us to significantly improve the manuscript's content and form.

This work has been partially supported by the REAKT-Strategies and tools for Real Time Earthquake RisK ReducTion FP7 European project funded by the European Community's Seventh Framework Programme (FP7/2007-2013) under grant agreement no. 282862.

References

Akkar, S. and Bommer, J. J.: Empirical prediction equations for peak ground velocity derived from strong-motions records from Europe and the Middle East, Bull. Seismol. Soc. Am., 97, 511–530, 2007.

Alcik, H., Özel, O., Apaydin, N., and Erdik, M.: A study on warning algorithms for Istanbul earthquake early warning system, Geophys. Res. Lett., 36, L00B05, doi:10.1029/2008GL036659, 2009.

Allen, R. M., Gasparini, P., Kamigaichi, O., and Böse, M.: The status of earthquake early warning around the world: an introductory overview, Seismol. Res. Lett., 80, 682–693, 2009.

Böse, M., Ionescu, C., and Wenzel, F.: Earthquake Early Warning for Bucharest, Romania: Novel and revised scaling relations, Geophys. Res. Lett., 34, L07302, doi:10.1029/2007GL029396, 2007.

Böse, M., Hauksson, E., Solanki, K., Kanamori, H., and Heaton, T. H.: Real-time testing of the on-site warning algorithm in Southern California and its performance during the July 29, 2008 M_{W} 5.4 Chino Hills earthquake, Geophys. Res. Lett., 36, L00B03, doi:10.1029/2008GL036366, 2009.

Bragato, P. L., Costa, G., Gallo, A., Gosar, A., Horn, N., Lenhardt, W., Mucciarelli, M., Pesaresi, D., Steiner, R., Suhadolc, P., Tiberi, L., Živčić, M., and Zoppé, G.: The Central and Eastern European Earthquake Research Network – CE^{3}RN, EGU General Assembly 2014, 27 April–2 May 2014, Vienna, Austria, 2014.

Carulli, G. B. and Slejko, D.: The 1976 Friuli (NE Italy) earthquake, Giornale di Geologia Applicata, 1, 147–156, 2005.

Colombelli, S., Amoroso, O., Zollo, A., and Kanamori, H.: Test of a Threshold-Based Earthquake Early Warning Using Japanese Data, Bull. Seismol. Soc. Am., 102, 1266–1275, doi:10.1785/0120110149, 2012..

Espinosa-Aranda, J. M., Cuellar, A., Garcia, A., Ibarrola, G., Islas, R., Maldonado, S., and Rodriguez, F. H.: Evolution of the Mexican Seismic Alert System (SASMEX), Seismol. Res. Lett., 80, 694–706, 2009.

Faenza, L., Lauciani, V., and Michelini, A.: Rapid determination of the shakemaps for the L'Aquila main shock: a critical analysis, Bollettino di Geofisica Teorica ed Applicata, 52, 407–425, doi:10.4430/bgta0020, 2011.

Giorgetti, F.: Isoseismal map of the May 6, 1976 Friuli earthquake, Boll. Geofis. Teor. Appl., 19, 707–714, 1976.

Goltz, J. D. L.: Introducing earthquake early warning in California: A summary of social science and public policy issues, technical report, Governor's Off. of Emergency Serv., Pasadena, Calif., 2002.

Gruppo di lavoro MPS: Redazione della mappa di pericolosita' sismica prevista dall'Ordinanza PCM 3274 del 20 marzo 2003, Rapporto conclusivo per il dipartimento di Protezione Civile, INGV, aprile 2004, Milano, Roma, 65 pp. + 5 appendici, available at: http://zonesismiche.mi.ingv.it/elaborazioni/ (last access: 31 March 2015), 2004.

Heinloo, A.: SeedLink design notes and configuration tips, http://geofon.gfz-potsdam.de/geofon/seiscomp.de/geofon/seiscomp/seedlink.html (last access: 31 March 2015), 2000.

Horiuchi, S., Negishi, H., Abe, K., Kamimura, A., and Fujinawa, Y.: An automatic processing system for broadcasting system earthquake alarms, Bull. Seismol. Soc. Am., 95 347–353, 2005.

Hoshiba, M.: Real-time correction of frequency-dependent site amplification factors for application to Earthquake Early Warning, Bull. Seismol. Soc. Am., 103, 3179–3188, doi:10.1785/0120130060, 2013.

Iannaccone, G., Zollo, A., Elia, L., Convertito, V., Satriano, C., Martino, C., Festa, G., Lancieri, M., Bobbio, A., Stabile, T. A., Vassallo, M., and Emolo, A.: A prototype system for earthquake early-warning and alert management in southern Italy, Bull. Earthq. Eng., 8, 1105–1129, doi:10.1007/s10518-009-9131-8, 2010.

Lancieri, M. and Zollo, A.: Bayesian approach to the real-time estimation of magnitude from the early P and S wave displacement peaks, J. Geophys. Res., 113, B12302, doi:10.1029/2007JB005386, 2008.

Lomax, A., Satriano, C., and Vassallo, M.: Automatic picker developments and optimization: FilterPicker – a robust, broadband picker for real-time seismic monitoring and earthquake early-warning, Seismol. Res. Lett., 83, 531–540, doi:10.1785/gssrl.83.3.531, 2012.

Luzi, L., Hailemikael, S., Bindi, D., Pacor, F., Mele, F., and Sabetta, F.: ITACA (ITalian ACcelerometric Archive): a web portal for the dissemination of Italian strong-motion data, Seismol. Res. Lett., 79, 716–722, 2008.

OGS: Bollettino della Rete Sismometrica del Friuli–Venezia Giulia e del Veneto, OGS (Istituto Nazionale di Oceanografia e di Geofisica Sperimentale), Centro di Ricerche Sismologiche, Udine, Italy, 1995–2013.

Pacor, F., Paolucci, R., Ameri, G., Massa, M., and Puglia, R.: Italian strong motion records in ITACA: overview and record processing, Bull. Earthq. Eng., 9, 1723–1739, doi:10.1007/s10518-011-9327-6, 2011.

Peng, H. S., Wu, Z. L., Wu, Y. M., Yu, S. M., Zhang, D. N., and Huang, W. H.: Developing a prototype earthquake early warning system in the Beijing Capital Region, Seismol. Res. Lett., 82, 394–403, 2011.

Picozzi, M., Colombelli, S., Zollo, A., Carranza, M., and Buforn, E.: A Threshold-Based Earthquake Early-Warning System for Offshore Events in Southern Iberia, Pure Appl. Geophys., doi:10.1007/s00024-014-1009-2, in press, 2014.

Picozzi, M., Emolo, A., Martino, C., Zollo, A., Miranda, N., Verderame, G., Boxberger, T., and the REAKT Working Group: Earthquake Early Warning System for Schools: A Feasibility Study in Southern Italy, Seismol. Res. Lett., 86, 398–412, doi:10.1785/0220140194, 2015a.

Picozzi, M., Zollo, A., Brondi, P., Colombelli, S., Elia, L., and Martino, C.: Exploring the Feasibility of a Nation-Wide Earthquake Early Warning System in Italy, J. Geophys. Res.-Solid, doi:10.1002/2014JB011669, in press, 2015b.

Satriano, C., Lomax, A., and Zollo, A.: Real-Time Evolutionary Earthquake Location for Seismic Early Warning, Bull. Seismol. Soc. Am., 98, 1482–1494, doi:10.1785/0120060159, 2008.

Satriano, C., Wu, Y.-M., Zollo, A., and Kanamori, H.: Earthquake early warning: Concepts, methods and physical grounds, Soil Dynam. Earthq. Eng., 31, 106–108, doi:10.1016/j.soildyn.2010.07.007, 2010.

Satriano, C., Elia, L., Martino, C., Lancieri, M., Zollo, A., and Iannaccone, G.: PRESTo, the earthquake early warning system for Southern Italy: concepts, capabilities and future perspectives, Soil. Dyn. Earthq. Eng., 31, 137–153, doi:10.1016/j.soildyn.2010.06.008, 2011.

SeisComP: Seismological communication processor, 2009, http://geofon.gfz-potsdam.de/geofon/seiscompS, last access: January 2010.

Steim, J. M. and Reimiller, R. D.: Timeliness of Data Delivery from Q330 Systems, Seismol. Res. Lett., 85, 844–851, doi:10.1785/0220120170, 2014.

Wu, Y. M. and Zhao, L.: Magnitude estimation using the first three seconds P-wave amplitude in earthquake early warning, Geophys. Res. Lett., 33, L16312, doi:10.1029/2006GL026871, 2006.

Zollo, A., Amoroso, O., Lancieri, M., Wu, Y. M., and Kanamori, H.: A threshold-based earthquake early warning using dense accelerometer networks, Geophys. J. Int., 183, 963–974, doi:10.1111/j.1365-246X.2010.04765.x, 2010.

Zollo, A., Colombelli, S., Elia, L., Emolo, A., Festa, G., Iannaccone, G., Martino, C., and Gasparini, P.: An Integrated Regional and On-Site Earthquake Early Warning System for Southern Italy: Concepts, Methodologies and Performances, in: Early Warning for Geological Disasters, Advanced Technologies in Earth Sciences, edited by: Wenzel, F. and Zschau, J., Springer, Berlin, Heidelberg, 117 pp., doi:10.1007/978-3-642-12233-0_7, 2014.

28

The impact of phenomena El Niño and La Niña and other environmental factors on episodes of acute diarrhoea disease in the population of Aguascalientes, Mexico: a case study

Martha Esthela Venegas-Pérez[1], **Elsa Marcela Ramírez-López**[1], **Armando López-Santos**[2], **Víctor Orlando Magaña-Rueda**[3], **and Francisco Javier Avelar-González**[1]

[1]Biochemical Engineering Department/Physiology and Pharmacology Department, Universidad Autónoma de Aguascalientes, Aguascalientes, México
[2]Unidad Regional Universitaria de Zonas Áridas, Universidad Autónoma Chapingo, Bermejillo, Durango, México
[3]Center for Atmospheric Sciences, Universidad Nacional Autónoma de México, Distrito Federal, México

Correspondence to: Elsa Marcela Ramírez-López (emramir@correo.uaa.mx)

Abstract. Acute diarrhoea diseases (ADDs) are one of the major health problems in Aguascalientes, Mexico. Due to the risk of significant increases of ADDs in the hot season, it has been necessary to determine the weather conditions that might lead to escalating ADD events. The effects of El Niño and La Niña phenomena on the morbidity rate of ADD (MRADD) in the State of Aguascalientes were determined during the period of 2000–2010. The MRADD was calculated from cases reported by the State Health Department. The Oceanic Niño Index (ONI) was obtained from the US National Oceanic and Atmospheric Administration. The impact of El Niño and La Niña on the MRADD was determined using the Pearson correlation coefficient and analysis of variance (ANOVA). The results gave a significant inverse correlation between El Niño phenomenon and MRADD ($r = -0.55$, $P = 0.001$), but a correlation was not observed on the La Niña phenomenon ($r = -0.022$, $P = 0.888$). Field data showed significant inverse influence of El Niño on MRADD for the years 2000–2010.

1 Introduction

Several authors have investigated the ways in which El Niño Southern Oscillation (ENSO) has occurred in Mexico. The main consequences of El Niño phenomenon in Mexico are the intensification of winter rainfall in the northwest and northeast of Mexico while decreasing toward the south. ENSO has produced colder winters and drier and warmer summers often producing severe droughts.

Several studies have shown a potential impact of ENSO on human health with vector-borne diseases (Schaffner and Mathis, 2014) and with acute diarrhoea diseases (ADDs) (Patz and Olson, 2006). These studies agree that the ENSO can increase the ambient temperature that could enhance the survival and persistence of many microorganisms causing ADDs because microorganisms can reproduce faster in warmer conditions.

Subsequent studies have shown that climatic factors significantly affect seasonal diarrhoea in susceptible populations (El-Fadel et al., 2012).

Global deaths of children aged less than 5 years from diarrhoea were estimated at 1.87 million (95 % confidence interval, CI: 1.56–2.19), approximately 19 % of total child deaths. WHO African and South-East Asia Regions combined showed 78 % (1.46 million) of all diarrhoea deaths occurring among children in the developing world; 73 % of these deaths occurred in 15 developing countries (Boschi-Pinto et al., 2008).

In Mexico, the recorded mortality rate of ADDs was 28 deaths in 2013 and 16 519 new cases per 100 000 children under one year of age were also reported. Moreover, 266 deaths were registered with a rate of 3.5 per 100 000 children among 1 to 4 years of age. The ADDs were the fifth cause of death in this age group.

In the state of Aguascalientes, ADDs were the fifth leading cause of infant death in the year 2013 (ISEA, 2014). It has been estimated that children under 5 years of age can experience between two to four diarrhoeal episodes per year.

According to Hernández (2002) and Ferrano et al. (2003) the age groups that are most affected by ADDs are the small children (under 5 years) and the elderly (over 60 years), who are more sensitive to suffer the effects due to excessive loss of electrolytes affecting the body during illness and can cause severe dehydration (Hernández et al., 2011).

The aim of this study was to determine the impact of El Niño and La Niña phenomena on the ADDs as well as other possible environmental factors during the period 2000–2010 in the population of the state of Aguascalientes, Mexico.

2 Data and methods

2.1 Area of study

The geographical unit of the study consisted of the State of Aguascalientes located in central Mexico
with geographic coordinates $22°27'36''$ N, $21°37'12''$ S of north latitude and $101°50'05''$ E, $102°52'41''$ W of west longitude (INEGI, 2014). It has a surface of $5616\,km^2$. The state of Aguascalientes represents 0.3 % of the Mexican surface area. Three major natural regions through its territory include the Sierra Madre Occidental, the Central Mesa or Plateau and Neovolcanic. The territory of the state is predominantly flat. The State is comprised of eleven municipalities with a total of 1 184 996 inhabitants (INEGI, 2010) concentrating 67 % of the population in the City of Aguascalientes. The climate prevailing in the state is semi-dry in 86 % of its territory, 14 % humid temperate is localized southwest and northwest of the state. The average annual temperature is 17–18 °C. The highest temperature (30 °C or more) occurs in the months of May and June and the lowest temperature (about 4 °C) in January. Rain is limited and occurs during the summer. Total annual precipitation is 526 mm (INEGI, 2010).

2.2 ADDs database

Information on the number of cases of more frequent and representative in the state of Aguascalientes was obtained from the Aguascalientes State Health Institute (ISEA, Spanish acronym). Data also included information on intestinal infections by other organisms and ill-defined with ICD 10 code: A04, A08-A09. The A04 code includes intestinal bacterial infections due *Escherichia coli*: enteropathogenic, enterotoxigenic, enterohemorrhagic and enteroinvasive, *Campylobacter Spp.*, *Yersinia enterocolitica*, and *Clostridium difficile*. The A08 code includes intestinal infections caused by viruses and other specified bodies: Rotavirus, Norwalk agent, and Adenovirus. The A09 code includes diarrhoea and gastroenteritis of presumed infectious origin. This information pertained for all the groups of 0–4 years, 5–59 years and above

60 years of age. Cases of ADDs included both men and women. The ADD rates were calculated dividing the numbers of episodes reported between the numbers of the annual population to express morbidity rates per 10 000 habitants. An ecological time-series study was also conducted.

2.3 El Niño and La Niña database

The information of the Oscillation El Niño Index (ONI) was obtained from the US National Oceanic and Atmospheric Administration (NOAA) for the period 2000–2010 (NOAA, 2014). This oceanic index defined the presence of El Niño with at least five successive overlapping three-month seasons experiencing sea surface temperature decreases of more than 0.5 °C; was defined as "La Niña" the consecutive presence of five or more values less than −0.5 °C.

2.4 Climate database

The daily weather variables of maximum temperature and precipitation were obtained from eighteen weather stations of the National Water Commission (CONAGUA, Spanish acronym). The period was from 2000 to 2010. Data quality control considered that 96 % of annual records were present in the time series of which was obtained the average of monthly maximum temperature and accumulated rainfall.

2.5 Statistical analysis

The statistical processing was performed with the statistical software Minitab® 16.2.3 (Minitab Inc., State College, PA). Descriptive statistics based on the calculation of measures of central tendency and the dispersion was performed. To correlate the phenomena under study with the rate of ADDs, the Pearson correlation coefficients were obtained, as well as to analyze the association between ADD with ONI and the number of days with temperature ≥ 30 °C and the number days with non-zero precipitation.

3 Results

3.1 Oscillation Niño Index and El Niño and La Niña phenomena

During the decade of the study, there was a total of four El Niño events, which were initiated between the months of May and August of the years 2002, 2004, 2006 and 2009, and extended to the autumn of the following year (2003, 2005, 2007 and 2010, respectively). According to the analysis of ONI, two of them were mild (2004 and 2006) and two with moderate intensity (2002 and 2009); ONI maximum values were reached in December 2009 (ONI = 1.6) as well as in November and December of 2002 (ONI = 1.3).

There was a total of five La Niña events. The first La Niña event began in July 1998 and continued until February 2001. The second and the fourth began in November and ended in March of the years 2005–2006 and 2008–2009. The third event occurred in August 2007 and continued until June 2008. The fifth event began in July 2010 and ended in April 2011. The latter phenomenon had strong intensity. The first and the third events varied in intensity between strong and moderate while the second and fourth events had weak intensities.

The lowest values of ONI were for the events that occurred in January and February 2000 (with values of -1.7 and -1.5 respectively), January and February 2008 and October, November and December 2010 (-1.5 in all cases).

3.2   Analysis of El Niño and La Niña phenomenon for those days with temperature above 30 °C and with nonzero precipitation

The average maximum temperature ranged between 20.1 and 32.4 °C, the average minimum temperature between 1.9 and 13.5 °C and the average rainfall from 0 to 252.2 mm. The warmest month was June and the coldest January. These results were based on the weather information for the state of a 31 year-period (1970 to 2010). As shown in Fig. 1, the number of days at or above temperatures of 30 °C in Aguascalientes, Mexico, showed a steadily and linear increase from 2000 to 2010 and the maximum values of days with temperatures at or above the 90th percentile (30 °C) were reached in 2009.

The correlation was sought between the phenomenon El Niño with the number of days with temperatures above 30 °C (90 percentile) and found that there was indeed significant inverse correlation ($r = -0.430$, $P = 0.014$). We did not find a correlation between phenomenon La Niña and the number of days with temperatures at or above 30 °C ($r = -0.228$, $P = 0.146$).

The rainfall pattern was other climate element in Aguascalientes that observed significant changes in atmospheric behaviour. Weather seasons were defined by the peculiarities of rainfall in the state: the rainy season which extended from May to October and the dry season from November to April (Fig. 2).

3.3   ADD analysis for children under 5 years of age

In the period 2000–2010, the average rate of ADD statewide was 20.07 episodes per 10 000 people. This ADD rate was similar to that presented in neutral conditions and during El Niño events (20.46 and 20.25 episodes per 10 000 inhabitants, respectively). During La Niña, there were 18.06 episodes per 10 000 inhabitants which was lower than the ADD rate obtained in neutral conditions or El Niño events, although no significant difference was found (Table 1).

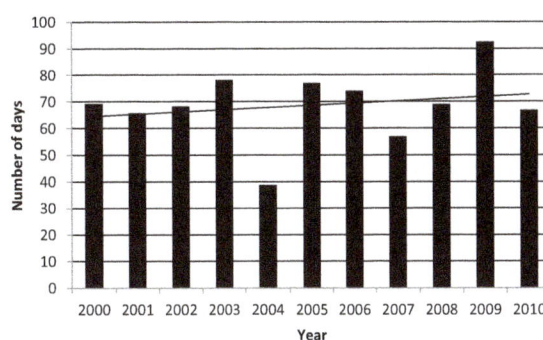

Figure 1. Number of days with temperatures above 30 °C (90th percentile) on Aguascalientes, Mexico.

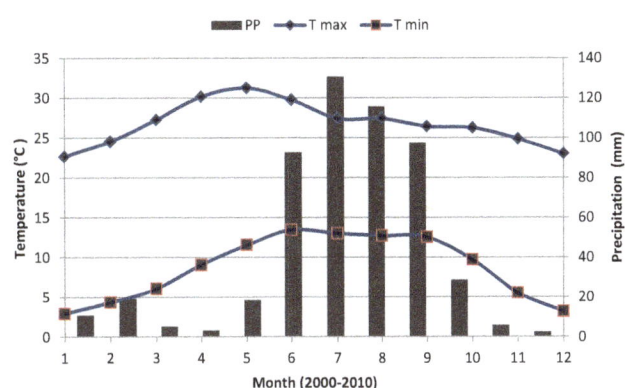

Figure 2. Climograph of Aguascalientes, Mexico for the period 2000–2010.

3.4   ADD Analysis for the population aged 5 and older

In the period 2000–2010, the morbidity of ADD varied between 107.40 and 25.90 episodes per 10 000 inhabitants. The highest rates were reported in June 2008 (70.16 events per 10 000 inhabitants) and from April to August 2010 (85.12, 103.32, 107.40, 95.98 and 76.53 episodes per 10 000 inhabitants, respectively).

During the observed phenomena, the rate of ADD in El Niño conditions was 44.37 episodes per 10 000 habitants. In neutral conditions, the ADD rate was 46.33 episodes per 10 000 inhabitants, and during La Nina phenomenon the ADD rate was 41.41 episodes per 10 000 inhabitants. To correlate the morbidity of ADD with ONI, a significant correlation (Table 2) was observed for the three age groups. In the case of La Niña and neutral conditions no correlation was found with ONI.

The incubation time that the ADD causing microorganism has before the disease symptoms appear are evident. There are time lags of several weeks, one month that may interfere with the interpretation of the quality of the relationship cause-effect between climate variables and ADD.

It is noteworthy to mention that when heavy rain falls after a prolonged drought, ADDs may significantly increase due

Table 1. Description of diarrhoea diseases by age and El Niño, La Niña and neutral from 2000 to 2010.

	El Niño phenomenon Mean ± SD	La Niña phenomenon Mean ± SD	Neutral Condition Mean ± SD	P value*
Children under 5 years	20.25 ± 6.37	18.06 ± 5.48	20.46 ± 4.50	0.000
People 5 to 59 years	44.37 ± 14.36	41.41 ± 13.78	46.33 ± 10.94	0.001
Adults over 60 years	3.86 ± 1.4	3.64 ± 1.23	4.0 ± 0.82	0.001

* Test of Levene

Table 2. Spearman's correlation analysis between diarrhoea diseases and El Niño, La Niña and neutral conditions in Aguascalientes, Mexico from 2000–2010.

	Lag		Children under 5 years	People 5 to 59 years	Adults over 60 years
La Niña phenomenon	0 month	Coefficient	0.361	0.154	0.101
		P value	0.019	0.329	0.523
	1 month	Coefficient	0.491	0.32	0.287
		P value	0.001	0.03	0.108
El Niño phenomenon	0 month	Coefficient	−0.648	−0.552	−0.453
		P value	0.000	0.001	0.009
	1 month	Coefficient	−0.482	−0.45	−0.263
		P value	0.001	0.051	0.019
Neutral Conditions	0 month	Coefficient	−0.079	−0.054	−0.041
		P value	0.59	0.711	0.778
	1 month	Coefficient	−0.034	0.058	0.100
		P value	0.817	0.691	0.493

to the presence of microorganisms in water bodies. Figure 3 shows a significant increase in the number of ADDs occurrences in 2010 most probably due to heavy rain events in 2008 and 2010 with a prolonged drought in between those years in 2009. The number of ADDs occurrences in 2010 was greater than in 2008 and 2009 (Table 3).

3.5 Analyses of the correlation between the number of days with temperatures over 30 °C, and El Niño and La Niña phenomena and the rate of ADD

By relating the phenomenon El Niño with the number of days with temperatures above the 90th percentile (30 °C), it was observed a significant inverse correlation with the increase of days. A significant direct correlation between the rates of ADDs were also determined (Table 4).

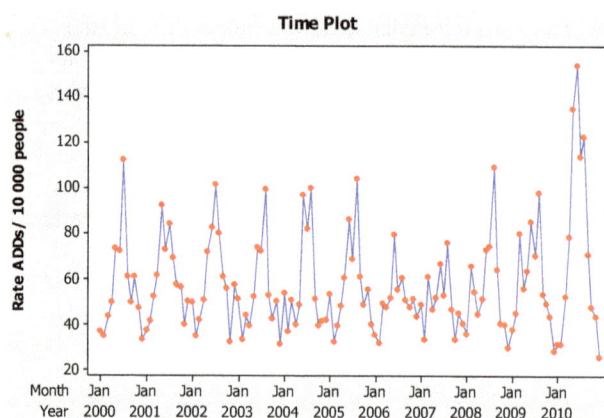

Figure 3. Time Plot rate ADDs.

4 Discussion

In the 2000–2010 period, a correlation of the number of days with non-zero precipitation with the morbidity of ADDs was determined for all three age groups (under 5, 5 to 59 and over 60 years). But no correlation was found with La Niña phenomenon, or under neutral conditions (Table 2).

Although these results do not imply a relationship cause-effect between climate variables and phenomena studied, the results suggest that both the disease and weather are seasonal events. These results also indicate that studies of the correlations between human diseases and weather are important and should be further investigated, as any variation of scale

Table 3. Correlations between diarrhoea diseases and ONI in Aguascalientes from 2009 to 2010 after one heavy rain (2009) a strong drought (2010).

	Lag		Children under 5 years	People 5 to 59 years	Adults over 60 years
La Niña Phenomenon	0 month	Coefficient	0.552	0.527	0.511
		P value	0.001	0.051	0.020
	1 month	Coefficient	0.620	0.451	0.370
		P value	0.008	0.060	0.010
El Niño Phenomenon	0 month	Coefficient	−0.82	−1.459	−0.326
		P value	0.004	0.182	0.058
	1 month	Coefficient	−0.626	−0.128	−0.003
		P value	0.053	0.724	0.090

Table 4. Spearman's correlation analysis between diarrhoea diseases and total days with temperatures above 30 °C and days with precipitation nonzero in Aguascalientes, Mexico from 2000–2010.

	El Niño Phenomenon number of days $T >= 30\,°C$		La Niña Phenomenon number of days $P_p > 0\,mm$		Neutral Condition number of days $P_p > 0\,mm$	
	Coefficient	P value	Coefficient	P value	Coefficient	P value
Children under 5 years	0.573	0.001	0.663	0.000	0.430	0.001
People 5 to 59 years	0.763	0.000	0.714	0.000	0.603	0.000
Adults over 60 years	0.800	0.000	0.626	0.000	0.620	0.000
condition	−0.339	0.018	0.251	0.031	0.029	0.468

in weather patterns would bring seasonal variations in epidemiological patterns.

Similar to other studies that have shown statistical correlations between El Niño and ADDs, the results from this study are in agreement with previous investigations. We observed a morbidity rate increased with an increase in the number of days with temperatures above 30 °C, which occurred under conditions of El Niño phenomenon causing the higher rate of morbidity than the state average under neutral conditions or La Niña phenomenon. This effect is consistent with that observed in the three groups of people in the state of Aguascalientes, Mexico.

Higher temperatures increase the risk of exposure of people to parasitic and bacterial diarrhoea and lengthen the survival of enterotoxigenic *Escherichia coli* bacteria in contaminated food (Black and Lanata, 1995; Checkley et al., 2000). Low temperatures increase the transmission of diarrhoea caused by viruses (Konno et al., 1995) and at intermediate temperatures (18–23 °C), children may be exposed to viral pathogens, bacteria, and protozoa (Checkley et al., 2000).

Something that is worth mentioning is the fact that approximately 96 % of treated and raw sewage generated by the various sectors of the State is directly discharged to the San Pedro River (Anónymous, 2007; Ramírez et al., 2007). In a previous study conducted by Guzmán-Colis et

al. (2011), it was found that water from the San Pedro River showed average concentrations of faecal coliform, about 3.16×106 MPN/100 mL, which exceeded four orders of magnitude the maximum limit.

Regarding the resistance of *Escherichia coli* than were present in this river; Ramírez-Castillo et al. (2013) found that 52 % of the strains were resistant to at least one antimicrobial agent, at least 37.3 % resistant to two agents, and 30.6 % were multi-resistant to multiple drugs. Twelve strains were also resistant to the fluoroquinolone and presented a multi-resistant phenotype.

It is important to mention that the waters of the San Pedro River are directly used for irrigation of vegetables, as well as a source of drinking water for animal's farms nearby to this river. The contamination of the San Pedro River can be a source of bacterial contamination for people and animals, either directly or indirectly through contact to water droplets that evaporate or by the direct consumption of vegetables.

Other studies have found a strain of *E. coli* O157: H7 in fruits and vegetables such as lettuce, radish and alfalfa (Bier, 1991; Como-Sabetti et al., 1997; Hilborn et al., 1999).

The ability of these bacteria to survive outside a host or in wet conditions was evaluated by various researchers to determine the possibility that they could survive in the atmosphere. For example, Rose et al. (1997) successfully isolated the bacterium *Escherichia coli* in various samples of urban dust in Mexico City and varied atmospheric conditions. Bautista-Olivas et al. (2013) found colonies of total coliforms and faecal coliforms in atmospheric water condensed in three areas of Mexico. Although the weather conditions of the state of Aguascalientes differ from the studied areas, these results suggest the *E. coli* bacteria can survive in the water from the atmosphere and environmental dust (under wet conditions and warm temperatures). This outcome suggests a potential for contamination and other pathogenic bacteria under these weather conditions.

Another probable line of transmission of *E. coli* bacteria could be the housefly (*Musca domestica*) as Béjar et al. (2006) showed that the housefly acts as a mechanical vector of enteropathogenic *Escherichia coli*. The presence of these bacteria was also diagnosed, among others, on dozens of species of cockroaches (Boschi-Pinto et al., 2008). The impact of these findings is enhanced by the fact that the propagation of these vectors with increased temperatures results in a greater chance of developing diarrhoea (Dakshinamurty, 1948; Abdullah, 1961).

The reproduction dynamics and mobility of insects along with the increase of days with temperatures above 30 °C can cause the appearance of new vector breeding sites on the banks of rivers and dams. Ambient temperatures exceeding the 90th percentile in rivers and dams can show a decrease in flow. Added heavy rains can cause the formation of new hatcheries that are quickly colonized. It is highly likely that these causes and effects have a high impact on increasing confirmed cases of ADD under El Niño in the population of the state of Aguascalientes.

On the other hand, the increase in the number of days with temperatures above the 90th percentile and behaviour patterns that are more common during warm season; such as increased water demand and little awareness hygiene practices climates (Black and Lanata, 1995; Checkley et al., 2000) and socioeconomic factors (Guerrant et al., 2002; Loyola and Soncco, 2007) such as access to water and sanitation services contribute to increased rates of diarrhoea.

The increase in morbidity of EDAs by *E. coli* in Aguascalientes was associated with an increase in days with higher ambient temperature above the 90th percentile, and these are exacerbated by the El Niño phenomenon.

Understanding the effect of climate variability on the epidemiology of the disease is necessary to take measures in the centre of health care to prevent or reduce the incidence of these diarrhoeal diseases and health education. Also, it is necessary to reinforce appropriate ADD preventive and treatment practices among parents and caregivers of children < 5 years of age.

Further studies for the collection of information such as social, economic, cultural and health areas and household conditions are needed. More studies are suggested to complete the total environment (socioeconomic, environmental, immunological and epidemiological) that accounts for disease dynamics in the study area and to be better prepared for future changes in climate caused by the intense El Niño events.

5 Conclusion

Meteorological and epidemiological data from the period 2000–2010 have shown a correlation between El Niño and ADDs, showing a morbidity rate increased by the increase in the number of days with temperatures above 30 °C, which occurred under conditions of El Niño phenomenon. But no correlation between rates of ADDs and La Niña or neutral conditions were found. The main constraint is associated with the lack of accurate and long term monitoring data on occurrences of ADD, especially related to differentiation among various causes, social and economic classes and environmental factors. Equally significant, the approach adopted in this study is based on the assumption that current associations will remain unchanged in the future which might introduce uncertainties because biological acclimatization as well as technologic and socio-economic developments will likely influence population vulnerability and exposure–response relationships.

Author contributions. M. E. Venegas-Pérez prepared the manuscript with contributions from all co-authors, E. M. Ramírez-López and V. O. Magaña-Rueda Co-tutors of dissertation and reviewers and contributors to the paper, A. López-Santos and F. J. Avelar-González assessors of dissertation and reviewers and contributors to the paper.

Acknowledgements. Authors acknowledge CONACyT for the scholarship to develop a PhD program in Biological Sciences. Special acknowledgements to SEMARNAT-CONACyT (S0010-2008-1) for the financial support to conduct this study.

References

Abdullah, M.: Behavioural effects of temperature on insects, Ohi J. Sci., 61, 212–219, 1961.

Anónymous: Comisión Nacional del Agua (CONAGUA), Estadísticas del agua en México, Secretaría de Medio Ambiente y Recursos Naturales, México, DF, 626 pp., 2007.

Bautista Olivas, A. L., Tovar Salinas, J. L., Mancilla Villa, Ó. R., Magdaleno Flores, H., Ramírez Ayala, C., Arteaga Ramírez, R., and Vázquez Peña, M. A.: Calidad microbiológica del agua obtenida por condensación de la atmósfera en Tlaxcala, Hidalgo y Ciudad de México, Rev. Int. Contam. Ambie, 29, 167–175, 2013.

Béjar, V., Chumpitaz, J., Pareja, E., Valencia, E., Huamán, A., Sevilla, C., Tapia, M., and Saez, G.: Musca domestica como vector mecánico de bacterias enteropatógenas en mercados y basurales de Lima y Callao, Revista Peruana de Medicina Experimental y Salud Publica, 23, 39–43, 2006.

Bier, J. W.: Isolation of parasites on fruits and vegetables, Sotuheast Asian, J. Trop. Med. Pub. Health., 22, 144–145, 1991.

Black, R. E. and Lanata, C. F.: Epidemiology of diarrheal diseases in developing countries, in: Infections of the gastrointestinal tract, edited by: Blaster, M. J., Smith, P. D., Ravdin, J. I., Greenberg, H. B., and Guerrant, R. I., Raven Press, New York, 13–16, 1995.

Boschi-Pinto, C., Velebit, L., and Shibuya, K.: Estimating child mortality due to diarrhoea in developing countries, B. World Health Organ., 86, 710–717, 2008.

Checkley, W., Epstein, L. D., Gilman, R. H., Figueroa, D., Cama, R. I., Patz, J. A., and Black, R. E.: Effects of El Niño and ambient temperature on hospital admissions for diarrhoeal diseases in Peruvian children, The Lancet, 355, 442–450, 2000.

Como-Sabetti, K., Allaire, S., Parrott, K., Simonds, C. M., Hrabowy, S., Ritter, B., Hall, W., Altamirano, J., Martin, R., Downes, F., Jennings, G., Barrie, R., Dorman, M. F., Keon, N., Kucab, M., Al Shab, A., Robinson-Dunn, B., Dietrich, S., Moshur, L., Reese, L., Smith, J., Wilcox, K., Tilden, J., Wojtala, G., Park, J. D., Winnett, M., Petrilack, L., Vasquez, L., Jenkins, S., Barrett, E., Linn, M., Woolard, D., Hackler, D. R., Martin, H., McWilliams, D., Rouse, B., Willis, S., Rullan, J., Miller, Jr., G., Henderson, S., Pearson, J., Beers, J., Davis, R., and Saunders, D.:: Outbreaks of Escherichia coli O157: H7 infection associated with eating alfalfa sprouts-Michigan and Virginia, June–July 1997, Morb. Mortal. Wkly. Rep, 46, 741–744, 1997.

Dakshinamurty, S.: The common House-fly, Musca domestica, L., and its behaviour to temperature and humidity, B. Entomol. Res., 39, 339–357, 1948.

El-Fadel, M., Ghanimeh, S., Maroun, R., and Alameddine, I.: Climate change and temperature rise: Implications on food-and water-borne diseases, Sci. Total Environ., 437, 15–21, 2012.

Ferrano, S., Vancampenhoud, M., and Troncone, A.: Diarreas agudas en la edad pediátrica, XIV Jornadas Nacionales de Infectología y IX Jornadas Nororientales, Puerto La Cruz, Venezuela, 2–3, 2003.

Guerrant, R. L., Kosek, M., Moore, S., Lorntz, B., and Brantley, R.: Lima Am. Magnitude and Impact of Diarrheal Diseases, Arch. Med. Res., 33, 351–355, 2002.

Guzmán-Colis, G., Thalasso, F., Ramírez-López, E. M., Rodríguez-Narciso, S., Guerrero-Barrera, A. L., and Avelar-González, F. J.: Evaluación espacio-temporal de la calidad del agua del río San Pedro en el Estado de Aguascalientes, México, Rev. Int. Contam. Ambie, 27, 89–102, 2011.

Hernández, F. M.: Diarrea aguda e infecciones respiratorias: caras nuevas de viejos conocidos, Revista de la Facultad de Medicina UNAM, 45, 103–109, 2002.

Hernández, C. C., Aguilera, A. M. G, and Castro, E. G.: Situación de las enfermedades gastrointestinales en México, Enfermedades Infecciosas y Microbiología, 31, 137–151, 2011.

Hilborn, E. D., Mermin, J. H., Mshar, P. A., Hadler, J. L., Voetsch, A., Wojtkunski, C., Swartz, M., Mshar, R., Lambert-Fair, M. A., Farrar, J. A., Glynn, M. K., and Slutsker L.: A multistate outbreak of Escherichia coli O157: H7 infections associated with consumption of mesclun lettuce, Arch. Intern. Med., 159, 1758–1764, 1999.

INEGI: Instituto Nacional de Estadística y Geografía INEGI Censo de Población y vivienda 2010, available at: http://www3.inegi. org.mx/sistemas/mexicocifras/default.aspx?e=_1, last access: 21 February 2014, 2010.

INEGI: Instituto Nacional de Estadística y Geografía, Anuario estadístico y geográfico de Aguascalientes, available at: http://internet.contenidos.inegi.org.mx/contenidos/productos/ /prod_serv/contenidos/espanol/bvinegi/productos/anuario_14/ 702825064853.pdf (last access: 2 March 2016), 2014.

ISEA: Instituto de Salud del Estado de Aguascalientes, Estadísticas – Principales Causas de Mortalidad Infantil 2008–2014, Secretaría de Salud, available at: http://www.aguascalientes.gob.mx/ isea/mortinfa.asp, last access: 5 September 2014.

Konno, T., Suzuki, H., Katsushima, N., Imai, A., Tazawa, F., Kutsuzawa, T., Kitaoka, S., Sakamoto, M., Yazaki, N., and Ishida, N.: Influence of temperature and relative humidity on human rotavirus infection in Japan, J. Infect. Dis., 147, 125–128, 1983.

Loyola, R. and Soncco, C.: Salud y Calidad del Agua en Zonas Urbanomarginales de Lima Metropolitana, Economía y Sociedad, Cies, 64, 80–85, 2007.

NOAA: National Oceanic and Atmospheric Administration/Weather Service/Climate Prediction Center/historical El Niño/La Niña/Cold and Warm Episodes By Season, available at: http://www.cpc.ncep.noaa.gov/products/analysis_monitoring/ ensostuff/ensoyears.shtml, last access: 18 August 2014.

Patz, J. A. and Olson, S. H.: Climate change and health: global to local influences on disease risk, Ann. Trop. Med. Parasit., 100, 535–549, 2006.

Ramírez, E. M., Avelar, F. J., Zaragoza, J., and Rico, R.: Estudio sobre los agentes, cargas contaminantes y toxicidad que afectan el cauce del río San Pedro en el municipio de Aguascalientes y zonas aledañas, Informe final, Centro de Ciencias Básicas, Universidad Autónoma de Aguascalientes, Ags., México, 218 pp., 2007.

Ramírez-Castillo, F. Y., González, F. J. A., Garneau, P., Díaz, F. M., Barrera, A. L. G., and Harel, J.: Presence of multi-drug resistant pathogenic Escherichia coli in the San Pedro River located in the State of Aguascalientes, Mexico, Frontiers in microbiology, 4, doi:10.3389/fmicb.2013.00147, 2013.

Rose, J. B., Lisle, J. T., and LeChevallier, M.: Waterborne cryptosporidiosis: incidence, outbreaks, and treatment strategies, in: Cryptosporidium and cryptosporidiosis, edited by: Fayer, R., CRC Press, Boca raton, 93–110, 1997.

Schaffner, F. and Mathis, A.: Dengue and dengue vectors in the WHO European region: past, present, and scenarios for the future, Lancet Infect. Dis., 14, 1271–1280, 2014.

29

An assessment of El Niño and La Niña impacts focused on monthly and seasonal rainfall and extreme dry/precipitation events in mountain regions of Colombia and México

María Carolina Pinilla Herrera[1,2] **and Carlos Andrés Pinzón Correa**[2]

[1]Centro de Investigaciones en Geografía Ambiental – UNAM, Morelia, México
[2]Fundación Natura, Bogotá, Colombia

Correspondence to: María Carolina Pinilla Herrera (omsha_ra@yahoo.com)

Abstract. The influence of El Niño and La Niña on monthly and seasonal rainfall over mountain landscapes in Colombia and México was assessed based on the Oceanic Niño Index (ONI). A statistical analysis was develop to compare the extreme dry/precipitation events between El Niño, La Niña and Neutral episodes. For both areas, it was observed that El Niño and La Niña episodes are associated with important increases or decreases in rainfall. However, Neutral episodes showed the highest occurrence of extreme precipitation/dry events. For a better understanding of the impact of El Niño and La Niña on seasonal precipitation, we did a compound and a GIS analyses to define the high/low probability of above, below or normal seasonal precipitation under El Niño, La Niña and cold/warm Neutral episodes. In San Vicente, Colombia the below-normal seasonal rainfall was identified during El Niño and Neutral episodes in the dry season JJA. In this same municipality we also found above-normal seasonal rainfall during La Niña and Neutral episodes, especially in the dry season DJF. In Tancítaro México the below-normal seasonal rainfall was identified during La Niña winters (DJF) and El Niño summers (JJA), the above-normal seasonal rainfall was found during La Niña summers (JJA) and El Niño winters (DJF).

1 Introduction

El Niño and La Niña events have been studied exhaustively and they are associated with the interannual extreme rainfall variability, especially in the tropical and sub-tropical regions of the Pacific basin (Zambrano, 1986; Ropelewski and Halpert, 1987; Philander, 1990; Allan et al., 1996; Manson and Goddard, 2001; Dewitte et al., 2013). The events of La Niña are associated with unusual cold temperatures in the equatorial Pacific Ocean, while El Niño is characterized by unusual high temperatures in the same region. Both of them have an important impact on precipitation patterns, which can increase or decrease the occurrence of extreme dry/precipitation events and therefore, affect primary economic activities such as fisheries and agriculture (Coelho and Goddard, 2009; Lavado-Casimiro et al., 2013; Jozami et al., 2015).

In Colombia, El Niño events increase the occurrence of extreme dry events over most the country, but especially from January to July in regions like the Caribbean and the Andes (Baldión and Guzmán, 1994; Guzmán and Baldión, 1997; Montealegre and Pabón, 2000; Poveda, 2004; Puerta and Carvajal, 2008; Ruíz and Pabón, 2013). On the other hand, La Niña increases the occurrence of extreme precipitation events from July to December with significant impacts on the Caribbean, the Andean and Amazonas (Cadena et al., 2006; Hurtado and González, 2012). Studies at regional level by Ramírez and Jaramillo (2009) and Poveda et al. (2001a, b) have reported different precipitation aspects during ENSO in the central mountain range of the country, known as central Andes; those studies found a prominent influence of ENSO on rainfall (decrease during El Niño and increase during La Niña) for DJF and JJA which are historically considered as the dry seasons.

Mexico exhibits a monsoonal climate with a rainy season during the summer months and a relatively dry (Escobar et al., 2001; Magaña et al., 2003; Pereyra et al., 2004). It has been demonstrated that El Niño events are related to intense droughts during spring and summer over the center and the south of the country, while Northern Mexico, the Pacific coast and the peninsula of Yucatan are exposed to rainfall in winter (Conde et al., 1997; Magaña et al., 2003). In the central and northern zones the cold fronts increase in winter, whereas summer droughts and fewer hurricanes are seen in the Caribbean and the Gulf of Mexico (Magaña, 1998; Badán, 2003; Granados et al., 2011). So, droughts in summer and cold fronts in winters are the meteorological events best known related to El Niño episodes (Magaña et al., 1998, 2003; Conde et al., 1997). For La Niña events in central Mexico has been reported intense droughts during winters and the increase of precipitation during summers. Nonetheless, Magaña et al. (2003) argued that the impact of El Niño and La Niña events on winter rainfall is not always the same. They attributed this to a southward shift of the Inter Tropical Convergence Zone (ITZC), more intense trade winds, a decreased number of tropical cyclones over the Intra Americas Seas (IAS) and reduced relative humidity, that may result in several and different regional/local patterns in the interseasonal climate variability related to ENSO.

There are significant advances in studies concerning the precipitation effects associated with ENSO and its impacts on extreme precipitation/dry events at national level and on both the Pacific coastal regions and the key coffee and maize cultivation zones in Columbia and Mexico (Magaña, 1998; Poveda et al., 2001a, b; Cadena et al., 2006; Conde and Saldaña, 2007; Puerta and Carvajal, 2008; Ramírez and Jaramillo, 2009; Granados et al., 2011; Ruíz and Pabón, 2013). Still, little is known about the impact of El Niño and La Niña events at local level in mountain landscapes, which are regions with relevant crop production and exposed to several climate variability processes. In particular, there are no studies about the influence of El Niño and La Niña on local seasonal precipitation over the mountainous northeast region of Colombia, which is the most important cocoa production area in the country. Similarly, no investigations have been done in the Purépecha Plateau of Mexico, the most important production area for avocados for export in México.

For both countries, several authors have recommended to include studies at regional or local scale for a better understanding of the influence of ENSO on monthly and seasonal rainfall patterns with socio-economic planning purposes (Montealegre and Pabón, 2000; Poveda, 2004; Ruíz and Pabón, 2013; Cadena et al., 2006; Conde and Saldaña, 2007; Nuñez and Treviño, 2013). This represents a challenge because climate information at local level is not accurate due to the low density of available meteorological/climate stations. However, data of monthly and seasonal rainfall are the only available data at local level and its analyses are relevant for perennial crop planning.

The aim of this study is to analyze how El Niño and La Niña episodes might affect the monthly and seasonal precipitation. We compared increases and decreases of extreme

dry/precipitation events during ENSO and Neutral episodes. Additionally, we studied the probability of occurrence of seasonal precipitation between El Niño, La Niña and Neutral episodes. We consider that for mountain landscapes with important rural economies, this kind of climate information can be used for the development of strategies to improve crop production and resource management.

2 Study locations and methods

The municipality of San Vicente de Chucurí is located in the central–western region of Santander, Colombia between the eastern mountain range and the valley of the Magdalena river, with altitudes between 300 and 3200 m a.s.l. (Fig. 1). In addition to the orographic patterns, the trade winds and the year-migration of the inter-tropical convergence zone (ITCZ) explain the local expression of the bimodal season in this region of Colombia. In San Vicente de Chucurí the annual precipitation varies from 1000 mm in the lowlands near the Magdalena Valley to 2500 mm in the higher mountain regions.

The municipality of Tancítaro, Michoacán state, is located in the physiographical province of the neo-volcanic axis of Mexico, specifically at sub- province of Tarasca, which is represented by volcanic mountain ranges with isolated volcanic strata, plains and basaltic plateaus (Fig. 2). Altitude is between 900 and 3200 m a.s.l. and the precipitation varies between 700 and 2000 mm year^{-1} depending on its orographic patterns and seasons. Since it is located in central western Mexico, this region reaches the minimum temperatures, medium humidity and droughts during the winter, while the most intense rainfalls and the highest temperatures and humidity occur mainly during the summer. Between June and November this region is exposed to tropical cyclones, which sometimes bring several rainfalls after the summer. In the highest altitudes, the temperature decrease around $-5\,°C$ between December and February, this may bring several freezing episodes. In the lowest altitudes the temperature could increase around 28 °C during summer time.

This research was based on available meteorological information. In the case of San Vicente, Colombia, the study was done with monthly precipitation data from fifteen meteorological stations at the IDEAM (Instituto de Hidrología, Meteorología y Estudios Ambientales de Colombia) for a period comprising 1970–2010. For Tancítaro, Mexico, the study was carried out with monthly precipitation data from the Servicio Meteorológico Nacional (Base de Datos Climatológica Nacional), which compiles the information of eight different meteorological stations around Tancítaro, for a period comprising 1952–2012.

The methodological approach included: (i) the quality control of datasets, (ii) the use of the ONI Index to classify the monthly accumulated precipitation by El Niño, La Niña or cold/warm Neutral events, (iii) a statistical analysis to report the extreme dry/wet months by El Niño, La Niña

Santander, Colombia

Figure 1. Location of Colombian study case, San Vicente de Chu-curi. Source: Convenio 46/3379, Fundación Natura – ISAGEN.

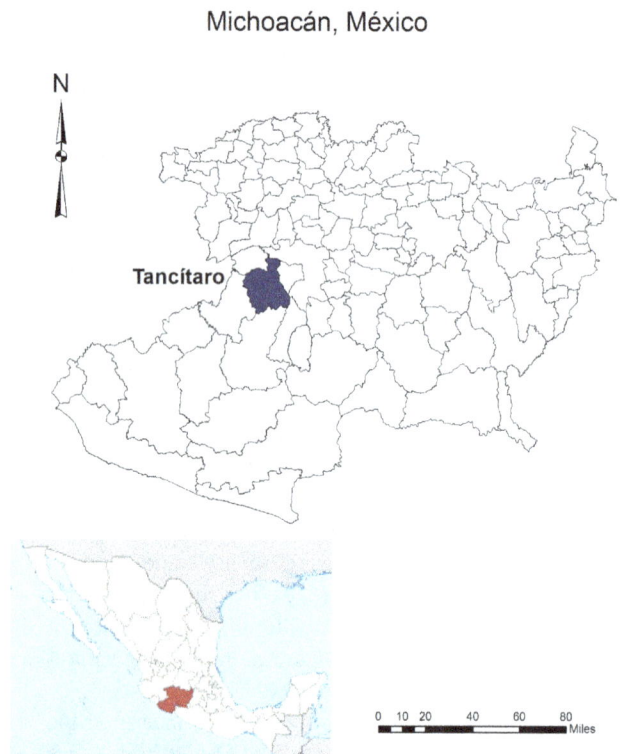

Michoacán, México

Figure 2. Location of Mexican study case, Tancítaro, México. Source: INEGI (2009).

or cold/warm Neutral events, and, (iv) a GIS-based geosta-tistical analysis of rainfall by El Niño, La Niña or cold/warm Neutral events per season.

For the quality control of datasets, the monthly accumu-lated precipitation records were homogenized using double mass cumulative techniques and missing data were filled by linear regression among stations, showing middle inter-station cross correlations. We expected high variability in the data from some meteorological stations since some failures have been reported on several devices and also with the data collection in Colombian and México. For these reasons we accepted P value $\alpha = 0.10/0.20$ and we rejected the records of some meteorological stations if they had at least 25 % of missing data (Valdivia et al., 2010, 2013), these included data from two meteorological stations in Colombia and one in México.

The data distribution and the months and years with rain-fall anomalies could be identify using the box plot analysis. The precipitation effects associated with El Niño, La Niña or Neutral episodes per month, were determined using the ONI index (Ramírez and Jaramillo, 2009; NOAA, 2013), we also compared the historical and monthly average values (Jozami et al., 2015).

The statistical analysis that reports the extreme dry/wet events was done using normalized differences and log trans-formation (Hamilton et al., 2012; Nuñez and Treviño, 2013), a comparison between the smaller number of El Niño-La Niña extreme episodes to the number of Neutral extreme episodes. For each month (classified by ONI index in El Niño, La Niña or Neutral episode) the precipitation values below 10th percentile (p10 dry extremes) and the precipi-tation values exceeding 90th percentile (p90 wet extremes) were identified and compared.

The GIS-based geostatistical analysis was based on:

a. The seasonal classification for both places:

   i. December-January-February: first dry season of the year (Colombia)–Winter (Mexico).

   ii. March-April-May: first rainy season (Colombia) – Spring (Mexico).

   iii. June-July-August: second dry season (Colombia) – Summer (Mexico).

   iv. September-October-November: second rainy sea-son (Colombia) – Autumn (Mexico).

b. A compound analysis by relative frequency (%) con-tingency tables per meteorological station was used to define the high/low probability of above, below or nor-mal seasonal precipitation under El Niño, La Niña, cold/warm Neutral events (Alfaro and Soley, 2009; Fallas and Alfaro, 2012). We used the ONI Index values to define cold/warm phases for Neutral episodes for years with a TSM anomaly $-0.4\,°C < T_{dep} < +0.4\,°C$ (which represent a relevant cold/warm condition) but not called El Niño or La Niña itself.

c. The selection of meterological stations with Chi-squared (X2) test: P value $\alpha = 0.05/\alpha = 0.10/\alpha = 0.20$

d. Frecuency histograms and map development with inverse-distance-weighted interpolation.

3 Results

3.1 San Vicente de Chucurí, Santander, Colombia

It was obtained fifteen Box-Plot diagrams that showed the record of the meteorological stations with a bimodal regime: two rainfall seasons between by two dry seasons. The highest rainfalls are in MAM and SON; and dry seasons are in DJF and JJA. Figures 3 and 4 show just two samples of the Box-Plot diagrams with the bimodal regime and the months and years with atypical precipitation records.

The ONI value analysis allows us to select El Niño events. From 1970 onwards fourteen El Niño events have occurred and six of them showed the highest ONI values related to dry events. The most significant impact on the local monthly precipitation occurred on September and October (Table 1). On other hand, from 1970 onwards there were fifteen events of La Niña and six of them indicated the highest ONI values related to increase of precipitation. In those years the highest recorded values occurred on August and October (Table 2).

To analyze the information by extreme dry/wet events, all monthly records were normalized and linked to El Niño, La Niña or Neutral episodes. Figure 5 shows that the highest occurrence of extreme precipitation events (above the 90th percentile) were registered during Neutral episodes. La Niña episodes, showed extreme precipitation events too but only in a few meteorological stations. The lowest occurrence of extreme precipitation events were during El Niño episodes. Checking the dataset, we found that months with more extreme precipitation events reported were August (on Neutral and La Niña episodes), October (on Neutral and La Niña episodes) and November (on Neutral and La Niña episodes). So, the extreme precipitation events during La Niña or Neutral episodes occurred mainly in August, October and November.

The highest occurrence of extreme dry events (below the 10th percentile) were reported also in Neutral episodes, but with high dispersal data. Extreme dry events occurred less frequently in El Niño and even less in La Niña episodes (Fig. 6). The months with more extreme dry events reported were September (on Neutral and El Niño episodes) and October (on Neutral and El Niño episodes). Therefore, the ex-

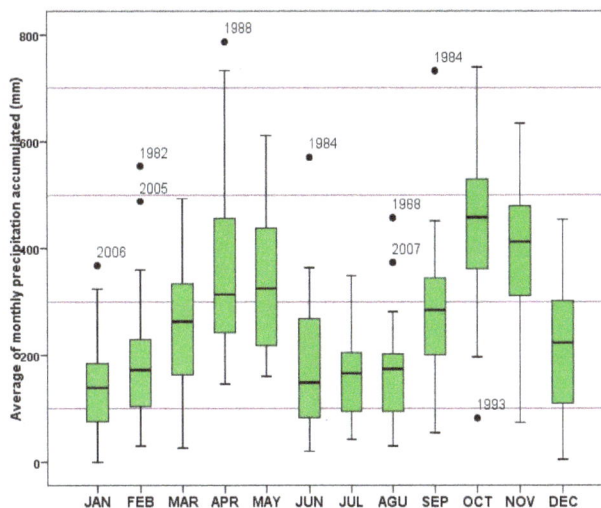

Figure 3. Monthly variability of precipitation (mm) at Puente La Paz meteorological station (1979–2010).

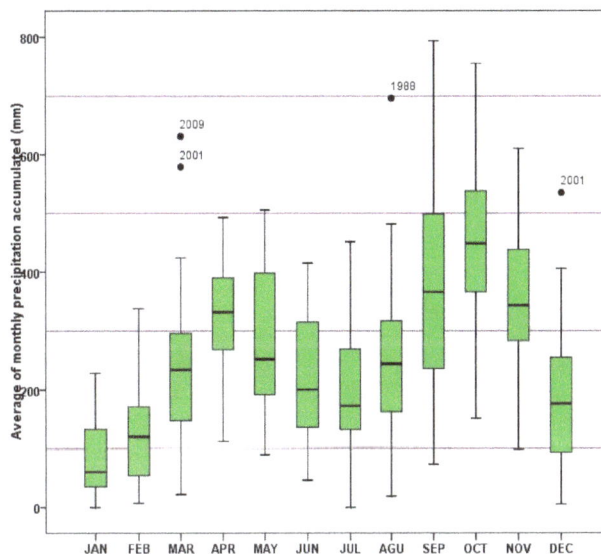

Figure 4. Monthly variability of precipitation (mm) at La Putana meteorological station (1973–2010).

treme dry events during El Niño or Neutral episodes occurred mainly in September and October.

As mentioned before, the compound analysis gave us the possibility to establish the high/low probability of above, below or normal seasonal precipitation under El Niño, La Niña, cold/warm Neutral episodes. The GIS-based analysis of these probabilities was done by frequency histograms and maps and both interpretations showed the same result. In this manuscript the Colombian case is only represented by the frequency histograms.

Figure 7 shows the probabilities of occurrence of above–normal, normal, and below–normal seasonal precipitation during El Niño, La Niña and cold/warm Neutral episodes.

Table 1. Comparison of the climatological monthly mean value vs. the precipitation value under El Niño episodes.

Meteorological station/month	Climatological monthly mean value	El Niño episode	ONI value	Precipitation value under El Niño episode	SE of the mean
Lebrija, September	130 mm	September, 1972	1.6	65 mm	0.33
Zapatoca, October	75 mm	October, 1973	1.8	38 mm	0.19
Girón, September	90 mm	September, 1982	1.9	20 mm	0.35
Betulia, October	450 mm	October, 1991	1.4	80 mm	1.85
Lebrija, September	145 mm	September, 1997	2.1	70 mm	0.38
San Vicente de Chucurí, October	280 mm	October, 2009	1.1	40 mm	1.2

SE: standard error.

Table 2. Comparison of the climatological monthly mean value vs. the precipitation value under La Niña episodes.

Meteorological station/month	Climatological monthly mean value	La Niña episode	ONI value	Precipitation value under La Niña episode	SE of the mean
San Vicente de Chucurí, October	240 mm	October, 1970	−0.8	580 mm	1.7
Girón, November	170 mm	November, 1974	−0.9	710 mm	2.7
Zapaptoca, August	100 mm	August, 1998	−1.0	275 mm	0.88
Los Santos, August	28 mm	August, 1999	−1.1	115 mm	0.44
Lebrija, October	145 mm	October 2007	−1.5	325 mm	0.9
Betulia, October	420 mm	October 2011	−0.8	790 mm	1.85

SE: standard error.

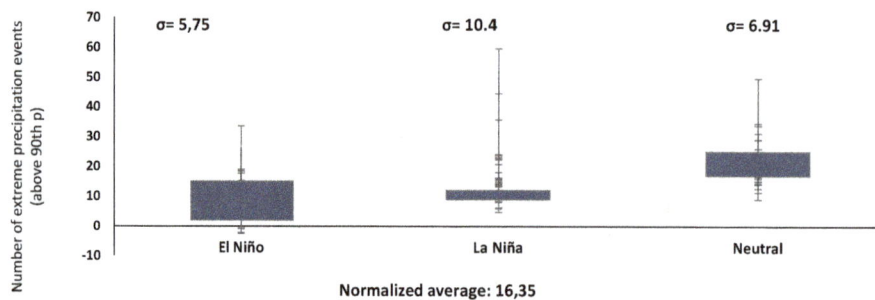

Figure 5. Normalized data of extreme precipitation events (above the 90th percentile) recorded in all meteorological stations around San Vicente de Chucurí, Colombia. Source: IDEAM.

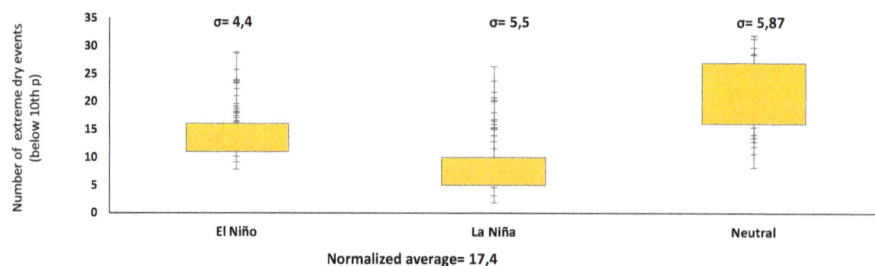

Figure 6. Normalized data of extreme dry events (below the 10th percentile) recorded in all meteorological stations around San Vicente de Chucurí, Colombia. Source: IDEAM.

The highest occurrence of the above-normal seasonal precipitation is in quarters DJF during La Niña and cold Neutral episodes (probabilities of 50 and 70 %) and JJA, SON in La Niña episodes (probabilities of 50 and 70 %). The highest oc-

currence of the below-normal seasonal precipitation is in JJA during El Niño (probabilities of 50 and 70 %), followed by SON during El Niño (probabilities of 50 and 70 %) and neutral cold episodes (probabilities of 30 and 50 %). The quar-

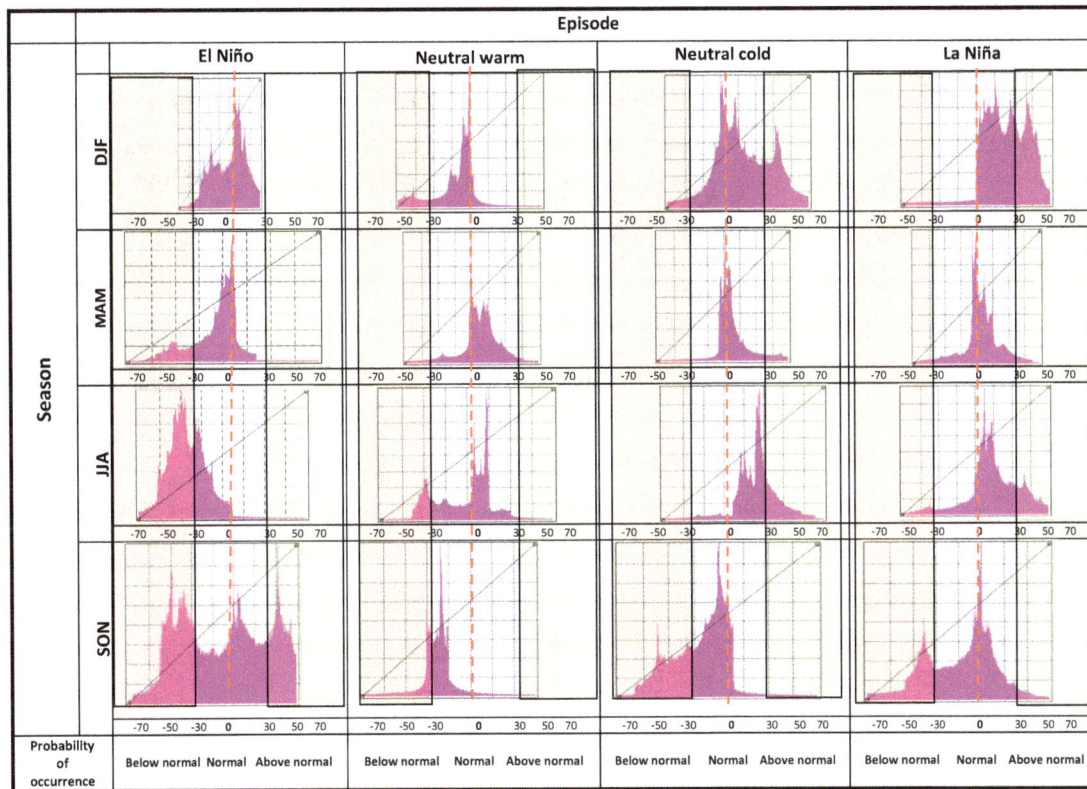

Figure 7. Histograms of probability of occurrence (%) of above–normal, normal, and below–normal seasonal precipitation during El Niño, La Niña and cold/warm Neutral episodes in Santander, Colombia. $\alpha = 0.20$ (4 meteorological stations for MAM; 3 meteorological stations for SON). $\alpha = 0.10$ (8 meteorological stations for MAM and 10 meteorological stations for SON, DJF, JJA). $\alpha = 0.05$ (5 meteorological stations for DJF and JJA).

ter MAM has the highest occurrence of the normal seasonal precipitation during El Niño, La Niña or cold/warm Neutral episodes.

This GIS-based analysis of rainfall reported a middle statistical significance because ten meteorological stations had $\alpha = 0.10$ for quarters SON, DJF and JJA. The highest statistical significance was obtained for quarters DJF, JJA and MAM ($\alpha = 0.05$) just in four meteorological stations. However, this geostatistical approach can be accepted to analyze historical records with at least 25 % of missing data (Valdivia et al., 2013).

Baldión and Guzmán (1994) as well as Puerta and Carvajal (2008) reported above-normal seasonal precipitation only in rainy seasons during La Niña events for the Caribbean and Andean zones. But our findings show that local precipitation in the mountainous northeast region of Colombia can significantly increase during dry seasons (especially on DJF) in La Niña episodes, as well as during rainy seasons in Neutral or La Niña episodes.

For the Caribbean region some works (Guzmán and Baldión, 1997; Ruíz und Pabón, 2013) described an important decrease in precipitation for DJF and JJA during ENSO (mainly El Niño episodes). Our results also reported a high

probability of occurrence of below-normal precipitation in those dry seasons during El Niño episodes. However, we found the same signal of below-normal seasonal precipitation for Neutral episodes.

The above-normal precipitation raifall in DJF during La Niña or cold Neutral episodes is especially critical for the emergency of diseases in the cocoa. The incidence of fungi such as *Monilia sp.* or *Phytoptora sp.* could increase and affect the most important harvest time of the year. The flowering cycles of the cocoa can also be affected by drought in JJA during El Niño events. If this information could be given to farmers and local institutions, they could develop strategies for fungi or crop management such as a frequent harvest of infected fruits or timely fumigations.

3.2 Tancítaro, Michoacán, Mexico

All the Box-Plot diagrams showed a monomodal regime characterized by a rainy season between May and October and a dry season between November and April. Figures 8 and 9 show just two samples of the Box-Plot diagrams with the monomodal regime and the months and years with atypical precipitation records.

Table 3. Comparison of the climatological monthly mean value vs. the precipitation value under El Niño episodes.

Meteorological station/month	Climatological monthly mean value	El Niño episode	ONI value	Precipitation value under El Niño episode	SE of the mean
Los Chorros de Varal, May	87 mm	May, 1998	2.3	0 mm	0.44
Los Chorros, Jun	125 mm	May, 1992	1.8	25.8 mm	0.5
Charapendo, July	99 mm	July, 1972	1.9	11.5 mm	0.44
Tanaco, August	49 mm	August, 2010	1.7	9.5 mm	0.24
Los Chorros, November	20 mm	November, 1977	1.6	127 mm	0.54
Paracuaro, December	4.6 mm	December, 1987	1.7	18 mm	0.07
Acahuato, January	19.5 mm	January, 2009	1.6	199 mm	0.9
Uruapan, January	46.8 mm	January, 1995	1.6	610 mm	2.82
Uruapan, February	9.9 mm	February, 1973	1.8	166.8 mm	0.78

SE: standard error.

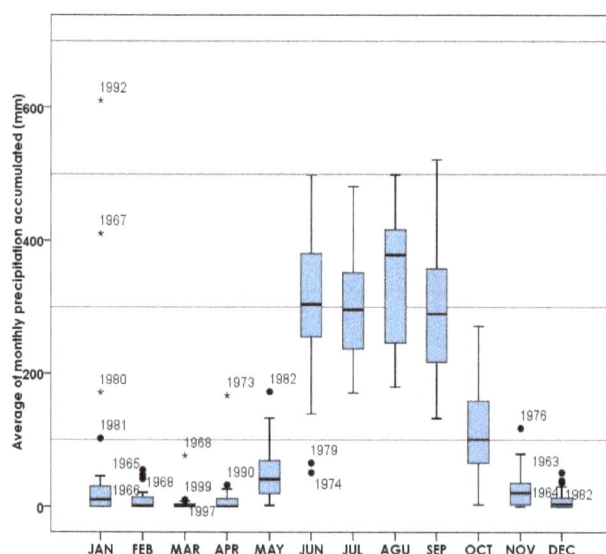

Figure 8. Monthly variability of precipitation (mm) at Uruapan meteorological station (1962–1999).

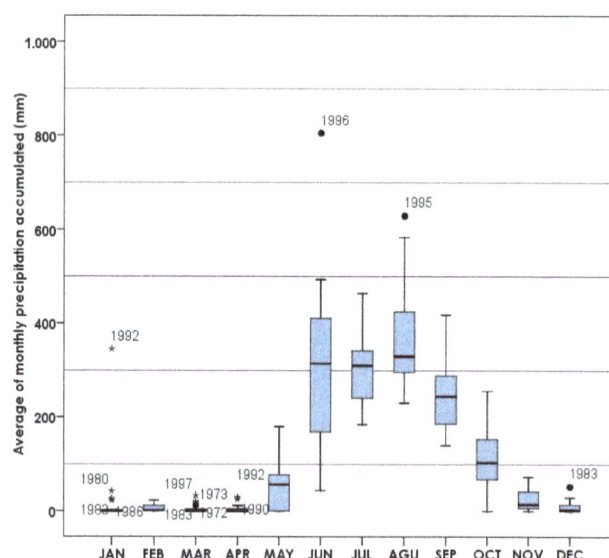

Figure 9. Monthly variability of precipitation (mm) at Peribán meteorological station (1969–1998).

Based on ONI Index we found precipitation increase by the strongest El Niño episodes during December, January and February, while May, Jun, July and August have been affected by a precipitation decrease (Table 3). During the strongest La Niña episodes, rainfalls between December and February showed a decrease and rainfalls from May to July showed an increase (Table 4).

The number of extreme dry/wet events recorded by monthly scale is shown in Fig. 10. The highest occurrence of extreme precipitation events (above the 90th percentile) were registered during in Neutral episodes, but the data series is highly dispersed. El Niño episodes had a fewer extreme precipitation events and La Niña had even less. Cheking all datasets, months with more extreme precipitation events reported were May and June on Neutral episodes and, December and February on El Niño episodes.

The highest occurrence of extreme dry events (below the 10th percentile) also were reported in Neutral episodes, but data dispersal is high (Fig. 11). Moreover the occurrence could change from one meteorological station to another depending on local conditions (such as orographical factors). The low dispersal data of La Niña events suggests that extreme dry events can affect the region and surroundings in a similar way. Months with more extreme dry events reported were January and February in Neutral episodes and May and June in La Niña ones.

The GIS-based analysis of the high/low probability of above, below or normal seasonal precipitation under El Niño, La Niña, cold/warm Neutral episodes for Tancítaro will be presented through probability maps (Fig. 12). The highest occurrence of the above-normal seasonal precipitation is in JJA during La Niña episodes (probabilities of 30 and 50 %) and MAM during cold Neutral episodes (probabilities of 30

Table 4. Comparison of the climatological monthly mean value vs. the precipitation value under La Niña episodes.

Meteorological station/month	Climatological monthly mean value	La Niña episode	ONI value	Precipitation value under La Niña episode	SE of the mean
Chorros de Varal, May	37.6 mm	May, 1988	−1.8	112.9 mm	0.38
Acahuato, May	23.5 mm	May, 2000	−1.6	103 mm	0.4
Paracuaro, June	170 mm	June, 1976	−1.9	509 mm	1.7
Charapendo, July	243 mm	July, 1985	−1.7	524 mm	1.41
Uruapan, August	221 mm	August, 1999	−1.5	484 mm	1.32
Uruapan, December	11.4 mm	December, 1999	−18	0 mm	0.06
Periban, January	21.7 mm	January, 1988	−1.7	0 mm	0.11
Acahuato, February	9.9 mm	February, 2011	−1.4	0 mm	0.05

SE: standard error.

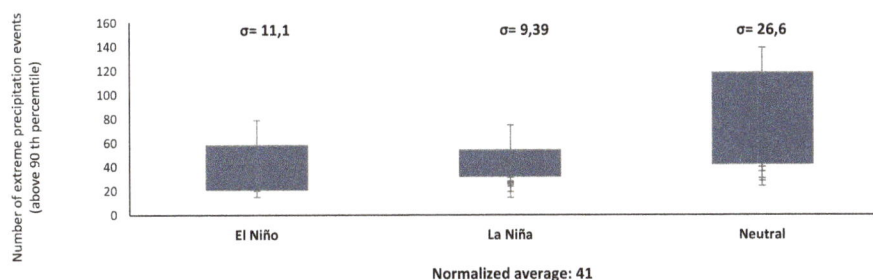

Figure 10. Normalized data of all extreme precipitation events (above the 90th percentile) in Tancítaro, Michoacán, Mexico. Source: Base de Datos Climatológica Nacional, SMN.

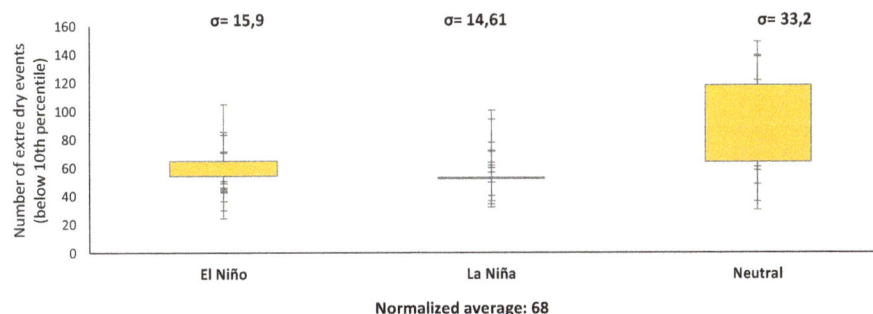

Figure 11. Normalized data of all extreme dry events (above the 10th percentile) in Tancítaro, Michoacán, Mexico. Source: Base de Datos Climatológica Nacional, SMN.

and 50 %). During La Niña episodes the occurrence of the below-normal seasonal precipitation is in quarters DJF and MAM (probabilities of 30 and 50 %). The highest occurrence of the normal seasonal precipitation is in quarters MAM during El Niño and warm Neutral, and SON during El Niño and La Niña episodes.

This GIS-based analysis of rainfall for Tancítaro reported a better statistical significance for quarters DJF and JJA because six meterological stations had $\alpha = 0.05$, six meterological stations had $\alpha = 0.10$ for quarter SON and five meterological stations had $\alpha = 0.10$ for quarter MAM.

Several authors had reported an increase of precipitation during El Niño winters and La Niña summers, and a decrease of precipitation in El Niño springs and La Niña winters for Central and southern México, for the state of Baja California and for the City of Tabasco (Magaña, 1998; Mosiño and Morales, 1998; Reyes and Troncoso, 2001; Pereyra et al., 2004). Our findings showed that the mountainous landscapes in the meseta Tarasca (Michoacán) follow this same signal in El Niño and La Niña episodes. However, we also found the occurrence of above-normal and below-normal seasonal precipitation for DJF and MAM in Neutral episodes.

The above-normal and the below-normal seasonal precipitation in quarter MAM during cold Neutral and La Niña episodes, respectively, are especially critical for the avocado production. These months are essential to complete the flow-

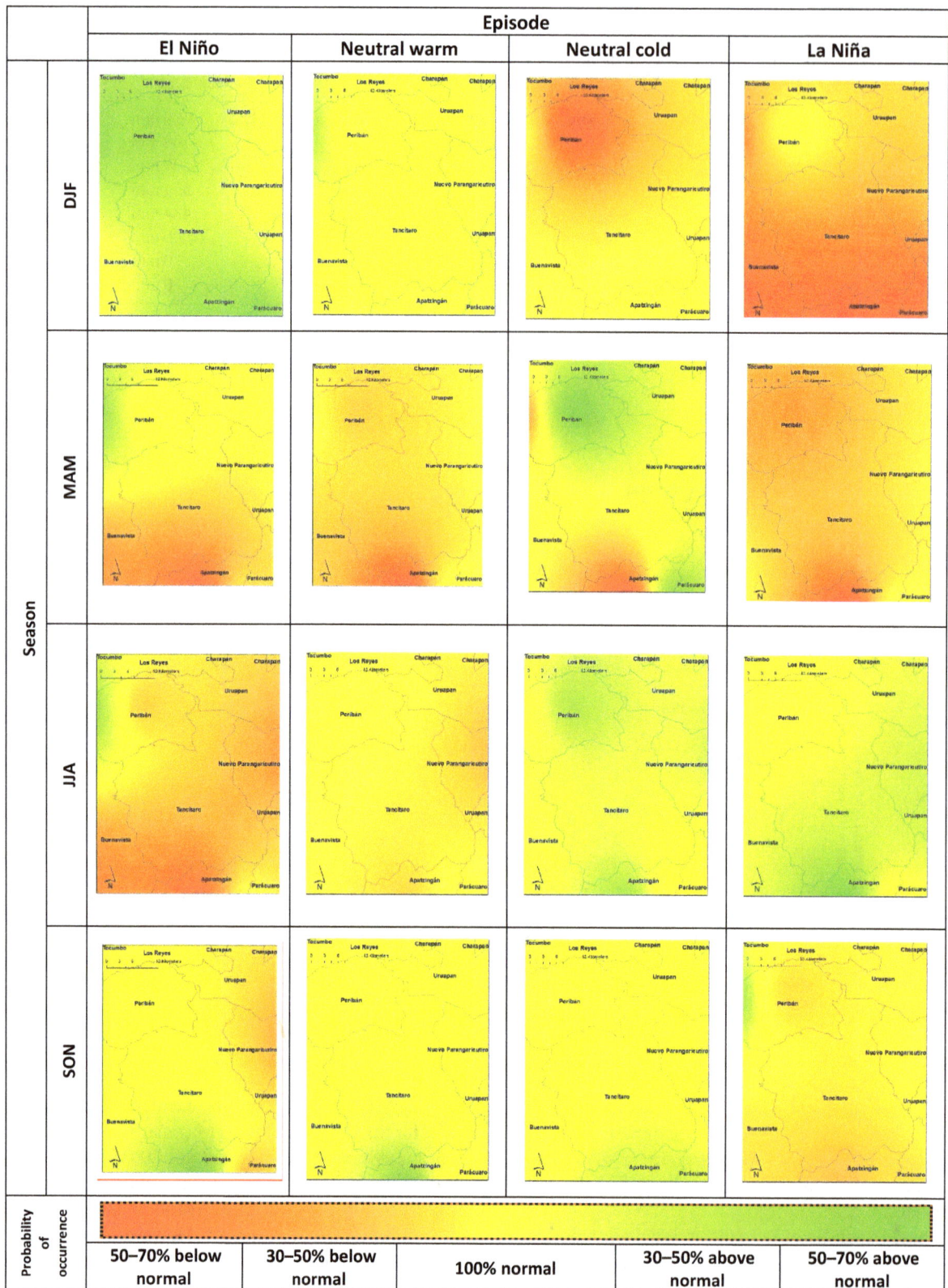

Figure 12. Maps of probability of occurrence (%) of above–normal, normal, and below–normal seasonal precipitation in El Niño, La Niña and cold/warm Neutral epidodes in Tancítaro, Mexico. $\alpha = 0.20$ (3 meteorological stations for MAM; 2 meteorological stations for SON). $\alpha = 0.10$ (5 meteorological stations for MAM, 6 meteorological stations for SON and 2 meteorological stations for DJF, JJA). $\alpha = 0.05$ (6 meteorological stations for DJF and JJA).

ering cycle and particularly for a local flowering episode called "La Loca", which is very important for farmers from lowlands in Tancítaro. With this information, the farmers can accomplish a timely crop irrigation and take actions against flower losses (there is local knowledge of several techniques).

4   Conclusions and perspectives

This study has shown the impacts of El Niño and La Niña episodes on local monthly and seasonal precipitation in both locations. The results are in agreement with other researches done at a national level and in different regions. However, our data adds new information concerning the occurrence of extreme dry/precipitation events and the probabilities of below-normal, above normal or normal seasonal rainfall in Neutral episodes.

We were able to recognized that in those mountainous regions, as mentioned Magaña et al. (2003), that ENSO impact can be sometimes weaker than intra-annual seasonality due to the local orographic factors, the intensity of hurricanes and tropical cyclone over the Intra Americas Seas (IAS), cold fronts, trade winds and the southward shift of the of ITZC. The study of precipitation effects associated with ENSO in mountain landscapes can be improved by analysis conducting within different elevation ranges like was presented by Pineda et al. (2013).

Agricultural institutions and farmers can benefit from this scientific information to identify climate risks on farms. If all agricultural stakeholders could be convinced to work together they may develop a wide-range of anticipatory and reactive management strategies in response to economic losses due to climate.

Local information about climate data is important, however, as we have seen in this study there can be a middle statistical uncertainty due to technical failures and unreliable data collection at meteorological stations. To overcome these barriers we need more state-of-the-art methodology and devices to provide additional and accurate local climate variability data.

For a suitable climate risk management (like phytosanitary warnings) it is also necessary to understand the perception of farmers on climate variability and climate risks, and also how they respond to the impact caused by climatic conditions.

Data availability

Climate data from this research are public accessible information from Instituto de Hidrología, Meteorología y Estudios Ambientales de Colombia (IDEAM) and Servicio Meteorológico Nacional. Databases from Santander, Colombia and Michoacán, México were purchased by projects; the information is property of ISAGEN-Fundación Natura and Centro de Investigaciones en Geografía Ambiental.

Author contributions. María Carolina Pinilla, Ecologist-PhD Candidate in Geography, designed the research, reviewed the methodological approach and results, prepared the manuscript and directed revisions in Spanish/English languages. Carlos Andrés Pinzón, Physicist-MSc., compiled data, conducted the statistical analysis and drafted the methodology and results.

Acknowledgements. The authors would like to recognize the total support of the following projects during this research: Convenio 46/3379 Fundación Natura Colombia-ISAGEN E.S.P.: "Programa para atender la percepción de la comunidad acerca de posibles cambios microclimáticos ocasionados por el embalse", and PAPIIT-UNAM IA300413 "Evaluación de la adaptación al cambio climático en comunidades rurales a través de su capacidad de respuesta en diferentes contextos geográficos en el estado de Michoacán". Special thanks to the Academic Writing team from UNAM Posgrado. The authors are grateful to Dra. María Pilar Cornejo and one anonymous reviewer whose valuable comments led us to the improvement of the different stages of this work.

References

Alfaro, E. and Soley, J.: Descripción de dos métodos de rellenado de datos ausentes en series de tiempo meteorológicas, Revista de Matemática: teoría y aplicaciones, 16, 60–75, 2009.

Alfaro, E., Soley, J., and Enfield, D.: Uso de una Tabla de Contingencia para Aplicaciones Climáticas, ESPOL publication, Guayaquil, Ecuador, 51 pp., 2003.

Allan, R., Lindsey, J., and Parker, D.: El Niño Southern Oscillation and climate variability, CSIRO Publication, 405 pp. 1996.

Badán, A.: The effects of El Niño in Mexico: A survey, Geofísica Internacional, 42, 567–571, 2003.

Baldión, J. and Guzmán, O.: Condiciones climáticas en la zona cafetera en los años 1991, 1992 y 1993 y su influencia en las cosechas de café, Cenicafe Avances Técnicos: 1–8, 1994.

Cadena, M., Pabón, J. D., Devis, A., Malikov, I., Reyna, J., and Ortiz, J.: Relationship Between the 1997/1998 El Niño and 1999/2001 La Niña Events and Oil Palm Tree Production in Tumaco, Southwestern Colombia, Adv. Geophysis, 6, 195–199, 2006.

Coelho, C. and Goddard, L.: El Niño–Induced Tropical Droughts in Climate Change Projections, J. Climate, 22, 6456–6476, 2009.

Conde, C. and Saldaña, S.: Cambio climático en América Latina y el Caribe: Impactos, vulnerabilidad y adaptación, Rev. Ambiente y Desarrollo, 23, 23–30, 2007.

Conde, C., Liverman, D., Flores, M., Ferrer, R., Araujo, R., Betancourt, E., Villareal, G., and Gay, C.: Vulnerability of rainfed maize crops in Mexico to climate change, Climate Change, 9, 17–34, 1997.

Dewitte, B., Bourrel, L., and Ambrizzi, T.: Editorial, Adv. Geosci., 33, 1–1, doi:10.5194/adgeo-33-1-2013, 2013.

Escobar, E., Bonilla, M., Badán, A., Caballero M., and Winckell, A.: Los Efectos del Fenómeno de El Niño en México, 1997–1998, 245 pp., 2001.

Fallas, B. and Alfaro, E.: Uso de herramientas estadísticas para la predicción estacional del campo de precipitación en América Central como apoyo a los Foros Climáticos Regionales – 1: Análisis de tablas de contingencia, Revista de Climatología, 12, 61–79, 2012.

Granados, R., Aguilar, G., Díaz, G., and Medina, M.: Alteraciones de los indicadores agroclimáticos en años con presencia del fenómeno El Niño en la región centro-occidente de México, Rev. Geográfica de América Central, 1–16, 2011.

Guzmán, O. and Baldión, J.: El evento cálido del Pacifico en la zona cafetera Colombiana, Cenicafe, 48, 141–155, 1997.

Hamilton, E., Eade, R., Graham, R., Scaide, A., Smith, D., Maidens, A., and MacLachlan, C.: Forecasting the number of extreme daily events on seasonal timescales, J. Geophys. Res., 117, D03114, doi:10.1029/2011JD016541, 2012.

Hurtado, G. and González, O.: Evaluación de la afectación territorial de los fenómenos El Niño/La Niña y análisis de la confiabilidad de la predicción climática basada en la presencia de un evento. Instituto de Hidrología, Meteorología y Estudios Ambientales de Colombia, 38 pp., 2012.

Jozami, E., Constanzo, M., and Coronel, A.: Influencia de "El Niño-Oscilación Sur" sobre las precipitaciones en Paraná y Lucas González (Entre Ríos, Argentina), Revista de Climatología, 15, 85–92, 2015.

Lavado-Casimiro, W. S., Felipe, O., Silvestre, E., and Bourrel, L.: ENSO impact on hydrology in Peru, Adv. Geosci., 33, 33–39, doi:10.5194/adgeo-33-33-2013, 2013.

Magaña, V.: Los impactos de El Niño en México, Universidad Nacional Autónoma de México y Secretaría de Gobernación, 229 pp., 1998.

Magaña, V., Pérez, J., and Conde, C.: El fenómeno de El Niño y la Oscilacion del sur (ENOS) y sus impactos, Ciencias, 14–18, 1998.

Magaña, V., Vázquez, J., Pérez, J., and Pérez, J.: Impact of El Niño on precipitation in México, Geofís. Int., 42, 313–330, 2003.

Manson, S. and Goddard, L.: Probabilistic precipitation anomalies associated with ENSO, B. Am. Meteorol. Soc., 82, 619–638, 2001.

Montealegre, J. and D. Pabón: La Variabilidad Climática Interanual asociada al ciclo El Niño-La Niña–Oscilación del Sur y su efecto en el patrón pluviométrico de Colombia, Meteorología Colombiana, 2, 7–21, 2000.

Mosiño, P. and Morales, T.: Los ciclones tropicales, El Niño y las lluvias en Tacubaya, D.F., Geofís. Int., 27, 61–82, 1998.

NOAA/National Weather Service: Cold & Warm Episodes by Season, Climate Prediction Center, available at: http://www.cpc.ncep.noaa.gov/products/analysis_monitoring/ensostuff/ensoyears.shtml, last access January, 2013.

Nuñez, D. and Treviño, E.: Spatial interpolation of monthly mean precipitation in the Rio Bravo/Grande basin, Tecnología y ciencias del agua, 4, 185–193, 2013.

Pereyra, D., Bando, U., and Natividad, M.: Influencia de La Niña y El Niño sobre la precipitación de la ciudad de Villahermosa, Tabasco, México, Universidad y Ciencia, 20 33–38, 2004.

Pineda, L., Ntegeka, V., and Willems, P.: Rainfall variability related to sea surface temperature anomalies in a Pacific-Andean basin into Ecuador and Peru, Adv. Geosci., 33, 53–62, doi:10.5194/adgeo-33-53-2013, 2013.

Philander, S. G.: El Niño, La Niña, and the Southern Oscillation, 1st Edn., Academic, San Diego, CA, 293 pp., 1990.

Poveda, G.: La hidroclimatología de Colombia: una síntesis desde la escala inter-decadal hasta la escala diurna, Rev. Academia Colombiana de Ciencias, 28, 201–222, 2004.

Poveda, G., Jaramillo, M., Gil, M., Quinceno, N., and Mantilla, R.: Seasonality in ENSO related precipitation, river discharges, soil moisture and vegetations index (NDVI) in Colombia, Water Resour. Res., 37, 2169–2178, 2001a.

Poveda, G., Rave, C., and Mantilla, R.: Tendencias en la distribución de probabilidades de lluvias y caudales en Antioquia, Meteorología Colombiana, 3, 53–60, 2001b.

Puerta, O. and Carvajal, Y.: Incidencia de El Niño-Oscilacion del Sur en la precipitación y la temperatura del aire en Colombia, utilizando el Climate Explorer, Ingenieria y Desarrollo, 23, 104–118, 2008.

Ramírez, V. and Jaramillo, A.: Relación entre el Índice Oceánico de El Niño y la lluvia en la región andina central de Colombia, Cenicafé, 60, 161–172, 2009.

Reyes, S. and Troncoso, R.: El Niño Oscilación del Sur y los fenómenos hidrometeorológicos en Baja California: el evento de 1997/98, Ciencia Pesquera, 15, 89–96, 2001.

Ruíz, A. and Pabón, D.: Efecto de los fenómenos de El Niño y La Niña en la precipitación y su impacto en la producción agrícola del departamento del Atlántico (Colombia), Rev. Colombiana de Geografía, 22, 35–54, 2013.

Ropelewski, C. F. and Halpert, M. S.: Global and Regional Scale Precipitation Patterns Associated with the El-Nino Southern Oscillation, Mon. Weather Rev., 115, 1606–1626, 1987.

Valdivia, C., Seth, A., Gilles, J., García, M., Jiménez, E., Cusicanqui, J., Navia, F., and Yucra, E.: Adapting to Climate Change in Andean Ecosystems: Landscapes, Capitals, and Perceptions Shaping Rural Livelihood Strategies and Linking Knowledge Systems, Ann. Assoc. Am. Geogr., 100, 818–834, 2010.

Valdivia, C., Thibeault, J., Gilles, J. L., García, M., and Seth, A.: Climate trends and projections for the Andean Altiplano and strategies for adaptation, Adv. Geosci., 33, 69–77, doi:10.5194/adgeo-33-69-2013, 2013.

Zambrano, E.: El fenómeno de El Niño y la oscilación del Sur, Acta Oceanográfica del Pacífico, 3, 195–203, 1986.

Permissions

The contributors of this book come from diverse backgrounds, making this book a truly international effort. This book will bring forth new frontiers with its revolutionizing research information and detailed analysis of the nascent developments around the world.

We would like to thank all the contributing authors for lending their expertise to make the book truly unique. They have played a crucial role in the development of this book. Without their invaluable contributions this book wouldn't have been possible. They have made vital efforts to compile up to date information on the varied aspects of this subject to make this book a valuable addition to the collection of many professionals and students.

This book was conceptualized with the vision of imparting up-to-date information and advanced data in this field. To ensure the same, a matchless editorial board was set up. Every individual on the board went through rigorous rounds of assessment to prove their worth. After which they invested a large part of their time researching and compiling the most relevant data for our readers.

The editorial board has been involved in producing this book since its inception. They have spent rigorous hours researching and exploring the diverse topics which have resulted in the successful publishing of this book. They have passed on their knowledge of decades through this book. To expedite this challenging task, the publisher supported the team at every step. A small team of assistant editors was also appointed to further simplify the editing procedure and attain best results for the readers.

Apart from the editorial board, the designing team has also invested a significant amount of their time in understanding the subject and creating the most relevant covers. They scrutinized every image to scout for the most suitable representation of the subject and create an appropriate cover for the book.

The publishing team has been an ardent support to the editorial, designing and production team. Their endless efforts to recruit the best for this project, has resulted in the accomplishment of this book. They are a veteran in the field of academics and their pool of knowledge is as vast as their experience in printing. Their expertise and guidance has proved useful at every step. Their uncompromising quality standards have made this book an exceptional effort. Their encouragement from time to time has been an inspiration for everyone.

The publisher and the editorial board hope that this book will prove to be a valuable piece of knowledge for researchers, students, practitioners and scholars across the globe.

List of Contributors

A. D'Alessandro
Istituto Nazionale di Geofisica e Vulcanologia, Centro Nazionale Terremoti, Rome, Italy
Università di Palermo, Dipartimento delle Scienze della Terra e del Mare, Palermo, Italy

I. Guerra and G. Stellato
Università della Calabria, Dipartimento di Fisica, Arcavacata di Rende (Cosenza), Italy

G. D'Anna and A. Gervasi
Istituto Nazionale di Geofisica e Vulcanologia, Centro Nazionale Terremoti, Rome, Italy

P. Harabaglia
Università della Basilicata, Scuola di Ingegneria, Potenza, Italy

D. Luzio
Università di Palermo, Dipartimento delle Scienze della Terra e del Mare, Palermo, Italy

D. Pesaresi
OGS (Istituto Nazionale di Oceanografia e di Geofisica Sperimentale), Trieste, Italy

J. Clinton
ETHZ, Zurich, Switzerland

H. Pedersen
RESIF, Grenoble, France

W. Li, Z. B. Wang, D. S. van Maren and H. J. de Vriend
Faculty of Civil Engineering and Geosciences, Delft University of Technology, Delft, the Netherlands

B. S. Wu
State Key Laboratory of Hydroscience and Engineering, Tsinghua University, Beijing, China

D. Idier
R3C, DRP, BRGM, Orléans, France

A. Falqués
Applied Physics Department, UPC, Barcelona, Spain

A. D'Alessandro
Istituto Nazionale di Geofisica e Vulcanologia, Centro Nazionale Terremoti, Rome, Italy
Università degli Studi di Palermo, Dipartimento di Scienze della Terra e del Mare, Palermo, Italy

D. Luzio
Università degli Studi di Palermo, Dipartimento di Scienze della Terra e del Mare, Palermo, Italy

G. D'Anna
Istituto Nazionale di Geofisica e Vulcanologia, Centro Nazionale Terremoti, Rome, Italy

Michaela Spiske
Universität Trier, Geozentrum, Behringstr. 21, 54296 Trier, Germany
Westfälische Wilhelms-Universität, Institut für Geologie und Paläontologie, Corrensstr. 24, 48149 Münster, Germany

L. Retailleau
Institut de Physique du Globe de Paris, Sorbonne Paris Cité, CNRS (UMS 7154), Paris, France CEA/DAM/DIF, F-91297 Arpajon, France

N. M. Shapiro
Institut de Physique du Globe de Paris, Sorbonne Paris Cité, CNRS (UMS 7154), Paris, France

J. Guilbert
CEA/DAM/DIF, F-91297 Arpajon, France

M. Campillo and P. Roux
Institut des Sciences de la Terre, CNRS, Université Joseph Fournier, Grenoble, France

A. Plüß and F. Kösters
Federal Waterways Engineering and Research Institute, Hamburg, Germany

T. Ahern, R. Benson, R. Casey, C. Trabant and B. Weertman
IRIS Data Management Center, 1408 NE 45th Street, Seattle, WA 98105, USA

S. Zen, G. Zolezzi and M. Tubino
Department of Civil, Environmental and Mechanical Engineering, via Mesiano 77, 38123, Trento, Italy

M.-T. Apoloner, G. Bokelmann and I. Bianchi
Department of Meteorology and Geophysics, University of Vienna, Vienna, Austria

E. Brückl
Department of Geodesy and Geoinformation, Vienna University of Technology, Vienna, Austria

H. Hausmann and R. Meurers
Zentralanstalt für Meteorologie und Geodynamik, Vienna, Austria

S. Mertl
Mertl Research GmbH, Vienna, Austria

M. Bès de Berc
Institut de Physique du Globe, UMR7516, Université de Strasbourg/EOST, CNRS, 5 rue René Descartes, 67084 Strasbourg, France

M. Grunberg and F. Engels
Réseau National de Surveillance Sismique, UMS830, Université de Strasbourg/EOST, CNRS, 5 rue René Descartes, 67084 Strasbourg, France

J. Bosboom
Faculty of Civil Engineering and Geosciences, Delft University of Technology, the Netherlands

A. J. H. M. Reniers
Faculty of Civil Engineering and Geosciences, Delft University of Technology, the Netherlands
Applied Marine Physics, Rosenstiel School of Marine & Atmospheric Science, University of Miami, USA

R. Mosquera, V. Groposo and F. Pedocchi
Instituto de Mecánica de los Fluidos e Ingeniería Ambiental (IMFIA), Facultad de Ingeniería, Universidad de la República, Montevideo, Uruguay

A. D'Alessandro and G. D'Anna
Istituto Nazionale di Geofisica e Vulcanologia, Centro Nazionale Terremoti, Italy

K. Valentine
Department of Earth and Environment, Boston University, 22015 Boston, USA
Boston College, 02467 Chestnut Hill, USA

G. Mariotti
Department of Earth and Environment, Boston University, 22015 Boston, USA
Massachusetts Institute of Technology, 02139 Cambridge, USA

S. Fagherazzi
Department of Earth and Environment, Boston University, 22015 Boston, USA

A. Anglade
Observatoire Volcanologique et Sismologique de Guadeloupe (OVSG/IPGP), Le Houëlmont 97113 Gourbeyre, Guadeloupe, French West Indies

A. Lemarchand, S. Tait, C. Brunet, A. Nercessian and F. Beauducel
Institut de Physique du Globe de Paris (IPGP), Paris, France

J.-M. Saurel and V. Clouard
Observatoire Volcanologique et Sismologique de Martinique (OVSM/IPGP), Morne des Cadets, 97250 Fonds Saint Denis, Martinique, French West Indies

M.-P. Bouin and J.-B. De Chabalier
Observatoire Volcanologique et Sismologique de Guadeloupe (OVSG/IPGP), Le Houëlmont 97113 Gourbeyre, Guadeloupe, French West Indies
Institut de Physique du Globe de Paris (IPGP), Paris, France

R. Robertson, L. Lynch, M. Higgins and J. Latchman
Seismic Research Centre (SRC/UWI), St. Augustine, Trinidad and Tobago, West Indies

G. Chatzopoulos, I. Papadopoulos and F. Vallianatos
Laboratory of Geophysics and Seismology, Technological Educational Institute of Crete, Chania, Greece

M.-T. Apoloner and G. Bokelmann
Department of Meteorology and Geophysics, University of Vienna, Vienna, Austria

I. Bianchi, M. T. Apoloner, E. Qorbani and G. Bokelmann
Institut für Meteorologie und Geophysik, Universität Wien, 1090 Wien, Austria

M. Anselmi
Sezione Sismologia e Tettonofisica, Istituto Nazionale di Geofisica e Vulcanologia, 00143 Roma, Italy

K. Gribovski
Institut für Meteorologie und Geophysik, Universität Wien, 1090 Wien, Austria
MTA CSFK Geodéziai és Geofizikai Intézet, 9400, Sopron, Csatkai E. u. 6–8, Hungary

S. C. Stähler
Dept. of Earth Sciences, Ludwig-Maximilians-Universität München, Theresienstrasse 41, 80333 Munich, Germany
Leibniz-Institute for Baltic Sea Research, Seestraße 15, 18119 Rostock, Germany

K. Sigloch
Dept. of Earth Sciences, University of Oxford, South Parks Road, Oxford, OX1 3AN, UK
Dept. of Earth Sciences, Ludwig-Maximilians-Universität München, Theresienstrasse 41, 80333 Munich, Germany

K. Hosseini
Dept. of Earth Sciences, Ludwig-Maximilians-Universität München, Theresienstrasse 41, 80333 Munich, Germany

W. C. Crawford, A. Mazzullo and M. Deen
Institut de Physique du Globe de Paris, Sorbonne Paris Cité, UMR7154 – CNRS, Paris, France

G. Barruol
Laboratoire GéoSciences Réunion, Université de La Réunion, Institut de Physique du Globe de Paris, Sorbonne Paris Cité, UMR7154 – CNRS, Université Paris Diderot, Saint Denis CEDEX 9, France

M. C. Schmidt-Aursch
Alfred Wegener Institute, Helmholtz Centre for Polar and Marine Research, Am Alten Hafen 26, 27568 Bremerhaven, Germany

M. Tsekhmistrenko
Dept. of Earth Sciences, University of Oxford, South Parks Road, Oxford, OX1 3AN, UK
Alfred Wegener Institute, Helmholtz Centre for Polar and Marine Research, Am Alten Hafen 26, 27568 Bremerhaven, Germany

J.-R. Scholz
Laboratoire GéoSciences Réunion, Université de La Réunion, Institut de Physique du Globe de Paris, Sorbonne Paris Cité, UMR7154 – CNRS, Université Paris Diderot, Saint Denis CEDEX 9, France
Alfred Wegener Institute, Helmholtz Centre for Polar and Marine Research, Am Alten Hafen 26, 27568 Bremerhaven, Germany

D. Pesaresi
OGS (Istituto Nazionale di Oceanografi e di Geofisic Sperimentale), Trieste, Italy

H. Pedersen
ESIF (Réseau sismologique & géodésique français), Grenoble, France

Y. Starovoit
CTBTO (Comprehensive Nuclear-Test-Ban Treaty Organization), Vienna, Austria

Jari Kortström, Timo Tiira and Outi Kaisko
Institute of Seismology, Department of Geosciences and Geography, University of Helsinki, Finland

F. Fuchs, P. Kolínský, G. Gröschl, M.-T. Apoloner, E. Qorbani, F. Schneider and G. Bokelmann
Department of Meteorology and Geophysics, University of Vienna, Althanstraße 14, UZA 2, 1090 Vienna, Austria

Olga Clorinda Penalba
Departamento de Ciencias de la Atmósfera y los Océanos (DCAO/FCEN), Universidad de Buenos Aires, Buenos Aires, C1428EGA, Argentina
Instituto Franco-Argentino para el Estudio del Clima y sus Impactos (UMI IFAECI/CNRS), Buenos Aires, C1428EGA, Argentina

Juan Antonio Rivera
Departamento de Ciencias de la Atmósfera y los Océanos (DCAO/FCEN), Universidad de Buenos Aires, Buenos Aires, C1428EGA, Argentina
Instituto Argentino de Nivología, Glaciología y Ciencias Ambientales (IANIGLA/CONICET), Mendoza, 5500, Argentina
Instituto Franco-Argentino para el Estudio del Clima y sus Impactos (UMI IFAECI/CNRS), Buenos Aires, C1428EGA, Argentina

Tito Maldonado
Centre for Natural Disaster Science, Uppsala University, Villav. 16, 752 36, Uppsala, Sweden
Department of Earth Sciences, Uppsala University, Villav. 16 752 36, Uppsala, Sweden
Center for Geophysical Research, University of Costa Rica, San Pedro de Montes de Oca, 11501-2060 San Jose, Costa Rica

Anna Rutgersson and Björn Claremar
Department of Earth Sciences, Uppsala University, Villav. 16 752 36, Uppsala, Sweden

Eric Alfaro
School of Physics, University of Costa Rica, San Pedro de Montes de Oca, 11501-2060 San Jose, Costa Rica
Center for Geophysical Research, University of Costa Rica, San Pedro de Montes de Oca, 11501-2060 San Jose, Costa Rica
Centre for Research in Marine Sciences and Limnology, University of Costa Rica, San Pedro de Montes de Oca

Jorge Amador
School of Physics, University of Costa Rica, San Pedro de Montes de Oca, 11501-2060 San Jose, Costa Rica
Center for Geophysical Research, University of Costa Rica, San Pedro de Montes de Oca, 11501-2060 San Jose, Costa Rica

M. Picozzi, L. Elia and A. Zollo
RISSC, Università "Federico II" di Napoli – AMRA, Naples, Italy

D. Pesaresi and M. Mucciarelli
CRS, OGS (Istituto Nazionale di Oceanografia e di Geofisica Sperimentale), Trieste, Italy

A. Gosar and M. Živčić
ARSO – Agencija Republike Slovenije za Okolje, Ljubljana, Slovenia

W. Lenhardt
ZAMG – Zentralanstalt für Meteorologie und Geodynamik, Vienna, Austria

Martha Esthela Venegas-Pérez, Elsa Marcela Ramírez-López and Francisco Javier Avelar-González
Biochemical Engineering Department/Physiology and Pharmacology Department, Universidad Autónoma de Aguascalientes, Aguascalientes, México

Armando López-Santos
Unidad Regional Universitaria de Zonas Áridas, Universidad Autónoma Chapingo, Bermejillo, Durango, México

Víctor Orlando Magaña-Rueda
Center for Atmospheric Sciences, Universidad Nacional Autónoma de México, Distrito Federal, México

María Carolina Pinilla Herrera
Centro de Investigaciones en Geografía Ambiental – UNAM, Morelia, México
Fundación Natura, Bogotá, Colombia

Carlos Andrés Pinzón Correa
Fundación Natura, Bogotá, Colombia

Index